THE Heavens & THE Earth

Dr. Marcus Ross
Dr. John Whitmore
Dr. Steven Gollmer
Dr. Danny Faulkner

Kendall Hunt
publishing company

Cover image © Shutterstock, Inc.

Kendall Hunt
publishing company

www.kendallhunt.com
Send all inquiries to:
4050 Westmark Drive
Dubuque, IA 52004-1840

Copyright © 2015 by Marcus R. Ross, Danny R. Faulkner, Steven M. Gollmer, John H. Whitmore

ISBN 978-1-4652-6385-8

Printed in the United States of America

CONTENTS

CHAPTER 3

The Earth's Rocks 55

CHAPTER 4

Plate Tectonics 83

CHAPTER 5

The Restless Earth 117

CHAPTER 6

Reading the Record of the Rocks................................149

CHAPTER 7

Earth's Geologic History: Two Contrasting Views 179

CHAPTER 8

Soils, Weathering and Mass Wasting 205

CHAPTER 9

Streams and Groundwater 231

CHAPTER 10

Glaciers and Deserts 263

CHAPTER 11

Earth Resources: Provisions from God's Creation...............293

CHAPTER 12

Oceans and Coastal Systems ...319

CHAPTER 13

Earth's Atmosphere 349

CHAPTER 18

Models of Cosmic Origin 489

PREFACE

In 2009, Marcus Ross and John Whitmore were having lunch at a Creation Geology Society meeting in Georgia, discussing the fact that they both used an excellent but old-Earth and evolution-oriented textbook for their Earth Science classes. With many other projects and responsibilities, the prospect of writing a textbook seemed far away indeed, a "someday" kind of goal. But by 2012 the continued growth of classes and programs at their and other Christian institutions made them reconsider the idea of writing a textbook in Earth Science with a specifically young-Earth perspective. Covering so many different topics in the Earth and space sciences meant that more expertise was needed than could be provided by two geologists, so they teamed with Steven Gollmer and Danny Faulkner to bring their knowledge of atmospheric and astronomical sciences to the project, as well as their deep commitment to Christ and passion for education.

The Heavens and the Earth is the result of our combined prayer, research, study, and time teaching in the college classroom, and it is our joy to present it to you. Each of us is a born-again follower of Jesus Christ. We acknowledge God as the great Creator of this universe, and we passionately advocate and seek to improve a young-Earth approach to understanding the creation, for it is with both the pages of scripture and the data of nature that we can discern best the history of the world from Creation Week to Noah's Flood to the world we live in today.

Our desire as authors is to craft a work that not only educates, but *edifies* the reader by showing how the creation testifies to and glorifies its Creator. Though we present many different topics, a common theme animates each and every chapter: God is the creator, sustainer, and provider of this world, and we wish to honor Him for His great work in all that we do. This Creator has made Himself known, not just through observation and understanding of the creation; more than that, He also reveals Himself to humanity in the pages of Holy Scripture and through His Son who dwelt among us. This great maker of all things is Yahweh, the God of Abraham, Isaac, and Jacob, and through His son Jesus Christ all things were made, and through whom we have forgiveness of sin and reconciliation with God.

Hallelujah!

Marcus R. Ross, Ph.D.

John H. Whitmore, Ph.D.

Steven Gollmer, Ph.D.

Danny Faulkner, Ph.D.

Marcus R. Ross

Since I was a small boy, two passions have always animated me: God and dinosaurs. And it was those two passions that set me on the path to becoming a Christian and a young-Earth creation paleontologist. That road was not always easy, but the challenge of reading the pages of scripture along with the layers of rock has a way of drawing you closer to God for understanding, wisdom, and patience. There are many things we still don't understand, but much also that has been learned in putting together a coherent and robust understanding of the creation and its history. My hope is that this book helps spark the imagination and creativity of more creation scientists-to-be, whose future contributions will vastly outshine our own.

My education, from kindergarten through doctorate, was in public schools and state universities where public displays of faith were not always well-received, and where young-Earth creation was never given a chance present its case. In fact, I did not spend a single day in a Christian school until I was hired as the first geology professor at Liberty University! Yet along the way I met many like-minded students, and became involved in creation groups at both Penn State University and the South Dakota School of Mines and Technology. Key to maintaining a strong faith were these like-minded students, supportive churches, and other geologists and Christians who had walked these roads before me, including Kurt Wise, Steve Austin, Art Chadwick, and my coauthor John Whitmore.

I owe my undergraduate and graduate advisors Roger Cuffey, Gale Bishop, and David Fastovsky an enormous debt of gratitude for the time an effort they spent training me in my studies. Each view the world and its history far differently than I, yet they each at times risked their own professional standing during my training, since I was known as a creationist to them and others while I was their student. They are men of extraordinary character, fairness, and class.

In 2002 I was joyously wed to Corinna, my greatest friend and companion. Her boundless enthusiasm and support for obtaining my postgraduate degrees and my work in creation geology seems at time too good to be true. She and our four children (Katriela, Micah, Daniel, and Sienna) sacrificed many hours so that I could dedicate myself to preparing this book, and for that I am grateful beyond mere words. And from the time I first learned the word "paleontologist" at age seven, my parents Michael and Lee-Ann, and my sister Jillian, have never ceased in their support and encouragement in my pursuits.

I am blessed beyond measure.

John H. Whitmore

I grew up in a home with two extraordinary parents and four brothers who taught me to love the outdoors, the woods, and Jesus Christ. My dad was an elementary teacher, my scout master, and the most influential teacher in my life. My love of paleontology blossomed because of two phenomenal Earth science teachers I had growing up in Ravenna, Ohio: Jerry Jividen and Rod Chlysta. As a result of that eighth grade class I decided to major in geology at Kent State University. One of my most respected professors at Kent refused to continue to teach me at the Master's level (because of my faith) and that led me to go to the Institute for Creation Research for further training in geology. I am grateful to people like Janet Ward, Sharon and Sam Goebel, and Carol and John Hood for sacrificially making my education at ICR possible. Sharon Goebel would later be my connection that got me started into teaching science and math at a small Christian school in Geauga County; a profession I never would have considered if not for her. Steve Austin was an incredible professor at ICR. He taught me how to make careful observations and introduced me to places like the Anza Borrego Desert, the Grand Canyon, and Mount St. Helens. Steve was the one who recommended that Cedarville College call a young geologist in 1991 to see if I might be interested in an Earth science teaching job. My doctoral training continued with Leonard Brand and Paul Buchheim at Loma Linda University, who continued to further challenge me and help me to grow in my understanding of science, geology, and faith.

Then, there is my lovely wife Jamie, whom I met nearly thirty years ago. On the very day we met she asked me about my dreams for the future. I told her that I wanted to start a geology major at a Christian College. I realized that there were only a handful of Christian geologists who took the Bible seriously and that there was a need to train young geologists to whom we could "pass on the torch." Developing a geology major at Cedarville has been only a partial fulfillment of that dream. The other part was to write a text that would help Christian students understand that geology is a worthwhile discipline and not just about identifying rocks and minerals. I wanted students to understand that Earth science is a powerful tool that we can use in helping us to understand the truth we find in scripture, especially when it comes to the Flood, the age of Earth, and evolution. It is my hope that this book will influence many young Christians to consider Earth science as a profession and will encourage them to solve many of the outstanding problems that we still have in the area of science and the Bible. I hope as you read this text, you will come to realize that Hannah's payer is true: "There is none holy like the LORD, For there is none besides You, Nor is there any rock like our God." I Sam. 2:2.

There are many more to whom I owe a huge debt of gratitude, but none more deserving than to Jamie and my six children. They have been more than patient with me as I traveled to "the ends of the Earth" studying geology. Those travels have certainly helped me understand the rocks and scripture better, but their sacrifice has been very great. Thank you my love; I am eternally grateful for your devotion. You mean the world to me and you are the most amazing woman I ever could imagine! "An excellent wife is the crown of her husband," Prov. 18:22.

Steven Gollmer

Growing up on a small farm in Illinois I gained an appreciation for God's creation and developed a love for math and science. My dad and mom were hard workers and made church a priority in our lives. In ninth grade my Earth science teacher asked how many students believed the Earth was 4.5 billion years old. I raised my hand with the majority of the class. To my embarrassment, my best friend didn't raise his hand, but indicated that he believed the Earth was thousands, not billions of years old. My embarrassment was not because he took that position, but because I didn't. Up to this point in my life I did not consider the relevance of the Bible to science. While I attended a summer Bible camp, I purchased Henry Morris' *The Bible and Modern Science*. This book built my confidence that the Bible was relevant to my study of science and that a Christian could honor God in the field of science.

Photo by Scott Huck

Through a series of God-directed circumstances, I ended up at Pillsbury Baptist Bible College after beginning my college career as a geology major. While there, I discovered a passion for teaching and a purpose for my life. Wayne Deckert, my science professor, exposed me to the creationist community and provided an opportunity to see A.E. Wilder-Smith as he made his last American speaking tour. This gentleman's demeanor and purposeful speech reinforced the value of thinking deeply about creation and mankind's current understanding of science. It seems things have come full circle, where I can pass on what I have learned about God and His creation. Cedarville University gives me an opportunity to do just that with likeminded colleagues and an excellent student body.

I have had numerous examples of good teachers and mentors in my life. My home church pastor, George Ford, provided many opportunities to be involved and teach while going through college. My parents as well as my sisters and their families have been supportive of someone they called "a professional student." My PhD advisor, Dr. Harshvardhan, demonstrated a mode of mentoring that I try to emulate, high expectations and patience. He emphasized the value of good writing and pushed me beyond what I thought I was capable of achieving. Most of all, I value the impact of Evelyn, my wife, on my life. She doesn't let me get by with sloppy thinking and keeps me grounded in the essentials of family and church. Her love is unconditional and she is the love of my life. My children are my joy and I thank them along with my wife for their patience while working on this project.

The value of my relationship with my Lord and Savior, Jesus Christ, cannot be expressed with mere words. He is the core of my life and provides the priorities by which all areas of my life work and makes sense. This centrality of Christ was not apparent to me in high school, although I was saved at an early age. However, His faithful work in my life and the words and writings of many of His servants have brought me to where I am today. Special thanks go to the Creation Research Society, the Creation Biology Society and the Creation Geology Society, which have provided supportive colleagues and kept me involved in creation research. I stand in awe of God's creation and am humbled by how little I know. My hope for this text is that it can provide, in some small way, what Henry Morris' writings began in my life: a confidence in God's Word, reasons for why I believe as I believe and a need to lovingly share with others. "But in your hearts honor Christ the Lord as holy, always being prepared to make a defense to anyone who asks you for a reason for the hope that is in you; yet do it with gentleness and respect." (I Peter 3:15)

Danny Faulkner

Astronomy has been a life-long passion for me. Even before I started school, I was fascinated with the stars. I'm different from most scientists in that I wasn't that interested in other sciences when I was growing up, but eventually I came to like them as well. My sophomore year of high school was a pivotal year in my development, because several important things happened that year. While I had the blessing of being born into a Christian home and was born again at age six, I rededicated my life in high school. By reading a book by Henry M. Morris, I became convinced of biblical creation and finally realized that my calling in life was to be an astronomer for God's glory.

To prepare for my life work, I enrolled at Bob Jones University, where I earned a Bachelor of Science degree in mathematics. From there I went to Clemson University, where I received a Master of Science degree in physics. After Clemson, I completed a Master of Arts and a Ph. D in astronomy from Indiana University. I was on the faculty at the University of South Carolina Lancaster for over 26 years, and upon my retirement received the title of Distinguished Professor Emeritus. In January 2013, I became the astronomer at Answers in Genesis, where I work in the Research Department, write planetarium shows for the Creation Museum, oversee Johnson Observatory at the Creation Museum, and write and speak for Answers in Genesis.

I have had the pleasure of using the facilities of some very nice observatories. Much of my research has been with eclipsing binary stars, leading to the publication of more than a hundred papers in various astronomy and astrophysics journals. I also have published much on astronomy in the creation literature, and so far I have published two books, *Universe*

by Design and *The New Astronomy Book*. I am on the Board of Directors of the Creation Research Society, and I also serve as the editor of the *Creation Research Society Quarterly*.

When I was a student at Bob Jones University many years ago, I took a few courses from George Mulfinger, including survey courses in astronomy and geology. Professor Mulfinger shared my passion for astronomy, and we developed a close relationship. He was the first creation scientist that I knew, and so he became a role model for me. Forty years ago, Professor Mulfinger was co-author of the first science book written especially for Christian high schools. It is fitting as I follow in his footsteps and participate as a co-author of the first Earth and space science textbook for Christian universities. It is sad that Professor Mulfinger did not live to see this day, but I am sure that he would be proud. It is my sincere desire that this textbook may inspire other Christian young people to take up the mantle from the authors of this work.

© Sergii Votit/Shutterstock.com

CHAPTER 1

THE EARTH IS
THE LORD'S

OUTLINE:

Many different components of Earth's systems find a point of interaction at the shoreline. Weathered rock materials make up the beach, which is worked over by waves of water driven by atmospheric winds that are generated by energy from the sun.

1.1 AN INTERDISCIPLINARY SCIENCE

The pages of this book provide an introduction to the **Earth Sciences**. These are a collection of different but related disciplines that together give us a unique understanding of the world in which we live and its context in space (figure 1.1). Each discipline focuses on the physical makeup and interactions of the planet, ranging from minerals to galaxies, and includes geology, hydrology, oceanography, meteorology, and astronomy.

Geology is the study of the solid earth, including rocks, minerals, erosion, volcanoes, plate tectonics, the planet's historical record, and many other areas. Geology has both present and past components, where geologists may be studying current materials and processes (called *physical geology*), or

attempting to determine events from the distant past (*historical geology*).

Hydrology is the study of water, and more specifically the storage and movement of water as it passes through different parts of the *hydrologic cycle* (see chapter 9). It has a strong focus on people's use of water, water pollution, and water management. Also connected to water is **oceanography**, the study of the world's oceans. This is one of the most broad-based disciplines, since it deals with the physics and chemistry of ocean water, its interactions with the atmosphere in producing currents and weather, geological connections to the ocean basins, and studies of the rich biological diversity of the marine world.

FIGURE 1.1. Earth Science incorporates the input from a wide variety of different fields of study.

The study of the atmosphere, weather, and climate is the discipline of **meteorology**. Also referred to as *atmospheric science*, meteorology is the area of science that we consult nearly every day when we check the weather report. Meteorology has a close association with oceanography, since the oceans cover 71% of our planet's surface and evaporate most of the water that later falls as precipitation.

A study of the Earth would not be complete without an understanding of our planet in the context of our place in space. **Astronomy** is one of the oldest of all the sciences, as people have been tracking the movements of the stars and planets for thousands of years. Astronomy is the study of the heavens and its bodies, from planets to galaxies and the universe as a whole. Like geology, astronomy has disciplines dedicated to both present and historical questions, such as measuring current solar output or developing historical models for the formation of solar systems and the universe.

1.2 MODELS OF EARTH AND ITS HISTORY

There are a wide variety of different perspectives about Earth's history, especially within Christianity. The four most common views are young-Earth creation, old-Earth creation, theistic evolution, and naturalistic evolution. The first three are perspectives held today by Christians, while the fourth denies either the existence or ability to know about God, and is therefore non-Christian. More specifically, **young-Earth creation** holds that 1) all of God's work of creation was accomplished in a six rotational-day period of time only a few thousands of years ago; 2) God directly and miraculously created "kinds" of organisms that have since diversified in a limited fashion; 3) humans were created separately and miraculously apart from other animals to be rulers over creation; and 4) a global flood destroyed the world and its inhabitants at the time of Noah, and is responsible for the majority of sedimentary rocks and fossils found on the planet's surface. Young-Earth creation is the longest-held view of creation within Christianity and Judaism, with the events and chronology accepted since Old Testament times, affirmed throughout the New Testament, advocated during the early Church period, and proclaimed in the Reformation. It continues to be strongly supported today. There are many reasons for this long-standing consistency. Chief among them is that the early chapters of Genesis (1-11) appear to record the events of creation, the Fall, and the flood as historical events, rather than non-historical myths or fables. Elsewhere in both the Old and New Testaments, the events and individuals are likewise treated as historical and important doctrines (including such issues as marriage, work, justice, and Christ's return) and are anchored in the events recorded in the texts.

The other three views are relative newcomers to Western thought and saw their rise to prominence begin in the late 1700s, largely in response to arguments for an ancient Earth and universe from geologists and astronomers. **Old-Earth creation** holds that the Earth is ancient (4.55 billion years), during which time God directly and miraculously created various "kinds" of organisms with limited evolutionary relationships, humans were new and miraculous creations by God, and Noah's flood was a local, rather than global event. **Theistic evolution** maintains that Earth is billions of years old, that God created life using only evolutionary processes, that humans share ancestry with apes, and that Noah's flood was a local event or perhaps a non-historical myth. **Naturalistic evolution** holds that God, if He exists, has played no scientifically discernable role in the formation of the universe and the evolution of life through time, and scripture has no bearing on matters of Earth history. Only natural processes working over immense periods of time are responsible for the present physical features of Earth and the evolutionary history of all its living and extinct inhabitants.

While learning the details of each of these views (and their many variations) is beneficial in light of the current diversity of opinion, this book focuses on the two most prominent views: young-Earth creation and naturalistic evolution. The former is the dominant view within Christianity in the United States and throughout Church history, and is the view held by the authors of this book. The latter is the view advocated in the majority of schools and universities of the world, and the view in which each of us as authors were trained as we pursued careers in science and education. Since both old-Earth creation and theistic evolution agree with naturalistic evolution on the timeframe and sequence of events (and also with biological evolution for theistic evolution), many of the issues covered in naturalistic evolution represent their views as well.

What we hope to do in the parts of the book where these issues are raised is to show that a young-Earth view is sound both scripturally and scientifically. Our goal is to discover the truth about our world and its history through God's leading (more on this below). We also wish to present the naturalistic evolution view fairly and honestly, so that you may properly understand the views held by a majority of scientists and why they believe that this model works. In comparing these very different views, we aim to be fair because we know and respect many scientists who hold to naturalistic evolution. Our arguments as creationists must be *for* young-Earth creation, rather than *against* evolution, because showing that someone else's idea is wrong doesn't mean that yours is right. Only by doing the harder work of successfully building models and testing hypotheses can young-Earth creationist scientists advance our understanding of God's magnificent world.

1.3 DOING GOOD SCIENCE

Science is a human endeavor, and like any human endeavor, it has its ups and downs. When done well, science leads us to great discoveries, advances, and insights. Its application through technologies and medicine can relieve us of grueling work and treat horrible diseases. Poorly-done science, however, slows our understanding and leads us along wrong paths, and science done with evil intent has caused much suffering. But good scientific work usually wins out because it is closer to the truth, results in a better understanding of our world, and brings us to a reverent and enthusiastic appreciation of God as the world's Creator and Sustainer.

So how do we do good science? There's no single formula, and as you'll see in the pages of this book, different branches even within the Earth sciences have different approaches because they are asking different kinds of questions. After all,

determining the history of a rock and forecasting tomorrow's weather are very different things indeed. But even with all the differences, there are some basic characteristics and approaches that scientists follow:

1) A scientist begins their inquiry because of past or current *observations* of the world around them. That is, something they've seen makes them curious.

2) These observations usually prompt some *questions* for why something is the way that it is, or what might happen to it under certain conditions.

3) The scientist thinks of a **hypothesis** (a possible explanation) that could answer the questions, or *predictions* about what one should discover if a set of observations is true.

4) There should to be a way to *test* that hypothesis/prediction against the existing observations and/or future observations.

5) The new observations will help to *evaluate* whether the hypothesis /prediction is successful. These observations can affirm or contradict the hypothesis/prediction, and often help the scientist to make adjustments and try again.

The steps above are often referred to as the *scientific method*, but we should really think of it as **scientific methods** (plural). These steps are not limited to the sciences, but the sciences as we consider them today (including astronomy, biology, chemistry, etc.) usually focus on evaluating data and discoveries from the natural world, and so are sometimes referred to as the *natural sciences*. Other areas like history or psychology are often referred to as *social sciences* due to their focus on people. *Engineering and technology* are usually reserved for areas of science that focus on the application of science to material inventions.

The natural sciences are capable of investigating events from both the past and the present and making predictions about the future, and each of the major branches of Earth science addresses these kinds of issues. Frequently scientists divide the types of questions into the **experimental sciences** and **historical sciences**. The experimental sciences are what we most often think of when we think of science: controlled conditions in which a hypothesis is tested in a repeatable fashion (figure 1.2). A geological example might ask the question, "What size particles can water flowing at 1 m/s move?" Using a flume, sluice, or other apparatus, a geologist can create a flow of 1 m/s and observe if silt, sand, small stones, or cobbles might be moved by the water. Another group could repeat this experiment at their own lab and re-test the results. The experimental sciences have the advantage of being able to test a single hypothesis many times over, to control for many different factors, and to adjust those factors precisely.

The historical sciences (figure 1.3) are different, because they seek to discover events that occurred just one time in the past; an event that cannot be repeated. For example, a paleontologist discovers a skeleton of *Tyrannosaurus rex*. The question, "What killed this *T. rex*?" is not subject to repeatable experimentation. Even if you could find another *T. rex* and kill it, that wouldn't tell you how the first one died. Instead, the paleontologist takes notes and samples of the fossil and the rocks to try to discern their history. Was the *T. rex* killed by another animal? Was it buried in a stream environment? Was it killed and buried during Noah's Flood? Working in the historical sciences usually means evaluating *multiple competing hypotheses* all at the same time, the same way a detective evaluates

FIGURE 1.2. A biologist working on bacteria in the lab (a) and a geologist measuring and mapping rocks (b) are examples of experimental science. From these observations, hypotheses and predictions can be made.

FIGURE 1.3. The observation of these tracks is part of experimental science, but determining if they came from a dinosaur, and what the dinosaur was doing at the time are historical science questions. Continued study and observations help us discover the most likely explanations from multiple competing hypotheses.

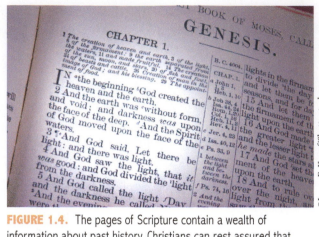

FIGURE 1.4. The pages of Scripture contain a wealth of information about past history. Christians can rest assured that this testimony from God is trustworthy, and we must use it to build accurate models of Earth history.

several murder suspects. Finding out important information, like an alibi, means that one of the suspects can be cleared of the crime. Discovering certain features of the rocks and fossils allow the paleontologist to exclude some of the hypotheses, leaving the more likely ones still standing. The approaches of both the experimental and historical sciences are necessary to understand our world and its history. The experimental sciences take some priority over the historical, because we can only make good inferences about the past if we have a good understanding of the present. For example, the flow-speed experiments might have a direct bearing on understanding the *T. rex* by helping the paleontologist interpret the sand and clay surrounding the fossil.

In historical studies of all kinds, reliable **eyewitness testimony** is important, because it brackets and narrows the possible explanations by providing information from someone who witnessed the event. Even the best historical science work, based on solid empirical studies, can be overturned if reliable testimony states that the event occurred differently than the interpretations and conclusions drawn from scientific investigation. This is why we authors hold to a young-Earth creation view of the world. God revealed to mankind the time and manner in which He created and judged the world (see esp. Gen. 1-11; figure 1.4), and these statements about the creation and its history provide us with accurate information. To ignore them would mean throwing out the *most reliable data* about the natural world, and doing so greatly reduces the likelihood of discovering the real history of God's creation.

The interplay between scientific methods with the Biblical record and its proper interpretation is a complex one, which is in part why there are so many views on origins among Christians, as outlined above. Dr. Leonard Brand provides a helpful diagram for how one might look at these interactions (figure 1.5). In it, Brand separates the science and religion domains, noting that they each operate according to their own sets of rules and principles (e.g., the methods of science vs. the methods of

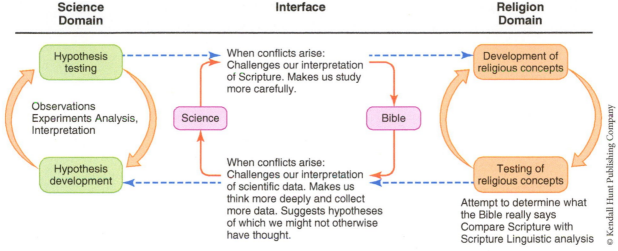

© Kendall Hunt Publishing Company

FIGURE 1.5. A model of science/religion interactions. Source: Brand (2009).

proper scriptural interpretation). However, Brand also shows that there is an **interface** (a point of contact and exchange) between them, where one domain can inform the other.

An example within young-Earth creation is the idea of a vapor canopy that was proposed to exist above the atmosphere at the beginning of creation, but later collapsed during Noah's flood to provide the rain for the flood. Early proponents of this idea saw that Genesis 1:6-7 (ESV) states:

> *⁶And God said, "Let there be an expanse in the midst of the waters, and let it separate the waters from the waters." ⁷And God made the expanse and separated the waters that were under the expanse from the waters that were above the expanse. And it was so.*

From these verses (religion domain), some creationists believed that the atmosphere divided two bodies of water, the oceans and some kind of water-vapor canopy above the atmosphere (interface). Creation atmospheric scientist Dr. Larry Vardiman spent many years working on computer models to see if this hypothesis could work (science domain), and the results were rather disappointing. It appears that the vapor canopy hypothesis simply doesn't hold water!

The failure of the vapor canopy hypothesis (science domain) has caused creation scientists to go back to scripture (through the interface) to see if a vapor canopy is really what these verses are talking about (religion domain), and today most young-Earth creationists do not hold to vapor canopy theory. Some believe that the "waters above" refer to waters at the edge of the universe, while others believe that the "waters above" is simply an ancient Hebrew idiom for rain. Both of these views lead to other hypotheses about the Earth, the universe, and the source of water for Noah's flood, and these are seeing much more success than vapor canopy had (see the discussion on "Catastrophic Plate Tectonics" in chapter 4).

What is instructive from this example is that the Bible *can* provide us with ideas about the world that we can then use to form scientific hypotheses and tests. The science and religion domains are different facets of the same, singular world that God has created. They are not separate realms of "fact" and "faith" or "physical" and "spiritual." In this book, we embrace the unity of creation and the rightful place of God to inspire our scientific exploration through scripture, inspiration, and the work of the Holy Spirit in the hearts and minds of scientists who ask Him for wisdom, guidance, and even for a good hypothesis!

When considering the Earth, scientists generally divide our world into four divisions, or **spheres**. Each of the spheres incorporates a variety of related processes and internal interactions.

1.4.1 Atmosphere

The outermost component of the Earth is the **atmosphere** (figure 1.6), which is the collection of gasses that surround the Earth and are held to it by gravity. The gasses that make up our atmosphere are primarily nitrogen and oxygen, with smaller contributions from argon, carbon dioxide, and several others (figure 1.7). Oxygen is particularly important because it provides us not only with the O_2 needed for our respiration, but also forms ozone (O_3) higher in the atmosphere, which filters out most of the ultraviolet radiation emitted by our sun. Other gasses, like water vapor, carbon dioxide,

and methane help retain thermal energy to keep the planet's temperature equitable; and nitrogen provides a gas that is transparent to visible light, allowing the sun's energy to reach the Earth's surface where it can be used by photosynthetic organisms and sustain life.

Scientists divide the atmosphere into four units based on whether temperatures are rising or falling (see Chapter 13 and figure 13.8). The upper boundary with space is gradational rather than sharp, and is placed at 100 kilometers (60 miles) above the surface. This is where a balance exists between the amount of energy coming into the Earth system from the sun and the amount leaving the Earth into space. The lowest portion of the atmosphere is where nearly all of our weather occurs.

Gasses are compressible (think of how much helium is actually in a tank for balloons), and the weight of the gasses higher in the atmosphere

Courtesy NASA

FIGURE 1.6. The moon rises over the Earth's atmosphere as seen in this image taken from the International Space Station. Nearly all of the planet's weather occurs in the lowest part of the atmosphere (called the *troposphere;* see chapter 13). The transition from blue to black, just above the moon, indicates the transition between Earth's atmosphere and space.

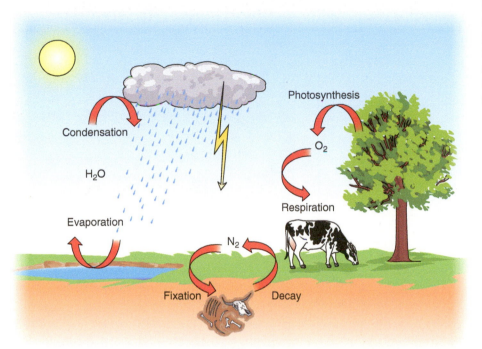

FIGURE 1.7. There are a variety of chemical cycling processes occurring in our atmosphere, including oxygen and nitrogen. These two gases make up approximately 99% of our planet's atmosphere.

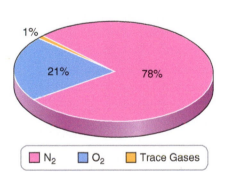

press down on the gasses near the Earth's surface. *Atmospheric pressure* (also called barometric pressure) is a measure of the amount of atmosphere over a particular place. The higher one rises into the atmosphere, the less gas is above them exerting pressure (figure 1.8). In fact, 50% of the atmosphere by weight is located within only 5.6 kilometers (18,000 feet or 3.4 miles) of the Earth's surface, and 90% is located within 16 kilometers (10 miles). This is why most aircraft have pressurized cabins: if you're flying at 34,000 feet (10.4 km or 6.4 miles), then the air pressure outside of the cabin is only 25% of air pressure at sea level. This means that there is only 25% of the oxygen available in the air outside compared to the amount of oxygen available at sea level. By pressurizing the cabin, sea-level atmospheric pressure and oxygen levels are maintained within the airplane.

1.4.2 Hydrosphere

All of the water on the planet's surface is part of the **hydrosphere**. Constantly shifting locations, the hydrosphere includes the water of the world's oceans, rivers, lakes, glaciers, atmosphere, and groundwater (figure 1.9). Various processes move the water from

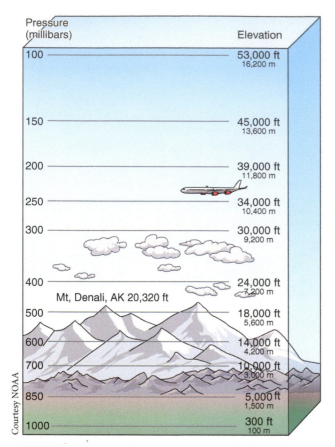

FIGURE 1.8. Atmospheric pressure decrease with altitude. At sea level, atmospheric pressure is 1013 millibars (mb).

FIGURE 1.9. The two largest reservoirs of water in the hydrosphere, the ocean and glaciers, meet in Glacier Bay, Alaska.

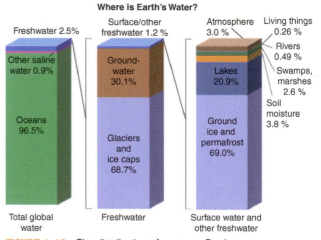

FIGURE 1.10. The distribution of water on Earth.

one place to another in a continuous recycling process that makes our planet so well-designed for life. It may be surprising that nearly 97% of all water of the hydrosphere is in the oceans (figure 1.10). More surprising, though, is that the remaining water is mostly located in glaciers and groundwater. Very little of the hydrosphere is stored in lakes, streams, and rivers, and only a tiny fraction is in the atmosphere, the source of all our precipitation! But while rivers and air don't carry much water at any given time, they move *huge* amounts of it on an annual basis. While not visible at the surface, the groundwater just below our feet is the most important source of water for humanity. Without it, much less of the globe would be populated, and much less agriculture could be sustained (figure 1.11).

FIGURE 1.11. Farming in arid regions is made possible by groundwater. A pivot irrigation system is responsible for the circular pattern of crops.

1.4.3 Geosphere

The **geosphere** refers to the solid, rocky components of the Earth, and is the largest of all the spheres. It begins at the Earth's surface of rock and sediment below the atmosphere or ocean, and extends to the center of the planet. Along the way, its composition and physical properties change significantly. Our direct experiences are limited primarily to the rocks of the continents and samples taken from the ocean basins. But through indirect methods, including data from earthquake waves, gravity anomalies, and magnetic field studies, geologists have come to an excellent understanding of the internal makeup of our planet as well. Here we will briefly survey the main components of the geosphere from surface to center, looking at both the chemical compositional differences that define the basic units of crust, mantle, and core, and also at the physical properties that cause these units (and the divisions within them) to behave differently, thus providing the picture of the Earth's interior we now have (figure 1.12).

The Crust. The outermost "skin" of the planet is called the **crust**. The crust ranges in thickness from 5-70 km. This may seem like a lot of rock (and it

is!), but in comparison to the entire geosphere, the crust is about as thick as the skin of an apple is to the whole apple. All told, the crust only makes up less than 1% of the Earth's volume. The rocks of the crust are cool or cold compared to deeper rocks, and they are also rigid and brittle. Rather than bending or flowing when subjected to stresses, they tend to fracture and break. Thus earthquakes are common in the crust, but do not occur in the deeper, hotter rocks of the planet.

Compositionally, we divide the crust into **oceanic crust** and **continental crust**. Oceanic crust covers about 60% of the Earth's surface and is made up of the igneous rocks basalt and gabbro (see chapters 3 and 4). Oceanic crust is usually less than 10 km thick, and its density is 2.9 g/cm³, or about 3 times the density of water (see box 1.1). These rocks are produced at the ocean ridges by volcanic activity. Continental crust is dominated by the igneous rocks granite and diorite (together called *granodiorite*), and are frequently covered by sedimentary rocks such as sandstone, limestone, and shale. Continental crust covers about 40% of the Earth's surface and is thicker than the oceanic crust, being about 30 km thick on average, and over 70 km thick in places like

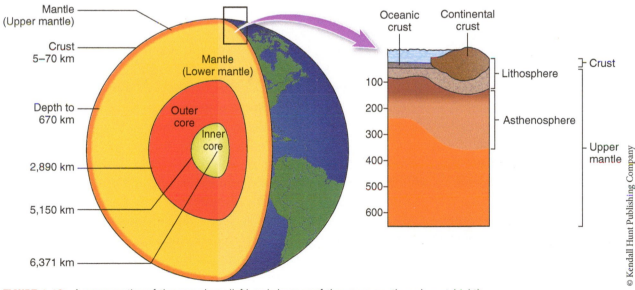

FIGURE 1.12. A cross-section of the geosphere (left) and close-up of the upper mantle and crust (right).

BOX 1.1

Density is the amount of mass an object has per unit volume, and is usually described in terms of g/cm³ (grams per cubic centimeter). This is an important unit for scientists, but is not commonly used in day-to-day life. So let's consider a comparison.

Water has a density of 1.0 g/cm³. A gallon of milk has about the same density of water, and weighs 8.3 pounds (3.8 kg). If we filled the same one-gallon container with granodirorite from the continental crust, it would weigh 22.5 pounds. Pretty heavy, but one gallon of basaltic oceanic crust would weigh a bit more: 24.2 pounds. Here are the average one-gallon weights for each of the geosphere regions described in the text:

Geosphere Unit	Density (g/cm³)	Weight (lb) of one gallon
Continental Crust	2.7	22.5
Oceanic Crust	2.9	24.2
Upper Mantle	3.9	32.5
Lower Mantle	5.0	41.7
Outer Core	11.0	91.8
Inner Core	13.0	108.5

the Himalayas. The granodiorite that makes up the continental crust is less dense (2.7 g/cm³) than the ocean crust, which is why the continents are mostly above sea level. While both continental and oceanic crust "float" above the mantle, the continental crust floats higher than the ocean crust.

THE MANTLE The next compositional unit below the crust is the **mantle**. At 84% of the planet's volume and 67% of the planet's mass, the mantle is easily the largest single component of the Earth. Due to the increasing pressure of the overlying rocks, the density of the mantle increases from 3.4 g/cm³ just below the crust to 5.6 g/cm³ near the core, and there is a gradual change in the dominant rock types, particularly between the upper and lower mantle. The *upper mantle* extends from the bottom of the crust to about 670 km deep, and is made mostly of the rock called peridotite, which is rich in iron and

magnesium. Its uppermost region is solid and brittle, like the crust, and the combined crust and uppermost upper mantle is called the **lithosphere** ("rocky layer"). The lithosphere is not a singular unit, but is actually composed of several large and many small slabs, called *plates*, which move with respect to one another. Their movements and interactions are the topic of *plate tectonics*, which is covered in chapter 4. Lithospheric plates are much thicker than just the crust. For example, lithosphere capped by oceanic crust is 70 km thick on average, and lithosphere capped by continental crust is typically 150 km thick, with some areas up to 280 km thick.

Below the lithosphere and extending to about 370 km deep is a unit of the upper mantle called the **asthenosphere** (meaning "weak layer"). While chemically similar to the mantle rocks in the lithosphere above, the higher temperatures allow the rocks of the asthenosphere to flow in a more plastic fashion, like a very thick Silly Putty™. The mechanical weakness of the asthenosphere allows the lithospheric plates ride over the top of the asthenosphere, and occasionally the plates get dragged down into and through the mantle itself. The remainder of the upper mantle is a transition zone with slightly stronger rocks due to the higher pressures.

The *lower mantle* extends from 670 km to 2,890 km deep. As the depth increases, the density, temperature, and strength of the mantle all increase as well. While varying from top to bottom, the lower mantle has an average density of about 5.0 g/cm³. The minerals have a higher amount of magnesium than the upper mantle, and though it still flows plastically, the lower mantle is more rigid than the upper mantle. Recent studies have also shown that oceanic lithosphere that has been dragged down through the mantle has come to rest near the core/mantle boundary.

THE CORE Beginning at a depth of nearly 2,900 km and continuing to the center of the planet is the **core**, a region composed mostly of iron with small amounts of nickel. It makes up about 15% of the planet's total volume, but over 30% of its mass (due to the high-density materials that comprise it). While the whole core is compositionally similar, it is divided into two units based on physical properties. The **outer core** is liquid, and the iron/nickel alloy flows with a viscosity similar to water but an extremely high density of nearly 11 g/cm³. The circulation of the metal is largely responsible for the Earth's magnetic field, which protects the planet from dangerous cosmic radiation emitted by our sun as well as distant stars and supernovae. The outer core extends from the boundary with the mantle to a depth of 5,150 km deep. The **inner core** is the final component of the geosphere, and is a solid rather than liquid due to the immense pressure of the geosphere surrounding it. The average density of the inner core is about 13 g/cm³, and it extends from the boundary with the outer core to the center of the planet at a depth of 6,371 km.

1.4.4 Biosphere

The **biosphere** includes all of the living organisms on the planet. It is easy and accurate to divide the spheres into the non-living (atmosphere, geosphere, hydrosphere) and living (biosphere) categories, since the living realm is so fundamentally different from non-living matter. After all, the "growth" of a mineral and the "growth" of a child are very different things. The former occurs solely due to the accumulation of similar chemical elements under the same conditions as the rest of the mineral, while the latter features far more dynamic levels of organization and development from tremendously varied inputs that are directed through much more complex biochemical processes.

Diverse and abundant life can be found in many parts of the globe, and the majority of the biosphere is located in the oceans. Here, sunlight penetrates through the first two hundred meters of water in a region called the *euphotic zone*. This is the portion of the ocean where there is enough light to permit photosynthesis. Microscopic marine photosynthetic organisms, called *phytoplankton*, form the base of the euphotic zone's food web, supporting the existence of countless species. Shallow continental shelves, coral reefs, and areas with nutrient-rich waters host the largest diversity of ocean life (figure 1.13a). In the deeper regions of the oceans where sunlight never penetrates (called the *aphotic zone*), a few unusual and hardy organisms make their living from organic material that falls down from the euphotic zone (anything from small organic particles to an entire whale carcass), or from alternate sources of energy, such as the heat and chemicals found at hydrothermal vents. Even there, where the water temperatures are near-freezing and the pressure is crushingly high, life forms such as crabs, worms, and microorganisms can survive.

Life on land is most diverse where temperatures are warm and water is abundant: tropical rain forests (figure 1.13b). In these environments, hundreds, and sometimes even thousands of different species of plants, animals, and fungus live in an area the size of a football field.

FIGURE 1.13a. Abundant sunlight, low sediment input, and warm, nutrient-rich waters make coral reefs (a) the most diverse of all marine ecosystems. Tropical rain forests, such as one in from Ecuador (b), host the largest species diversity on Earth.

Life also exists deep in the continental interiors and in sediments below the ocean floor. In fact, perhaps 40% of all life is found *within* rocks! Whether in tropical rain forests or in deserts, temperate forests, or arctic tundra, the animals and plants of this world have spread far and wide, fulfilling God's commandment after the Flood to "be fruitful and multiply on the earth" (Gen. 8:17, ESV).

1.5 EARTH SCIENCE: SYSTEMS AND SYNTHESIS

Each of the spheres described above contains its own set of processes and actions. For example, the hydrosphere involves evaporation and precipitation; the atmosphere transmits and retains different forms of energy; the surface of the geosphere is shaped by the movements of plates; the biosphere is characterized by food webs. Yet at the surface of the earth, each of the spheres interacts with the others in complex and dynamic ways, operating in concert to make our planet unique in its ability to maintain a world where life is not just possible, but capable of thriving.

Earth scientists discovered that by studying both the spheres and their interactions, they could get a better picture of the Earth as a **system**. We can think of a system as a series of interactive components working together to form a complex and interdependent unit. We frequently use the word *system* to describe a group of equipment. Your audio system may include a receiver, amplifier, bass unit, and a series of speakers. Not all of these components have to be operating at the same time (the back surround speakers, for example), and each component has its own set of internal components to make it work properly (e.g., woofers and tweeters for the speakers). Taken together, the entire system works because each of the components contributes to producing a rich audio experience.

Earth systems science, then, is an interdisciplinary approach to trace and understand the systems and subsystems of the planet. Let's take an example of a simple interaction between the spheres, taken from the list given above: evaporation and precipitation are part of the hydrosphere because they involve the water on the planet. But they are also part of the atmosphere, because as water is evaporated from the oceans, it converts from liquid water to water vapor, which is a gas. At certain concentrations and temperatures, the atmosphere can no longer retain water vapor, which will then crystallize as ice to form clouds, or condense into droplets and fall as precipitation. So the two

FIGURE 1.14. Soil represents a complex interface among Earth's various spheres. Weathered rocks and minerals, air and water contribute to soil. Animals and plants work to aerate the soil, provide organic materials and nutrients, and help break up the bedrock below the soil.

most basic components of the *hydrologic cycle* (the set of processes that move water around the earth) involve two of the spheres.

A more complex interaction involving all of the spheres can be found in soils (figure 1.14). Soil is composed primarily of weathered rock materials (geosphere), decaying organic material (biosphere), and frequently contains small amounts of water (hydrosphere) flowing through open spaces filled with air (atmosphere). But the interactions continue beyond just the mere components: tree roots and water break apart and degrade the rocks below the soil by both physical and chemical processes. Worms, ants, and other burrowing organisms aerate the soil by tunneling through it. This increases the permeability of the soil, accelerating soil development even further. Sediment and organic material deposited during a river flood provide additional nutrients to the soil and foster plant growth.

Energy sources are needed to drive Earth's various systems, subsystems, and cycles. In the case of the hydrologic cycle and soil development, the major source of energy is the sun. Its energy warms the oceans and land (creating evaporation and weather patterns) and provides energy for plants via photosynthesis, which then provides energy for other organisms. These hydrologic, atmospheric, and biological systems are all driven by the sun's emission of radiation to our planet.

The other major source of energy for Earth's systems comes from the planet itself. As discussed above, the interior of the planet is very hot. This thermal energy comes from the planet's initial formation during Creation Week as well as from the decay of radioactive materials in the mantle and crust during Noah's Flood and through to today. The flow of energy from the hot interior to the cool exterior is expressed at the planet surface in volcanic eruptions, geysers, and plate tectonic movements.

1.6 DOMINION, STEWARDSHIP, AND THE ENVIRONMENT

So far, our focus in this chapter has been on surveying the basics of Earth Science, including basic methods, components, and interactions. But why? For what ultimate purpose should we spend time learning about our planet? Two reasons stand out above all others. First is that understanding our Lord's world may help us understand Him better as the creator and sustainer of all things. This leads to thanksgiving, reverence, and praise to God, who alone is worthy of our worship. The second is that after creating the first man and woman, God gave them the privilege of becoming the caretakers of His magnificent creation. In Genesis 1:28 (ESV), God spoke to them saying "Be fruitful and increase in number; fill the earth and subdue it. Rule over the fish in the sea and the birds in the sky and over every living creature that moves on the ground." Genesis 2:15 (ESV) also states "the LORD God took the man and put him in the Garden of Eden to work it and take care of it." These verses help to define the **dominion mandate** of mankind. From the beginning, the expectation was that mankind would serve as a ruler over creation, as a representative of God on Earth (this is in part what being "made in the image of God" means). Man is to rule over all creation, and to do so in a way that cares for God's world, recognizing that it is His creation, not ours. By birthright, this duty falls to each and every one of us, and learning about this world will aid us in fulfilling our God-given tasks as stewards.

So what do we have to manage? For Adam and Eve, their role as stewards was to tend the Garden of Eden, to bring forth its abundance and bounty. Thinking about this for a moment, it means that the Garden would not simply provide Adam and Eve with everything they desired. They actually had to *work* for things to happen correctly; it was the plan for humanity from the beginning. Granted, their work was not difficult or overly laborious; the curse on the ground after Adam and Eve's sin made their formerly pleasant work turn difficult and hard. Sin now pervades the world and has affected our ability to properly manage the planet as God's stewards. Yet our role has not changed; we are not relieved from our duties simply because we do not do them properly.

Stewards are caretakers. They are not the owner or ruler of an area, but they are put in charge by the owner/ruler and act on their behalf. So we are to act as God's representatives as caretakers over His world and use its resources wisely. A **resource** is any material that can be used by people, usually in reference to earth materials, organisms, land, and energy resources. Scientists and educators usually divide the resources into **renewable** and **non-renewable resources** (figure 1.15). Renewable resources are those which can be replenished at about the same rate in which they are used. Wood for construction, water from dams, and sunshine for solar energy are all examples of renewable resources. Non-renewable resources are those that exist in fixed quantities. These include rock and mineral resources, such as building stone and ores, or fossil fuels like petroleum and natural gas.

This division is a useful one, but it has its limitations. Some renewable resources are actually quite rare and may be non-renewable in certain cases. For example, water resources are abundant in the eastern United States but very limited in the west and southwest areas of the country. Water from streams and underground aquifers must be strictly managed, often as if they were non-renewable, because the rate of replenishment is much slower than the rate of extraction. On the other hand, some of the non-renewables can be treated as effectively inexhaustible: the vast abundance of granite, limestone, and other important rock and mineral materials mean that there is no real limit on their availability. Their limitations are based on acquiring mining rights and permits, the presences of particular types of rocks in an area, and on the technology that determines how deep a quarry can go before having

FIGURE 1.15. There are a wide variety of resources, from physical materials to food to land to energy. Renewable resources, such as trees harvested by the logging industry (a) can be replenished over time, while non-renewable resources like ores for making metals; (b) the Fimiston Open Pit Gold Mine in Australia exist in fixed quantities.

to close and reclaim the land. In other cases, non-renewable resources can be recycled, as in the case of many metals. While this does not increase their abundance or make them renewable, efficient recycling can greatly extend a non-renewable resource.

As defined above, a *resource* are those things that are useful to people. In terms of economic forecasting a resource represents the highest possible amount of that material that may be available in an area. A **reserve** is a resource that has been determined to exist or has a good likelihood of existing in the resource area, and a **proved reserve** is a reserve that is known to exist and can be recovered economically. Each term becomes more narrowly defined, and the potential recovery value gets smaller. For example, let's say that a state geological survey estimates that a shale deposit contains a large amount of natural gas (methane), and reports the estimated volume of gas at 1 trillion cubic feet. This is the amount of the resource that exists. An energy company specializing in natural gas then estimates that the amount of gas that they could possibly extract from the whole resource area is 100 billion cubic feet (the reserve value), and they expected to actually recover 10 billion cubic feet (the proved reserves). Proved reserves can change with the price of the commodity: if the price goes up, the more expensive sources

of the commodity become possible to extract, and the amount of proved reserve increases. Likewise, if the price goes down, companies will shift to fewer and more profitable sources, and the amount of proved reserve decreases. When world leaders, corporations, and reporters discuss the world's natural resources, it is helpful to keep these terms in mind.

Stewardship is not only about resource acquisition, it is also about care of the environment from which the resources come. We must realize that there is *no way* to use resources without affecting the environment. All actions have consequences that are good, bad, or (more often) some combination of the two. Since we are commanded to manage this planet with an aim towards human benefit, we cannot view all use of resources to be bad or wrong. That leads us to worship the creation instead of its Creator. Human beings are not alien invaders to this world; the world exists because God wanted to create us, and this is our home. So our job of stewardship is one of wisdom, where we recognize needs or desires, and bring them about while treating the world as God intended us to. The effects of sin and the Fall hamper this, but through prayer, study, and God's inspiration, we can do better. Consider the following two examples:

The Cuyahoga River flows through northern Ohio, entering Lake Erie at Cleveland. In 1969, an oil

FIGURE 1.16. The Cuyahoga River in northern Ohio was once so polluted that it occasionally caught on fire, such as the photo above from 1952 (upper). The water was so polluted that in the 1980s, ecological surveys often came back with only a handful of fish (not fish species, just a few *total* fish). Today, the water quality is much improved (lower), and as many as 40 fish species now make the upper Cuyahoga River their home.

slick from chemicals dumped in the river by industries caught fire. This was not the first time that the Cuyahoga River had caught fire; there had been several before, beginning in the early 1900s. But news of this particular fire went national when it was featured in *Time* magazine (which actually showed a picture from a previous fire in 1952, not the 1969 fire; figure 1.16), and the Cuyahoga River became a call for a national program to clean up the polluted waterways of the United States. What emerged in 1972 was the Clean Water Act, which allows the Environmental Protection Agency (EPA) to monitor and regulate water quality. A lack of proper care for the creation resulted in a *river* catching on fire. The recognition of this poor stewardship motivated people to action and regulation, which substantially improved the quality of the river and lake ecosystems in the United States.

A second example involves quarries that produce rock materials for roads, concrete, and other types of construction materials. Before establishing a particular quarry in Lynchburg, Virginia, a local quarry company (the Boxley Company) needed to properly address a stream flowing through what would become the main pit of the quarry. This small stream feeds into another stream, which eventually flows into the James River. The quarry spent time planning out an alternate route for the stream, and worked on building a new stream channel that would go around the quarry and re-enter its existing channel further downstream (figure 1.17). By creating small meanders and planting appropriate vegetation, Boxley crafted an equivalent channel that bypassed the quarry zone. Once the quarry is exhausted and closed, Boxley has plans to convert the quarry to a reclaimed lake. While the landscape will be permanently altered (there was no lake there before, and all actions have consequences), the area and its environment will be properly managed.

Courtesy Josh Wilson Boxley Company

FIGURE 1.17. This small stream was carefully repositioned around a quarry site owned by the Boxley Corporation outside of Lynchburg, Virginia.

1.1 An Interdisciplinary Science

- Earth science comprises a number of disciplines: geology, hydrology, oceanography, meteorology (or, atmospheric science), and astronomy.
- Geology is divided into physical and historical branches. Physical geology deals with present-day events, materials, and processes, while historical geology attempts to reconstruct past geological events and conditions.

1.2 Models of Earth and Its History

- Four major views of evolution are held today: young-Earth creation, old-Earth creation, theistic evolution, and naturalistic evolution. The first three are positions held by Christians, while naturalistic evolution is agnostic or atheistic.
- Young-Earth creation is the historic position of Christianity and Judaism. It follows a close reading of the Bible, and affirms that the earth and universe are only a few thousands of years old, and that a catastrophic flood at the time of Noah formed much of the sedimentary rocks and fossils.
- Young-Earth creation and naturalistic evolution are the most dissimilar of the four positions, and are contrasted at times in this book.
- A case for any of the origins positions must be made on its own merits. They cannot advance our understanding if they are only negative critiques.

1.3 Doing Good Science

- A variety of scientific methods are used by Earth scientists to investigate the planet, its components, interactions, and history.
- Experimental sciences are those which rely on repeatable and controllable tests of hypotheses.
- Historical sciences investigate past events that cannot be recreated in the same way that experimental sciences can. Historical sciences attempt to narrow answers among multiple competing hypotheses.
- Reliable eyewitness testimony serves to inform us on historical matters.
- The Bible and science are different domains with different methods, but they can inform each other to provide insights and hypotheses across an interface.

1.4 The Earth: Components and Context

- There are four major divisions (spheres) to Earth: atmosphere (gas), hydrosphere (water), geosphere (solid earth), and biosphere (life).
- The atmosphere is composed almost entirely of just three gasses: nitrogen, oxygen, and argon.
- The bulk (97%) of the hydrosphere is composed of saline water, most of which is in the oceans. Glaciers and groundwater are the major reservoirs of fresh water.

- The geosphere is divided into three primary units of different chemical composition: the crust, mantle, and core. The crust and upper mantle form the lithosphere and "plates", which ride over top of the asthenosphere. The core is made of iron and nickel, and the liquid outer core transmits Earth's magnetic field.
- Both in the oceans and on land, the biosphere is most productive in regions of abundant nutrients, water, and sunlight.

1.5 Earth Science: Systems and Synthesis

- Systems are composed of several interactive components that together form a complex unit.
- The Earth acts as a system, with numerous interfaces between the various subsystems and cycles (e.g., soils and the hydrologic cycle).

1.6 Dominion, Stewardship, and the Environment

- The dominion mandate of Christian theology affirms that humans were created by God in order to serve as stewards over the planet.
- Our planet has many different types of resources. Some of them are renewable, while others are non-renewable.
- The effects of sin have often hampered our ability to act as proper stewards, but with prayer and careful study, we can manage our planet honorably before God.

KEY TERMS

Asthenosphere	Hypothesis
Astronomy	Interface
Atmosphere	Lithosphere
Biosphere	Mantle
Core (outer and inner)	Meteorology
Crust (continental and oceanic)	Naturalistic evolution
Dominion mandate	Oceanography
Earth Science	Old-Earth creation
Earth systems science	Proved reserve
Experimental science	Reserve
Eyewitness testimony	Resource (renewable and non-renewable)
Geology	Scientific methods
Geosphere	Spheres
Historical science	System
Hydrology	Theistic Evolution
Hydrosphere	Young-Earth creation

1. Describe the contributions and interactions of the Earth's spheres at the beach/shoreline.

2. Complete the following diagram by labeling the lettered components of the geosphere, and the depths (in km) of each boundary.

3. An atmospheric scientist is working on determining the relationship between temperature, moisture content, and ice crystal formation by manipulating an artificial atmosphere in the laboratory. Is this research experimental or historical science? Explain your choice.

4. What are the two sources of energy that drive Earth processes? What are two examples of these processes at work?

5. Provide a brief comparison of young-Earth creation and naturalistic evolution concerning the following issues: use of the Bible, age of the earth and universe, and ancestry of life forms.

6. What verses from the Book of Genesis contribute to the dominion mandate? How are the terms and language used in these verses similar? How are they different?

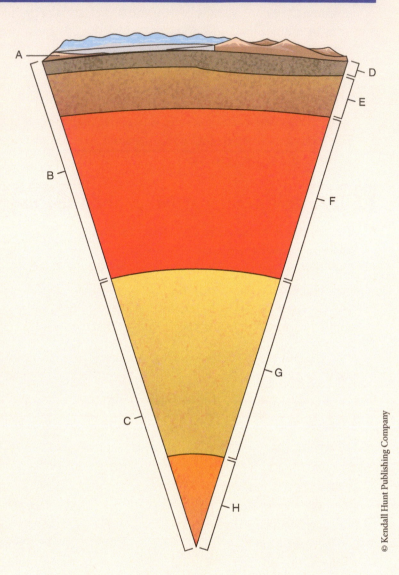

© Kendall Hunt Publishing Company

FURTHER READING

Young-Earth Creation

Brand, Leonard. *Faith, Reason, and Earth History*, 2nd edition. Andrews University Press. 2009.

Garner, Paul. *The New Creationism*. Grace Distributions. 2009.

Wise, Kurt. *Faith, Form, and Time*. B&H Books. 2002.

Christianity and the Environment

The Cornwall Alliance: www.cornwallalliance.org

Evangelical Environmental Network: www.creationcare.org

REFLECT ON SCRIPTURE

Read Genesis 1, then Psalm 104. These are two beautiful descriptions of God's creation in historical narrative and poetic styles, respectively. Write down how the two are similar in their descriptions, terms, and order of events. Also consider their differences in style and in their discussion of the creation.

CHAPTER 2

MINERALS

OUTLINE:

© DimasEKB/Shutterstock.com

A quarry in Crete, Greece, where the mineral gypsum is mined. Inset: A crystal of selenite (a form of gypsum) and drywall, one of the products made from gypsum.

27

2.1 INTRODUCTION

Minerals are all around us, and often in unexpected places. The metals and gems that make up jewelry are easy to spot, as are the beautiful crystals seen in museum collections. But look closer, and you will find minerals nearly everywhere. A handful of dirt contains sand-sized grains of quartz and fine clays. The computer chips inside our televisions, phones, tablets, and microwaves are made up of metals (like gold and copper) and silica, which comes from pure quartz sand. The interior walls of our homes, made from drywall or plaster, contain minerals of gypsum. Our sidewalks are made with calcite from limestone, and our batteries are made with exotic elements, such as lithium and yttrium, mined from only a few locations on Earth. With over 4,900 minerals known (and counting), God has provided us with an amazing and diverse array of minerals!

2.2 WHAT EXACTLY IS A MINERAL?

People use the word "mineral" to describe several different things. In areas of food and health, "minerals" are nutrients that are needed for our body's proper functions (usually various ions, like potassium, iron, and zinc). In the mining industry, "mineral" resources often refer to any material extracted from rocks, such as sand, limestone, and iron ore. We also think of gemstones and beautiful crystals as "minerals" when shopping for jewelry.

Geologically speaking, a **mineral** *is a naturally occurring object that is a crystalline solid, generally inorganic, and has a definite chemical formula* (figures 2.1 and 2.2). Unsurprisingly, *mineralogy* is the study of minerals, and a *mineralogist* is the one who studies them. In thinking about this geological definition of a mineral, it is helpful to break down each of the components. We refer to these characteristics shared by all minerals as the **mineral attributes**:

1. *Naturally occurring*—To be a mineral, the object must form without aid or intervention by human activity, and samples must be found in a geological setting. While many minerals can be reproduced in laboratories, these are termed *synthetic minerals*, to distinguish them from their counterparts in nature.

2. *Crystalline*—All minerals have a regular and repetitive framework that results from an organized set of atoms (figure 2.3). This ordered structure controls the shape that mineral crystals display. Materials like opal that are *amorphous* (they have no ordered atomic structure) are termed *mineraloids*.

3. *Solid*—An object must be solid at Earth's surface in order to be a mineral. When frozen, ice can be considered a mineral, but in liquid form it cannot.

4. *(Generally) Inorganic*—Minerals are typically understood as being solely the products of physics and chemistry, without the direction of biological agents. However, many organisms regularly create orderly crystalline structures (such as the shells of clams and corals) that are, for all intents and purposes, minerals, and are occasionally important in rock formation. We may refer to these as *biominerals*.

FIGURE 2.1. *A few types of minerals* (a) Aquamarine, a form of beryl. (b) Garnet is a deep red mineral that can form in sheet-like structures or 12-sided balls. (c) Sulfur is yellow and emits an odor when scratched. (d) Azurite is a deep blue mineral that forms crusts.

5. *Definite chemical formula*—each mineral is characterized by a set of elements that allow the mineral to be written as a chemical formula. Some minerals' formula is simple, such as SiO_2 (quartz) and $NaCl$ (halite), while others like $K(Mg, Fe)_3AlSiO_{10}(OH)_2$ (biotite mica group) are far more complex. In some cases, different elements can be included in the same spot in the mineral without changing the mineral's overall structure (such as the "Mg, Fe" in biotite, meaning there can be either magnesium

or iron in this spot). It is the types of atoms and their bonding that control the crystalline structure.

How, then, is a rock different from a mineral? In most cases, a **rock** *is an aggregate (a collection) of one or more minerals.* In other cases, a rock may be composed of mineraloids (such as volcanic glasses) or solid organic matter (such as coal seams). Minerals, then are the building blocks of rocks (figure 2.4).

FIGURE 2.2. *Some non-mineral Earth materials*
(a) Obsidian is a non-crystalline material.
(b) Granite is a rock made of visible mineral grains. (c) Coal is made from plant material.
(d) Crude oil (petroleum) is liquid and ultimately made from plankton.

FIGURE 2.3. (a) The mineral halite (salt). (b) The crystal framework of halite.

Salt   NaCl

- Na
- Cl

FIGURE 2.4. Left: An outcrop of the igneous rock gabbro. Right: A close-up shows that the gabbro is made up of many individual mineral crystals.

2.3 ATOMS AND BONDING: FOUNDATIONS OF MINERALS

2.3.1 Elements, Atoms, and Electrons

Since minerals are a product of the bonding and regular arrangement of atoms, it is useful to briefly discuss these fundamental components of matter. First, we can define an **atom** as *the smallest particle of matter that cannot be split into simpler substances by chemical processes*. Atoms are composed of three basic materials (figure 2.5): **protons** are positively charged particles that are confined to the atom's central region, called the **nucleus**; **neutrons** are also in the nucleus, but have no charge; **electrons** are extremely small, negatively charged particles that orbit the nucleus. Because the charges of protons and electrons are opposite but equal in strength, in a neutral atom the number of electrons matches the number of protons.

The **periodic table** of the elements (figure 2.6) organizes atoms according to the number of protons in the nucleus, which is called the **atomic number**. All of the atoms with the same number of protons are representatives of one **element**, a material of characteristic physical and chemical properties that cannot be broken down into simpler substance by chemical processes.

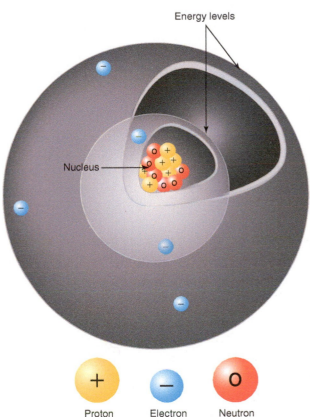

FIGURE 2.5. Basic structure of an atom. Positively-charge protons (yellow) and non-charged neutrons (red) occupy the nucleus while negatively-charged electrons (blue) orbit the nucleus at various distances.

Periodic Table of the Elements

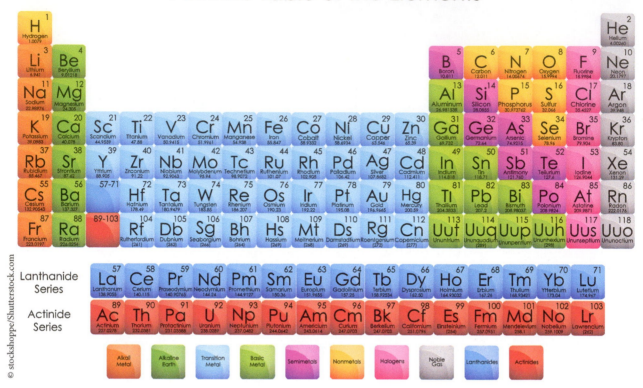

FIGURE 2.6. The periodic table of the elements.

For example, all atoms with six protons in their nucleus are carbon atoms, and all atoms with seven protons in their nucleus are nitrogen atoms. They will each have neutrons in their nucleus as well, and be surrounded by six and seven electrons, respectively.

The total of all protons AND neutrons is the **atomic mass** of the atom. On the periodic table, the atomic number is listed above the element's symbol, and the atomic mass is below. Notice that the atomic number is represented by a whole number while the atomic mass has a decimal with numbers following. This is because the number of protons defines an element, while the number of neutrons varies from one atom to another of the same element and is not a fixed number all the time. Atoms of the same element that have different numbers of neutrons are called **isotopes**, and they exist in different abundances within nature.

Carbon, for example, exists in three isotopes: $^{12}_{6}C$, $^{13}_{6}C$, and $^{14}_{6}C$ (figure 2.7). Because they are all carbon atoms (chemical symbol "C"), each has six protons in their nucleus (shown with the subscript: $_{6}C$). Due to having six, seven, or eight neutrons in the nucleus, the atomic mass can be twelve, thirteen, or fourteen (shown with the superscript: ^{12}C, etc.). The ratios of the isotopes have been discovered by scientists and are used to determine the average atomic mass of the element on Earth, which is represented in the decimal number on the periodic table. Since $^{12}_{6}C$ is by far the most common isotope of carbon, the atomic mass of the element is very close to 12.

The electrons of an atom orbit the nucleus. It is easiest to think of these electrons like planets orbiting a star, though this concept is a bit inaccurate. Electrons are free to orbit over various distances away from the atom's nucleus, but the laws of physics only allow so many of them to be in any given distance

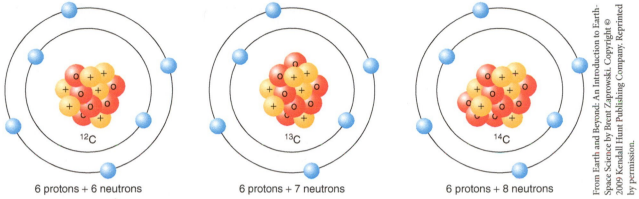

From Earth and Beyond: An Introduction to Earth-Space Science by Brent Zaprowski. Copyright © 2009 Kendall Hunt Publishing Company. Reprinted by permission.

6 protons + 6 neutrons	6 protons + 7 neutrons	6 protons + 8 neutrons

FIGURE 2.7. Three isotopes of carbon.

from the nucleus (figure 5). But when it comes to linking atoms together to form **compounds**, the outermost electrons (called *valence electrons*) control what types of bonds, and how many, will form.

2.3.2 Types of Bonds

In the outermost portion of the electron cloud, often called the *valence shell*, eight electrons can be held (the exception to this is hydrogen and helium, which can hold at most two electrons). The **octet rule** states that atoms are most stable when their valence shell is filled with eight electrons. Since only the noble gasses possess a full valence shell by themselves, following the octet rule is done via **chemical bonds**, which

link atoms by gaining, losing, or sharing electrons. The three most common types of chemical bonds are ionic, covalent, and metallic. A fourth type, the hydrogen bond, is a bit different but very important in mineralogy.

Ionic bonds are formed when one atom transfers one or more electrons to another atom. The loss or gain of the electron(s) creates an imbalance in the number of electrons compared to the protons in the nucleus, and the once-neutral atom becomes a positively- or negatively charged **ion** (figure 2.8). Like in love, opposites attract, so a negatively-charged ion (called an *anion*) is attracted to a positively-charged ion (called a *cation*), and the two stick together like magnets (figure 2.9). Ionic bonds can form

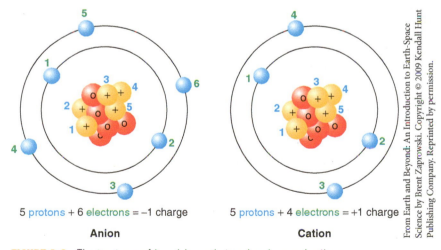

5 protons + 6 electrons = –1 charge	5 protons + 4 electrons = +1 charge
Anion	**Cation**

From Earth and Beyond: An Introduction to Earth-Space Science by Brent Zaprowski. Copyright © 2009 Kendall Hunt Publishing Company. Reprinted by permission.

FIGURE 2.8. The two types of ions (charged atoms): anions and cations.

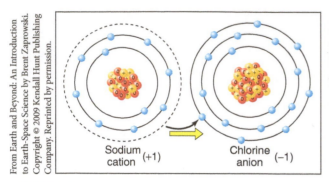

FIGURE 2.9. Salt (Sodium chloride or NaCl) forms when a positively-charged sodium ion is attracted to a negatively-charged chlorine ion. This is also known as ionic bonding.

FIGURE 2.10. The mineral halite (rock salt) is formed by ionic bonds between sodium and chlorine.

between single-element ions, or groups of atoms that together have a charge (figure 2.9).

The most familiar material displaying ionic bonding is common table salt, geologically known as halite (figure 2.10). Halite is a bond of positive sodium ions (Na^+) with negative chlorine ions (Cl^-). Sodium has only one electron in its valence shell. Losing it is energetically favored, and so it quickly escapes under many different conditions. In contrast, chlorine has seven electrons in its valence shell, and filling it with a stray electron is likewise energetically favored. The result is two roughly equal and oppositely charged ions, which quickly form an ionic bond to form an electrically neutral molecule.

$$Na^+ + Cl^- \rightarrow NaCl$$

Covalent bonds involve two or more atoms sharing each other's electrons in order to fill their valence shells. These bonds are much stronger than ionic bonds. In water (figure 2.11), two hydrogen atoms each need one additional electron to fill their valence shell (which can hold only two electrons). Oxygen has six electrons in its valence shell, and needs two electrons to fill its octet. By sharing an electron with each hydrogen (which also share one electron each from the oxygen atom), all three atoms fill their valence shell through covalent bonds.

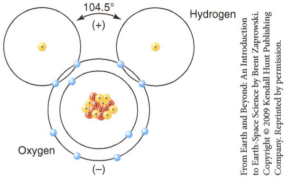

FIGURE 2.11. A water molecule, displaying covalent bonding and polar structure. Covalent bonds between hydrogen and oxygen in a molecule of water. The angle between the hydrogen nuclei is 104.5°. The polar nature of water is due to the proximity of all the electrons to the oxygen, making that region is negatively charged, while the region near the hydrogen nuclei is positive.

Metallic bonds, like their name suggests, are found among metals, such as copper, iron, and aluminum. In metallic bonding, the valence electrons are shared among neighboring atoms, but not so strongly as in covalent bonds (figure 2.12). Instead, the electrons are relatively free to move among the three-dimensional structure of the solid. Because the electrons can move from atom to atom, metals can conduct heat and electrical charges easily, and be hammered into sheets and drawn out into wire. These properties have made metals into very sought-after materials, recovered from various rocks and minerals around the world.

CHAPTER 2: Minerals

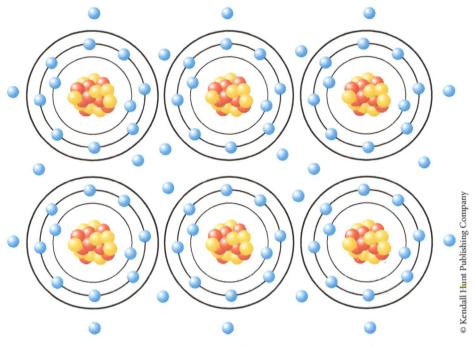

FIGURE 2.12. Metallic bonds. Valence electrons are able to move easily from one metal atom to another.

Hydrogen bonds are loose connections between weakly charged regions of overall neutral molecules. The arrangement of bonds in and around the physical shape of some molecules results in portions of that molecule where hydrogen atoms are exposed at edges, and their electrons bonded toward another atom in the molecule.

Water once again serves as an example (figure 2.11). Notice that the atoms of the water molecule are not oriented in a straight line. Instead, the hydrogen atoms are "bent" at an internal angle of 104.5°. This is because the electrons are pulled closer to the oxygen atom than to the hydrogens, and the clustering of the negatively charged electrons around the oxygen pushes the hydrogen atoms toward one side of the molecule. This gives the area around the oxygen a negative charge, and the exposed, positively-charged hydrogen nuclei give the area around the hydrogens a positive charge.

The positive and negative regions, or "poles", result in water being a *polar molecule*. These regions are attracted to opposite charges, such as those from other water molecules. This causes water molecules to weakly "stick" to each other along positive-negative attractions, and is the reason for water's surface tension (figure 2.13). It also allows ionically bonded molecules, such as halite, to break apart and dissolve easily in water (figure 2.14).

Hydrogen bonds are important in minerals, where they often play a role in minerals like micas and clays in controlling the ways these minerals combine or break along their surfaces.

2.3.3 The Making of Mineral Crystals

Crystals form by the accumulation of ions. In magmas (chapter 3), for example, many different ions and compounds are found within the liquid rock. As these ions and molecules interact, they begin building mineral crystals according to the types of bonds that they can make, and the sorts of atoms present in the magma. Once a certain kind of mineral begins to form, only the same molecules as the original "seed" crystal will join the growing crystal.

FIGURE 2.13. The "pond skater" insects can walk on water due to surface tension.

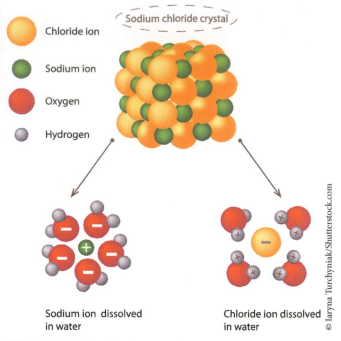

Sodium chloride crystal

Chloride ion

Sodium ion

Oxygen

Hydrogen

Sodium ion dissolved in water

Chloride ion dissolved in water

FIGURE 2.14. How halite (rock salt) dissociates among polar water molecules.

So if oxygen (O) begins joining with silicon (Si) in a 2:1 ratio to form a quartz crystal, most other types of atoms are excluded because they won't connect with the established pattern of silicon and oxygen atoms.

In a way, a crystal's atomic structure is like a scaffold or framework (figure 2.15). This process, and the others in igneous, sedimentary and metamorphic rocks, is responsible for forming the many types of minerals that make up the rocks of our planet.

FIGURE 2.15. (a) A quartz crystal (b) The Washington Monument. The scaffolding around it was needed for repairs after a magnitude 6.0 earthquake in 2011.

Once we have decided that what we are looking at is a mineral (it possesses all of the *mineral attributes* defined earlier), our next question is determining what particular mineral it is. As of 2014, the International Mineralogical Association lists over 4,900 different minerals, with more being discovered all the time. The vast majority of these minerals are extremely rare and only form under very specific circumstances. A few dozen minerals are common, and just nine make up 90% of the rocks we find on the surface of the Earth!

To determine the identity of a mineral, we need to look at its **mineral properties**. These are the unique physical characteristics that a mineral possesses, which fall into categories of optical, shape, mass-related, mechanical, and other properties. By observing and testing for these properties, you can determine the identity of a mineral.

2.4.1 Optical Properties

The first thing that anyone looking at a mineral notices is its *color*. Then we observe that light is reflected off of the mineral's surface in some way, which we call the mineral's **luster**. Luster is often grouped into two large categories, *metallic* and *non-metallic* (figure 2.16). In the metallic category, light is reflected off of the surface in a way that mimics the look of metals like

© papa1266/Shutterstock, Inc.

© Allocricetulus/Shutterstock, Inc.

© Siim Sepp/Shutterstock, Inc.

© Brian C. Weed/Shutterstock, Inc.

© Imfoto/Shutterstock, Inc.

© Only Fabrizio/Shutterstock, Inc.

FIGURE 2.16. Mineral lusters. Top row from left to right, pyrite, galena, and copper display a metallic luster. Bottom row from left to right, some non-metallic mineral lusters: glassy (ruby), earthy (kaolin), pearly (muscovite mica).

gold, iron, or copper. Non-metallic lusters are a wide variety of appearances, including dull, earthy, glassy, pearly, resinous, or satiny.

One other helpful visual clue is called **streak**. This is the color of a mineral when it is powdered, usually by scratching it on a ceramic tile (figure 2.17). A mineral's streak color is very consistent, which is very useful if the mineral's overall color is variable.

A few transparent minerals are capable of splitting light into two beams, creating a *double refraction* of an image (figure 2.18).

2.4.2 Shape Properties

Minerals take on particular shapes as they grow by the accumulation of ions. If given enough space, they will form consistent geometric shapes, called

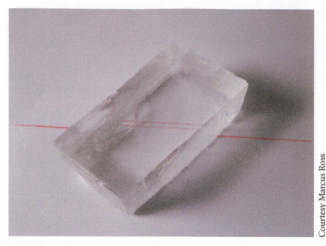

FIGURE 2.18. Double refraction by a clear calcite crystal.

FIGURE 2.17. Streak is the color of a mineral when powdered, and is most useful in identifying metallic minerals. Hematite (top two minerals) leaves a brick-red streak on a ceramic tile, regardless of the sample's color. Pyrite (lower right) has a dark gray streak despite the mineral's golden color, while copper (lower left) leaves a streak identical to its color.

the **crystal form** or **habit** of the mineral. Some minerals will form blades, prisms, or cubes, while others might be platy or crusts (figure 2.19). The flat surfaces formed as crystals grow are called *crystal faces.*

Sometimes the same materials can form a variety of different minerals or crystal forms, due to differences in the conditions of formation. These different minerals are called **polymorphs**. For example, two mineral polymorphs of carbon are diamonds and graphite. In diamonds, the carbon atoms are arranged in tetraheda, whereas in graphite they are arranged in sheets. These structural differences result in both one of the hardest and softest minerals on Earth, both from the same single element!

2.4.3 Mass-Related Properties

The **specific gravity** of a mineral is the density of that mineral divided by the density of water at 4°C. Since it is a density (in g/cm³) divided by another density (also in g/cm³), specific gravity is a unitless value. For example, the specific gravity of galena (an ore of lead, and one of the densest minerals) is 7.58, while the specific gravity of quartz is 2.65.

(a)

(b)

(c)

(d)

(e)

FIGURE 2.19. Various crystal forms, also known as habits. (a) cubic, (b) prism, (c) dendritic, (d) pyramidal, and (e) dodecahedral.

2.4.4 Mechanical Properties

Many mineral properties are connected to mechanical behavior; that is, how the mineral behaves under stress. A mineral's **hardness** measures its resistance to being scratched. Geologists often use a scale developed by German mineralogist Friedrich Mohs (figure 2.20). Mohs' scale is based on ten minerals (figure 2.21) which are relatively common. The scale is relative, and neither linear nor exponential. Using a variety of simple and common materials and tools that have a known hardness (such as a knife blade, glass plate, or penny), you can make a rough but accurate determination for a mineral's hardness. For example, if a mineral can be scratched by a penny (hardness of 3) but not by your fingernail (hardness of 2.5), then the mineral's hardness is somewhere in between these. If two minerals are very similar, more specialized laboratory techniques and equipment are needed to determine their absolute hardness.

Tenacity refers to the behavior of a mineral when it is broken or deformed (including bent, stretched, pounded, etc.). Common terms to describe a mineral's tenacity include *flexible, brittle, elastic, malleable, ductile, and sectile.*

If a mineral is broken, the crystalline framework will allow breaks to follow particular paths, or none at all. When a mineral breaks along a plane of weakness, it produces a flat surface called a **cleavage plane**. Minerals can possess zero cleavage planes or several (figure 2.22). Cleavage planes are always parallel to each other, but they may or may not be parallel to any of the crystal faces of the mineral's crystal form/habit. Depending on how rough or flat the surfaces are when broken, a mineral's cleavage may be described as poor, fair, good, or excellent.

If a mineral breaks and shows no evidence of parallel cleavage planes, the break is considered **fracture** (figure 2.22). Fractures can fall under several categories that describe their appearance, such as *splintery, irregular,* or *conchoidal.*

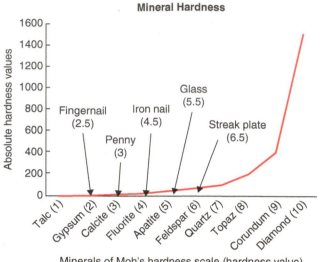

Mineral Hardness

FIGURE 2.21. Mineral hardness according to Mohs' scale (horizontal) and absolute hardness (vertical scale). Useful testing items are noted according to their Mohs hardness value.

FIGURE 2.20. Geologist Friedrich Mohs invented a useful scale of mineral hardness.

FIGURE 2.22. Cleavage and fracture among various minerals. (a) Chalcedony (a variety of quartz) displays smooth, rounded *conchoidal fracture*. (b) Irregular fracture in hematite. (c) Muscovite mica shows excellent cleavage in one direction, creating flat sheets. (d) Potassium feldspar displays two planes of cleavage at right angles to each other. (e) Flourite displays a cubic habit (green), but octagonal cleavage (purple).

FIGURE 2.23. The many cuts of diamonds all follow the mineral's cleavage planes.

The most familiar experience we have with mineral cleavage is in jewelry. The various cuts of diamonds, emeralds, and other precious gems are determined by the number and types of cleavage planes that the mineral possesses (figure 2.23).

2.4.5 Other Properties

While all minerals display cleavage or fracture (or some combination), and each has a color and a specific gravity, there are some properties that are rarer but very useful. Some minerals, such as kaolin and sulfur, emit an *odor* when scratched (earthy and rotten-egg, respectively). Others have distinct *tastes*, such as the salty flavors of halite (NaCl) and sylvite (KCl).

Some common minerals display *magnetism*, in which they are attracted to a magnetic field (figure 2.24). A few minerals will react with acids. For example, when dilute hydrochloric acid (HCl) is applied to calcite and dolomite (the main minerals in limestone and other carbonate sedimentary rocks [see chapter 3]), the reaction releases gasses, and the mineral displays *effervescence* (figure 2.25).

FIGURE 2.24. The mineral magnetite can attract iron-bearing objects, such as iron filings.

FIGURE 2.25. Carbonate minerals, including those used in antacids, react with acid to make carbon dioxide bubbles.

2.5 CLASSIFYING MINERALS INTO GROUPS

With over 4,900 known minerals (and counting!), it is perhaps surprising that this wide diversity actually falls into just a few major groups, yet classifying minerals is quite straightforward once their chemistry is known.

Since minerals grow by the accumulation of ions, it is the ions that help us create a classification system. In particular, the negatively charged anions define the mineral groups, and the positively charged cations determine the particular minerals within the group.

So what materials are available with which to build minerals? Out of all the elements on the periodic table, a mere eight make up over 98% of all materials in the crust. And of those eight, oxygen (46.6%) and silicon (27.7%) far surpass all others (figure 2.26).

2.5.1 Silicate Minerals

The vast majority of minerals of Earth's crust are **silicate minerals**, such as quartz, feldspars, olivine, and micas. These minerals are very different from one another in their mineral properties (color, cleavage, density, etc.), but all of them share a common building block: the **silicon-oxygen tetrahedron**. The silicon-oxygen tetrahedron is an anion made of a central atom of silicon covalently bonded to four oxygen atoms. The silicon's octet is filled, and each oxygen atom can form one additional bond (–1) with another atom, giving us the anion SiO_4^{4-} (figure 2.27). From this building block, 92% of the minerals of Earth's crust are made.

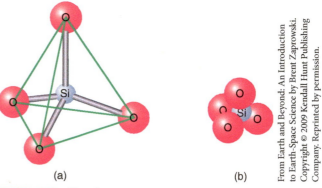

From Earth and Beyond: An Introduction to Earth-Space Science by Brent Zaprowski. Copyright © 2009 Kendall Hunt Publishing Company. Reprinted by permission.

FIGURE 2.27. The silicon-oxygen tetrahedron in (a) ball-and-stick model; and (b) large-ball model. The covalent bonds fill the silicon's octet, but the oxygen atoms each have one more bond available.

Relative Abundance of Elements in Earth's Crust

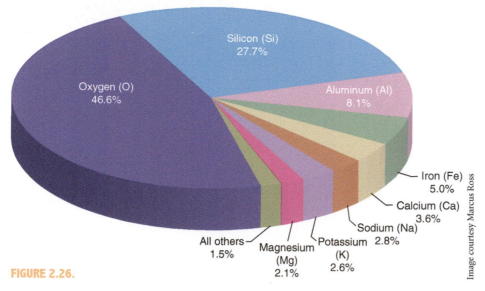

Image courtesy Marcus Ross

FIGURE 2.26.

The silicate minerals are subdivided based on how the silica-oxygen tetrahedra are arranged with other ions and with each other. The five major groupings of tetrahedra are a) *isolated*; b) *single chain*; c) *double chain*; d) *sheet*; and e) *framework silicates* (figure 2.28).

One of the most abundant minerals on Earth is olivine, whose form as a gemstone is the deep-green peridot. In it the *isolated tetrahedral* are separated from each other by iron and/or magnesium ions, and the tetrahedral do not share any bonds among

FIGURE 2.28.

Mineral Group	Silicate Structure	Common Mineral
Orthosilicates	Isolated tetrahedra 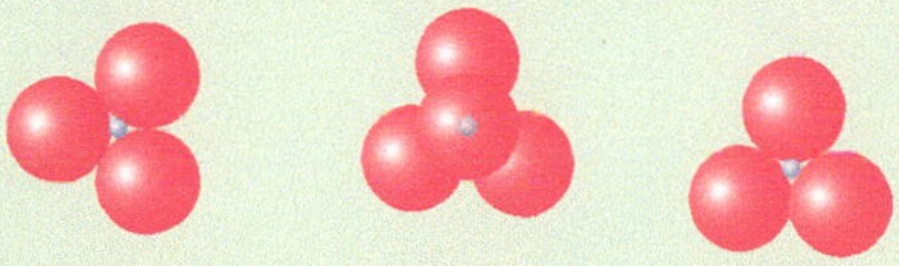	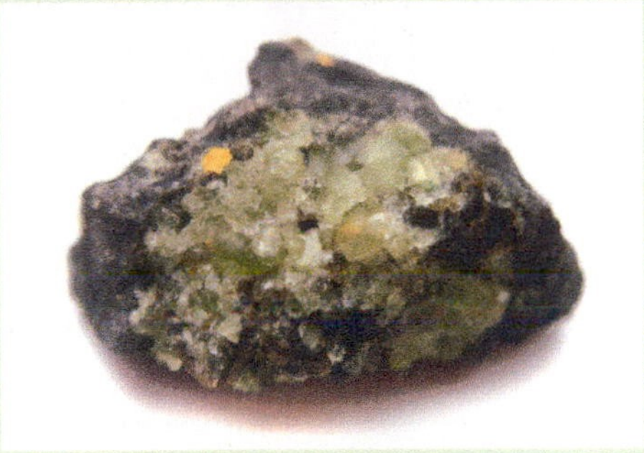 Olivine $(Mg,Fe)_2SiO_4$
Chain silicates	Single chains 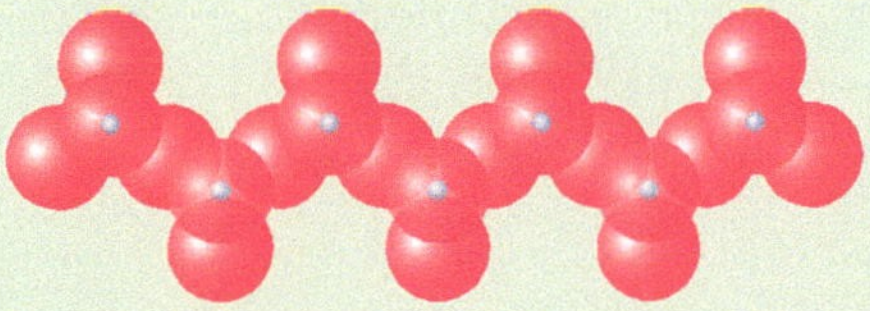	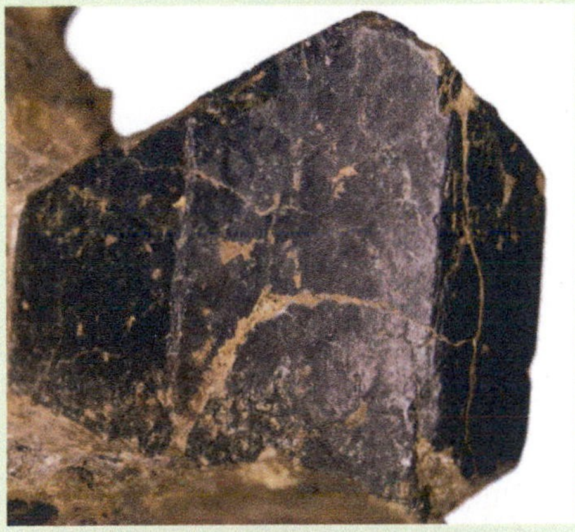 Augite $(Mg,Fe)SiO_3$
	Double chains 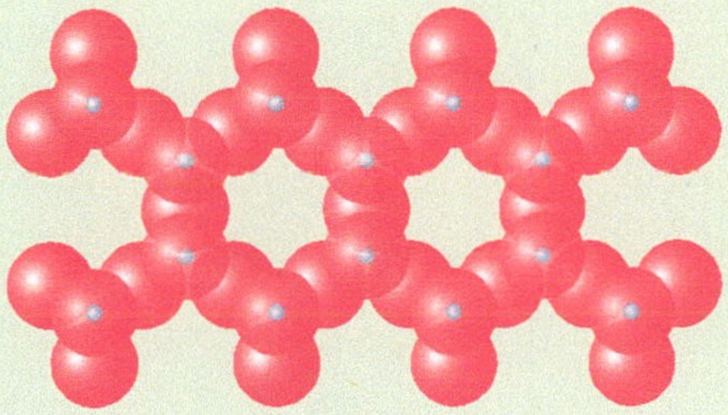	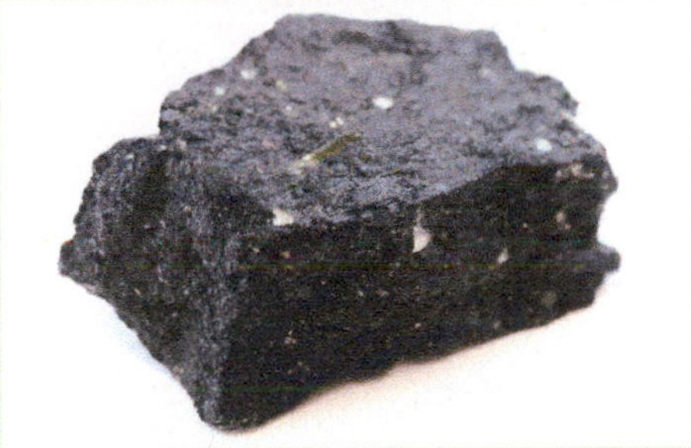 Hornblende $Ca_2(Fe,Mg)_5Si_8O_{22}(OH)_2$

(*Continued*)

FIGURE 2.28. *(Continued)*

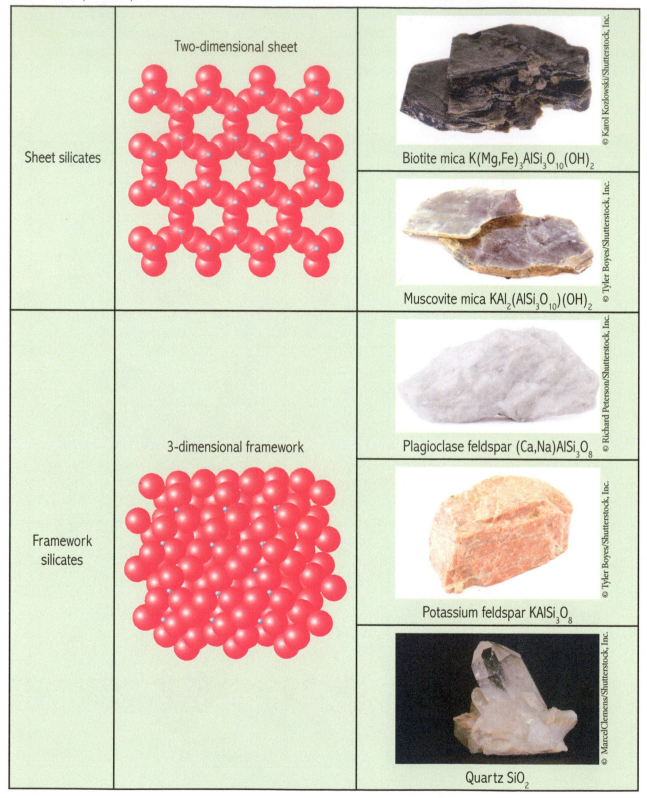

Sheet silicates	Two-dimensional sheet	Biotite mica $K(Mg,Fe)_3AlSi_3O_{10}(OH)_2$ © Karol Kozlowski/Shutterstock, Inc.
		Muscovite mica $KAl_2(AlSi_3O_{10})(OH)_2$ © Tyler Boyes/Shutterstock, Inc.
Framework silicates	3-dimensional framework	Plagioclase feldspar $(Ca,Na)AlSi_3O_8$ © Richard Peterson/Shutterstock, Inc.
		Potassium feldspar $KAlSi_3O_8$ © Tyler Boyes/Shutterstock, Inc.
		Quartz SiO_2 © MarcelClemens/Shutterstock, Inc.

their oxygen atoms. Garnet and topaz are also in this silicate mineral group.

In the dark-green to black mineral augite (a member of the pyroxene group), the tetrahedra form a *single chain* by sharing two oxygen atoms with neighboring tetrahedra. Hornblende, a dark gray to black member of the amphibole group, is composed of tetrahedra that alternately share 2 or 3 of their oxygen atoms with other tetrahedra, forming a *double chain* structure.

Sheet silicates form when three of the oxygen atoms of a tetrahedron are shared with neighboring tetrahedra. The result is a two-dimensional network of tetrahedra and other ions. Micas (such as dark-colored biotite and light-colored muscovite) and clay minerals are part of this silicate group.

In the *framework silicates*, the tetrahedra share each of their four oxygen molecules with neighboring tetrahedra, creating a three-dimensional network of atoms. In quartz (SiO_2), each of the oxygen ions bond to a silicon ion in a 2:1 ratio throughout the crystal. The feldspar minerals incorporate aluminum with either potassium or calcium and sodium in definite but somewhat interchangeable proportions. This is why feldspar is considered a group of minerals by geologists: the properties of the feldspar minerals are all very similar, but made from slightly different sets of ions, much like a chocolate chip cookie might have milk chocolate, dark chocolate, or white chocolate chips, or some combination of each.

2.5.2 Non-silicate Minerals

The remaining 8% of the minerals that make up the Earth's crust are collectively called **nonsilicate** minerals, but this is not a formal group. It simply refers to all the minerals that are not characterized by the silica-oxygen tetrahedron anion. Instead, there are several groups of nonsilicates, and most are characterized by their anion (figure 2.29).

Halides are characterized by anions from the halogen group (charge of −1), such as Cl^- and F^- and Br^-. These anions are bonded to metal cations, such as Na^+ and K^+. Common halides include halite (NaCl) and fluorite (CaF_2).

Oxides have a free oxygen anion (O^{2-}) bonded to a metal cation. The mineral hematite (Fe_2O_3) is a mineral form of iron rust, and magnetite (Fe_3O_4) is a magnetic oxide mineral.

Sulfides have a free sulfur anion (S^{2-}) bonded to a metal cation. The metallic minerals galena (PbS), pyrite (FeS_2; a.k.a. fool's gold), and chalcopyrite ($CuFeS_2$) are sulfides and are important metal ore resources.

Sulfates have an SO_4^{2-} anion bonded to a metal cation. They are frequently formed at Earth's surface by precipitation out of a water solution. Gypsum ($CaSO_4 \cdot H_2O$), which is used in making drywall and plaster of Paris, is an important sulfate mineral.

After the silicates, *carbonates* are the second most abundant mineral group. They are defined by the CO_3^{2-} anion, bonded to a metal cation. Calcite ($CaCO_3$) and dolomite ($CaMg[CO_3]_2$) are abundant minerals that make up limestone and other carbonate rocks (chapter 3).

Native elements exist as minerals of a single element. Diamond and graphite are native elements of carbon, and sulfur forms native element crystals and rock deposits, especially around volcanoes. Copper, gold, and silver may form as native metal elements.

2.6 MINERAL USES: PRACTICAL PLAYERS TO PRECIOUS GEMS

There's a phrase in the mining community: "If you can't grow it, you have to mine it." Think about it for a moment as you look around the room: what materials by you right now started off as physical Earth resources, made from rocks, minerals, and fossil fuels? From plastics (oil) to concrete (rock) to

FIGURE 2.29. Several of the many non-silicate minerals. Top row (left to right): halite, hematite, and galena. Middle row: pyrite, gypsum (selenite), and calcite. Bottom row: gold and magnetite.

drywall (mineral), chances are you are surrounded by mined materials!

From the perspective of human history, it took very little time after Eden for mankind to explore and make use of mineral resources. At a most basic level, stone tools may have been used for farming by Adam and Eve either before or after their removal from the Garden of Eden. Onyx and gold are mentioned prior to the Fall (see Genesis 2:12), and by the seventh generation from Adam via Cain, Tubal-Cain had a reputation as a skilled metal craftsman. So God made mineral resources available from the beginning, with the intention that we would use them as part of our role is to act as stewards of this world (cf. Gen. 1:26). Long after the creation, the Flood's profound effect on Earth's surface (Chapter 4) had a dramatic impact on the location and availability of many mineral resources, both in destroying some and creating many others (see Box 2.1).

We can divide mineral resources into a few broad categories, and also leave the formal definition of minerals to one side. In industry terms, **mining** is the recovery of solid metallic or nonmetallic resources from rock or loose sediment. Thus mining includes excavating coal seams, rock quarries, salt mines, gravel pits, and digging for precious gems. The extraction of oil and natural gas is not considered part of the mining industry, and will be discussed in Chapter 11. There you will also be introduced to the effects of fossil fuels, mining, land use, and other human activities on the Earth system.

Metallic resources come from both the native elements (which were discovered first) and from **ores**: metal-rich minerals and rocks (figure 2.30). By mining and processing ores, metals can be extracted and purified. Nearly all of the metals we use come from ores, rather than from native elements.

Nonmetallic resources include a wide variety of important rock and mineral materials used in many different applications. For building materials, large slabs of rock, such as granite and marble are cut out of the earth by fast-moving cables, torches, or water jets. Clay, gravel, gypsum, and limestone are all very important raw materials for the production of building materials like bricks and cement (see page 26). Rocks rich in phosphorous are often mined to make fertilizers for agricultural use (figure 2.31).

Precious and semiprecious gems are those minerals (and a few other substances) that are often used in jewelry. Diamonds, ruby and sapphire (both forms of corundum), and emerald (a form of beryl) top the list for the most sought-after gems. Other common gemstones include garnet, tourmaline, aquamarine (another form of beryl), and the many colorful varieties of quartz (figure 2.32).

© Lee Prince/Shutterstock, Inc.

FIGURE 2.30. Arial view of the Kennecott Bingham Canyon Mine, the largest man-made excavation on Earth. Products include copper, gold, silver, and molybdenum.

FIGURE 2.31. A large phosphate quarry and phosphate-rich rock (inset).

FIGURE 2.32. The many varieties of quartz. Top (left to right): citrine, amethyst, and smoky quartz. Bottom: rose and milky quartz.

CHAPTER 2: Minerals

BOX 2.1

Precious Gems from Noah's Flood

Noah's Flood dramatically altered the world, as massive geological processes literally moved continents across the Earth's surface. But in spite of its destruction (indeed, *because* of it), we now have access to some amazing minerals. Diamonds, rubies, sapphires, and emeralds are the most sought-after and expensive gem-quality minerals. And the conditions needed to form them fit with our current view of how the Flood operated.

Diamonds are mineral forms of carbon, and they must be formed at great depths in the Earth (about 100 miles/160 km!). Carbon from the pre-flood ocean crust was dragged deeper into the Earth at subduction zones (see chapter 4), where high temperatures and pressures force the carbon atoms to create the four-sided covalent bonds that make diamonds so hard (10 on Mohs hardness scale). But to get to the surface, a diamond needs to take an express-speed trip to the surface. A special, gas-injected magma called a *kimberlite* must pick up and the diamonds from 160 km deep and bring it to the surface *in only one day*, or else the diamond structure will degrade into the much less-prized form of carbon, graphite. Late in the Flood and afterward, the kimberlites brought the diamonds quickly to the surface.

Diamonds may get the most attention, but rubies are actually the most expensive gemstones. More common than rubies, sapphires are much less expensive. Both rubies and sapphires are variations of the mineral corundum, and extremely hard mineral (9 on Mohs hardness scale). Only the clearest forms of red or blue corundum are considered rubies and sapphires, respectively. To form, both rubies and sapphires need very

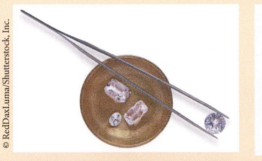

Many minerals formed during Noah's flood, including some of the most sought-after gems: diamond, ruby, sapphire, and emerald.

high temperatures and pressures, such as those found deep in the Earth's crust and where continents collided during the Flood. The new gems are then brought to the surface by basalt lavas and/or erosion.

Emerald is a green form of the mineral beryl, and is the third most valuable gemstone after diamond and ruby. Though still a very hard mineral (Mohs hardness of 7.5-8), it is the softest of this group, and the only one prized more for the richness of its color than the clarity of the mineral. Emeralds may form in several ways, but all require water and high temperatures and pressures from mountain-building, and each was present during the Flood.

For more information on how these gems formed during the Flood, see the articles by geologist Dr. Andrew Snelling in "Further Readings."

2.1 Introduction

- Minerals are important components of daily life.
- There are over 4,900 known minerals.

2.2 What Exactly Is a Mineral?

- A mineral is a naturally occurring, crystalline, solid, inorganic material with a distinct chemical formula.
- Some biological materials, such as the shell material of coral and clam shells, are also considered minerals and are called biominerals.
- Rocks are aggregates (collections) of at least one type of mineral, and often many types of minerals.

2.3 Atoms and Bonding: Foundations of Minerals

- Atoms are the building blocks for all matter, including minerals. Their basic components are protons and neutrons within the nucleus, with electrons orbiting the nucleus.
- Chemical bonds form among the valence electrons of atoms. The bonds may be covalent, ionic, metallic, or hydrogen.

2.4 Determining Mineral Identities

- A variety of mineral properties are used to identify particular minerals. The most important include color, streak, luster, hardness, crystal form, and cleavage.
- Some mineral properties, such as taste, smell, or reaction to acid are rare but very diagnostic for particular minerals or mineral groups.

2.5 Classifying Minerals into Groups

- Minerals are divided primarily into silicate and non-silicate groups. The silicates all possess the silicon-oxygen tetrahedron as a basic building block, arranged in isolated tetrahedra, chains, sheets, or three-dimensional networks.
- Non-silicate minerals s are primarily grouped according to their anions (negatively-charged components of the mineral compound). Major non-silicate groups include carbonates, halides, oxides, sulfides, and native elements.

2.6 Mineral Uses: Practical Players to Precious Gems

- Mining is the primary means of acquiring mineral resources.
- Metallic mineral resources are called ores. Non-metallic mineral resources are used in a wide variety of industrial, agricultural, and commercial areas.

Atom
Atomic mass
Atomic number
Chemical bond
Cleavage (and cleavage plane)
Compound
Covalent bond
Crystal form (or habit)
Electrons
Element
Fracture
Habit
Hardness
Hydrogen bond
Ion
Ionic bond
Isotopes
Luster

Metallic bond
Mineral
Mineral attributes
Mineral properties
Mining
Neutron
Non-silicate minerals
Octet rule
Periodic table
Polymorphs
Proton
Rock
Silicate minerals
Silicon-oxygen tetrahedron
Specific gravity
Streak
Tenacity

REVIEW QUESTIONS

1. Why is coal not considered a mineral? Which attribute(s) is it lacking?
2. Describe how covalent and ionic bonds differ, and give an example of a molecule for each type of bond.
3. Quartz and halite (salt) are both clear minerals. What are some of the properties that might distinguish them?
4. Based on the distribution of elements in the Earth's crust (figure 2.26), would you expect silicates or non-silicates to be the most abundant group of minerals?
5. Look at the chemical formulas for the silicate minerals in figure 2.28. What metals are common in the top four vs. the bottom four minerals shown? What colors dominate the top four vs. the bottom four minerals?
6. People in the mining industry often say "If you can't grow it, you have to mine it." Look around you. What are three or four objects that required mined resources? What types of mined resources are used in them?

FURTHER READING

Chesterman, C.W., and Lowe, K.E. *National Audubon Society Field Guide to North American Rocks and Minerals*. Knopf Publishers. 1979.

Nesse, W.D. *Introduction to Mineralogy*, 2nd ed. Oxford University Press. 2011.

www.mindat.org (a huge online database of minerals, collecting localities, etc.)

Snelling, Andrew. "Emeralds—Treasures from catastrophe." *Answers Magazine*, vol. 6, no. 4, p. 72–76. 2011. Available online at https://answersingenesis.org/geology/rocks-and-minerals/emeralds-treasures-from-catastrophe/

FOCUS ON SCRIPTURE

Many different minerals are mentioned throughout the Bible. Aaron's breastplate held twelve gems for the twelve tribes of Israel (Exodus 28:17–20), and the foundations of the walls of the New Jerusalem are made of twelve precious stones (Revelation 28:19–20). Read the description of the New Jerusalem in Revelation 28, and look up a few of these minerals.

54

Photo by John Whitmore

CHAPTER 3

THE EARTH'S ROCKS

OUTLINE:

The Grand Canyon is one of the best places in the world to study the diverse rock types that we have on earth. One can find all the major rock types (igneous, sedimentary and metamorphic) and gain an appreciation for how they are distributed in the earth's crust. Most of the rocks in this view are marine sedimentary rocks which blanket the continents—even at this elevation of more than a mile above sea level. Some of these rock layers can be traced across the entire continent of North America.

Rocks are aggregates of one or more minerals. They can be made of a single mineral or a whole collection of minerals. For example, the rock limestone is made only of the mineral calcite. The rock granite is made of a whole collection of minerals: quartz, potassium feldspar, biotite, plagioclase and sometimes others. There are some things that we call "rocks" even though they may not strictly fit into this definition. Do you remember the five-part definition of a mineral (Chapter 2)? Coal is usually considered a rock, but it is made up of organic material so it does not strictly fit the definition of a rock (because minerals can't be organic).

Rocks are important both for economical and scientific reasons. For example, rocks can provide data to help us evaluate the theories of evolution or creation. Rocks contain things such as fossils and radioactive elements which are often used for geological dating. Rocks contain important economic minerals like iron ore, copper, gold and coal (figure 3.1). Rocks are also important because of various fluids they contain such as drinking water, crude oil and natural gas—substances that we routinely use every day in many different ways. For example, crude oil is not only used for gasoline, it is used in the manufacture of many chemicals, including plastics.

Geologists typically recognize three categories of rocks: **igneous**, **sedimentary** and **metamorphic** (figure 3.2, 3.3). Igneous rocks crystallize from liquid rock, like the granite mentioned previously. Sedimentary rocks consist of small particles or pieces of other rock, like the particles of sand that make up sandstone. These types of rocks are usually easily identified because of the layers contained within them. Metamorphic rocks are rocks that become changed by heat and/or pressure. They start off as one type of rock and become something

FIGURE 3.1. The Bingham Canyon copper mine, near Salt Lake City, Utah, is the world's largest manmade excavation at ¾ of a mile deep and 2 ¾ miles in diameter! The mine produces about 300,000 tons of copper, 500,000 ounces of gold 4,000,000 ounces of silver and several other precious metals each year. Since the 1870's, when the mining began in this area, about 8,000 vertical feet of material has been removed.

FIGURE 3.2. Deep in the Grand Canyon igneous and metamorphic rocks can be found. The pink granites are known as *intrusive* igneous rocks because the magma that formed them was injected into the dark green metamorphic rocks that were there first. All of these processes happened deep underground, probably early during the creation week described in Genesis 1.

FIGURE 3.3. This is a series of sedimentary rocks at the top of the Grand Canyon. Most sedimentary rocks have the flat layers because they were formed in the bottom of an ocean during Noah's flood.

The Three Types of Rocks:

Igneous
Sedimentary
Metamorphic

else during the metamorphic process. For example, heat and pressure applied to the sedimentary rock limestone can turn it into the metamorphic rock marble.

3.2 THE ROCK CYCLE

Rocks can change from type to type by the processes of **weathering**, **metamorphism**, melting and **recrystallization**. The **rock cycle** (figure 3.4) is a schematic drawing that shows how these processes change rocks from one type to another. Weathering can occur through physical or chemical processes (see Chapter 8) which causes rocks to be broken into smaller pieces or can even dissolve the rock altogether. Notice from the diagram that any type of rock (igneous, metamorphic or sedimentary) can be affected by weathering processes. Broken up pieces of loose rock and minerals are called **sediment**. Sediment particles can be small in size like mud or sand, or larger like gravel or boulders. When sediment is lithified or hardened, it becomes sedimentary rock. **Lithification** involves three steps. First the sediment must be compacted. **Compaction** reduces the amount of empty space between the grains and packs the grains closer together. Secondly, **dewatering** goes hand in hand with compaction. Dewatering simply means that water is squeezed out of the rock, like pressing on a wet sponge. Thirdly, the grains must be cemented together. This process is called **cementation**. A variety of substances can "glue" or

cement the rocks together (figure 3.5). Most commonly the loose grains are cemented together with the minerals quartz, calcite, or hematite. Together compaction, dewatering, and cementation turn the loose sediment into a sedimentary rock.

Any type of rock can be changed by heat and pressure because the minerals present can become unstable, recrystallizing to accommodate new temperatures and pressures. Recrystallization is the result of hot fluids interacting with the mineral ingredients of the rock to produce new minerals. It is important to realize that during the process of metamorphism that the rock does not melt. *All of the changes take place while the rock is in its solid state.* If enough heat and pressure is applied to the rock so that it melts, we are no longer in the metamorphic realm; we enter into the igneous realm. Three types of metamorphic rocks are generally recognized depending on the reason for the recrystallization. Heat and pressure create **regional metamorphic rocks**. This is the most common type of metamorphic rock and includes such things as slate (sometimes used as shingles on roofs). If heat is the main cause of change, the rock is referred to as a **contact**

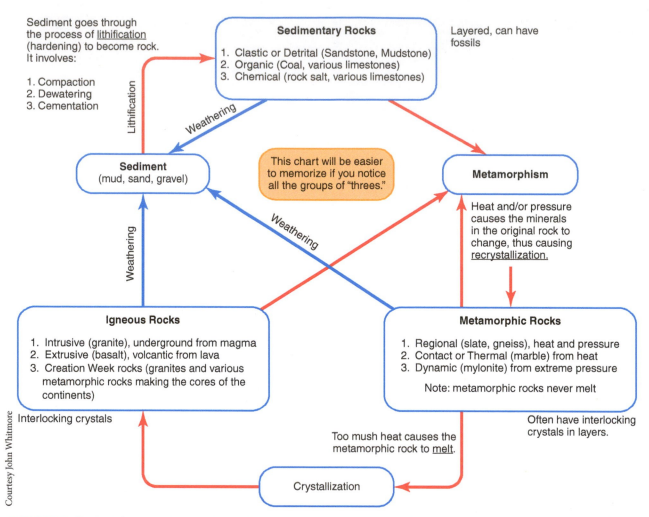

Sediment goes through the process of <u>lithification</u> (hardening) to become rock. It involves:

1. Compaction
2. Dewatering
3. Cementation

Sedimentary Rocks

1. Clastic or Detrital (Sandstone, Mudstone)
2. Organic (Coal, various limestones)
3. Chemical (rock salt, various limestones)

Layered, can have fossils

Lithification

Weathering

Sediment
(mud, sand, gravel)

This chart will be easier to memorize if you notice all the groups of "threes."

Metamorphism

Heat and/or pressure causes the minerals in the original rock to change, thus causing <u>recrystallization.</u>

Weathering

Weathering

Igneous Rocks

1. Intrusive (granite), underground from magma
2. Extrusive (basalt), volcanic from lava
3. Creation Week rocks (granites and various metamorphic rocks making the cores of the continents)

Interlocking crystals

Courtesy John Whitmore

Metamorphic Rocks

1. Regional (slate, gneiss), heat and pressure
2. Contact or Thermal (marble) from heat
3. Dynamic (mylonite) from extreme pressure

Note: metamorphic rocks never melt

Often have interlocking crystals in layers.

Too mush heat causes the metamorphic rock to <u>melt.</u>

Crystallization

FIGURE 3.4. The Rock Cycle, showing the relationships among igneous, sedimentary and metamorphic rocks. Note that every type of rock can be weathered into sediment, and that any type of rock can be changed by heat and pressure into a metamorphic rock. Metamorphic rocks can be heated so much that they melt, and when the liquid rock cools, it crystallizes into an igneous rock.

or thermal metamorphic rock. Marble is a common type of this rock. This typically happens when hot liquid rock bakes or cooks rock that it comes into contact with, but cannot melt. Metamorphism by pressure only happens rarely and is most commonly found in fault zones and is called **dynamic metamorphism**. Mylonite is a type of rock that is often found there.

If enough heat and pressure are present to melt the rock, the rock turns into **magma**. If the liquid rock erupts from a volcano we then refer to it as **lava** (figure 3.6). The only difference between magma and

lava is whether the liquid rock is above ground or below ground. The crystallization of magma or lava produces an igneous rock. We recognize three types of igneous rocks (in keeping with our pattern of three's). **Intrusive igneous** rocks form when magma cools and crystallizes underground. This process usually produces fairly large crystals that can be viewed with the naked eye. **Extrusive** or **volcanic igneous** rocks generally cool quickly above ground. This process usually produces rocks that have crystals too small to be viewed with the naked eye. A third category that we may consider here are **Creation Week**

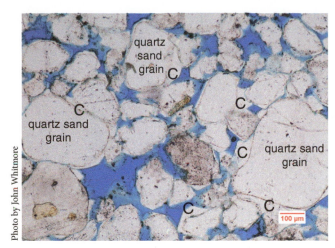

FIGURE 3.5. This is what a sandstone looks like under the microscope. When the sample is prepared, a blue epoxy is forced into the air spaces of the rock; thus the blue areas represent the empty spaces between the quartz sand grains. The sand grains are held together by quartz cement (some of which is labeled "C"). Note the 100 μm scale. 1,000 μm = 1mm, thus this image is only about 2mm wide!

rocks (figure 3.2). In reading the creation story in Genesis chapter 1, there appears to be some significant geologic activity that took place during the first three days of creation as the oceans and continents were formed. The cores or the foundations of the continents are often made of granite (figure 3.7). The continents appeared on the third day of creation as they rose up out of the ocean waters (Genesis 1:9-10).

From a conventional geological perspective, the rock cycle is understood to be a never-ending cycle that occurs over long periods of time. From

FIGURE 3.6. Lava, which flows out of a volcano, produces the extrusive (or volcanic) igneous rocks such as basalt.

a biblical perspective, we see three major times in earth history when the rock cycle was probably most active:

CREATION WEEK First, during the initial three days of creation, there must have been tremendous geological activities to form the cores of the continents. As we look deep into the continents we often find a morass of igneous and metamorphic rocks that are truncated by later processes. Thus these rocks must be the oldest on earth (figure 3.2). They lack fossils, so they must have originated before the first

---

Genesis 1:9-10

[9]Then God said, "Let the waters below the heavens be gathered into one place, and let the dry land appear"; and it was so. [10]God called the dry land earth, and the gathering of the waters He called seas; and God saw that it was good.

Times in Earth History when the Rock Cycle was probably most active:

1. First 3 days of the Creation Week
2. During Noah's flood
3. Early post-flood times

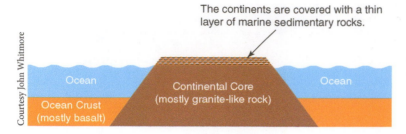

The continents are covered with a thin layer of marine sedimentary rocks.

FIGURE 3.7. A cross-section of what a typical continent looks like today. The cores of the continents are primarily made of granite, while the ocean floor primarily consists of basalt. The continental cores were made early in the creation week and rose up out of the oceans on the third day of creation. The continents and oceans are covered with a thin veneer of marine sedimentary rock that was deposited on them during and after Noah's flood.

life on earth (which was created on days 3, 5, and 6). There is evidence of igneous intrusion, metamorphism and some sedimentary processes in these rocks. As the continents rose on the third day, water likely ran off of the continents causing tremendous amounts of weathering, and erosion that caused thick non-fossiliferous sedimentary deposits.

Noah's Flood Secondly, during Noah's Flood the Bible tells us that the oceans covered the continents. We think there was a tremendous amount of weathering, sedimentary processes, igneous and metamorphic activity during this event. It is very clear from a straightforward reading of Genesis 6–9 that this was a world-wide event that was started by *all* of the "fountains of the great deep" breaking open. We think this was a period in Earth history where the rock cycle was very active for about one year; Noah and his family were on the ark for 371 days.

The Post-Flood Period Thirdly, in the days, months, and years immediately following the Flood, there must have been great geological activity. All of the marine sediments that were deposited during the Flood were uplifted to re-form the continents (figure 3.7). These sediments would have been water saturated which is one of the key factors that initiates large landslides and other mass movement catastrophes (see Chapter 8). The immediate post-Flood rocks also contain much evidence of volcanic activity and mountain building. These processes would have tapered down from the Flood to the slower rates that we observe today.

3.3 IGNEOUS ROCKS

Igneous rocks form from the cooling of molten rock, and all igneous rocks are *crystalline*. That is, they are formed from interlocking crystals that fit together much like the pieces of a puzzle (figure 3.8). There are no empty spaces between the crystals; all the space becomes filled as the molten rock crystallizes and turns into a solid rock. Usually in intrusive rocks (figure 3.8) the crystals are big enough to see with the naked eye (usually > 1mm). In extrusive rocks, the crystals are usually quite small and they can only be viewed with the aid of a microscope (usually < 1mm). Note that metamorphic rocks are also crystalline, but will have some important differences from igneous rocks.

We cannot directly observe the processes that form intrusive igneous rocks since it occurs deep underground. However, we can observe the rocks after they have cooled and then have been exposed by erosion or brought to the surface by mountain building processes. We can more easily observe the formation of extrusive igneous rocks when volcanoes erupt, since those processes occur at the Earth's surface.

FIGURE 3.8. The interlocking crystals of the intrusive igneous rock granite fit together much like the pieces in a puzzle. Each different color is a different mineral crystal.

3.3.1 Classification and Identification of Igneous Rocks

The basic types of igneous rocks are shown in table 3.1 and in figure 3.9. When classifying an igneous rock you need to ask two basic questions: 1) What is the texture of the rock, and 2) What is the

When classifying an igneous rock ask yourself these two questions:
1. What is the texture of the rock?
2. What minerals are present (or what is the rock's color)?

mineral composition or color of the rock? If you can answer these two questions, you should be able to come up with the correct name for the rock.

The first step in igneous rock identification is to determine the texture (figure 3.10). Igneous rock textures refer to the size of the minerals or other physical attributes of the rock. Textures are listed vertically, on the left side of table 3.1. **Pegmatites** have crystals that are greater than 30 mm in size. Rocks with **phaneritic** textures have crystals 1–30 mm in size. **Aphanitic** textures are when most crystals are too small to identify with the naked eye (< 1 mm). This type of texture is common in many

Granite   Diorite   Gabbro

Rhyolite   Andesite   Basalt

FIGURE 3.9. The basic types of igneous rocks. Compare these basic types with the names in Table 3.1.

Pegmatitic

Phaneritic

Aphanitic

Porphyritic

Glassy

Vesicular

Pyroclastic

FIGURE 3.10. The basic textures of igneous rocks. Compare these textures with Table 3.1.

volcanic rocks. **Porphyritic** textures are a combination of phaneritic and aphanitic textures; these rocks have two distinct crystal sizes which results when erupting lava already has crystals in it. Glassy textures result when volcanic material cools so fast that minerals do not have a chance to form making the rock appear as "glass." Obsidian is the most common example of this rock. Native Americans often used this material to make knives, arrowheads and spears because when it breaks, it has very sharp edges. **Vesicular** textures are the result of volcanic gases escaping from cooling lava, leaving many tiny holes. Because of its many holes, pumice is so light that it will often float in water (figure 3.11). A **pyroclastic** texture is usually the result of explosive volcanism. The rock characteristically contains many fragments of broken volcanic rock.

The second step is to determine either the color or the mineral composition of the rock. Basic colors and mineral compositions are listed horizontally, along the top of the table 3.1. Use the mineral composition of the rock if you have a pegmatitic or phaneritic texture (larger crystals). It is only appropriate

to identify the rock by color for the volcanic textures. Often only a few minerals are needed to identify a rock. For example a dark-colored rock with a porphyritic texture containing crystals of plagioclase would be basalt porphyry. A rock with a pegmatitic texture containing the minerals biotite, quartz and K feldspar would be granite pegmatite.

3.3.2 The Cooling and Crystallization of Igneous Rocks

As a general rule, volcanic rocks have smaller crystals (aphanitic textures) and intrusive igneous rocks have larger crystals (phaneritic or pegmatitic textures). Often we think this is the case because intrusive rocks have more time to cool than extrusive ones. One of the classic arguments against a six- to ten thousand year old Earth has been that large igneous bodies (like a huge granite intrusion) take far too long too cool for a young earth to be possible. However, evidence is mounting that even intrusive granite pegmatites can cool within a matter of weeks. Laboratory studies have indicated that large

TABLE 3.1.

The basic kinds of igneous rocks. If the rock has large crystals (>1 mm) use the intrusive igneous rock textures for a name, and for very large crystals (>30 mm), use the term "pegmatite." If the crystals are small (< 1mm), use the volcanic rock textures and identify the rock by its overall color. The minerals or the overall color can also be used to roughly estimate the original chemical composition of the magma.

Color → (very general)		Felsic (light)	Intermediate (gray)	Mafic (dark)	Ultramafic (very dark)
Composition → (typical minerals present)		Hornblende (small %) Biotite Na plagioclase K feldspar Quartz	Hornblende Ca plagioclase Na-Ca plagioclase No K feldspar No quartz	Olivine Pyroxene Ca plagioclase	Olivine Pyroxene
Texture	Intrusive	Pegmatite	Granite Pegmatite	Diorite Pegmatite	Gabbro Pegmatite	
		Phaneritic	Granite	Diorite	Gabbro	
	Volcanic	Aphanitic	Rhyolite	Andesite	Basalt	Peridotite
		Porphyritic	Rhyolite porphyry	Andesite porphyry	Basalt porphyry	
		Glassy	Obsidian			
		Vesicular	Pumice		Scoria	
		Pyroclastic	Volcanic tuff (frags < 2 mm) or volcanic breccia (frags > 2 mm)			

Types of magma: Felsic magmas | Intermediate magmas | Mafic magmas | Ultramafic magmas

Chemistry of the magma:

← Increasing K and Na

Increasing Ca, Fe and Mg →

75% silica ← Increasing silica → 45% silica

crystals, reaching pegmatitic sizes, can grow within hours to days. Andrew Snelling has a nice summary of this research in his paper on "Catastrophic Granite Formation" in the *Answers Research Journal*. The process (figure 3.12) involves the continuous flow of water already present in the molten rock and in the surrounding rock it intrudes, working to cool the rock while at the same time allowing larger crystals to grow (the water increases crystal growth rates). While we still don't have all the answers in this area, young-Earth creationists have made excellent progress providing several mechanisms and evidences for rapid cooling of large igneous bodies.

Early in the twentieth century, geologist N.L. Bowen began a series of experiments where he melted the ultramafic igneous rock peridotite, then cooled it to various temperatures to determine which minerals formed. From these experiments, it became apparent that all of the minerals in an igneous rock do not crystallize at the same temperature, but they crystallize over a wide range of temperatures. Bowen recognized several patterns which are summarized in a diagram known as **Bowen's Reaction Series** (figure 3.13). It turns out that two factors control final mineral composition of an igneous rock: temperature and the magma's chemistry.

The minerals in figure 3.13 are organized into three groups: a discontinuous branch, a continuous branch, and a cluster of quartz, potassium feldspar, and muscovite mica. The minerals at the top of the

diagram crystalize first and the minerals at the bottom of the diagram crystallize last as temperatures drop. Olivine and calcium-rich plagioclase form first, at about 1,200° C, while quartz is the very last mineral to form at about 600° C.

Along the discontinuous branch, olivine forms first. As the magma cools, pyroxene (augite) begins to form by using both the remaining atoms in the melt as well as the atoms in the olivine. So as more pyroxene is formed, there is less and less olivine. Likewise, as temperatures continue to drop, the pyroxene is replaced by amphibole (hornblende), which is later replaced by biotite mica. Thus as temperatures cool, the silica tetrahedrons form more and more complex structures at the expense of the minerals that had cooled previously, and typically only two close members of the discontinuous branch might be found in an igneous rock. So why don't all igneous rocks just have biotite? The answer lies in the chemistry of the original magma: magmas with very little silicon (Si) cannot make the more complex silicate structures like amphibole and biotite, they might only form olivine and amphibole (look back at chemical formulas for these minerals in figure 2.19). The amount of silicon

FIGURE 3.11. Is this eleven year old boy really that strong? He is lifting a large piece of pumice that probably weighs no more than 30 pounds. Because of its vesicular texture, pumice is so light that it will float in water. Notice the skill of the boy: he can lift the rock and blow a bubble with his chewing gum at the same time! Mono Lake area, California.

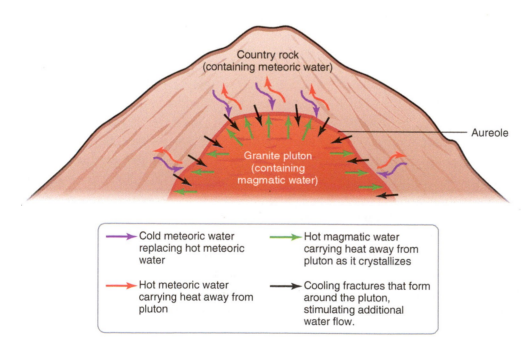

→ Cold meteoric water replacing hot meteoric water	→ Hot magmatic water carrying heat away from pluton as it crystallizes
→ Hot meteoric water carrying heat away from pluton	→ Cooling fractures that form around the pluton, stimulating additional water flow.

FIGURE 3.12. The cooling processes involved in a large granite body. Both magmatic and other water help transfer the heat from the pluton to the surrounding rocks.

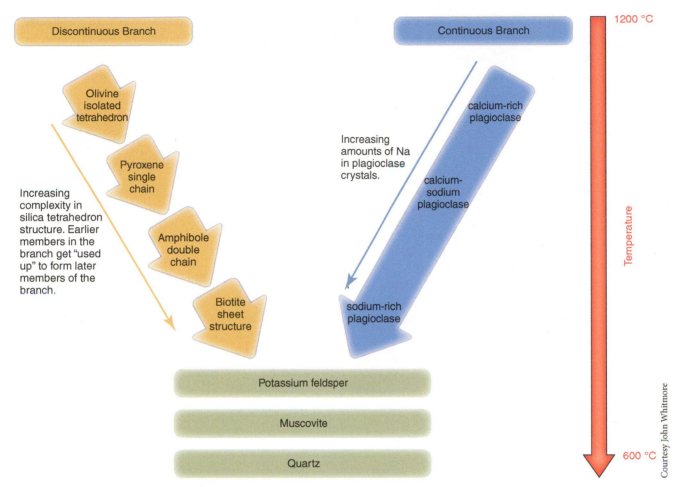

FIGURE 3.13. Bowen's Reaction Series.

Discontinuous Branch

Olivine isolated tetrahedron

Pyroxene single chain

Amphibole double chain

Biotite sheet structure

Increasing complexity in silica tetrahedron structure. Earlier members in the branch get "used up" to form later members of the branch.

Continuous Branch

calcium-rich plagioclase

calcium-sodium plagioclase

sodium-rich plagioclase

Increasing amounts of Na in plagioclase crystals.

Potassium feldsper

Muscovite

Quartz

1200 °C

Temperature

600 °C

(and some other elements) limits what minerals can form along the discontinuous branch, which is also seen by reading the mineral compositions of igneous rocks in table 3.1 from right to left.

The continuous branch of Bowen's Reaction Series is a little different. At the highest temperatures, calcium (Ca) prefers to react to form plagioclase crystals over sodium (Na). The first crystals of plagioclase to form are very rich in calcium. As temperatures drop, the calcium in the melt gets used up and sodium begins to become incorporated into the plagioclase crystals. At higher temperatures, the plagioclase will be 100% calcium plagioclase. As the temperature drops the plagioclase crystals will be 90% calcium and 10% sodium plagioclase.

At intermediate temperatures the mix will be 50% calcium and 50% sodium plagioclase. At the lowest temperatures, the plagioclase crystals will be 100% sodium plagioclase. Thus, there is a continuum from calcium rich crystals to sodium rich plagioclase crystals. Again, consider table 3.1. Note that calcium rich plagioclase occurs on the right, intermediate plagioclase occurs in the middle and sodium rich plagioclase occurs on the left of the diagram. In table 3.1, temperatures decrease from right to left.

If liquid rock is still present below 700° C the magma will contain high amounts of potassium, sodium and silica. In this case, these ingredients will crystallize to form potassium feldspar (orthoclase), muscovite and finally quartz. Quartz is the last mineral to crystallize. It will only form if there is still molten silica tetrahedrons remaining that have not been used to make other minerals. Quartz is a framework of pure silica tetrahedrons.

3.3.3 Useful Types of Igneous Rocks

Igneous rocks have many uses. Granite is commonly quarried, cut, and polished to make upscale countertops, flooring, tomb stones, monuments and building fascia (figure 3.8). Diamonds, garnets, rubies and many other gem stones are sourced from igneous rocks. Diamonds, which are made of pure carbon, require extreme temperatures and pressures to crystallize. These temperature and pressures exist 150-500 km below the earth's surface. If the magma from these areas reaches the Earth's surface, it forms a very narrow conduit called a **kimberlite pipe** (see figure 3.14). Diamonds can be found in the cooled magma. In areas that have ongoing igneous activity close to the surface, the heat energy can be tapped as an alternative energy source for heating

homes and buildings. Many important ore minerals are also associated with igneous rocks such as chrome, platinum, tin, copper and gold (figure 3.1). Igneous rocks are also important because they are the source of most radioactive dates (discussed in Chapter 6).

FIGURE 3.14. A diamond mine in a kimberlite pipe, Siberia. Kimberlite pipes are igneous eruptions at the earth's surface that originate from very deep within the earth (150–500 km). If the melt contains carbon, diamonds will crystallize under the high temperatures and pressures. Kimberlite pipes were named after the town of Kimberly, South Africa, the location of another diamond mine.

3.4 SEDIMENTARY ROCKS

Sedimentary rocks are some of the most easily recognized and most familiar of all rock types. They are made from small pieces of sediment which are compacted and cemented to form sedimentary rocks (review section 3.2 on the rock cycle). Since the continents are mostly above sea level, it seems curious that the vast majority of sedimentary rocks that we find on land is of marine origin. But if most sedimentary rocks were formed during Noah's Flood (figure 3.7), then the widespread marine sediments on top of the continents makes sense. Sedimentary rocks are familiar and easily recognized because they usually form flat layers, like those seen in the Grand Canyon (figure 3.15). If we could remove soil and vegetation from the continents, and cut deep canyons into them, then

FIGURE 3.15. Flat, continuous, uniformly thick layers are one of the most conspicuous features of sedimentary rocks. The continents are covered with marine sedimentary rocks containing billions of fossils, and some of these layers can be traced all the way across North America.

most of the earth's land surfaces would consist of layered sedimentary rocks. These rocks contain many important industrial and energy resources as well as many clues about Earth's past.

3.4.1 The Classification of Sedimentary Rocks

Depending on the origin of the sediment, three types of sedimentary rocks are generally recognized: clastic, chemical, and organic sedimentary rocks.

Clastic (sometimes called **detrital**) **sedimentary rocks** are made of pieces of other rock. The rock is named according to the grain size of the pieces clay particles are the smallest recognized particles at < 1/256 mm, which are too small to see, even with a standard microscope. Rocks composed entirely of clay-sized particles are referred to as **shale**. Silt particles are larger (1/256-1/16 mm), but most of them will still be too small to see, even with magnification. A rock made completely of silt-sized particles is called a **siltstone**. Because clay and silt particle sizes can't easily be seen, a technique that can be used in the field (with caution) is to chew a small piece of the rock. If the rock is gritty, it is siltstone. If it is smooth and pasty, it is shale. Both siltstone and shale are **mudrocks** (a broader category of fine-grained rocks that also include mudstones and claystones). **Sandstone** is a bit easier to identify. Sand particles are 1/16 to 2 mm in size and although very tiny, most sand particles can be seen with the naked eye. If the rock has a combination of larger clasts (> 2 mm) and smaller matrix grains (< 2 mm) filling in the spaces between the larger clasts, the rock is referred to as a **conglomerate** or **breccia**. Conglomerates have rounded clasts, and breccias have more angular clasts. The different types of clastic sedimentary rocks are described in table 3.2. (see also figure 3.16).

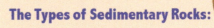

The Types of Sedimentary Rocks:

Clastic (or detrital)
Chemical
Organic

TABLE 3.2.

Classification of clastic (or detrital) sedimentary rocks. The rocks are classified according to the grain size of their particles.

Particle Size (mm)	Particle Name	How to identify	Rock Name
< 1/256	clay	Will not be able to see clay particles with magnification (10-40 x). Pasty if chewed in mouth.	shale
1/256 – 1/16	silt	Some of the larger silt particles will be visible with magnification (10x). Gritty if chewed in mouth.	siltstone
1/16 – 2	sand	Should be able to make out most sand particles with naked eye, but they will be tiny. Can easily see with magnification (7-10x).	sandstone
Mix of above	—	Usually mostly clay and silt particles with some sand.	mudstone
2 – 64	pebble	Pebble-sized clasts with clay to sand matrix (measure with a ruler).	conglomerate (if rounded clasts) breccia (if angular clasts)
64 – 256	cobble	Cobble-sized clasts with clay to sand matrix (measure with a ruler).	
> 256	boulder	Boulder-sized clasts with clay to sand matrix (measure with a ruler).	

In **chemical sedimentary rocks**, the sediment is created by some type of chemical process that happens within water. The easiest example to understand is rock salt, which can form as a salty lake dries up and leaves sodium chloride (halite) behind. Most of the examples of chemical sedimentary rocks can be identified by various physical properties of minerals that you learned in Chapter 2 like testing

FIGURE 3.16. Examples of various clastic (detrital) sedimentary rocks.

Rock Type	Hand Sample	Outcrop Photos
Shale	 © Siim Sepp/Shutterstock, Inc.	 © sigur/Shutterstock, Inc.
Siltstone	 © Tyler Boyes/Shutterstock, Inc.	 Photo by John Whitmore
Sandstone	 © michal812/Shutterstock, Inc.	 © Johnny Adolphson/Shutterstock, Inc.
Conglomerate (rounded clasts)	 © sonsam/Shutterstock, Inc.	 © Zbynek Burival/Shutterstock, Inc.
Breccia (angular clasts)	 © Siim Sepp/Shutterstock, Inc.	 © Siim Sepp/Shutterstock, Inc.

FIGURE 3.17. Various types of chemical sedimentary rocks.

Rock Type	Hand Sample	Outcrop Photos
Rock Salt	 © Sstiling/Shutterstock, Inc.	 © Stanislav Bokach/Shutterstock, Inc.
Gypsum	 © farbled141555412/Shutterstock, Inc.	 © Sementer/Shutterstock, Inc.
Travertine Limestone	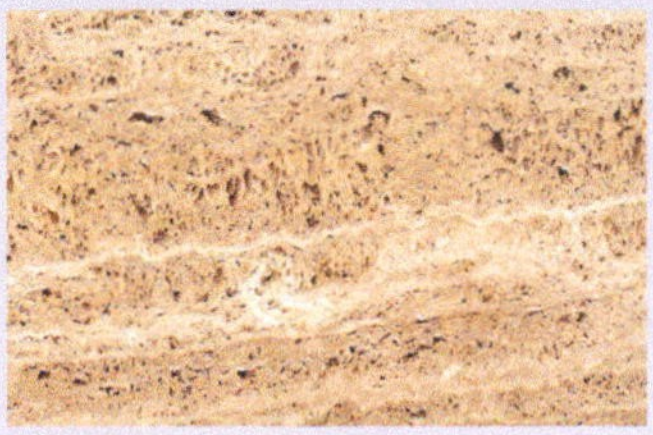 © Photographee.eu/Shutterstock, Inc.	 © Andrea Willmore/Shutterstock, Inc.
Dolomite	© Siim Sepp/Shutterstock, Inc.	 Photo by John Whitmore
Chert	 © vvoe/Shutterstock, Inc.	 Photo by John Whitmore

with hydrochloric acid, taste, cleavage, or hardness. Essentially each of the chemical sedimentary rocks (figure 3.17) are large-scale examples of various non-silicate minerals, so you can use the mineral properties to identify them.

Salt deposits such as gypsum ($CaSO_4 \cdot H_2O$) and halite ($NaCl$) are quite common in the sedimentary rock record. For example, there are extensive and thick deposits of salt below the states of Ohio, Michigan, and Louisiana. The Zechstein Salt in Europe covers 2.5 million square kilometers (1/3 the size of the United States), and its thickness exceeds 600 m in some places! Many of these deposits are commercially mined for halite. Gypsum deposits are mined to make drywall (also called gypsum board, sheetrock, wallboard, or gyprock). After processing, the gypsum is put between two sheets of thick paper and used to make interior walls.

Today, salt deposits can form in desert settings such as the Bonneville Salt Flats in Utah, or the Dead Sea which separates the countries of Jordan and Israel (figure 3.18). It is probably because we can actually observe halite deposits forming by evaporation in places like Death Valley, that we have become conditioned to believe that almost all salt deposits form in this way.

However, there are new studies which point away from modern evaporative models as the best explanation for salt deposits in the sedimentary record, especially the large-scale salt deposits mentioned above. The new hypotheses involve the salts (like halite and gypsum) being deposited on the sea floor rapidly in association with volcanic activity. Normally, sodium chloride comes apart in water creating positively charged sodium and negatively charged chlorine ions. This happens because water (H_2O) is a polar molecule and the sodium and chlorine atoms disperse to the charged polar ends of the water molecule (see Chapter 2.3 and figure 2.14). However, when water is heated to exceptionally high temperatures and is contained under very high pressures (like on the deep ocean floor) it will

FIGURE 3.18. Halite deposits on the floor of Death Valley, California (Badwater Basin), formed by the evaporation of shallow, mineral-rich water that occasionally occupies the valley bottom. Rocks formed in this way are known as "evaporites."

not boil. Instead, the water becomes what we call supercritical. When this happens, water becomes a nonpolar molecule (figure 3.19), and the dissolved sodium and chlorine will become attracted to each other and form an instant salt (NaCl) deposit. Many components to this model of salt precipitation were in place during Noah's flood, but are challenging to envision in old-Earth scenarios.

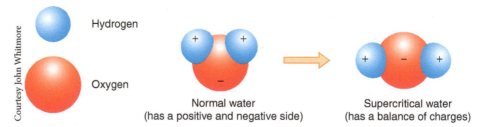

Hydrogen

Oxygen

Normal water
(has a positive and negative side)

Supercritical water
(has a balance of charges)

FIGURE 3.19. The formation of supercritical water. Under normal conditions, water is a polar molecule with positive and negative sides. When water is superheated, and forced to stay in the liquid phase (because of high pressures), it changes its shape and becomes "supercritical." In this arrangement, the molecule is nonpolar, or has a balanced set of charges. In this condition, dissolved ions like sodium and chlorine will immediately precipitate out of the supercritical water, forming a large salt deposit.

CHAPTER 3: The Earth's Rocks

Limestone is made of the mineral calcite ($CaCO_3$) and dolostone (sometimes simply called dolomite) is made of the mineral dolomite ($CaMg[CO_3]_2$). These types of rocks are generally called carbonates because of the carbonate (CO_3^{-2}) building block that both rocks have in common. Chemical processes are usually responsible for the production of these rocks, but sometimes the rocks are clastic, biological, or a combination of clastic/chemical/biological components. The wide variety of limestone types (figure 3.20) attests to the many sources and processes that form limestone.

When geologists study rocks, they usually go to places where they can actually see the rocks forming today so they can get some insights into how the rock formed in the past. For example, if we want to study limestone formation, we might go to a place like the Bahama Islands where limestone is currently forming in abundance in the ocean. However, when it comes to dolomite (or dolostone) we have a problem. It has been estimated that about 10% of all sedimentary rocks are dolomite, yet dolomite can nowhere in the world be found forming in large quantities, and laboratory studies to artificially make dolomite under current ocean conditions have all but failed. This has often been called the *dolomite problem* in geochemistry textbooks. What is clear is that dolomite formed in a marine setting and probably at rather high temperatures and pressures, but these conditions don't normally occur in the oceans today. In Chapter 4, we describe the hypothesis of catastrophic plate tectonics (CPT). If such a process happened during Noah's flood, conditions for dolomite formation may have been more favorable for dolomite to precipitate out of sea water than either current conditions or various old-Earth hypotheses.

Chert is a type of silica-rich type of quartz mineral which is closely related to flint, jasper, and agate. It can often be found as nodules in many limestones. For reasons that are not completely understood, the chert

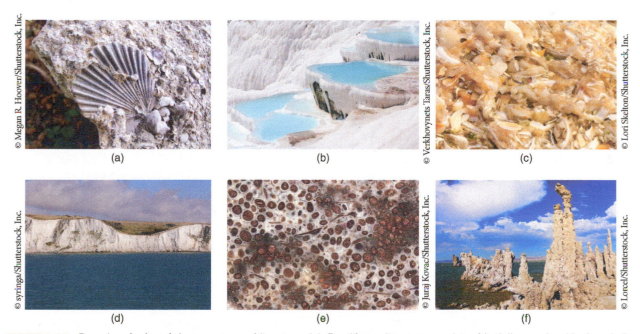

FIGURE 3.20. Examples of a few of the many types of limestone. (a). Fossiliferous limestone consists of both lime mud and broken shells. (b). Travertine limestone often forms in the vicinities of springs, often producing beautiful terraces. (c). Coquina is almost exclusively made up of broken shells. (d). The Chalk Cliffs of Dover are made up of millions of small organisms with calcite shells. (e). Oolitic limestone has the appearance of many small balls about the size of BB's, which formed as calcium carbonate accumulates around a small fragment of fecal material or a sand grain that is rolled back and forth in the surf. (f). Tufa is an extremely porous type of limestone that can also form in the vicinity of springs.

can replace the organic shelly material in the limestones such as clam shells or corals turning the calcite skeletons of the organisms into chert (figure 3.21).

The particles that make up **organic sedimentary rocks** have living things as their origin. For example some limestones are made almost completely of broken shells; called a coquina (figure 3.20. If the limestone is made partly of shells and carbonate sand or mud, then we would call the limestone a *fossiliferous limestone*. We can also have fossiliferous sandstones and shales. In some mudstones, you can smell oily **kerogen** in broken pieces of the rock; kerogen is a waxy oil derived from the soft tissues of living things. These rocks are referred to as *oil shales*. It is thought most of the oil in the Green River Formation of Colorado, Utah, and Wyoming accumulated from the remains of algae that lived in the water of the large post-Flood lakes. During Noah's flood there was a tremendous amount of organic material, from many different kinds of sources that was buried and now has turned into natural gas, oil, and coal deposits (more about these in Chapter 18).

Coal is an organic sedimentary rock that is almost made entirely of tree bark and other organic remains from plants. Dr. Kurt Wise proposed a theory in 2003 that most of the coal we find in the Paleozoic coal measures was made during Noah's Flood as large floating forests were destroyed and buried. We no longer have these types of ecosystems on earth, but from the unusual plants and animals that we often find in coal measures, this hypothesis makes a lot of sense. Coal comes in three basic varieties. **Lignite**, or brown coal, has the lowest percentage carbon and has the least energy per unit mass. **Bituminous coal** has a greater energy density, is black, and typically burns with higher heat energy and less smoke. **Anthracite coal** is the highest grade of coal and is the only coal considered a metamorphic, rather than sedimentary, rock. It often has a black glassy appearance, much like obsidian. Most of the anthracite reserves in the United States are in the eastern Appalachians of Pennsylvania and New York. This type of coal has a high carbon content and burns very hot with few pollutants. Its primary use is in metallurgy; however, most of the major reserves have been extracted and many mines are now closed.

3.4.2 Sedimentary Structures

Sedimentary rocks contain clues just like the scene of a crime, and it is the job of the geologist to collect the various clues and then make an interpretation about the rock's origin and history, including its original environment, source of materials, temperatures, currents, etc. **Sedimentary structures** are important clues that geologists use to make these types of interpretations. Sedimentary structures are features that are made within the rocks while they are in the process of formation, so they provide the record of the processes that made the rock.

Many geologists have done experimental work creating many different types of sedimentary structures. Videos of some of these experiments can be found posted on the internet. **Planar bedding**, or horizontal layering is one of the most recognizable characteristics of sedimentary rocks. It can form under a variety of different conditions, but most commonly it forms when sediment is carried and then deposited in water. **Cross-beds** (figure 3.22) form as the result of sediment avalanching down the faces of dunes, which form angled bedding instead of planar bedding. Whether the dunes are small or many meters high, the process is similar.

FIGURE 3.21. A horn coral in the Redwall Limestone that has been completely replaced by chert. Jeff Canyon, northern Arizona

Photo by John Whitmore

FIGURE 3.22. Very large cross-beds in the DeChelly Sandstone, DeChelly National Monument, Arizona. Cross-beds often appear as angled layers in between horizontal surfaces (called bounding surfaces). They can form in a variety of settings including desert dunes, deltas and large underwater dunes called sand waves.

For example, small-scale **ripples** (figure 3.23) can contain cross-beds much like larger dunes. Depending on the types of ripples (there are many) geologists can get some clues about the direction and velocity of the water when the ripples formed.

Mud cracks and **rain drop prints** can often be found in modern settings like deserts (figure 3.24). But when supposed examples from the rock record are examined, there are some important differences (see the article by Whitmore (2009) in the references). **Fossils** are the remains of organisms in the rock record and are another important clue for past conditions. To become fossilized, animals or plants must be buried quickly before they decompose. Fossils are important whether they are body fossils or footprints that have been left behind. Fossils are important indicators as to the conditions under which rocks formed in the past. So fossils, whether they are body fossils of an actual organism or footprints left behind as an animal walked through the mud are important indicators as to how the rocks they are contained in formed.

Some sedimentary structures give us clues as to the condition of a lower layer of sediment when a layer above was deposited. **Load casts** are made when the layers below are relatively soft and the upper layer pushes down into it. These structures are typically formed when sediments are accumulating very rapidly. **Sand injectites** (figure 3.25) occur when a fluid mix of sand and water is forced into other rock units. In the Grand Canyon, sand injectites occur at the base of the Coconino Sandstone; where the sand has been injected downward into the Hermit Formation (a shale). According to old-Earth views of geologic history, the Coconino Sandstone was deposited about 275 million years ago during the Permian period (see Chapter 7), and the earthquake which caused the liquefaction of the sand moved about 50 million years ago (seen today as the Bright Angel Fault). If conventional dating is accurate, the sand in the Coconino had to remain soft for about 225 million years in order for it to become liquefied during the earthquake, even while many other rock units were deposited on top of it over supposedly millions of years. The sand injectites demonstrate that this is a very unlikely

FIGURE 3.24. Mud cracks and rain drop prints found beside a small stream in Capitol Reef National Park, Utah. These features can often be found in modern settings, but examples from the rock record remain rather dubious, in our opinion.

FIGURE 3.23. Several types of small ripples exposed (in cross-section) in a sand bank along the Colorado River, Arizona.

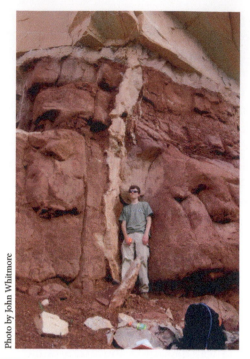

FIGURE 3.25. A sand injectite penetrating from the Coconino Sandstone (above) into the Hermit Shale (brownish-red rock below), Dripping Springs Trail, Grand Canyon National Park, Arizona. In order for this sand injectite to form, the sand in the Coconino Sandstone had to be full of water and unlithified during the time of an earthquake.

scenario, because they can only be explained if there was only a short period of time between the deposition of the Coconino and the formation of the injectites, not millions of years.

3.4.3 Important Sedimentary Rocks

Sedimentary rocks are valuable in several different ways. First, they are the source for many of the earth's coal, oil, and natural gas reserves. Organic material was concentrated and deposited in sedimentary rocks, primarily during the flood as microorganisms, animals, and plants were buried during the catastrophe. Other deposits that also primarily formed during the flood are rock salt (NaCl, halite) and gypsum ($CaSO_4 \cdot H_2O$). Salt has many uses including table salt, road salt, an industrial source of chlorine, animal hide tanning, textile dyeing and the extinguishing of class D fires. Gypsum is mined for use in making wall boards, commonly referred to as "drywall," and also for plaster of Paris and Portland cement. Limestone is typically quarried and crushed, then used for road bases, construction aggregates and fill material, as well as being a major component of cement. Certain types of shales with special clay minerals are often used in the manufacture of bricks and ceramics. Most of the water that we drink every day comes out of the ground. This water is typically pumped out of sedimentary rocks like sandstones. Sedimentary rocks also help us answer important questions about Earth history. Sedimentary rocks contain fossils so we can use them to test various hypotheses about Noah's Flood, evolution, and the age of the Earth.

3.5 METAMORPHIC ROCKS

Of the three different types of rocks, metamorphic rocks often have the most different appearance. Like igneous rocks they are crystalline (having interlocking crystals), but as you will see, they often look very different from igneous rocks. Metamorphic rocks often occur in the cores of mountain ranges or deep in canyons. To become exposed, they usually must either be pushed up from deep within the earth, or a canyon must be cut deep into the earth, or both.

Metamorphosis is a common concept within biology; we think about caterpillars turning into butterflies or tadpoles turning into frogs. The term simply refers to incredible change that takes place. The end metamorphic rock product often doesn't look anything like the igneous or sedimentary rock with which the process was started. In Romans 12:2, Paul uses the Greek word *metamorphoo*, (often translated "transformed") to talk about the changes that occur

3.5.1 The Classification of Metamorphic Rocks

Metamorphic rocks usually come in one of three different types: regional, contact (or thermal) and dynamic. Regional metamorphic rocks are the most common type and when they occur they are quite extensive. They are formed as both heat and pressure cause mineral changes within the rock. These types of rocks often occur in the cores of mountain ranges or are found deep below the surface in the continental core (figure 3.7). Because these rocks undergo so much pressure in the process, when the new minerals grow, they grow perpendicular to the pressure. Rocks that form in this manner are **foliated**; that is, they display groups of parallel layers, banding patterns, or lines. If you see a rock that is both crystalline and layered, it is likely a regional metamorphic rock that is exhibiting foliation (figure 3.26). Don't get the layering caused by foliation confused with the layering in sedimentary rocks. Recall that metamorphic rocks have interlocking crystals, whereas most sedimentary rocks have pore spaces in between the grains (figure 3.5).

Regional metamorphic rocks come in four varieties depending on the intensity of the metamorphism. To show what happens to a rock when it is metamorphosed, we will metamorphose a mudstone (a sedimentary rock) as an example (figure 3.27). When heat and pressure is applied to the mudstone, the mineral grains in the mudstone will begin to change. These changes take place while the rock is hot, but long before it melts. In

in a person as our minds are changed as a result of the work that Christ has done in us. The metamorphosis that Christ does in our minds and lives when we trust Him, is truly remarkable! Like the metamorphosis in caterpillars, tadpoles, and rocks, the change that occurs in a true believer is complete; the latter is unrecognizable from the former.

As we learned in the rock cycle, heat and pressure are the most common factors that change rocks. During the process of metamorphism, the original crystals of the rock become unstable and change into crystals that are more stable under the new pressure and temperature regimes. Regional metamorphic rocks are the most common and are formed by both heat and pressure. Contact or thermal metamorphic rocks are formed by large amounts of heat typically from a nearby magma chamber. Dynamic metamorphic rocks are formed by large amounts of pressure, for example in a fault zone. Not only do heat and pressure play large roles in the metamorphic process, but fluids are important as well. Hot fluids not only can provide the heat for metamorphism to occur, but they can also act as catalysts for new mineral growth, introduce new chemical ingredients to the system, or take other ingredients away. Hot fluids are primarily in the form of water. Many granite magmas have significant amounts of water contained within them. As the magma cools, the hot water is released causing metamorphism of the surrounding rock.

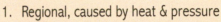

The Three Types of Metamorphic Rocks

1. Regional, caused by heat & pressure
2. Contact (or thermal), caused by heat
3. Dynamic, caused by pressure

FIGURE 3.26. Regional metamorphic rocks are foliated, and often appear to be layered. The new metamorphic minerals (like micas) grow in the direction of least resistance.

fact, all the changes that occur in a metamorphic rock take place while the rock is still solid. If the rock becomes so hot that it melts and recrystallizes, it is no longer a metamorphic rock; it becomes an igneous rock. It is kind of difficult to imagine how so many changes can take place while the rock is still a solid, but remember that hot fluids flowing through the rock are important in bringing change as well. Heat and pressure cause unstable conditions to the minerals that make up the mudstone. So, the minerals begin to change to accommodate the new temperature and pressure conditions. When the minerals grow, they grow perpendic-

ular to the pressure being applied to the rock. At first, the minerals are so tiny that they require a special microscope to be seen. If the metamorphism stops at this phase, the rock is called a slate. Slates have a dull luster because the minerals are so small. If metamorphism continues, the minerals grow slightly larger and the rock becomes a phyllite. In a phyllite, the crystals are still tiny, but you might be able to make a few of them out with your naked eye or with a small magnifying glass. A phyllite has a shiny texture compared to the slate because the minerals are slightly larger and better reflect light. If metamorphism continues, the minerals will continue to grow larger until they can be seen (and identified) with the naked eye. Metamorphic rocks in this category are called schist. The word *schist* usually does not stand alone. The schist needs to be named according to the minerals it contains. Examples might be talc schist, chlorite schist, muscovite schist, muscovite-garnet schist, etc. If metamorphism continues further, the schist will change into a gneiss (pronounced "nice"). A gneiss is characterized by crystals that are just as big as those found in a granite. In fact many types of gneiss look like granites; with the exception of

FIGURE 3.27. Microscopic thin sections (upper row) showing the changes that take place in crystal size as a shale is turned into various metamorphic rocks and then eventually a granite. Note the increase in crystal size as increasing amounts of metamorphism take place. The lower row shows hand-sized samples of the same rocks. The scale bars are 500μm in length, or 0.5mm.

Mine shaft

Bass Limestone

Contact metamorphism

Diabase sill

Photo by John Whitmore

FIGURE 3.28. A metamorphic aureole is a baked zone next to an igneous intrusion. In this case, near the bottom of the Grand Canyon, a molten diabase sill was injected below the Bass Limestone. As a result, the limestone was metamorphosed and created the aureole (the white zone) above the diabase sill. The mineral asbestos was created in the process and was commercially mined in the past.

the dark-light banding that is common in gneiss. Sometimes gneiss can partially melt. When this happens, the rock becomes a migmatite.

Contact (or thermal) metamorphic rocks form as liquid rock from an igneous pluton (figure 3.12) bakes the surrounding area (figure 3.28). The baked zone is where the rocks become metamorphosed. The baked zone is known as an **aureole**. Imagine a lasagna that is left in the oven too long or is cooked at too high of a temperature. The top, bottom and sides of the lasagna will be burnt black, but the inside will be red and juicy. The metamorphic aureole is like the black burned crust around the outside of the lasagna (except the heat is coming from the other direction). Contact metamorphic rocks are usually very local in

nature; they don't occur over extremely large areas like the regional metamorphic rocks. If the pluton is small, the aureole might only be a centimeter or less in thickness. Larger intrusions can have much thicker aureoles.

Occasionally, if small amounts of pressure are involved, contact metamorphic rocks might exhibit some foliation. The most common types of contact metamorphic rocks (figure 3.29) are marble (changed from limestone and/or dolomite), quartzite (changed from quartz sandstone) or anthracite coal (changed from bituminous coal). Quartz sandstone and quartzite are both made of quartz; limestone and marble are both made of calcite. In these rocks the changes occur by

FIGURE 3.29. Some common types of contact metamorphic rocks: quartzite, marble and anthracite.

the minerals growing larger and becoming more tightly interlocked with each other (figure 3.30).

Dynamic metamorphic rocks are the least common types of metamorphic rocks. They occur in extremely local areas. Dynamic metamorphic rocks are caused by mineral changes that take place due to extremely great pressures, like we find in the midst of an earthquake fault zone. In these situations, the common metamorphic rock type is mylonite.

It is now believed that that widespread regional metamorphism can take place only within a few days if the right conditions are present. Dr. Andrew Snelling's work from the Great Smoky Mountains

along the Tennessee-North Carolina border and from New South Wales, Australia shows that metamorphism happened in these areas in less than two weeks during the time of the Flood!

3.5.2 Useful Metamorphic Rocks

There are many metamorphic rocks that are quite useful to man. Many precious metal ore deposits (copper, gold, silver, etc.) have their origin as hot fluids carry metals away from magmas. The metals can't fit into the crystal structure of the igneous rocks, so they are concentrated and carried away from plutons. As hot water cools in the fractured metamorphic rock around the pluton, the metals come out of solution and are deposited in the surrounding rock. Some of the most important metal deposits in the world form as a result of this, such as the Bingham Canyon copper mine, near Salt Lake City, Utah (figure 3.1). Many precious gem stones also form during metamorphism; gems like garnet, turquoise, ruby, sapphire and beryl. Some metamorphic rocks are useful for other purposes. Marble is soft enough that it can be carved into statues or large slabs to build buildings. Slate can be used for roofing or flooring. Anthracite is a type of metamorphic coal that burns very hot and clean. Talc can be used for talcum powder, cosmetics, or as a lubricant and filler in quality paints. Graphite makes an excellent dry lubricant which can be purchased at the auto parts store to lubricate the door locks on your car, for example.

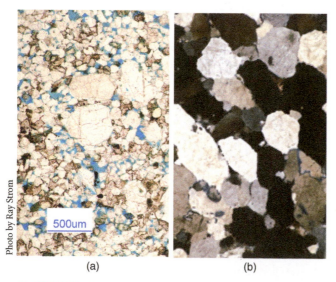

FIGURE 3.30. A quartz sandstone (a) turns into a quartzite (b) when heat is applied. Notice in image B that the grains fit tightly against one another like puzzle pieces.

1. The "rock cycle" illustrates how rocks may change back and forth from igneous, sedimentary and metamorphic rocks. The processes in the rock cycle were more active at some points in Earth history than others (like early in the creation week and the flood).
2. Igneous rocks form from magma or lava. Some igneous rocks were likely formed by processes early in the creation week.
3. Most igneous rocks can be identified by texture and mineral composition.
4. Igneous intrusions can be concordant (sills, laccoliths) or discordant (dikes, stocks, batholiths).
5. As magma (or lava) cools it doesn't all crystallize at once. It goes through a process of crystallization called Bowen's Reaction Series.
6. Sedimentary rocks are classified according to the size of the particles contained within them.
7. Many (but not all) sedimentary rocks were made during Noah's flood; the fossils they contain demonstrate that the rocks formed quickly.
8. Sedimentary structures and various minerals within sedimentary rocks help to discern under what kinds of conditions the rocks were formed under. By studying sedimentary structures we can learn some things about the rate and processes that led to the formation of the rock.
9. Metamorphic rocks are rocks that have been changed as a result of heat and/or pressure.
10. Most metamorphic rocks can be classified by the amount of foliation they exhibit.

KEY TERMS

Anthracite coal
Aphanitic
Aureole
Bituminous coal
Bowen's reaction series
Breccia
Cementation
Chemical sedimentary rock
Clastic (detrital) sedimentary rock
Compaction
Conglomerate
Contact (ort thermal) metamorphic rocks
Creation week rocks
Cross-bedding
Detrital sedimentary
Dewatering

Dynamic metamorphic rock
Extrusive igneous rock
Foliation
Fossil
Igneous rock
Intrusive igneous rock
Kerogen
Kimberlite pipe
Lava
Lignite coal
Lithification
Load casts
Magma
Metamorphic rock
Metamorphism
Mud cracks

Mudrock	Sand injectite
Organic sedimentary rock	Sandstone
Pegmatite	Sediment
Phaneritic	Sedimentary rock
Planar bedding	Sedimentary structures
Porphyry (porphyritic texture)	Shale
Pyroclastic	Siltstone
Rain drop prints	Thermal (or contact) metamorphic rock
Recrystallization	Vesicular
Regional metamorphic rock	Volcanic igneous rock
Ripples	Weathering
Rock cycle	

REVIEW QUESTIONS

1. Why is the study of rocks important?
2. Be able to sketch the rock cycle (figure 3.4) without referring to the diagram!
3. Be able to sketch and identify the different types of plutons from photographs.
4. What are the main criteria in the identification of igneous rocks? Can you sketch out a diagram for the identification of igneous rocks?
5. Can you explain Bowen's Reaction Series to someone who is unfamiliar with it?
6. Why are radiohalos important in a biblical understanding of earth history?
7. If you heated up a granite rock and melted it, what would melt first? Second? Third?
8. What are the main factors used in classifying sedimentary rocks?
9. What are some of the different types of limestone? How are they similar? How are they different?
10. List some examples of sedimentary structures and explain why they are important to geologists to study.
11. List some important rock types and their uses (igneous, sedimentary and metamorphic).
12. What is the primary way metamorphic rocks are classified?

FURTHER READING AND REFERENCES

- Arvidson, R.S., and Mackenzie, F. T., 1999. The dolomite problem; control of precipitation kinetics by temperature and saturation state. *American Journal of Science*, v. 299, p. 257–288.
- Hovland, M., Rueslåtten, Johnsen, H.K., Kvamme, B., and Kuznetsova, T. 2006. Salt formation associated with sub-surface boiling and supercritical water. *Marine and Petroleum Geology*, v. 23, p. 855–869.

- Snelling, A.A. 2008. Radiohalos in the Cooma Metamorphic Complex, New South Wales, Australia: the mode and rate of regional metamorphism. In A. A. Snelling (Ed.) (2008). *Proceedings of the Sixth International Conference on Creationism* (pp. 371–387). Pittsburgh, PA: Creation Science Fellowship and Dallas, TX: Institute for Creation Research.
- Snelling, A.A. 2008. Catastrophic granite formation: Rapid melting of source rocks and rapid magma intrusion and cooling. *Answers Research Journal*, v. 1, p. 11–25.
- Whitmore, J.H. 2009. Do Mud Cracks Indicate Multiple Droughts During the Flood?, in: M.J. Oard and J.K. Reed (eds.), *Rock Solid Answers*, Master Books, Green Forest, Arkansas, p. 167–183.
- Whitmore, J.H. 2007. Should fragile shells be common in the fossil record? *Answers* v. 2(2), p. 78–81.
- Whitmore, J.H. and Strom, R. 2010. Sand injectites at the base of the Coconino Sandstone, Grand Canyon, Arizona, *Sedimentary Geology*, v. 230, p. 46–59.
- Whitmore, J.H., Forsythe, G., and Garner, P.A. 2015. Intraformational parabolic recumbent folds in the Coconino Sandstone (Permian) and two other formations in Sedona, Arizona (USA). *Answers Research Journal*, v. 8, p. 21–40.
- Wise, K. P. 2003. The pre-Flood floating forest: A study in paleontological pattern recognition. In R. L. Ivey, Jr., editor, *Proceedings of the Fifth International Conference on Creationism*, pp. 371–381, Pittsburgh, Pennsylvania: Creation Science Fellowship.

WEBSITES TO VISIT

- http://www.learner.org/interactives/rockcycle/index.html An interactive website that will help you learn and understand some of the basics of the rock cycle. Includes some quizzes. (Annenberg Foundation).
- http://geology.com/rocks/ Igneous, metamorphic and sedimentary rock identification (includes photos and descriptions).
- https://www.youtube.com/watch?v=cJo0fTpJypg A flume experiment illustrating how cross-beds can form.

SCRIPTURES FOR FURTHER STUDY

- Romans 12:1-2. The word *transformed* comes from the Greek work *metamorphoo*. What is Paul teaching about the "metamorphism" that happens in our lives when we become Christians?
- Genesis 1:9-10. The origin of the continents on the third day of creation.
- Genesis 6-9. This passage discusses Noah's flood which we believe accounts for most of the fossiliferous marine sediments that cover every continent (see figure 3.11).
- Using a concordance, find scripture passages in the Old and New Testament using the term "rock." Note that God is often referred to as our "Rock."

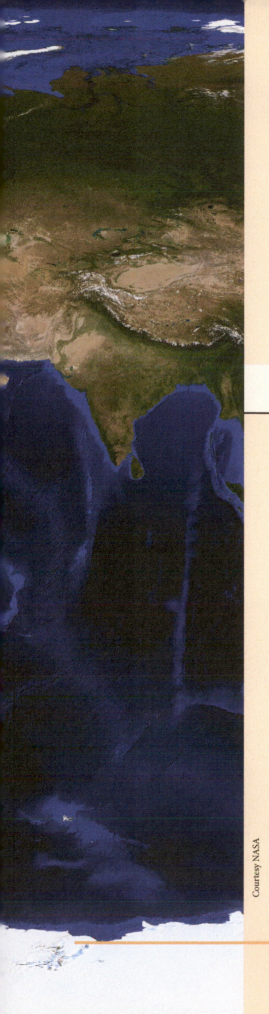

Courtesy NASA

CHAPTER 4

PLATE TECTONICS

OUTLINE:

The continents and oceans of the world contain evidence of having moved in the past, and continue moving today. Image courtesy NASA's Visible Earth.

Ever remember looking at a globe as a child and noticing how South America and Africa seemed to fit together like pieces of a puzzle? Our ability to recognize the seemingly good fit of South America with Africa is easily taken for granted. We live in a world where the geography of the continents and coastlines has been known for centuries, and modern satellite imagery can be viewed with the click of a mouse or a swipe of the smartphone. But it wasn't always this way. The first really good maps of the eastern coasts of the New World were made in the late 1500s and early 1600s as explorers and colonists raced to claim lands for their European mother countries. It appears that the first person to notice the close association of the coastlines of South America and Africa was the Flemish cartographer Abraham Ortelius, who suggested that North and South America were ripped away from Europe and Africa in the past by earthquakes and floods.

Many other scientists and cartographers would also puzzle over the fit of the continents,

including leading thinkers such as Francis Bacon and Benjamin Franklin. The first scientist to put together a full argument for why these continents shared physical similarity was Antonio Snider-Pellegrini, a geographer who likewise noticed the close fit of South America and Africa. He also knew of various rocks and fossils of Pennsylvanian layers (see chapters 6 and 7) that were common to both North America and Europe. Putting these and other evidence together in 1858, Pellegrini argued that all of the continents were once part of an original, single continent which in the past had broken apart (figure 4.1). Not only that, but Pellegrini believed that the mechanism that drove these continents apart were catastrophic, with Noah's Flood being the most recent major event. Sadly, Pellegrini's ideas fell on deaf ears because at the time geologists had abandoned a young-Earth view of history and catastrophic explanations in geology were no longer in favor (see chapter 6).

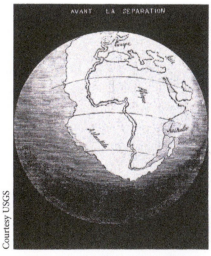
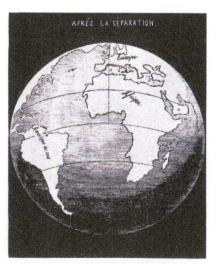

Courtesy USGS

FIGURE 4.1. Pellegrini's illustration of the continents before and after Noah's Flood. Note how similar Pellegrini's pre-Flood continent is to the supercontinent Pangaea in figure 2.

4.2 ALMOST THERE: CONTINENTAL DRIFT

It would take several more decades before Alfred Wegener, a gifted and daring German scientist, would renew a case for moving continents. Wegener's specialty was meteorology and the study of glaciers, but he had a keen mind for geology. In 1915, Wegener published his first scientific paper in which he argued that the continents had slowly and gradually moved over time. Despite having some of the same arguments, there's no evidence that Wegener knew of Pellegrini's work. He followed this up with a much larger work which was widely translated in 1924, and argued that all of the continents were once part of a singular and massive supercontinent he called **Pangaea** ("all the Earth", figure 4.2). According to Wegener, Pangaea split first into northern and southern portions called **Laurentia** and **Gondwana**, which later split to form today's continents as they slowly moved to their present locations over eons of geological time.

4.2.1 Wegener's Evidence for Continental Drift

Wegener called this process **continental drift**. To make this case, Wegener marshalled an impressive set of observations.

FIT OF THE CONTINENTS As with others before him, Wegener noted that the eastern coasts of the Americas and the western coasts of Europe and Africa fit with almost puzzle-piece accuracy. He believed that this was a result of forces that split these continents apart. Most geologists of Wegener's time believed that the continents did not move horizontally, and no one could directly measure any horizontal movement at the time.

MATCHING GEOLOGICAL UNITS Wegener observed that similar rock units that were found on continents from either side of the Atlantic Ocean. Both in terms

of the lithology (physical makeup of the rocks) and their position in the rock record, deep crystalline rocks from Brazil matched those from southern Africa. In North America, the rocks of the Appalachian Mountains were the same relative age as mountains in eastern Greenland, Scotland, Scandinavia, and also the Atlas Mountains of Morocco. Wegener used these clues to piece together the fit of North America against Europe and Africa (figure 4.3), forming a single and immense mountain chain linking each of these mountain chains.

DISTRIBUTION OF FOSSILS Pellegrini had previously noticed similar plant fossils between Europe and North America. Wegener concentrated on the southern continents, where a variety of plant and animal fossils could be seen from South America, Africa, Antarctica, Australia, and India (figure 4.4). By uniting these continents into a large southern region of Pangaea (Gondwana), the geographic range of each organism was continuous over land, with no major oceans between them.

DISTRIBUTION OF LIVING ORGANISMS Why are marsupials (pouched mammals) largely restricted to Australia and South America? Wegener believe that these continents maintained a close connection with each other after the breakup of Pangaea, and that marsupials had evolved on them without contact with placental mammals (mammals with a womb) because their continents were separated from North America, Europe, and Asia by oceans. Wegener thought that continental drift could explain the **biogeography** of both fossils and living plants and animals.

ANCIENT CLIMATES Wegener discovered that there were many records of presumed glacial activity in locations that today simply cannot support glaciers, such as Africa and southern India (now located in tropical areas) or the hot, dry deserts of western

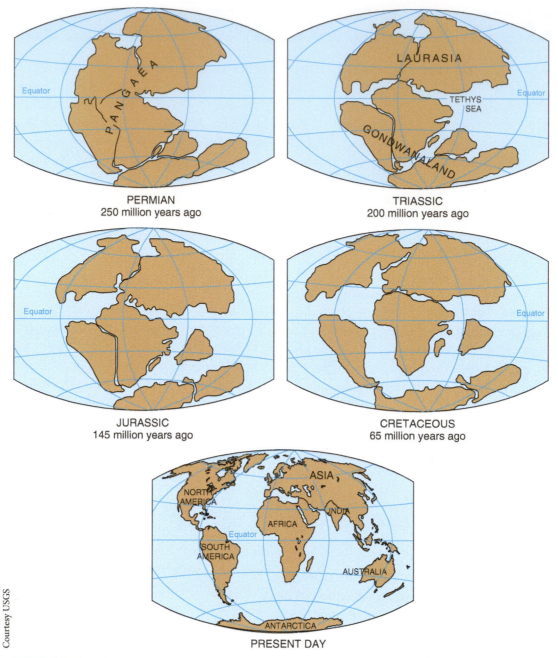

FIGURE 4.2. A modern reconstruction of Pangaea and its later split and movement over an old-Earth timeframe.

Courtesy USGS

Australia. When glaciers flow over land, they often pick up pieces of rock and drag them across the land surface (see chapter 10). This produces marks that point in the direction of glacial movement. By combining the southern continents, the striations all seemed to point away from southeastern Africa, which Wegener believed was once positioned over the South Pole, allowing for the development of glaciers in its center that then flowed out in many directions over the other continents (figure 4.5).

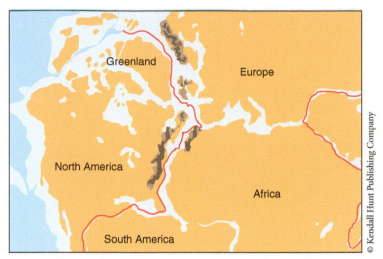

FIGURE 4.3. When put together as Pangaea, mountain chains on different continents (such as the Atlas, Appalachian, and Scandinavian) form a single line.

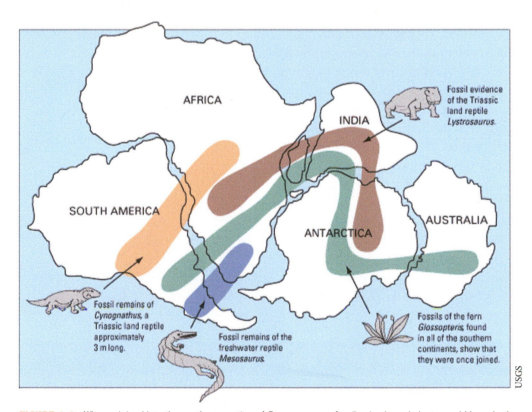

FIGURE 4.4. When rejoined into the southern portion of Pangaea, many fossil animals and plants would have had a continuous range across continents. These included the semi-aquatic reptile *Mesosaurus*, the small carnivorous reptile *Cynognathus*, the herbivorous reptile *Lystrosaurus*, and the semitropical plant *Glossopteris*.

4.2.2 Wegener's Problem of Mechanism

So with all of this evidence marshalled for continental drift, why was Wegener's idea rejected by most members of the geological community? Wegener had developed a very comprehensive, but still circumstantial case for continental drift. There were no measurements of continents moving today, and indeed, our common experience is to think that the continents are not moving (they certainly don't *feel* like they are moving!). Many of Wegener's arguments also had alternate hypotheses or challenges from a stable-continent perspective, such as land-bridge connections to account for fossil distributions or difficulties in explaining non-collision mountain belts like the Rockies.

The bigger problem was in Wegener's proposed mechanisms for how the continents were supposed to drift over time. Wegener initially thought that perhaps the continents drifted by plowing through the ocean floor due to forces from the rotation of Earth. This proved physically impossible; the forces needed for the continents to plow through the ocean basins were far too large to overcome. Wegener dropped this explanation, but had few compelling alternatives, such as gravitational or tidal forces, which likewise failed. Despite his tireless work in promoting and arguing for his theory of continental drift, Wegener was largely ridiculed for his efforts. Yet he would ultimately be

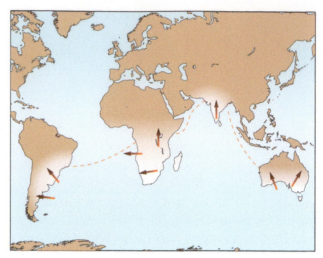

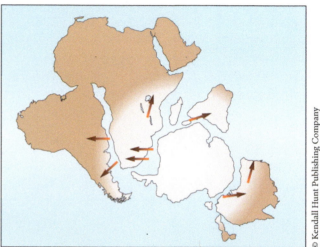

FIGURE 4.5. Wegener argued that striations in rocks on different continents were formed by glaciers that once covered southern Gondwana.

vindicated by a series of discoveries made just a few decades after his death.

4.3 THE PLATE TECTONICS REVOLUTION

With Wegener's ideas effectively defeated by the 1930s, the idea of continental drift was largely ignored or worse, ridiculed. This changed after World War II, particularly during the late 1950s and throughout the 1960s. As the evidence piled up, it took only about 10 years for geologists to discover the keys to moving both the continents *and* the oceans, and transform Wegener's cast-off idea of continental drift into perhaps the most important and comprehensive systems of geology: plate tectonics. Two areas of study paved the way: paleomagnetism and ocean surveying.

4.3.1 Magnetism and the Rock Record

Compasses have been used for direction and navigation since at least the 11th century in China. Whether ancient or modern, a magnetic compass uses a magnetic material (such as the mineral magnetite) that can orient itself with respect to the earth's magnetic field. This orientation occurs because the Earth's magnetic field is a **dipole**, meaning that the magnetic field has two distinct zones, north and south. This field is sustained by fluid motion in the outer core, resulting in a flow of magnetism out of the southern polar region, flowing north and returning to the core through the northern polar region (figure 4.6). The magnetic north and south poles are not directly over the geographic north and south poles (which are defined by rotation on Earth's axis), but are often a few hundred kilometers away.

COMPASSES IN THE ROCKS By the 1800s, scientists had discovered that certain rocks strongly affected compasses, and that magnetic fields parallel to Earth's surface were effectively "frozen" in recent lava flows. This occurs when a magma or lava cools to form an igneous rock (see Chapter 3). If the magnetite is one of the igneous rock minerals, once its temperature drops below 580° C, it can be influenced by Earth's magnetic field. This temperature is called the *Curie point*. As the lava solidifies, the magnetite becomes locked into place by the other minerals that grew around it. The magnetite's orientation is thus a magnetic signature for the rock, indicating its position on the planet at the time of formation. This is known as **paleomagnetism**.

POLAR WANDER In the 1950s, new inventions in detecting magnetic fields were applied to rock units, and their signatures could be more fully understood. The record of paleomagnetism seemed to indicate that the magnetic north pole had not always been in the same place. Scientists called the looping path that the magnetic north pole seemed to take over time "**polar wander**" But even more interesting was the fact that the polar wander path derived from

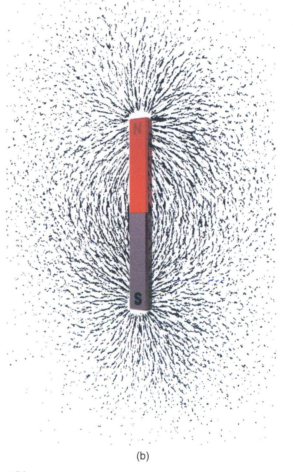

(a)

(b)

FIGURE 4.6. Earth's dipole magnetic field (a) is similar to that of a bar magnet (b).

igneous rocks in North America was different from the polar wander path derived from igneous rocks in Europe! Both paths started in the same place (near the geographic north pole at present), but then diverge further and further among older and older rocks (figure 4.7a).

The solution to the problem of these differing polar wander paths was not that the magnetic field had moved, but that the *continents* had. Only when the continents were reconfigured could the polar wander paths point towards a single pole at other times in Earth history (figure 4.7b). The paleomagnetic studies of the 1950s and 1960s on continental rocks confirmed Wegener's hypothesis of continental drift, though the problem of a mechanism to move them still remained.

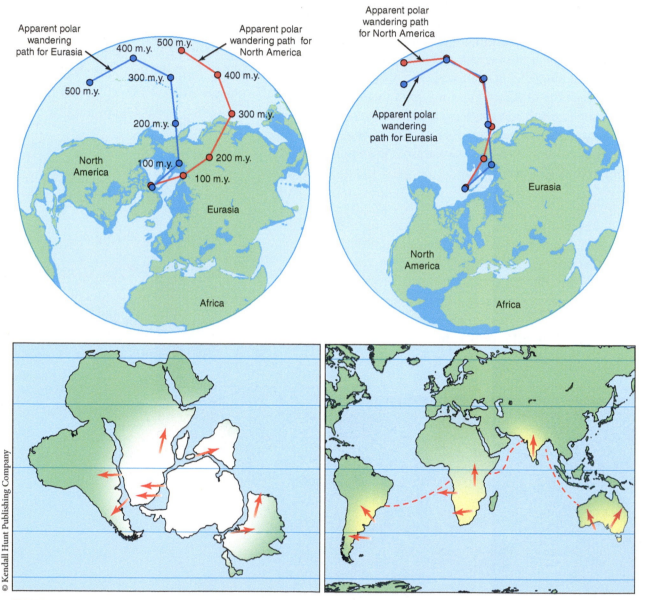

FIGURE 4.7. Polar wander curves (a) do not align when the continents are fixed, but (b) properly align when continents are placed in historical plate tectonic locations.

MAGNETIC REVERSALS One further paleomagnetic issue proved crucial in the development of plate tectonics. Beginning in the early 1900s, some geologists noticed that the orientation of magnetic mineral grains was opposite to the current magnetic field. That is, instead of the magnetite in the lava pointing towards the magnetic north pole, it was pointing south. Little progress was made until after World War II, when geologists became interested in these rocks and their backward-pointing internal compasses.

What became apparent was that Earth's magnetic field had undergone a **magnetic reversal** in the past, a period of time when the magnetic field flipped its orientation. The field had emerged from the north magnetic pole and flowed into the south magnetic pole, opposite of the current orientation. Moreover, this was not a singular event: there are literally hundreds of reversals recorded in the rocks of the planet. Rocks that were formed during periods where the magnetite grains point north (as they do today) are called "normal" periods, while those which point south are assigned to "reverse" periods.

4.3.2 Ocean Features and Composition

Mapping of the ocean floor, especially during and after World War II, produced the final pieces of the plate tectonics puzzle. These included the discovery of ocean ridges, deep-ocean trenches, the nature of the ocean crust and its sediments, and magnetic signatures in the ocean crust.

THE CONTINENTAL SHELF AND SLOPE The science of determining water depth is called **bathymetry**. The first major research vessel to explore the oceans was a British ship, the *H.M.S. Challenger*, from 1873–1876. From this journey, the crew of the *Challenger* was able to determine the broad outline of the **continental shelf**, where water depths were typically less than 150 meters. The continental shelf extends from a few tens to a few hundreds of kilometers further out to sea and its shape can vary a bit from the coastline (figure 4.8). At the end of the shelf is a steep drop-off into the **abyssal plain** at the **continental margin** (the edge of a continent), and water depths plunged to two thousand meters or more. With the development of sonar in the 1920s, many governments began to aggressively measure the seafloor in their own territorial waters and beyond.

OCEAN RIDGES The journey of the *Challenger* also identified a large mountainous structure through the middle of the Atlantic Ocean. With the advent of sonar and its extensive use by the U.S. Navy during World War II and the early Cold War, a picture emerged of

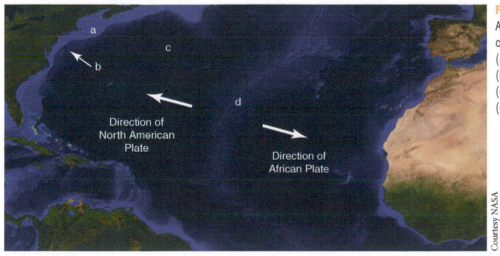

FIGURE 4.8. The North Atlantic, showing important ocean features:
(a) the continental shelf;
(b) the continental margin;
(c) the abyssal plain; and
(d) the mid-Atlantic ridge.

Direction of North American Plate

Direction of African Plate

Courtesy NASA

an immense linear mountain chain now known as an **ocean ridge** (marked by a *d* in figure 4.8). Stretching from north of (and through!) Iceland and nearly to Antarctica, the Mid-Atlantic Ridge gently rises, peaks, and falls across more than 1,000 km, and stands 2,000 to 3,000 meters above the level of the ocean basin. The ridge is cut and shifted by many faults, and a large depression, called a **rift**, is in its center. The rift is volcanic and produces a large number of shallow and relatively small-magnitude earthquakes.

Looking at figure 4.8, notice that the Mid-Atlantic Ridge has a shape that mirrors the western continental margins of Europe and Africa, and the eastern continental slopes of North and South America, and is a far better match than the modern coastlines which inspired Pellegrini and Wegener to propose moving continents.

DEEP-OCEAN TRENCHES Another unusual discovery by the *Challenger* survey was the existence ultra-deep **ocean trenche**s (figure 4.9). While the average depth below sea level in the abyssal plain is about 2–3 km, these narrow zones can plunge to depth of 11 km. While there are a few trenches in the Atlantic (most are near the Caribbean islands), they are far more numerous along the margins of the Pacific Ocean. Among the best-known are the Peru-Chile Trench and the Mariana Trench.

OCEAN CRUST AND ITS SEDIMENTS Beginning in 1968, the research vessel *Glomar Challenger* (named in honor of the earlier *H.M.S. Challenger*) was sent out to get samples of the sediments and rocks of the ocean floor using a large on-board drilling rig. The *Glomar* scientists made two crucial discoveries. First, the ocean crust was composed everywhere of basalt, a rock dominated by the iron- and magnesium-rich mafic minerals with a density of 2.9 g/cm^3. The ocean crust was not made of the less-dense (2.7 g/cm^3), silica-rich igneous and

Courtesy NASA

FIGURE 4.9. The Peru-Chile Trench appears as a dark-blue zone on South America's west coast. It is nearly 6,000 km (3,700 miles) long and has a maximum depth of over 8,000 m (26,400 ft).

metamorphic rocks that make up the continents. The second discovery was that the further away from the ocean ridges samples were taken, the sedimentary cover over top of the basalt crust became progressively thicker, and included geologically older sediments as well (figure 4.10).

MAGNETIC SIGNATURES OF THE OCEAN CRUST In the early years of the Cold War, research ships towed magnetometers to measure changes in the magnetic field strength of the ocean basalts (and also to find any nearby submarines). Looking across sections of the ocean ridges, scientists discovered that the ocean crust on either side of the ridge showed a nearly identical pattern of normal and reversed paleomagnetism, parallel to the ocean ridge axis (figure 4.11).

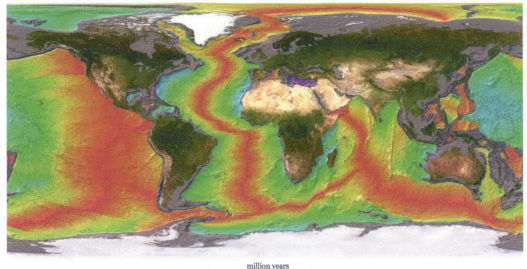

FIGURE 4.10. Age of the seafloor according to old-Earth estimates. Note that the youngest material is found at the ocean ridge, and that older material is found far from the ridge. Arrows indicate the direction of motion of the plates.

Courtesy NOAA

million years

0 20 40 60 80 100 120 140 160 180 200 220 240 260 280

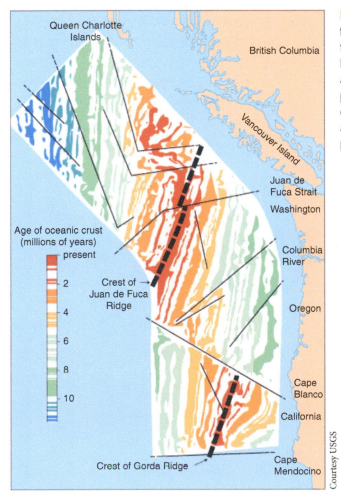

Queen Charlotte Islands

British Columbia

Vancouver Island

Juan de Fuca Strait

Washington

Age of oceanic crust (millions of years)

present

Columbia River

2

Crest of Juan de Fuca Ridge

4

Oregon

6

8

Cape Blanco

10

California

Crest of Gorda Ridge

Cape Mendocino

Courtesy USGS

FIGURE 4.11. Paleomagnetic features at the Juan de Fuca ridge off the coast of Oregon, Washington, and British Columbia. The normal (colors) and reverse (white) patterns are parallel to the ridge axis, and provided evidence for seafloor spreading. The ages (in millions of years before present) are old-Earth time estimates.

4.3.3 Putting It All Together

From the late 1950s through the 1960s, the discoveries described above were falling into place, and it became clear that Wegener was actually closer to the truth than anyone had ever thought. The continents did move, but they did not plow through the ocean floor like an icebreaker. Different forces, involving the formation, movement, and destruction of ocean crust, were behind the shifting position of many large and small pieces of Earth's crust.

SEAFLOOR SPREADING Geologist and Rear Admiral of the U.S. Navy Harry Hess was instrumental in tying together the many components of the ocean floor described above, weaving them into a single hypothesis now called **seafloor spreading**. Hess theorized that the ocean crust was like a conveyor belt: new ocean crust formed from rising magma at the ocean ridges. Since Hess believed that this is where two pieces of ocean crust spread apart, these regions are also known as **spreading centers**. In the areas where deep-ocean trenches were found, Hess argued that the ocean crust dipped downward at the deep-ocean trenches and traveled back into the mantle in a process called

subduction. Subduction meant that the ocean crust was recycled: new material at the spreading center replaced the material destroyed at subduction zones (figure 4.12).

Hess' hypothesis of seafloor spreading was confirmed with the discovery of the mirrored and alternating pattern of magnetism found in the ocean crust parallel to the ridge axis, and the evidence of progressively deeper and older sediments away from the ridge axis (figures 4.10 and 4.11). If new crust was formed at the ridge axis, as it cooled the grains of magnetite would orient with the magnetic field. As the crust moved away from the ridge, new crust would form at the ridge, and when the magnetic field reversed the new magma would pick up a reversed magnetic signature. Further from the ridge, the older, solid crust would retain its original magnetic signature.

Likewise with the sediment cover. When the new ocean crust is freshly formed, there are no sediments above it. As the crust moves away from the ridge, progressively more sediment is deposited. The further away from the ridge, not only has more sediment been deposited, the further regions also have older sediments above the crust compared to younger parts of the crust nearer to the ridge axis.

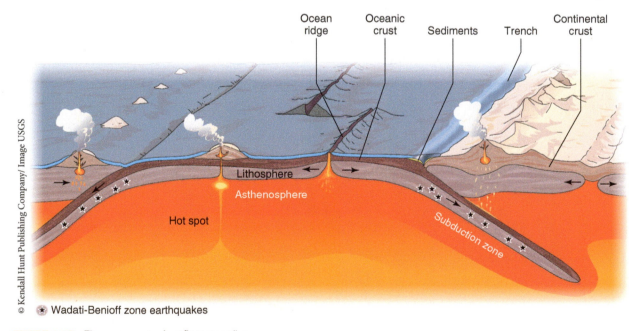

FIGURE 4.12. The components of seafloor spreading.

Hess' concept of moving ocean crust tied together many different sets of facts into a single mechanism.

SUBDUCTION ZONE EARTHQUAKES Stating that subduction was occurring at the deep-ocean trenches may have been reasonable, but the hypothesis needed additional confirmation. This came by the way of earthquakes. Earthquakes were already known to occur in and nearby the trenches, but in the 1960s more accurate and numerous seismometers confirmed the presence of long, steeply angled regions of earthquake activity near the trenches. Called **Wadati-Benioff zones** (figure 4.12) in honor of their discoverers, these earthquake regions extend from the trench surface downward at angles ranging from 30°–70°. The linear nature of these earthquakes give geologists the location of the subducting ocean crust, which generates earthquakes as it slides against another piece of crust and the upper mantle, and always moving away from the ocean ridge that formed it.

PLATES AND PLATE BOUNDARIES IDENTIFIED From Hess and others we now know that the surface of Earth is divided into a series of **plates** that move with respect to each other. These plates are rigid, and deform by cracking or breaking to produce earthquakes. The plates are composed of both crust and some of the upper mantle, which move together as a single unit. This combination of crust and upper mantle is called the **lithosphere**, which means *layer (or region) of rock*. Ocean crust is quite thin (7–10 km), so the lithosphere capped by it is about 100 km total depth. Continental crust is much thicker (30–70 km) than ocean crust, resulting in a lithosphere that is 150 km deep on average, and can be up to 280 km deep under the largest mountain chains. The plates move over a section of the upper mantle called the **asthenosphere** (*"weak layer/region"*), where the rocks are so close to their melting temperatures that they will not break, but instead are able to flow despite being solid, somewhat similar to Silly Putty™.

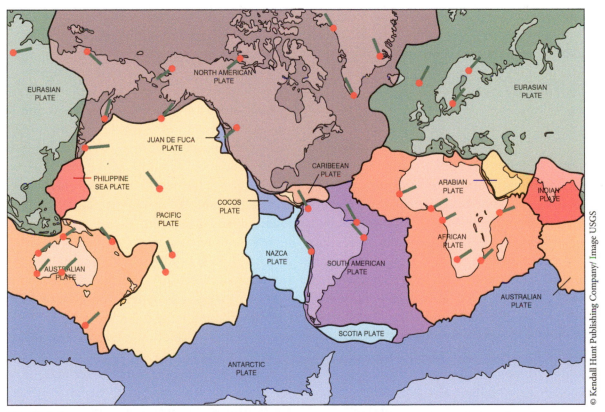

FIGURE 4.13. The direction of motion among Earth's major plates as determined by GPS measurements. Regions of significant earthquake and/or volcanic activity help identify the plate boundaries.

There are fourteen major plates and many other smaller ones as well. Figure 4.13 illustrates the major plates and their direction of motion. It is the distribution of earthquakes, ridges, and subduction zones that geologists use to determine plate boundaries. In looking at figure 4.13, you will also notice that many plates include both continental *and* oceanic crust. The Eurasian and African plates, for example, extend to the Mid-Atlantic Ridge, where they are met by the North and South American plates. In these plates, the continental shelves tend to be wide, and they are characterized by a **passive margin**: the continental and oceanic crusts are part of the same plate, and their

contact is not characterized by earthquakes and volcanic activity. At **active margins**, the edge of the continent coincides with a plate boundary, and earthquake and/or volcanic activity is common. Active margins characterize the land around the Pacific Ocean.

As the lithospheric plates move, they carry whatever type of crust happens to be above them. Because our understanding of the forces involved is different from Wegener's views of continental drift, (where only the continents moved) we use the term *plate tectonics* to refer to the components and mechanisms that move lithosphere plates around Earth's surface.

4.4 A CLOSER LOOK AT PLATE BOUNDARIES

Since the plates move horizontally over Earth's surface, they interact with each other at plate boundaries in three ways. They can a) move away from each other; b) move towards each other; or c) move past each other. Geologists refer to these movements as *divergent*, *convergent*, or *transform* plate boundaries. Each type of interaction produces a characteristic set of physical features.

DIVERGENT BOUNDARIES When two plates move in opposite directions from each other, they are separated by a **divergent boundary**. The most common form of divergent boundary is the ocean ridge which is illustrated in profile in figure 4.14. Two plates, both capped by ocean crust, diverge at the central rift, where magma rises to form new lithosphere capped by basaltic ocean crust. The constant extensional (opposite-direction) forces form a large number of faults and fractures as the two plates diverge, and the buoyant pool of magma below keeps the region near the rift elevated, forming the mountainous ridge system seen in figure 4.8.

Another type of divergent boundary is found within a continent, called a **continental rift**. Here the situation is more complex than at ocean ridges,

as the forces that produce a continental rift can continue through a number of stages (figure 4.15), ultimately forming a new ocean basin. The first stage occurs when extensional forces begin to pull one continent in opposite directions. These forces begin to stretch the continent, resulting in earthquakes along developing fault lines. While the whole region is lifted upward by rising magma below, the rocks nearest the center of the rift slide down the vertical faults, forming an elongated *rift valley*. Some volcanic or geothermal activity begins to occur.

In the second stage, the continental crust has been sufficiently thinned that the rift valley has widened and deepened. Volcanic structures and activity are common in the rift valley, as are rivers and lakes. In the third stage of development, the extensional forces have separated the continent into two pieces, and magma rises more freely in the valley center. Now below sea level, ocean water may invade the region, creating a long, linear sea. Over time, the extensional forces continue to pull the continents away from each other, and the linear sea forms into an ocean basin. The former rift valley has transitioned into an ocean ridge. Various stages of the continental rift ocean ridge are seen along

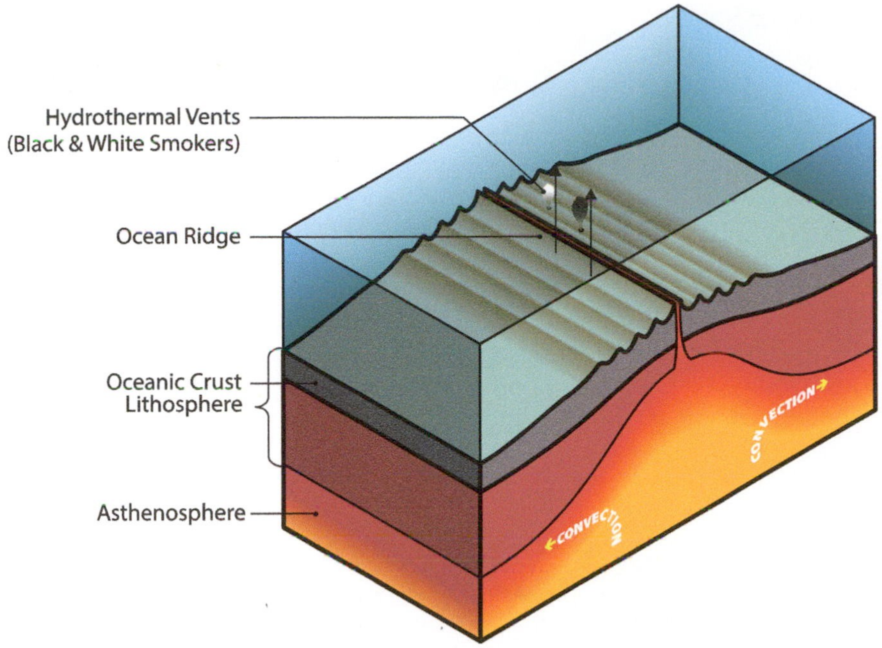

Zern Liew/Shutterstock.com

FIGURE 4.14. A cross-sectional profile of an ocean ridge.

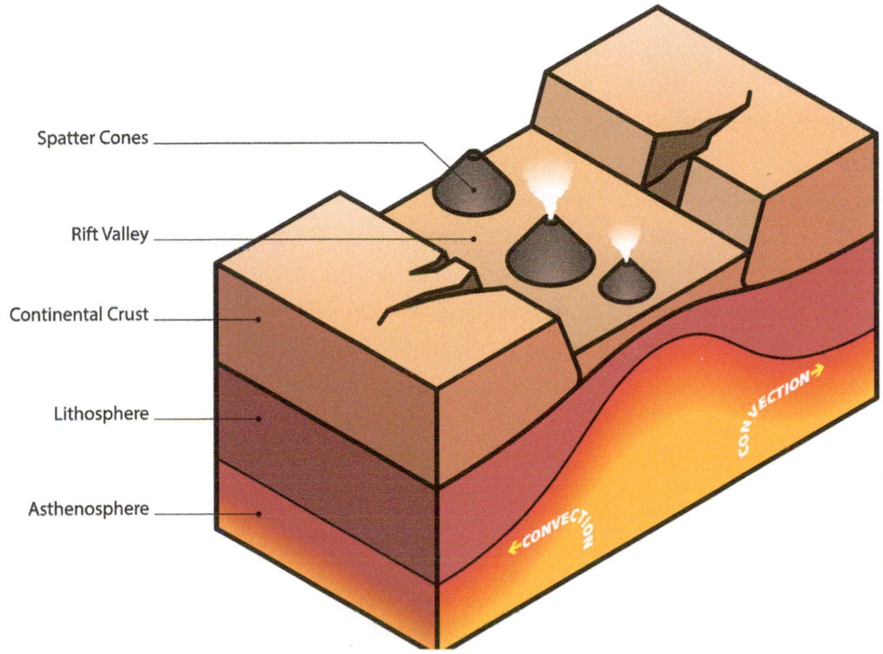

Hydrothermal Vents
(Black & White Smokers)

Ocean Ridge

Oceanic Crust
Lithosphere

Asthenosphere

CONVECTION

CONVECTION

FIGURE 4.15. Extensional forces (arrows) separating a continent produce a continental rift zone.

Spatter Cones

Rift Valley

Continental Crust

Lithosphere

Asthenosphere

CONVECTION

CONVECTION

Zern Liew/Shutterstock.com

the East Africa rift (figures 4.16), a complex region characterized by wide rifted valleys where numerous active volcanoes (such as Mt. Kilamanjaro) and large lakes (like Lake Victoria) stand out as dominant features of the landscape. The East African Rift is the boundary between two sub-plates on one continent: the Nubian plate to the west and the Somali plate to the east.

CONVERGENT BOUNDARIES In the case of two plates moving towards one another, a **convergent boundary** is seen. With two types of crust atop the lithospheric plate, there can be three kinds of convergent boundaries: continental-continental, continental-oceanic, and oceanic-oceanic.

At a *continental-oceanic convergence* (figure 4.17), the density of the ocean crust is greater than the

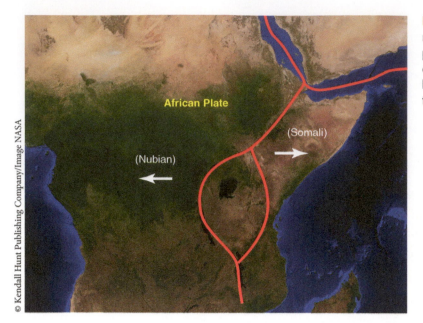

African Plate

(Somali)

(Nubian)

FIGURE 4.16. The East Africa rift system (EAR; red line) is a continental rift that splits the African plate into the Nubian and Somali sub-plates. The EAR connects to the north to a triple junction including two linear seas with ocean crust and small ocean ridges: the Red Sea and the Gulf of Aden.

continental crust, forcing the plate capped with ocean crust to subduct below the plate capped with continental crust. The oceanic plate plunges down into the mantle, forming an ocean trench and producing the line of earthquakes of the Wadati-Benioff zones. At depths of 65-130 km, parts of the subducting ocean plate begin to melt, forming a magma that rises upwards into the continent. This can emplace a massive amount of new igneous rock (typically intermediate/andesitic in composition; see Chapters 3 and 5), and often produces volcanoes at the surface. The Cascades and Andes mountains of North and South America are examples of volcanic mountain ranges formed by oceanic-continental convergence, called **continental arc volcanoes**.

At an *oceanic-oceanic convergence* (figure 4.18), two plates each capped with oceanic crust move toward one another. One of the two plates has ocean crust that is slightly colder and denser than the other due to being further from its magma source at a distant ridge. This denser plate will subduct, producing similar features to those seen in the continental-oceanic convergence: a deep-ocean

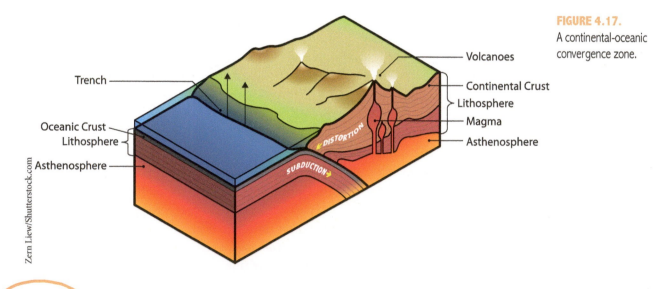

Trench

Oceanic Crust
Lithosphere
Asthenosphere

DISTORTION

SUBDUCTION

Volcanoes

Continental Crust
Lithosphere
Magma
Asthenosphere

FIGURE 4.17. A continental-oceanic convergence zone.

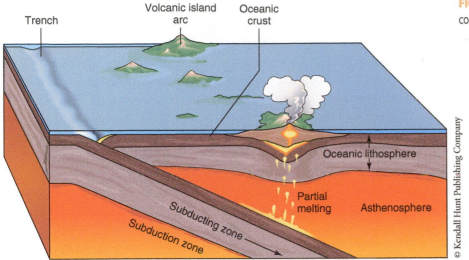

© Kendall Hunt Publishing Company

FIGURE 4.18. An oceanic-oceanic convergent boundary.

Labels on figure: Trench, Volcanic island arc, Oceanic crust, Oceanic lithosphere, Partial melting, Asthenosphere, Subducting zone, Subduction zone

trench, numerous earthquakes, and partial melting of the subducting plate. Because the plates are under the ocean, when the rising magma penetrates the overlying plate it produces underwater volcanoes of intermediate/andesitic composition, which over time may grow large enough to become volcanic islands. These systems of underwater and exposed volcanoes are called **volcanic island arcs**, because of the semi-circular shape that they have at the surface. Volcanic island arcs are common features around the Pacific Ocean's "ring of fire", and include New Zealand, Japan, and the Philippines.

The Caribbean Islands are an example from the Atlantic Ocean.

When two plates capped with continental crust collide in a *continental-continental convergence* (figure 4.19), neither of the plates can subduct into the mantle, since both are made up of low-density granitic rock. When this occurs, the force of the impact of these two continental plates results in a very large mountain belt, characterized by deformed and faulted rocks, but with little volcanic activity because the continental rocks are not extensively melted. The Alps and Himalayas are the best modern example of

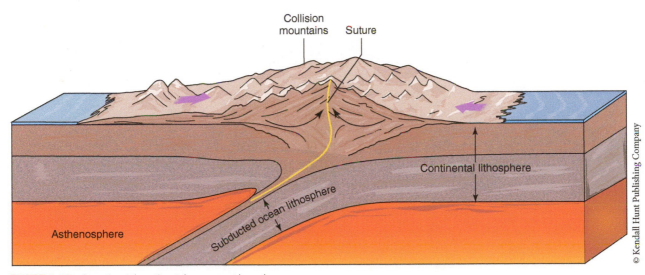

Labels on figure: Collision mountains, Suture, Continental lithosphere, Asthenosphere, Subducted ocean lithosphere

© Kendall Hunt Publishing Company

FIGURE 4.19. A continental-continental convergent boundary.

a continental-continental convergence. In fact, the Himalayas are still growing, because India continues to move northward into Asia. As a testimony to the continued and at times violent nature of this collision, a major earthquake struck Nepal on April 25, 2015. The quake registered 7.8 on the moment magnitude scale and surface shaking of up to IX on the modified Mercalli scale (see Chapter 5 for a discussion of these scales). Thousands died and many more were injured. The quake actually released enough pressure to cause Mt. Everest to drop 2.8 centimeters! When North America collided with Africa and Europe during the formation of Pangaea, the Atlas, Appalachian, Scottish and Scandinavian mountains were all formed through this same type of collision (figure 4.3).

TRANSFORM BOUNDARIES The final type of plate boundary are **transform boundaries**, which form where two plates move alongside and past each other, rather than directly towards or away from each other. Much like traffic on either side of the yellow line in the middle of a road, the plates are moving in opposite directions without either splitting apart or colliding along the boundary. There are two main areas where transform boundaries are observed. The most common are those found near the ocean ridges; rarer but more famous for their impact to humans are those that cut across continents.

Look again at figure 4.9. You will notice that the ridge is frequently offset by lines oriented perpendicular to the ridge axis, and generally follow the shapes of the continental margins on either side of the Atlantic. These lines are known as **fracture zones** (figure 4.20), and actually connect the offset ridge segments. While the fracture zones as a whole are often hundreds to thousands of kilometers long, there are only earthquakes in the *active zone* located between offset ridges. There, two plates are moving in opposite directions, and earthquakes are common, shallow, and low-magnitude as the plates slide against each other. Once past the active zone, the fracture continues, but only within one plate. These *inactive zones* have ocean crust on either side of the fracture, but they are moving in the same direction, providing no friction for earthquakes to occur.

Transform boundaries are much rarer in continental crust, but their effects on humans are far greater than those in the fracture zones. The best-studied continental transform boundary is the San Andreas fault, found in western California. A spreading center that separates the Baja Peninsula from mainland Mexico converts into a transform boundary that stretches from Baja to Fort Bragg, about 250 km north of San Francisco (see figure 4.22). This transform boundary includes the famous San Andreas Fault, though the boundary is made up of hundreds of other active faults as well. Along the San Andreas, the rocks of California

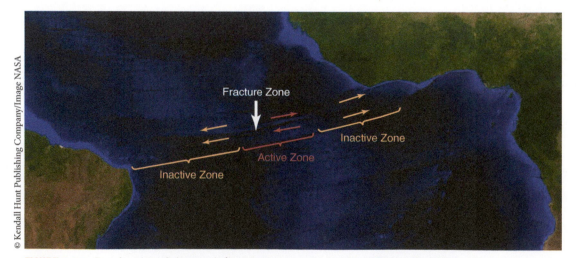

FIGURE 4.20. Transform boundaries seen in fracture zones near the Mid-Atlantic ridge in the southern Atlantic Ocean.

to the east are part of the North American Plate, and are moving westward. On the west side of the fault, the rocks are actually attached to the Pacific Plate, which is moving northwest with respect to North America. This means that San Diego, Los Angeles, and San Francisco aren't actually on the North American Plate!

Another important continental transform is found in the Middle East. The Dead Sea Transform (figure 4.21) separates the African Plate (Nubian portion) from the Arabian Plate. The Jordan River flows through this valley, forming the border between Israel and Jordan. To the south, the Dead Sea Transform connects to the spreading center in the middle of the Red Sea. This region has obviously played an important role in the history of Israel, whose people escaped Pharaoh's army across the Red Sea (a divergent boundary) and re-entered their homeland by crossing through the Jordan River (a transform boundary). Centuries later, Jesus would be baptized in this same river, and leading up to His

FIGURE 4.21. The Dead Sea Transform (yellow line and arrows) connects to the divergent Red Sea spreading center (gray line and arrows).

return, he prophesized that many earthquakes would occur (Box 4.1). It's not surprising, then, that geologists have discovered that this is a fault-dominated land which will shake at Jesus' return!

BOX 4.1   JESUS' RETURN

In Matthew 24 (and paralleled in Mark 13 and Luke 21), Jesus describes the events leading up to His return, including "earthquakes in various places" (verse 7). Earthquakes that are part of the time of tribulation preceding Jesus' return are described in detail as part of the judgments in Revelation 6, 8, 11, and 16. Zechariah 14 foretells the great earthquake that accompanies Jesus' return to Earth, which will split the Mount of Olives in two!

4.5 HOT SPOTS AND PLATE RATES

In addition to the many features of the oceans and continents described earlier, there is one more important feature that allowed geologists to test the ideas of plate tectonics. **A hot spot** is *a plume of magma originating in the lower mantle and rising upward into or through the crust.* As the hot spot rises and contacts the lithosphere, the hot magma forces its way up through cracks in the rock, and erupts at the surface. Thus, a hot spot forms volcanoes on the

crust above it. Figure 4.22 shows the distribution of several of the nearly 100 hot spots currently known.

It appears that hot spots are stationary, or nearly so within their mantle location. This means that as a plate moves over a hot spot, the hot spot can produce a series of volcanoes, one after the other. This is the case with the most famous hot spot chain, the Hawaiian Islands (figure 4.23). All of the eight major Hawaiian Islands are volcanic in origin, and

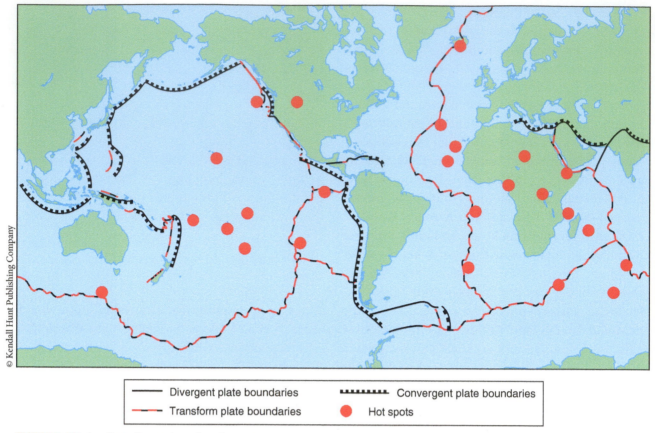

| —— Divergent plate boundaries | ▪▪▪▪▪▪ Convergent plate boundaries |
| — Transform plate boundaries | ● Hot spots |

FIGURE 4.22. Locations of several prominent hot spots.

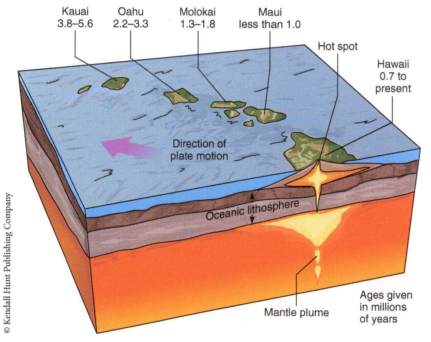

Kauai
3.8–5.6

Oahu
2.2–3.3

Molokai
1.3–1.8

Maui
less than 1.0

Hot spot

Hawaii
0.7 to
present

Direction of
plate motion

Oceanic lithosphere

Mantle plume

Ages given
in millions
of years

FIGURE 4.23. The Hawaiian Islands represent a series of volcanoes formed as the Pacific Plate migrated northwest over a hot spot. Dates provided are according to old-Earth views. Young-Earth geologist believe that these islands formed after Noah's Flood, and became larger due to slowing plate motion.

formed after the Flood was over but only the island of Hawai'i (the "Big Island") is active. The others are now extinct and have moved west/northwest, carried away from the hot spot by the migration of the Pacific Plate. Following the west/northwest trend, there are numerous other extinct volcanic structures under the ocean surface, some of which have coral reefs built on top of the submarine volcanoes. Further out, this chain abruptly shifts to the north, with numerous submarine extinct volcanoes (called the Emperor Seamounts) leading all the way to a subduction zone near the Aleutian Islands of Alaska. All together, the Hawaiian Island chain and Emperor Seamounts span about 6,000 km. These hot spot-generated features have been used by geologists to infer plate movement direction and speed.

With the development of the Global Positioning System (GPS), plate motions can now be directly measured in the present. These tools were unavailable to Wegener in the 1920s or Hess in the 1960s, but they now confirm that the plates are moving according to the expectations of plate tectonics theory. The plate directions shown on Figure 4.13 are based on the velocities of dozens of GPS sites, with thousands more located around the world. We know that current plate velocities vary around the globe from only a few mm/yr up to 20 cm/yr.

4.6 NOAH'S FLOOD: WHERE SCRIPTURE MEETS GEOLOGY

To this point, we have considered the evidence that the Earth's surface is composed of lithospheric plates capped by oceanic and/or continental crust, and that there is sufficient evidence to convince us that these plates have moved significant distances throughout Earth's history. As young-Earth creationists, these evidences require a historical account and mechanism that can satisfactorily explain them while remaining faithful to the foundational truths of the Biblical record.

Does the Bible specifically state that the continents and oceans have either moved or remained unmoved over time? That answer is "no"; the Bible makes no direct claim for either situation. In terms of geology, is the evidence for long-distance, horizontal movements of the continents and ocean basins compelling? We believe that answer is "yes." Given this, we seek to discover how plate tectonic movements can be understood within the young-Earth framework provided by the Bible. Surprisingly, the answer helps us understand the likely mechanism behind Noah's Flood.

4.6.1 Scriptural Framework

The account of Noah is found in Genesis chapters 6-9, and chapters 7 and 8 describe the event that we commonly call "Noah's Flood." There are over 200 accounts of a large-scale flood from different peoples and cultures around the world (which lends helpful extra-biblical evidence for its occurrence), and the account found in Genesis is the most realistic and detailed of all. In following the days and months outlined in the account, the Flood lasted about 371 days from the time Noah enters the ark to the time he and the animals leave it (compare Gen. 7:11 and 8:14-16). This is much longer than the commonly assumed duration of 40 days, which is actually just the first phase of the Flood (Gen. 7:17).

Several terms and phrases used in Genesis 6-9 point to Noah's Flood being a unique event that was global in scope, affecting the entirety of both the physical Earth and all the organisms (including humans) living on it. These include:

Mabbul—This Hebrew word translated is as "flood" or "deluge" in Genesis 6–11 and also in Psalm 29:10. In Greek, the term *kataklysmos* is used in the Septuagint (the Greek translation of the Hebrew Old Testament) and in the New Testament. Both *mabbul* and *kataklysmos* are used only for Noah's Flood; it is a singular event unlike any other in history.

"All" and *"every"*—These terms are used over 60 times in the Flood account to describe the scope and extent of the destruction of the world. Constant and repetitive, they are an emphasis to drive home to the Israelites that indeed *all flesh* was destroyed. This included not only all humans (whose wickedness brought judgment upon themselves) but also all the land dwelling, air-breathing animals: domestic livestock, wild animals, birds, and "creeping things."

The waters rose—This phrase is a continuous reminder throughout Gen. 7 that the waters of the *mabbul* are drowning the entire planet, not just a localized region. By covering all of the high hills by more than 6 meters (20 feet), the flood left no place for land-dwelling, air-breathing animals or humans to escape (see Gen. 7:20-23; I Pet. 3:20).

But God remembered Noah—The first words of Gen. 8 function as both the focal and hinge point for the entire Flood account. It is the focus because God's "remembering" of Noah is God acting upon the covenant to save Noah as promised in Gen. 6:8. It is the hinge because the first half of the Flood account (Gen. 7) is filled with rising water and death; the second half (Gen. 8) sees lowering water and the promise of life.

With these and the other details presented in the Biblical text, creationists recognize that the Flood occurred at a particular time in Earth history, had a specific duration, covered the entire planet, and killed off vast numbers of humans, plants, and animals. We now turn to a possible mechanism that fits these parameters and can also explain the physical evidence of dramatic plate motions discovered by geologists.

4.6.2 Catastrophic Plate Tectonics

For many decades, young-Earth creationists simply did not have an adequate scientific model of Noah's Flood. Admittedly, the Bible's record of the flood does not include many clues about the mechanisms that caused it. We are told that "the fountains of the great deep burst open, and the floodgates of the heavens were opened" (Gen. 7:11, NIV), and we are given details regarding rain and rising and falling water levels, but little else. After all, Noah couldn't see down to the lithosphere from the ark!

The prospects for a useful and comprehensive Flood theory began changing in 1994, when a group of six creation scientists proposed **catastrophic plate tectonics** (or CPT; see Austin *et al.* in the Further

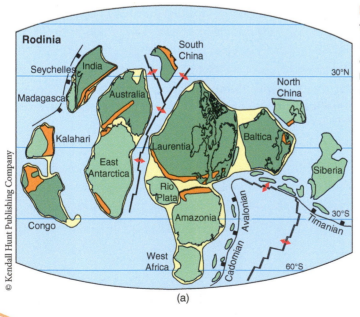

FIGURE 4.24. Three stages of Earth's tectonic history is a biblical perspective: (a) The pre-Flood (Precambrian) continental assemblage known as Rodinia; (b) the mid-Flood supercontinent of Pangaea; and (c) modern world geography.

FIGURE 4.24. (Continued)

The formation of Pangaea early in the Flood.

(b)

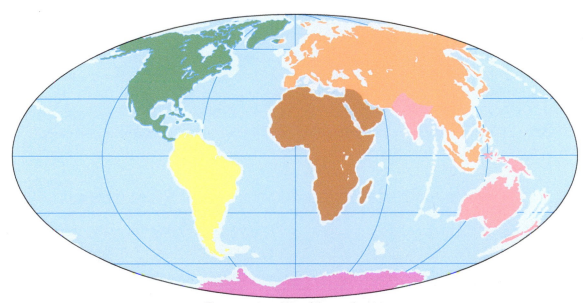

The current position of the continents.

(c)

Reading section). This theory argued that very rapid movements of the plates occurred during Noah's Flood, and indeed was the major physical mechanism that God used to destroy the world. It remains the best comprehensive explanation for the many physical features on the Earth that point towards large-scale plate motions while remaining faithful to the Biblical record of Earth history.

The CPT model begins with the Earth structured similar to today, comprised of an inner and outer core, mantle, and crust. The pre-Flood world had oceanic and continental crust for the oceans and the land. The configuration of the continents, however, was dramatically different than the present: perhaps organized as a supercontinent known as Rodinia (figure 4.24a). The ocean crust was cold and dense, similar to its composition today far from ocean ridges (which probably did not exist at the beginning of creation). Given that God called his creation "very good" at the end of creation week, we doubt that there were many large volcanic structures, earthquakes, or other geologically produced natural hazards before the Flood.

The Flood begins when the ocean crust alongside the continents fractures and begins to sink. This occurs because dense ocean crust (such as that seen at the subduction zone by Japan) is actually denser than the underlying mantle, creating an unstable arrangement. As the pre-Flood ocean lithosphere sinks into the mantle, the friction between them creates large amounts of heat, which makes the mantle less viscous and allows the ocean slab to subduct even faster. This generates even more heat and allowing for more and faster subduction, resulting in a positive feedback loop. Dr. John Baumgarder, who pioneered the computer modeling of CPT, calls this "runaway subduction", and may be part of the "fountains of the great deep burst open" described in Gen. 7:11. According to Baumgardner's modeling, the ocean lithosphere was moving at a few *kilometers per hour* both at the surface and through the mantle during the Flood. This is much, much faster than the current few cm/yr we see today, and can move the oceans and continents thousands of miles during the year-long period of the Flood.

As the subducting slabs travel down through the mantle, two things are occurring. At the surface, rocks (including ocean crust) are brittle, and as the subducting slab moves deeper into the mantle, some portion of the ocean crust still at the surface breaks, forming a huge rift exposing the upper mantle. This exposure causes the mantle to partially melt, forming magma for new ocean crust to fill in the gap. This is the beginning of ocean ridges, which being the process of seafloor spreading to replace ocean lithosphere lost to subduction. As the new magma comes in contact with ocean water, it creates a massive, linear geyser around the world that blasts superheated water from the oceans high into the atmosphere, where it cools and condenses as a drenching downpour of rain (figure 4.25). This begins the intense rainfall described when "the floodgates of heaven were opened" (Gen. 7:11), which is described right after the breaking of the fountains of the great deep. Deeper in the mantle, as the cold lithosphere sinks, warmer regions of the mantle rise and create a convection current, similar to the movement of boiling water in a pot.

FIGURE 4.25. The eruption of water at the beginning stages of the Flood.

Back at the Earth's surface, the continents cannot subduct (they are too light), but they are strongly affected by these events. The subduction of ocean lithosphere, formation of seafloor spreading zones, and convection of the mantle tear Rodinia apart, and begin moving the pieces around the Earth's surface. At some point early in the Flood, these broken continental fragments recombine to form Pangaea, producing the mountain belts, sedimentary basins, and other features that help geologists reconstruct its shape and location (figure 4.24b). Then, perhaps a month or two after Pangaea's formation, continued mantle convection, subduction, and continental rifting split Pangaea apart into new fragments that will become today's continents. These fragments shifted and moved closer to their present-day position through the remainder of the flood, gradually slowing but continuing to move in the centuries after the Flood until they reached their present-day locations (figure 4.24c).

Most of the description of the Flood in Gen. 7 and 8 concerns the rising and falling of water. How did that happen according to CPT? There are several components, and the following four mechanisms described below and illustrated in figure 4.26 show how water from the ocean basins caused global flooding over all of the shifting continents during the Flood:

Geyesers at ocean ridges—as discussed above, the fracturing of the ocean crust during the flood resulted in the formation of the ocean ridge system. Rising magma contacting the ocean created a massive, linear geyser of superheated water, which cooled and condensed to form torrential rains (Hebrew: *geshem*).

Subduction at continental margins—where the pre-Flood ocean lithosphere subducted along continental margins, friction between the plates caused the continents to be pulled downward towards the mantle. While they did not sink into the mantle, this downward pull allowed ocean water to invade far into the continents and hundreds of meters deep.

Formation of new ocean crust—while the pre-Flood ocean crust was cold and dense, the formation of ocean crust at the new ocean ridges produced much warmer and more buoyant

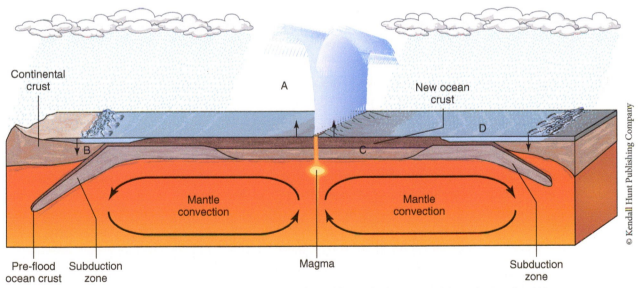

FIGURE 4.26. A cross-section through the crust and upper mantle during the Flood, showing the principle mechanisms by which ocean water covers over the continents: (a) rainfall from mid-ocean ridge geysers; (b) drag-down of continents at subduction zones; (c) new, thicker ocean crust; and d) thermal expansion of ocean water.

ocean crust during the Flood. This effectively lifted the elevation of the ocean basin by as much as one *kilometer* or more, causing huge volumes of ocean water to advance over the continents.

Thermal expansion of ocean water—as the water of the oceans was warmed by volcanic activity during the Flood, it expanded like all other substances do when heated. This effect is the least important of all, having very little influence on flooding the continents. The expansion rate of water is small (0.000214/°C), and even multiplied by the enormous volume of ocean water, would likely result in less than a 10m rise.

Combined, these factors were easily sufficient to move large volumes of ocean water over the continents to erode and cover "all the high hills" of the pre-Flood world as recorded in Genesis. How did the Flood stop? That answer comes from the eventual shutdown of runaway subduction. As new ocean crust was formed and pulled away from the ocean ridges, it eventually came to subduction zones. While early in the Flood runaway subduction and mantle convection could continue to draw down this lighter, more buoyant new ocean crust, at some point its lower density overcame these forces and the ocean crust began resisting subduction. Once that happened, subduction and seafloor spreading rates began to dramatically slow. The massive geysers would eventually come to a stop, and the continents would rise back up as slowing subduction speeds reduced their downward pull on the continental margins, allowing water to drain off the continents. Cooling of the ocean crust would also cause it to contract, lowering the sea floor level and increasing the volume of water in the oceans. The waters for the Flood came from the oceans to flood the continents, then returned to new ocean basins at the Flood's end, where they remain to this day, as promised by God to never again destroy the world with a flood (Gen. 9:8-17, Ps. 104:8-9; see Box 4.2.).

A theory such as CPT is helpful in understanding the movement of the plates as they occurred during Noah's flood and in the years following it. But a good scientific theory should be not only descriptive, but also predictive. CPT has many significant challenges, particularly in dealing with how much heat would be generated by all of the proposed tectonic activity, how it relates to accelerated nuclear decay (discussed in Chapter 6, and which adds even more heat to the global system), and whether there are alternate, Flood-based interpretations of geological features that seem to indicate a non-marine origin (such as the evidence for glaciers used by Wegener and others to reconstruct Pangaea). These are all areas that challenge CPT and any other model of the Flood.

Yet despite challenges, CPT has also seen powerful confirmation of important parts of the theory. Data regarding Earth's magnetic field and its history of reversals show that reversals can occur very quickly. Evidence in support of this comes from both ocean crust basalts and terrestrial lava flows. As mentioned earlier, the ocean crust records magnetic field reversals in parallel bands opposite each side of the spreading center (figure 4.11). The details are a bit more curious, however because the normal and reverse polarity bands show a mottled pattern of magnetic field orientation within the rock, rather than a solid and uniform block of north- or south-oriented magnetic orientations. That is, within a normal polarity segment of the crust there are many spots and areas that show a reverse polarity, and vice versa. This indicates that the rock far away from the spreading center had not completely cooled by the time the reversal occurred. Most of the rock cooled to record one polarity, while some parts remained hot enough that when a reversal happened, they picked up the opposite polarity. In a young-Earth and CPT view, this would be expected, because the ocean crust is forming quickly and reversals are happening frequently. In an old-Earth plate tectonics view, the basalt far from the ridge would have cooled

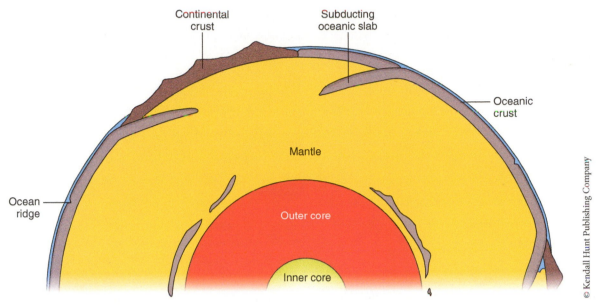

FIGURE 4.27. A cross-section through the geosphere, showing subducting oceanic lithosphere and the remains of pre-Flood ocean lithosphere located in a "plate graveyard" near the core-mantle boundary.

entirely, with no mottling except right near a reverse zone while the lava was erupting.

Terrestrial, post-Flood lava flows from Oregon show us that the reversals themselves occur quickly. Creationist D. Russell Humphreys predicted in 1986 that if reversals happened in a young-Earth setting, then thin, fast-cooling lava flows might "capture" the evidence of the reversal by recording different magnetic directions in different parts of the lava (fast-cooling outer part vs. slower-cooling inner parts). Three years later, old-Earth geologists discovered that a thin lava flow recorded a 90° shift in polarity that would have taken only 15 days to form. The same group of researchers again described even faster shift rates in 1995. These point strongly to rapid changes in the magnetic field as predicted by young-Earth models, but competing old-Earth models expect these changes to happen over *thousands of years*, not a few *weeks*!

A different type of confirmation came from seismic studies of the Earth's interior. Chapter 5 discusses how earthquake waves are used to determine the materials and features of the crust, mantle, and core. The theories of runaway subduction and CPT require that large slabs of the cold, pre-Flood ocean lithosphere sank down through the mantle, traveling towards the outer core. If Noah's Flood occurred only a few thousands of years ago, then there has not been enough time for these slabs of lithosphere to become the same temperature as the mantle around them. In 1995, just one year after CPT was presented, old-Earth geologists discovered large, cold regions of the deep mantle near the outer core, just as CPT would expect (figure 4.27). This was surprising to many old-Earth geologists, whose own models for plate tectonics kept the subducted ocean lithosphere in the upper mantle. But even when adjusting their model (as a good scientist does when faced with contradictory data), old-Earth views still have trouble explaining why the slabs are cold, since they are supposed to have been subducting and warming in the mantle over tens to hundreds of millions of years. Here the young-Earth and CPT views are much more consistent with the data we know, and provide us with confidence that CPT, though still incomplete, is on the right path to a better understanding of Earth history.

Regardless of the actual mechanism by which the flood occurred, the result of the flood was a completely different world. Noah and his family entered the ark seeing a world similar to, but corrupted after, the initial creation; they left the ark to find new continents, new oceans, new valleys and mountain chains (such the Rockies, Himalayas, and the "mountains of Ararat" where the ark landed; figure 4.28). *Everything* is different now.

FIGURE 4.28. Massive mountain chains, such as the Rocky Mountains, did not exist before the Flood. Rather, rocks formed during creation week and earlier in Noah's Flood were lifted up to form these mountains later in the Flood, and also during the post-flood period.

BOX 4.2 GOD'S ENDURING PROMISE

After Noah left the ark, God speaks with him and makes a covenant with both humans and all other animals that He will never again destroy the world with a flood. This is seen in Genesis 9:8-17 and Psalm 104:8-9 (ESV).

Genesis 9:8-17.

[8]Then God said to Noah and to his sons with him, [9]"Behold, I establish my covenant with you and your offspring after you, [10]and with every living creature that is with you, the birds, the livestock, and every beast of the earth with you, as many as came out of the ark; it is for every beast of the earth. [11]I establish my covenant with you, that never again shall all flesh be cut off by the waters of the flood, and never again shall there be a flood to destroy the earth." [12]And God said, "This is the sign of the covenant that I make between me and you and every living creature that is with you, for all future generations: [13]I have set my bow in the cloud, and it shall be a sign of the covenant between me and the earth. [14]When I bring clouds over the earth and the bow is seen in the clouds, [15]I will remember my covenant that is between me and you and every living creature of all flesh. And the waters shall never again become a flood to destroy all flesh. [16]When the bow is in the clouds, I will see it and remember the everlasting covenant between God and every living creature of all flesh that is on the earth." [17]God said to Noah, "This is the sign of the covenant that I have established between me and all flesh that is on the earth."

Psalm 104:5-9

[5] He set the earth on its foundations, so that it should never be moved.

[6] You covered it with the deep as with a garment; the waters stood above the mountains.

[7] At your rebuke they fled; at the sound of your thunder they took to flight.

[8] The mountains rose, the valleys sank down to the place that you appointed for them.

[9] You set a boundary that they may not pass,so that they might not again cover the earth.

CHAPTER 4: Plate Tectonics

CHAPTER 4 IN REVIEW: PLATE TECTONICS

4.1 Early Clues

■ Maps of the New World produced in the 1500s and 1600s provided the first clues of plate movements, because the eastern coastlines of North and South America appear to "fit" with the western coastlines of Europe and Africa.

4.2 Almost There: Continental Drift

■ German meteorologist Alfred Wegener developed the concept of continental drift.

■ Wegener argued that all the continents were once assembled into a single supercontinent called Pangaea.

■ Wegener's evidence for Pangaea included the fit of the continents, matching geological units, the distribution of fossils, the distribution of living organisms, and ancient climates.

■ Continental drift was ultimately rejected by the geological community at the time.

4.3 The Plate Tectonics Revolution

■ Discoveries about the ocean during and after World War II proved important in the development of plate tectonics.

■ Two facets of paleomagnetism (the record of magnetic orientation in Earth's rocks) are important in tracking plate motion over time: apparent polar wander and magnetic reversals.

■ Oceanic crust is basaltic in composition. Ocean features such as the continental shelf and margin, abyssal plain, ocean ridges, trenches, and the patterns of ocean sediments all point towards plate motion over time.

■ Plate tectonics theory successfully combined Wegener's continental drift with seafloor spreading.

4.4 A Closer Look at Plate Boundaries

■ Plate boundaries may be divergent, convergent, or transform. Different types of landforms and other surface features are associated with each type of boundary.

■ Plates move away from each other at divergent boundaries and are regions where new lithosphere is produced

■ Plates move towards each other at convergent boundaries and are regions where lithosphere is destroyed. They are characterized by volcanic arcs (either oceanic or continental) or large collisional mountain belts.

■ Plates move past each other along transform boundaries. Most are found near ocean ridges, while some cut across continents.

4.5 Hot Spots and Plate Rates

- A hot spot is a plume of magma originating in the lower mantle and rising upward into or through the crust.
- Hot spots are stationary, and may be used as fixed points to measure the rate of plate motion as the plate moves over the hot spot.

4.6 Noah's Flood: Where Scripture Meets Geology

- The Flood account of Genesis 6-9 provides information about the destruction of the world at the time of Noah.
- Numerous scriptural terms and descriptions point to a global, rather than local flood. These include unique terms for the Flood, superlative language, and the record of its extent and duration.
- Catastrophic Plate Tectonics (CPT) is a young-Earth creation hypothesis for rapid plate motions during Noah's Flood.
- The waters for Noah's Flood came primarily from the ocean, and returned to the ocean when the Flood was finished. Mechanisms to accomplish this include massive geysers, subduction, the formation of new ocean crust, and thermal expansion of ocean water.

KEY TERMS

Abyssal plain
Active margin
Asthenosphere
Bathymetry
Biogeography
Catastrophic Plate Tectonics (CPT)
Continental arc volcanoes
Continental drift
Continental margin
Continental rift
Continental shelf
Convergent boundary
Dipole
Divergent boundary
Fracture zone
Gondwana
Hot spot
Laurentia

Lighosphere
Magnetic reversal
Ocean ridge
Ocean trench
Paleomagnetism
Pangaea
Passive margin
Plate(s)
Plate tectonics
Polar wander
Rift
Seafloor spreading
Spreading center
Subduction
Transform boundary
Volcanic island arc
Wadati-Benioff zone

1. What were Wegener's main lines of evidence for Pangaea and continental drift? What was the major problem that he faced in presenting his argument?

2. Look at the introductory figure at the beginning of this chapter, and observe the shape of the continental shelves on either side of the Atlantic Ocean. Is the "fit" of the continents better or worse than looking at modern shorelines? Why do you think that is the case?

3. Describe what paleomagnetism is in your own words.

4. You are vacationing on a beautiful island in the Pacific, and notice that the mountain in the center of the island is a large andesitic volcano. At what type of plate boundary do you find yourself? What other types of features are likely to be in the nearby area?

© Kendall Hunt Publishing Company

5. What type of plate boundary is most likely represented at the long, linear sea (A) in the image above?

6. What Hebrew and Greek terms are used for Noah's Flood in the Bible? What are two of the other scripture-based reasons for why the Flood is global?

7. How long did the Flood last?

8. According to Catastrophic Plate Tectonics, what are the four ways that water flooded the world?

9. How have discoveries in paleomagnetism and seismology (the study of earthquakes) provided evidence for rapid magnetic field reversals and rapid plate motions?

FURTHER READING

Austin, Steven A., *et al.* "Catastrophic Plate Tectonics: A Global Flood Model of Earth History." *Proceedings of the Third International Conference on Creationism*. 1994. Available online at http://static.icr.org/i/pdf/technical/Catastrophic-Plate-Tectonics-A-Global-Flood-Model.pdf

Frisch, Wolfgang, Martin Meschede, and Ronald C. Blakey. *Plate Tectonics: Continental Drift and Mountain Building*. Springer. 2011.

Snelling, Andrew. *Earth's Catastrophic Past*, volumes I and II. Master Books, 2014.

Snelling, Andrew. Can catastrophic plate tectonics explain Flood geology? *The New Answers Book 1*, pp. 186–197, Master Books, 2006.

United States Geological Survey: www.usgs.gov

Wise, Kurt. *Faith, Form, and Time*. B&H Press, 2002.

FOCUS ON SCRIPTURE

Read Genesis 1:1–2:3, then 7–8. Create a list of terms that are found in common between the two, such as groups of animals, components of the Earth, and descriptions of the events taking place. What is the scope of both accounts? That is, does the language center on global or local actions and events? Finally, read II Peter 3, a New Testament passage that refers back to the Flood. What Christian theological doctrine is supported by the historical reality of the Flood? Does this doctrine concern few, many, or all people? How might this passage affect your understanding of the Flood account?

CHAPTER 5

THE RESTLESS EARTH

OUTLINE:

Rainer Albiez/Shutterstock.com

Earthquakes and volcanoes are some of the most devastating catastrophes that mankind can experience. Here is a nighttime eruption of the Italian volcano Stromboli.

5.1 INTRODUCTION

Earthquakes and volcanoes are two of Earth's most scary and devastating natural disasters. Both of these phenomena are related to the structure and interior of the earth and both give us clues about the nature and inaccessible depths of the earth. Some of the most devastating earthquakes that have happened in earth history, happened during Noah's Flood as "all the fountains of the great deep burst open" (Genesis 7:11). These earthquakes certainly must have been on the floor of the ocean, within minutes causing giant tsunamis that devastated the coastal areas of the earth. The events of the book of Amos are identified as happening two years prior to "the earthquake" (Amos 1:1). Several earthquakes are mentioned in the New Testament including one at the time of the crucifixion (Matthew 27:51) and during the resurrection (Matthew 28:2). An earthquake happened when Paul and Silas were in prison that caused all the doors to open and their chains to fall off (Acts 16:26). Great earthquakes were predicted by Jesus to be part of the signs of the end times (Matthew 24:7; Revelation 16:17-21).

Volcanic eruptions are not mentioned as frequently in Scripture, but they are certainly implied in many places such as Revelation 16. Although it was certainly a miraculous event, the destruction of Sodom and Gomorrah seems as though it was some type of volcanic activity (Genesis 19). The Ark landed in the volcanic mountains of Ararat (Genesis 8:4).

The Greatest Earthquake on Earth

Revelation 16:17-21.[17] Then the seventh *angel* poured out his bowl upon the air, and a loud voice came out of the temple from the throne, saying, "It is done."[18] And there were flashes of lightning and sounds and peals of thunder; and there was a great earthquake, such as there had not been since man came to be upon the earth, so great an earthquake *was it, and* so mighty.[19] The great city was split into three parts, and the cities of the nations fell. Babylon the great was remembered before God, to give her the cup of the wine of His fierce wrath.[20] And every island fled away, and the mountains were not found.[21] And huge hailstones, about one hundred pounds each, came down from heaven upon men; and men blasphemed God because of the plague of the hail, because its plague was extremely severe.

Genesis 7:11 In the six hundredth year of Noah's life, in the second month, on the seventeenth day of the month, on the same day all the fountains of the great deep burst open, and the floodgates of the sky were opened.

In Psalm 104:32, the psalmist describes God as the only one who has power to touch the mountains and cause them to smoke.

Earthquakes and volcanoes help us understand much about the interior of the earth. Man has never drilled a hole or dug a mine shaft through the crust of the earth, even though the crust of the earth is (comparatively) not very thick. When earthquakes happen, they produce energy, called *waveforms*, which travel through rocks. Some of these waveforms travel through the earth and come out the other side; others bounce off layers within the earth and come back to the surface. By studying patterns of waveforms, we can infer the thickness and composition of the interior of the earth. Volcanic eruptions bring magma (and sometimes rock) to the earth's surface that is often sourced from very deep within the earth (sometimes from hundreds of kilometers down). Studying this material can give us more direct clues as to the composition of the deeper parts of the earth.

There are three main types of forces within the earth: **compressional, tensional,** and **shear forces** (figure 5.1). Solid rock can withstand very large compressional forces without changing much. On the other hand hard rock does not handle tensional and shear forces very well. For example, imagine taking a cement block which is connected by chains in between two tractors. As the tractors begin to pull against each other (tensional force), the block will quickly fail. On the other hand, the block could probably easily support the weight of the two tractors (compressional force) if you were able to balance the weight of tractors on top of the cement block. In the solid earth, we often see both bends and breaks; we refer to them as **folds** and **faults**. When we see folded rock we typically think of forces bending the rock while it was still relatively soft. When we see faulted rocks, we know the rock was hard enough to break by compressional, tensional, or shear forces.

Let us look at an example that causes some problems for a conventional interpretation of earth history. Carbon Canyon is a side canyon of the Grand Canyon. In this area are some fantastic folds (figure 5.2) in the Tapeats Sandstone which was deposited about 520 million years ago according to the conventional time scale. It is believed the Tapeats was folded when the entire area was uplifted about 60 million years ago (supposedly 460 million years *after* its formation according to old-Earth estimates!). We think the sandstone was deposited during the Flood and was still relatively soft when the area was uplifted about a year or so later. Since not much time had passed, the sandstone easily bent instead of broke. The dilemma for the conventional geologist is: How could these sandstones possibly remain soft for almost one half billion years? The folds in these rocks indicate their deposition, uplift and folding were all closely timed events. The presence of the folds eliminate millions of years of time within this section of rocks.

5.2.1 Types of Folds

When sedimentary rocks are deposited, they are laid down as flat, horizontal layers. This is often referred to as the **Law of Original Horizontality** (see Chapter 6). But sometimes, when we drive through a road cut, we see that the rocks are not

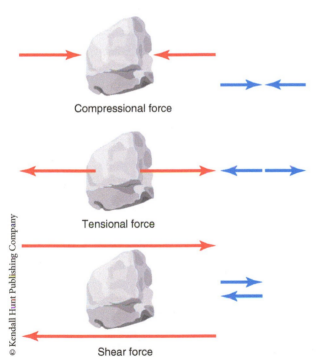

Compressional force

Tensional force

Shear force

© Kendall Hunt Publishing Company

FIGURE 5.1. Three different types of forces can be applied to rocks: compressional, tensional and shear.

FIGURE 5.2. Folds in the Tapeats Sandstone, Carbon Canyon, Grand Canyon, Arizona. Originally, the layers in the sandstone were flat. They have been folded, so that some of the layers on the right side of the photo are nearly vertical. Notice the people for scale in the lower right. These rocks had to bend while they were still soft.

horizontal, but that they are bent into various types of folds. There are two basic shapes, and a number of more complex patterns. A "U"-shaped fold is called a **syncline** (figure 5.3) and an upside down "U" is called an **anticline** (figure 5.4).

A road cut or cliff face is only a two dimensional exposure, while the folds have three dimensional shapes. Sometimes, large-scale, three-dimensional folds develop that can be many tens of kilometers in size. The two basic types are domes and basins (figure 5.5).

A **dome** is like an overturned soup bowl, while a **basin** is just the opposite. Sometimes domes, basins or the three-dimensional shapes of anticlines and synclines can make up entire mountain ranges, or whole series of mountain ranges. For example, the Black Hills of South

FIGURE 5.3. Sedimentary strata are deposited horizontally in flat lying beds. These rocks began as horizontal, but have been folded into a "U" shape. Sidling Hill, Maryland, along Interstate 68.

FIGURE 5.4. A fold that has the appearance of an arch is called an anticline. Anza Borrego Desert, southern California.

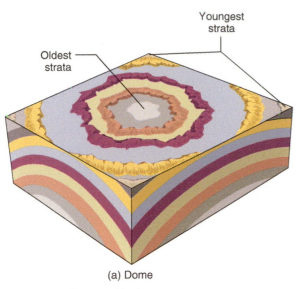

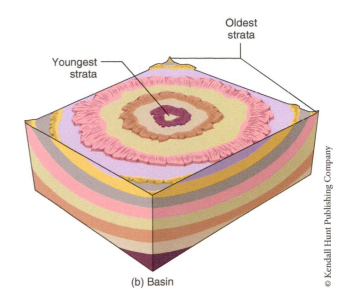

© Kendall Hunt Publishing Company

(a) Dome   (b) Basin

FIGURE 5.5. Domes and basins.

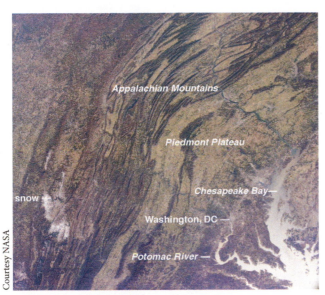

Courtesy NASA

FIGURE 5.6. The "v" shaped mountain ridges of the Appalachians are formed from plunging anticlines and synclines.

Dakota is a giant dome that has been eroded, while many of the Appalachian Mountains in Pennsylvania are giant anticlines and synclines (figure 5.6).

5.2.2 Types of Faults

When forces are applied to hard rocks, the rocks will break instead of bending, resulting in a fault. Faults are often easy to pick out in a road cut or cliff face because two different types of rock will be juxtaposed (figure 5.7). If the rocks haven't moved too much, it is easy to figure out which way the rocks have moved.

During the gold rush days of the mid-1800's, miners recognized that gold and other minerals sometimes accumulated in quartz veins associated with faults. The miners would dig their shafts and tunnels along the faults and collect the gold ore. This was before the days of electricity, so the miners would hang lanterns on the ceilings of the mine tunnels. Thus, the upper part of the fault (the top of the mine tunnel) became known as the **hanging wall** or **hanging block**. The bottom part, where the miners walked, became known as the **footwall** or **footblock**. Faults exhibiting vertical movement are still described using this terminology (figure 5.8).

There are three basic kinds of faults. The movement in **dip slip faults** is vertical and causes the rocks on each side of the fault to move up and down in relationship to each other. In other words, the hanging wall will move either up or down compared to the footwall. The movement in **strike slip faults** is horizontal. In other words, if you are standing on the surface, the rocks on each side of the fault zone will

Photo by John Whitmore

FIGURE 5.7. Sometimes where a fault occurs, two rocks can be found sitting beside each other that don't match. Here, the Moenkopi Formation (left), which usually sits on top of the Kaibab Limestone (right) has been shoved down, while the Kaibab has been pushed up. US89A, northern Arizona.

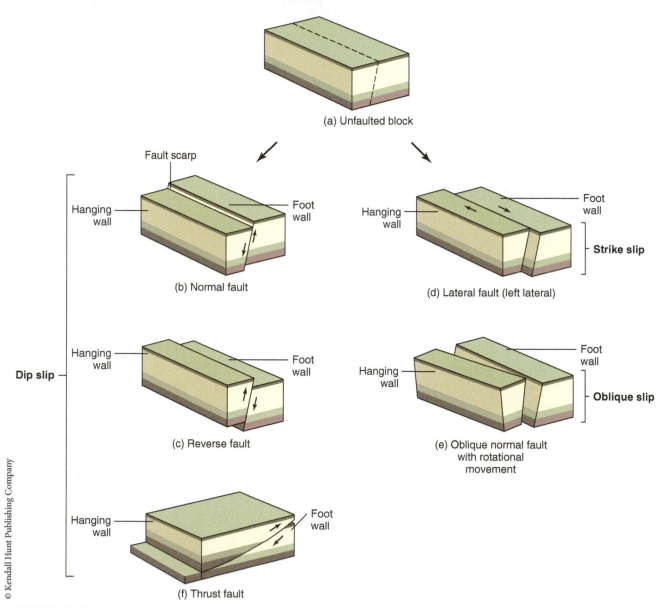

(a) Unfaulted block

Fault scarp

Hanging wall — Foot wall

(b) Normal fault

Hanging wall — Foot wall

Strike slip

(d) Lateral fault (left lateral)

Dip slip

Hanging wall — Foot wall

(c) Reverse fault

Hanging wall — Foot wall

Oblique slip

(e) Oblique normal fault with rotational movement

Hanging wall — Foot wall

(f) Thrust fault

© Kendall Hunt Publishing Company

FIGURE 5.8. The basic kinds of faults caused by the three different types of geological forces (tension, compression, and shear).

CHAPTER 5: The Restless Earth

move horizontally in relationship with one another. **Oblique slip faults** are a combination of dip slip and strike slip. In other words, both vertical and horizontal movements take place.

To make things simpler, the movement in dip slip faults is usually described as which way the hanging wall has *apparently* moved. Often times it is difficult to tell if the hanging wall moved, the footwall moved, or if there has been a bit of movement in both. If the hanging wall appears to have been shifted downward with respect to the footwall, we refer to the fault as a **normal fault**. If the hanging wall appears to have been shifted upward with respect to the footwall, we refer to the fault as a **reverse fault**. If a reverse fault has a particularly shallow angle (~10° or less), it is referred to as a **thrust fault**. See figure 5.8 for each of these.

5.3 EARTHQUAKES

Earthquakes happen when stresses build up in the earth and rocks fracture. This process causes much shaking and vibration, which if large enough can have a severe impact on people and their structures. When people think about earthquakes in the United States, they usually think of California; but there are even more earthquakes in Alaska. California is much more populated, so even small earthquakes that occur there tend to make the news. If you look at the "top 20 earthquake list" from the United States, surprisingly you can see that California starts showing up at number 13.

Deadliest earthquake lists (such as those listed at the USGS website) are quite sobering. The most devastating earthquakes to humans aren't necessarily the biggest ones. The earthquake that happened in Haiti in 2010 was certainly large (7.0 on the moment magnitude scale), but more importantly it occurred in an area of extreme poverty. Poorly-constructed buildings collapsed killing many people immediately, and aftermath was just as bad. The people of Haiti had very little infrastructure to be able to deal with all of the injuries, so many people died from very treatable injuries and diseases. Events like this provide an incredible opportunity for Christians and geologists to serve the poor and to share the gospel of Christ.

5.3.1 Earthquake Generation

As mentioned above, earthquakes occur because stresses build and cause rocks to break. The **elastic rebound theory** explains why moving rocks produce an earthquake. Imagine breaking a dry wooden stick. At first the stick will bend as force is applied to it. However, the stick can only bend so far before it snaps. Rocks will bend slightly when tectonic forces are applied to them, but they will not bend far. Rocks can only take so much pressure before they snap, causing an earthquake (figure 5.9). The larger the force involved, the bigger the earthquake will be. When the earthquake happens, energy in the form of **seismic waves** propagates from the **focus** of the earthquake (figure 5.10), much like what happens when you throw a rock into a pond. Typically, the focus of an earthquake is underground. A line can be

Top 20 list of earthquakes in the US. See following website: http://earthquake.usgs.gov/earthquakes/states/10_largest_us.php

Largest earthquakes in the world since 1900. See following website: http://earthquake.usgs.gov/earthquakes/world/10_largest_world.php

Elastic rebound theory

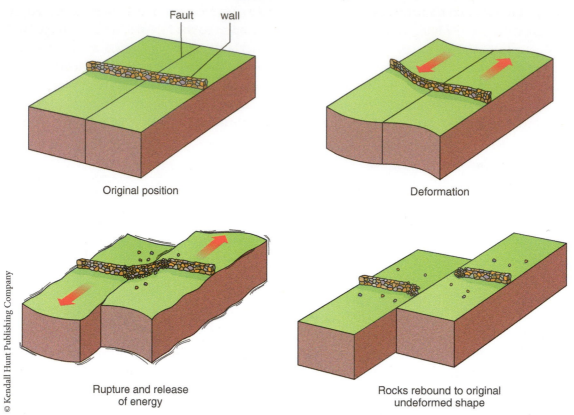

Fault wall

Original position

Deformation

© Kendall Hunt Publishing Company

Rupture and release
of energy

Rocks rebound to original
undeformed shape

FIGURE 5.9. The elastic rebound theory helps us understand how stresses build up and then are suddenly released within the earth.

drawn from the focus of an earthquake upwards to a point on the Earth's surface directly above the focus, called the **epicenter**. Typically when earthquakes are reported by the media, the epicenter and the **moment magnitude scale** value (the energy of the earthquake) are reported. The epicenter is usually

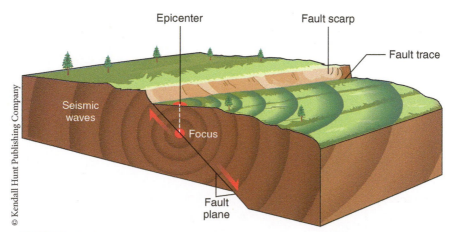

Epicenter

Fault scarp

Fault trace

© Kendall Hunt Publishing Company

Seismic
waves

Focus

Fault
plane

FIGURE 5.10. Earthquake waves (or seismic waves) travel away from the focus of an earthquake much like waves travel away from a stone that is thrown into a pond. The epicenter is the location on Earth's surface directly above the focus.

CHAPTER 5: The Restless Earth

described as a number of miles/kilometers away from a town or other geographic feature.

5.3.2 Seismic Waves

Wave energy from an earthquake can travel along the surface of the earth (**surface waves**) or the energy can travel down into the earth (**body waves**). There are two types of surface waves and two types of body waves. Body waves travel faster than the surface waves so if you are in an earthquake, you will usually feel the body waves first, followed shortly by the surface waves—unless you are very close to the epicenter then you will probably feel everything at once.

The two types of body waves are **primary waves** and **secondary waves**. *Primary waves* (figure 5.11; called "P" waves for short) are compressional waves,

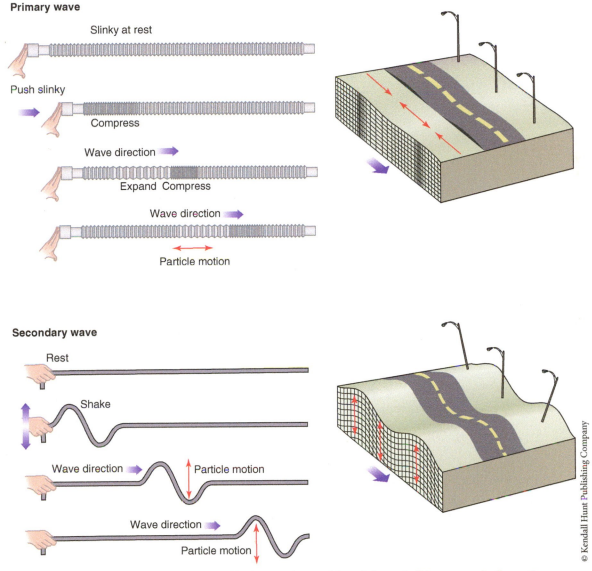

FIGURE 5.11. Representations of the two kinds of body waves that travel through the earth. Primary waves (or P waves) are compressional waves and behave much like the slinky with zones of compression followed by zones of expansion. Secondary waves (or S waves) are transverse wave types and behave much like a wave form traveling down a long rope after it is shaken.

© Kendall Hunt Publishing Company

causing the rock to expand and contract as they travel. Sound waves travel in much the same way. If you pluck a Slinky® a compressional wave will travel back and forth along the spring. P waves travel faster than any other type of earthquake wave, so they will be the first waves from an earthquake to arrive at a recording station, or the first waves you will feel if you are in an earthquake. P waves can travel through all types of mediums (air, liquid, and solids). *Secondary waves* ("S" waves) travel second fastest and are transverse wave types. If two people hold a rope or long spring at each end, and then it is "flicked" a transverse wave will travel back and forth. These types of waveforms can only travel through solids. During large earthquakes, P waves will travel all the way through the earth and come out on the other side, while S waves will not. This implies that there are liquid layers deep within the earth that absorb the S waves.

Surface waves are the slowest, but do the most damage because they cause the rigid surface of the earth to flex. The two wave types travel about the same speed and were named after geologists who studied and described them. **Love waves** cause the surface rocks to move back and forth and to have a lateral motion with respect to how the wave is propagating (figure 5.12). Thus, pipes in a wall might bang back and forth against the inside of a wall or a bridge might be shifted off its foundation because of Love waves. **Rayleigh waves** cause a rolling motion within the earth, causing the earth's rigid surface to move up and down much like ocean waves (figure 5.12). In fact, during large earthquakes you might be able to see these types of waves move across large flat surface like fields or parking lots!

5.3.3 Measurement of Earthquakes

A **seismometer** is an instrument that can detect an earthquake. The Chinese developed the earliest seismometers (figure 5.13), which consisted of a vase that loosely held balls in the mouths of dragons. There

Love waves
The ground moves side to side

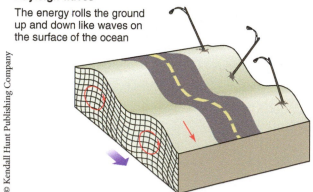

Rayleigh waves
The energy rolls the ground up and down like waves on the surface of the ocean

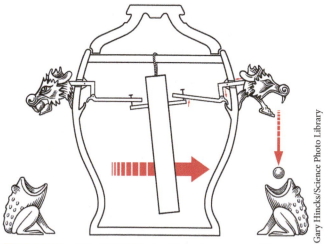

© Kendall Hunt Publishing Company

Gary Hincks/Science Photo Library

FIGURE 5.12. A representation of how Love waves and Rayleigh waves deform the surface of the earth as they travel away from an earthquake epicenter.

FIGURE 5.13. A illustration that shows how the an ancient Chinese seismometer worked. It was invented in 132 AD by Zhang Heng. Often the devices were intricately fashioned.

the rotating paper drum, it creates a **seismogram**, the paper record of an earthquake. Today, most seismograms are produced and recorded digitally on a computer, and recording and reporting of seismic events is almost instantaneous. You can see earthquake data reported within minutes on the USGS website.

Modern methods to measure earthquakes are based on measuring either the earthquake's magnitude (amount of energy released along the fault) or its intensity (level of shaking and damage at the surface). While different, both are useful in measuring and describing an earthquake. The first accurate magnitude scale, the **Richter Scale**, which was developed in 1935 by California geologist Charles Richter (figure 5.15).

On the seismogram (figure 5.15), P waves, S waves and surface waves are usually separated from

was a mechanism that released the balls and allowed them to fall into frog mouths at the slightest shaking. They developed instruments like this because they thought if they could detect small earthquakes, they might be able to see some patterns before a larger and more destructive earthquake occurred. Today, **seismographs** (figure 5.14) are used to both detect and record earthquakes. Early- to mid-1900s seismographs had a rotating drum of paper that a pen could mark on during the shaking of an earthquake. The instrument was anchored on bedrock, which is more sensitive to slight shaking. As the pen makes marks on

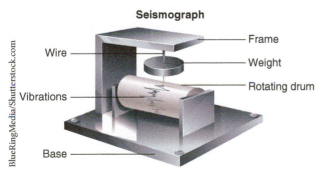

Seismograph

Wire — Frame
Weight
Vibrations — Rotating drum
Base

BlueRingMedia/Shutterstock.com

FIGURE 5.14. A seismograph is an instrument that can both detect and record an earthquake. A pen records the earthquake vibrations on a rotating drum of paper. Modern instruments record all of the data digitally on a computer.

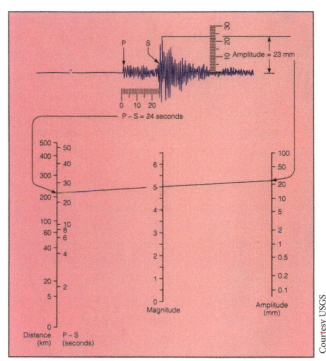

Courtesy USGS

FIGURE 5.15. A diagram showing how the Richter scale magnitude number of an earthquake is calculated. The time difference between the arrival of the P and S waves is measured from the seismogram and then plotted on the vertical scale on the left. The greatest amplitude is measured and then plotted on the vertical scale on the right. A line is drawn between the left and right scales which intersects the middle scale. The Richter magnitude is then read from the middle scale; which is a "5" in this case.

one another. This is because the waves travel at different velocities and they arrive at a recording station in the order of their velocity. The time difference between the arrival of the first P wave and the arrival of the first S wave indicates how far away the earthquake was. To get a magnitude scale number for an earthquake, the distance to the seismic event is the first thing that must be determined. Then a measurement is made of the highest peak recorded on the seismogram. These two numbers are then plotted on a graph and a line is drawn which will intersect the Richter scale magnitude of the earthquake. Notice that scales are logarithmic, not linear. You can't think of a Richter 5.0 earthquake as just being a little bigger than a Richter 4.0 earthquake; it is actually about *32 times larger* in terms of energy released! Likewise, a Richter 6.0 earthquake is about 1024 times larger than a 4.0 (32 x 32 = 1024).

While this is the most well-known scale, it was actually replaced in the 1980s by the **moment magnitude scale**, and this is the magnitude number that is reported by the USGS to news outlets. The moment magnitude scale is nearly identical to the Richter, but provides better estimates of large-magnitude earthquakes.

Because the Richter scale did not always adequately describe the damage done in a particular area, the **Mercalli scale** was developed (figure 5.16). This is an intensity scale based on actual damage reports and human sensitivity to the earthquake, and ranks earthquakes in Roman numerals from I to XII. "I" is the smallest earthquake, generally not noticed by people. Earthquake damage on this scale does not begin to occur on this scale until IV or V. Absolute, devastating damage is reserved for XII on this scale. The Mercalli scale is useful for assessing historical accounts of earthquakes that occurred prior to the invention of seismometers. An example would be the Charleston, South Carolina earthquake from 1886 (figure 5.17). One limitation of the Mercalli scale is that it cannot be used for remote areas where no humans are present. For example, you couldn't use it for an earthquake occurring in Greenland or Antarctica.

5.3.4 Types of Earthquake Damage

Probably the most apparent type of damage to most people during a major earthquake is structural damage to buildings and structures, or falling objects (unattached bookshelves falling over, dishes falling out of the cupboard, etc.). Things like power lines, utility poles, roof overhangs, brick walls, chimneys and building fascia can fall causing potential harm to people. During the 1989 Loma Prieta earthquake near San Francisco a portion of the double-decker Nimitz Freeway collapsed, trapping and killing many people between the decks of the highway.

Fire can be a serious concern during a major earthquake. Rigid natural gas and water lines can rupture; gas can explode and the fire department can do nothing when water lines have also ruptured. Fire caused some of the greatest damage during the 1906 and 1989 earthquakes in the San Francisco area.

There are many other types of damages that can happen during earthquakes that often have much more devastating effects. **Tsunamis** happen when

Some Recent Deadly Earthquakes		
Year	Place	Deaths
1970	Peru	70,000 (landslide)
1976	China	243,000
1990	Iran	50,000
2004	Sumatra	228,000 (tsunami)
2005	Pakistan	86,000
2008	China	88,000
2010	Haiti	316,000
2011	Japan	21,000 (tsunami)

Magnitude / Intensity Comparison

The following table gives intensities that are typically observed at locations near the epicenter of earthquakes of different magnitudes.

Magnitude	Typical Maximum Modified Mercalli Intensity
1.0 – 3.0	I
3.0 – 3.9	II – III
4.0 – 4.9	IV – V
5.0 – 5.9	VI – VII
6.0 – 6.9	VII – IX
7.0 and higher	VIII or higher

Abbreviated Modified Mercalli Intensity Scale

I. Not felt except by a very few under especially favorable conditions.

II. Felt only by a few persons at rest, especially on upper floors of buildings.

III. Felt quite noticeably by persons indoors, especially on upper floors of buildings. Many people do not recognize it as an earthquake. Standing motor cars may rock slightly. Vibrations similar to the passing of a truck. Duration estimated.

IV. Felt indoors by many, outdoors by few during the day. At night, some awakened. Dishes, windows, doors disturbed; walls make cracking sound. Sensation like heavy truck striking building. Standing motor cars rocked noticeably.

V. Felt by nearly everyone; many awakened. Some dishes, windows broken. Unstable objects overturned. Pendulum clocks may stop.

VI. Felt by all, many frightened. Some heavy furniture moved; a few instances of fallen plaster. Damage slight.

VII. Damage negligible in buildings of good design and construction; slight to moderate in well-built ordinary structures; considerable damage in poorly built or badly designed structures; some chimneys broken.

VIII. Damage slight in specially designed structures; considerable damage in ordinary substantial buildings with partial collapse. Damage great in poorly built structures. Fall of chimneys, factory stacks, columns, monuments, walls. Heavy furniture overturned.

IX. Damage considerable in specially designed structures; well-designed frame structures thrown out of plumb. Damage great in substantial buildings, with partial collapse. Buildings shifted off foundations.

X. Some well-built wooden structures destroyed; most masonry and frame structures destroyed with foundations. Rails bent.

XI. Few, if any (masonry) structures remain standing. Bridges destroyed. Rails bent greatly.

XII. Damage total. Lines of sight and level are distorted. Objects thrown into the air.

From The Severity of an Earthquake, USGS.

FIGURE 5.16. An abbreviated modified Mercalli intensity scale. The intensity scale shows how humans perceive the earthquake and how much damage is done to manmade structures.

© Kendall Hunt Publishing Company/Image USGS

an earthquake occurs on the ocean floor as a result of dip slip faulting, where the water column above the fault is pushed upwards (figure 5.18). During the great Sumatran earthquake of 2004, the ocean floor was vertically displaced as much as 20 m along a distance of 150 km! This produced a tsunami that swept across the Indian Ocean and hit many islands and coastal villages without warning; many were so far away that they did not even feel the 9.1 magnitude earthquake. About 228,000 people died. Tsunami waves (mistakenly called "tidal waves") usually do not have a large wave height in the open ocean. It is

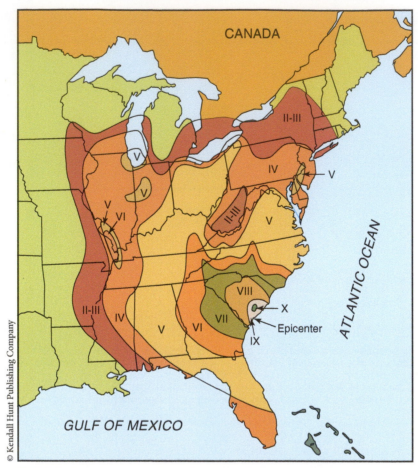

© Kendall Hunt Publishing Company

FIGURE 5.17. A Mercalli intensity map for the 1886 Earthquake that happened in Charleston, South Carolina. Note how the intensities can vary from place to place.

AP Photo/Kyodo News

FIGURE 5.18. The great Japanese tsunami of 2011. 19,000 people died as a result of the 9.0 moment magnitude scale earthquake and subsequent tsunami.

only as they arrive on the continental shelf and are focused into narrow bays that they grow in height. There is a tsunami warning system in the Pacific Ocean as a result of the devastating earthquakes and tsunamis following the 1960 Chile and 1964 Alaskan earthquakes. But there was no warning system in the Indian Ocean. Special buoys can be placed in the ocean to measure changes in water pressure as the result of a tsunami and then send a signal to a satellite for early warning.

5.4 THE INTERIOR OF THE EARTH

Recall from Chapter 1 that the earth has a radius of 6,371 km and is composed of three major parts: the **crust** (5 to 70 km thick), the **mantle** (2,890 km thick) and the **core** (with a radius of 3,471 km; figure 5.19). In reality, we really don't know much at all about the interior of the earth. Miners and drillers have never been able to penetrate past the crust, despite some incredible attempts to do so. The deepest hole ever drilled reached a depth of 12.3 km before the operation had to stop due to hot borehole temperatures (180°C). So how do we know anything at all about the deep parts of the earth if they are so inaccessible?

Three pieces of information help us best understand the deep interior of the earth. First of all, volcanoes are usually sourced from deep within the crust and mantle of the earth. Lava and **xenoliths** give us clues as to the chemical makeup of the material below the crust. Xenoliths ("foreign rocks") are solid pieces of rock that get incorporated into lava as it moves to the surface. Certain types of minerals are stable only under specific temperatures and pressures. So by studying the minerals and chemistry of lava and xenoliths, we can make inferences on the conditions deep within the earth.

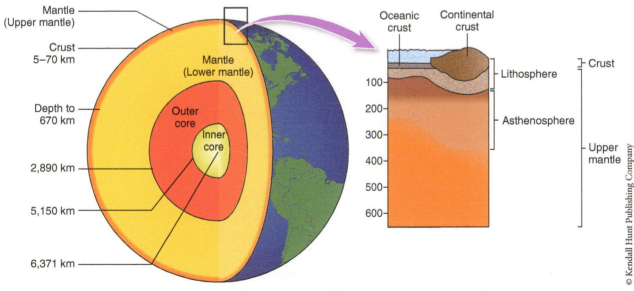

FIGURE 5.19. The structure of the interior of the earth showing the detail of the earth's crust and upper mantle.

Secondly, the earth is a giant magnet, and something within the earth must be generating the earth's magnetic field. Additionally, the earth has a certain mass, which is able to be estimated by how the earth behaves gravitationally with the sun, moon and other planets. In order to explain the magnetism and the mass, a core of iron and nickel makes the most sense, and these materials are abundant in the stony meteorites of our solar system.

Thirdly, earthquake body waves (P and S waves) travel through the earth and are capable of coming out through the other side. As they travel through different densities and encounter the curvature of the earth, they change velocities and refract (bend). Based upon the pattern of when and how the waves come out on the other side of the earth, we can make some important inferences (figure 5.20).

As P waves travel down through the crust, they suddenly increase in velocity at about 100 km below the surface. This change was first discovered by Croatian geologist Andrija Mohorovičić, and the crust/mantle is technically known as the Mohorovičić Discontinuity (or **Moho**). Within the upper mantle is a plastic-like zone called the **asthenosphere**. Everything above the asthenosphere is called the **lithosphere**, and includes both the uppermost upper mantle and the crust above it. The lithosphere is rigid, and it is only here that earthquakes can occur.

Remember that S waves cannot travel through liquids; so based on the pattern of "missing" S waves on the opposite side of the earth from an earthquake's origin (figure 5.20a) we can infer that the outer core of the earth is a liquid. P waves can travel through the liquid, but also get refracted and create their own pattern of missing waves (figure 5.20b). Interior to the outer core is the inner core, and the further refraction of P waves indicates that the inner core is a solid portion of the earth. Thus from volcanoes, magnetics, and seismic waves, we can discover the various regions of Earth's interior and their properties.

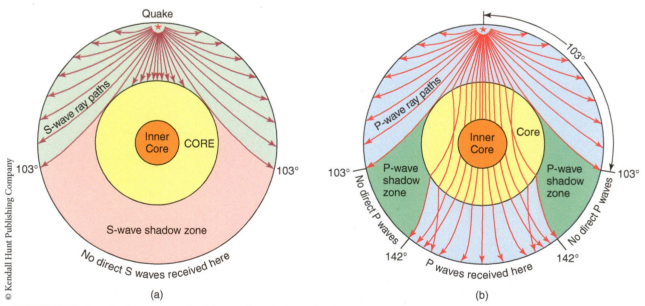

FIGURE 5.20. S waves do not penetrate all the way through the earth (a). Based on their behavior, we can infer that the outer core of the earth is comprised of liquid rock. P waves can penetrate liquids, but are refracted (b).

Volcanoes are some of the most amazing geological displays of power that can be witnessed. Even relatively small volcanic eruptions can release the energy equivalent to man-made nuclear bombs—every second of their eruption! Earthquakes generally both precede and accompany volcanic eruptions due to vast amounts of magma explosively moving at supersonic velocities just below the volcano. Many volcanoes reach high into the earth's atmosphere and as a result are snow covered. If a volcano like this erupts, the melted snow and ice can cause flooding and mudflows that can extend far beyond the immediate blast zone of the volcano. Volcanoes can develop steep slopes which make them prone to massive landslides, even if the volcano is inactive.

Volcanoes occur in many places around the earth (figure 5.21), but the most significant ones occur in two areas: The first is the "Ring of Fire" which surrounds the Pacific Ocean. These include many volcanoes in the Andes Mountains of South America, the Cascades and Aleutian Islands of North America, and volcanoes in New Zealand and Japan. The other most active belt goes through the Mediterranean Sea and Asia Minor. These include volcanoes like Mt. Vesuvius and Mt. Ararat.

5.5.1 Volcanic Materials

Volcanoes can emit several types of products: liquid rock (lava flows), hot pieces of solid rock (**pyroclastic material**) and various gases. Gas is released during all volcanic eruptions; it may surprise you that water is by far the most common type of gas released. Other types of gas can include various

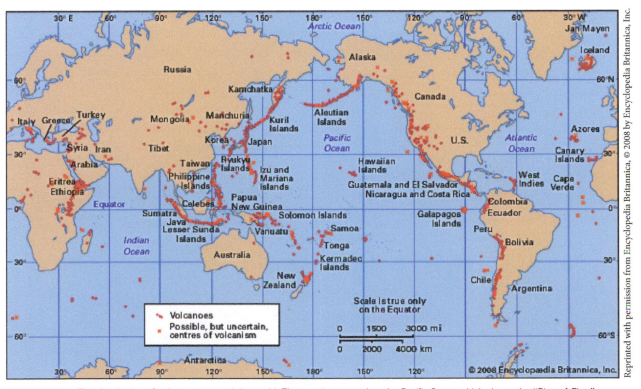

Reprinted with permission from Encyclopedia Britannica, © 2008 by Encyclopedia Britannica, Inc.

FIGURE 5.21. The distribution of volcanoes around the world. The map is centered on the Pacific Ocean which shows the "Ring of Fire."

sulfurous gases and carbon dioxide. Gases vent forcefully out of **fumaroles** if hot volcanic rock is present just below the surface (as in Yellowstone National Park). Gases are often poisonous because they do not contain any oxygen. Sometimes volcanic gases produce a phenomenon called **vog** (volcanic smog) composed of sulfurous gases that are released from large volcanoes. On the "big island" of Hawaii, Mt. Kilauea's gasses can result in hazardous air warnings hundreds of kilometers to the north on the other islands.

Pyroclastic material (or **tephra**) consists of hot (pyro) pieces of solid rocks (clastic). Pyroclastic deposits closely resemble sedimentary rocks because the material is thrown up into the air and then settles back down to the earth (figure 5.22). If the pieces are small (sand-sized) the material is called **volcanic ash**. Small pea-sized rocks are referred to as **lapilli** and larger rocks are referred to as **volcanic bombs.** Lapilli and bombs usually land quite close to the volcano, but volcanic ash can be forced very high into the atmosphere and be carried great distances by the jet stream. Some airliners have had problems with volcanic ash getting into their engines, so commercial airline traffic generally steers wide of an erupting volcano. Pyroclastic rocks are usually filled with holes due to the large amount of gas that was present in the frothy lava, just before the eruption.

Dark-colored pyroclastic material is generally called **scoria**; lighter colored material is called **pumice**. Some pumice is so full of holes that it is light enough to float in water!

Geologists recognize several types of lava. Some lava is smooth and looks kind of like taffy; it is called **pahoehoe** [pah-hoh-ey-hoh-ey]. **Aa** [ä, ä] is rough, jagged lava (figure 5.23). Both terms are Hawaiian in origin. Here is an easy way to remember which lava is sharp and angular: What words would come out of your mouth if you walked across aa without any shoes or socks? Aa!

5.5.2 Volcanic Dangers

What goes up must come down. When a volcano throws pyroclastic material up into the air, the larger particles free-fall back to the earth (sand sized fragments and larger). This hot material often falls back directly onto the volcanic cone, and then proceeds to violently avalanche down the slope of the volcano. This is known as a "glowing avalanche" or **nuée ardente**, French for "burning cloud" (figure 5.24). This is the primary and most serious danger of living close to a volcano. The avalanches can come down the volcanoes at great velocities (up to 700 km/hr), have temperatures up to 1,000° C, and extend many kilometers from the foot of the volcano. They totally extinguish and bury all living things in their path. The deposit that is formed can be tens to hundreds

FIGURE 5.22. Tephra (or pyroclastic material) consists of hot solid material that is erupted into the atmosphere. It can be as small as volcanic ash or consist of larger blocks and bombs. This is the eruption of the Icelandic volcano Eyjafjallajokull.

FIGURE 5.23. Aa lava is rough and jagged (left) while pahoehoe lava is smooth like taffy (right).

onime/Shutterstock.com

FIGURE 5.24. A pyroclastic flow (or *nuée ardente*) descending down the flank of a volcano in Indonesia. These events can travel hundreds of kilometers an hour and are one of the most considerable volcanic dangers.

Courtesy NOAA

FIGURE 5.25. In June of 1991 the Philippine volcano Mt. Pinatubo erupted blanketing the area with volcanic ash. When monsoon rains fell, the volcanic ash mixed with water to create massive mudflows. This photo shows the destruction of a bridge when more than five meters of mud flowed down the river.

of meters thick and blanket hundreds to thousands of square kilometers of landscape.

Lava flows generally move slowly enough that they do not cause loss of life during an eruption. However, people cannot move their homes or all of their possessions if a lava flow approaches. **Lahars** or **mudflows** can move much faster and can extend tens to hundreds of kilometers away from the volcano. During a large volcanic eruption in places like the Cascades, hot volcanic ash melts snow and ice and the resulting water mixes with ash to form a giant flood of mud. As was true in the 1980 eruption of Mt. St. Helens, mudflows usually cause the most serious dollar damage during an eruption. In 1991, Mt. Pinatubo had a much larger eruption than Mt. St. Helens, covering the region with a thick blanket of ash. Because the volcano is in the tropical Philippine Islands, it was not covered with snow and ice. Instead, several months after the eruption, the monsoon rains began, causing large amounts of ash to be turned into mud causing severe lahars in the area (figure 5.25).

Substantial landslides can occasionally happen during some volcanic eruptions. The largest landslide ever witnessed by man occurred in 1980 when Mt. St. Helens erupted; the landslide had a volume of 2.8 km³ and covered an area of 60 km² to an average depth of 46m! In 1883 the Indonesian volcano Krakatoa erupted. During the eruption, part of the volcano slid into the ocean creating a tsunami that killed approximately 35,000 residents in nearby islands. With the advent of detailed seafloor mapping, geologists have discovered that there have been many landslides in the Hawaiian islands, and it is estimated that over *half* of the island of Oahu has slipped into the ocean! One of the largest of the slides is the Nuʻuanu Slide. It was over 32 km wide and carried huge pieces of rock 200 km away from the island, out onto the ocean floor. One of the identified boulders is almost 30 km long and 1.5 km thick!

5.5.3 Types of Volcanoes

In general, there are three main types of volcanoes which can be distinguished from each other by their sizes, slope angles, and composition (lava flows, pyroclastics, or both). **Cinder cones** (figure 5.26) are

FIGURE 5.26. SP Crater cinder cone in the San Francisco Volcanic Field of northern Arizona. Note the dark colored lava flow that has come out of the base of the cone and flowed toward the top of the photo.

the smallest type of volcano; usually less than 1,000 m and most are much smaller. Their eruption cycle lasts a few decades, and eruptions consist mostly of

FIGURE 5.27. Mt. Ararat (elevation 5,137 m) is a composite volcano near the city of Yerevan in Turkey which last erupted in 1840. Genesis 8:4 says the Ark landed "on the mountains of Ararat" not on Mt. Ararat itself.

loose pyroclastic debris (ash and hot cinders) and infrequent lava flows.

Composite cones (also called *stratovolcanoes*) are some of the best known volcanoes on earth, not only because of their abundance, size and beauty, but because many of them are notorious. Their eruptions are often sudden and violent, sending volcanic ash high into the atmosphere, decimating the landscape around the volcano and destroying every bit of life in the vicinity. They get their name from the fact that both pyroclastic and lava eruptions produce alternating layers (a *composite*) of ash and lava flows. The volcanoes in the Cascade Range, extending from Northern California into southern Canada, are the best known within North America. Composite cones continue around the rim of the Pacific Ocean as the "Ring of Fire," because they are formed from magmas produced from the subduction zones found around the Pacific margin. The Aleutian Islands, Japan (Mt. Fuji), the Philippines (Mt. Pinatubo), many of the southwestern Pacific island chains (Krakatoa, Tambora), New Zealand, and volcanoes along the western coast of South and Central America are all composite cones. In fact, the extrusive igneous rock andesite (Chapter 3) gets its name from the Andes Mountains of Chile. In the Mediterranean and Asia Minor examples include Mt. Vesuvius, Mt. Etna and Mt. Ararat (figure 5.27).

Shield volcanoes are the largest volcanoes on earth by volume (figure 5.28). Hawaii's Mt. Kilauea is perhaps the world's most active volcano, continuously erupting since 1983. Shield volcanoes get their name because of their broad and low-sloping profile (think of Captain America's shield). They have shallow slopes (usually < 15°) because they

FIGURE 5.28. Mauna Loa is one of the earth's largest volcanoes. Its summit is 4,169 m above sea level. If it were removed from the ocean and placed beside Mt. Everest, it would be taller and much, much wider!

primarily emit high-temperature, low-viscosity lava that flows easily over the landscape until it cools.

5.5.4 Caldera Formation

When a volcano erupts, it generally empties a magma chamber that is present somewhere below the summit of the volcano. If the chamber is close to the surface, the summit of the volcano will collapse into the chamber after the eruption is over creating a large circular crater called a **caldera** (figure 5.29). This is a common feature at the summit of all different types of volcanoes. Shortly after Noah's Flood was over, one of the largest volcanoes on earth erupted. It was in the area that we now call Yellowstone National Park in the northwest corner of Wyoming. The eruption was so explosive that there is little left. However, remnants of the circular caldera can still be found (figure 5.30). It is on the

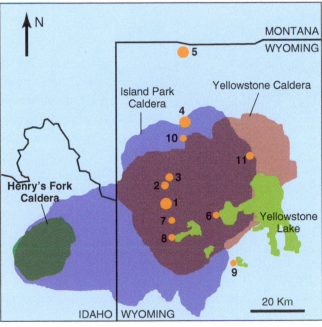

FIGURE 5.29. Perh Crater Lake, Oregon is the result of caldera formation after the eruption of Mt. Mazama. After the collapse of the volcano, the crater filled with water. Wizard Island is a small volcano that developed after the caldera formed.

FIGURE 5.30. The huge Yellowstone Caldera is what remains from a massive volcano that erupted after the Flood and spread its ash across the western half of North America. It is part of several calderas in the area (a). Hot springs, geysers, and other thermal features can be found on the floor of the caldera. In the photo (b), the dark rim of the caldera can be seen in the background.

Geothermal Area

1. Upper Geyser Basin
2. Midway Geyser Basin
3. Lower Geyser Basin
4. Norris Geyser Basin
5. Mammoth Hot Springs
6. West Thumb Geyser Basin
7. Lone Star Geyser Basin
8. Shoshone Geyser Basin
9. Heart Lake Geyser Basin
10. Gibbon Geyser Basin
11. Mud Volcano/ Sulfur Caldron Area

(a)

(b)

FIGURE 5.30. (*Continued*)

floor of the Yellowstone caldera that most of the hot springs, geysers, and other thermal features can be found in the Park. There is still hot rock and magma below this volcano! Yellowstone is one of many extremely large "super volcanoes" that erupted after Noah's Flood (figure 5.31).

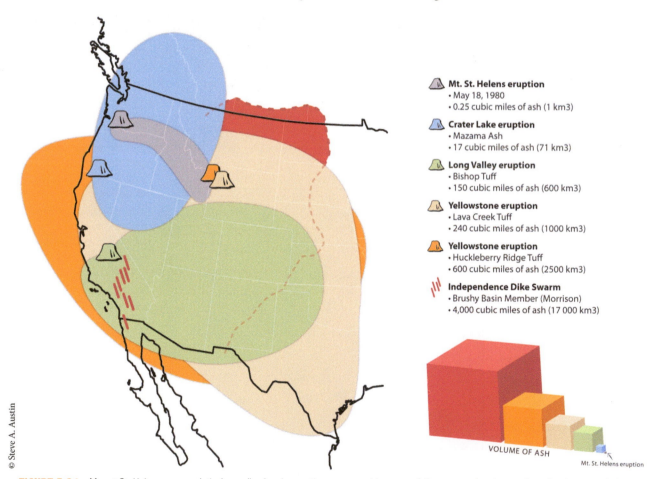

Mt. St. Helens eruption
- May 18, 1980
- 0.25 cubic miles of ash (1 km3)

Crater Lake eruption
- Mazama Ash
- 17 cubic miles of ash (71 km3)

Long Valley eruption
- Bishop Tuff
- 150 cubic miles of ash (600 km3)

Yellowstone eruption
- Lava Creek Tuff
- 240 cubic miles of ash (1000 km3)

Yellowstone eruption
- Huckleberry Ridge Tuff
- 600 cubic miles of ash (2500 km3)

Independence Dike Swarm
- Brushy Basin Member (Morrison)
- 4,000 cubic miles of ash (17 000 km3)

VOLUME OF ASH

Mt. St. Helens eruption

FIGURE 5.31. Mount St. Helens was a relatively small volcanic eruption compared to some of the super volcanic eruptions that happened after Noah's Flood, like Yellowstone.

5.5.5 Lava Floods

Sometimes volcanic lava eruptions are so wide-spread and voluminous that it is difficult to identify where the initial volcano(s) was. This is the case of the Columbia River Plateau Basalts in the northwestern part of the United States (figure 5.32). Here, multiple lava flows can be found up to 2,000 m thick and covering many hundreds of thousands of square kilometers. Based on their stratigraphic position, most of these lava flows probably happened near the end of the Flood or shortly after. The early post-Flood times were some of the most volcanically active times that we have had in earth history.

FIGURE 5.32. The Columbia River Basalts cover much of the states of Washington and Oregon and represent massive outpourings of lava in the northwestern part of the United States shortly after Noah's Flood.

5.6 CASE STUDY: MT. ST. HELENS

In March of 1980 a series of small earthquakes began to occur under the Mt. St. Helens volcano in the southwest corner of the state of Washington (figure 5.33). The volcano is one of dozens of composite volcanoes that are part of the Cascade Range. Later that month, a steam explosion left a huge crater in the glacier-covered summit of the 2,950 m mountain. This was the first activity since it last erupted in 1857. Over 10,000 earthquakes rocked the volcano leading the major eruption on May 18th. By that time the slope of the volcano was rising as much as 15m per day as magma and gasses built up inside the volcano!

On the morning of May 18th, a magnitude 5.1 earthquake was recorded as occurring below the

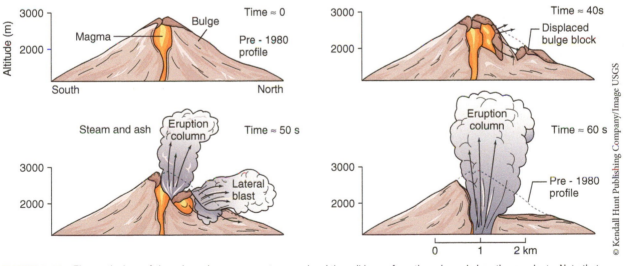

FIGURE 5.33. The north slope of the volcano became over steepened and then slid away from the volcano in less than a minute. Note that there was both a lateral blast and a vertical blast of pyroclastic material.

Courtesy USGS

FIGURE 5.33. (*Continued*)

volcano, and about 15 seconds later, the volcanic eruption began with the entire north slope of the volcano sliding away. As the north side moved, it allowed for a lateral blast, and then a vertical blast out of the volcano. The violent eruption continued for 9 hours, putting ash 24 km high into the atmosphere in just 15 minutes.

Part of the landslide entered into nearby Spirit Lake causing a 250 meter tall tsunami that stripped the hillsides of all its trees. Over one million trees were swept into the lake and created a huge floating log mat, parts of which have floated for more than 35 years. Pyroclastic flows followed the landslide and lateral blast covering 16 km² of area just north of the volcano. These flows moved at velocities of up to 130 km/hr and made deposits up to 37 m deep. As volcanic ash mixed with the melting glaciers on the mountain, mudflows began to travel down rivers

that drained the north side of volcano at speeds up to 80 km/hr. Many roads were washed out, 27 bridges were destroyed and nearly 200 homes were leveled or buried.

Many geologists have studied the eruption of Mt. St. Helens, but in particular Dr. Steve Austin has made important contributions. As a result we have learned seven major things that were rather unexpected.

1. *Finely laminated deposits don't always have to be made in quiet conditions.* Normally, when a geologist observes finely laminated deposits (figure 5.34) they have been trained to think about quiet, still conditions that allow sediments to accumulate grain by grain—like a deep lake or ocean bottom. However, the violent pyroclastic flows made paper thin layers in a matter of *seconds*.

1980 Mt. St. Helens Eruption Facts:

- Violent VEI 5 Eruption on May 18, 1980
- Pre-eruption elevation: 2,950 m
- Post-eruption elevation: 2,549 m
- 57 people died
- Eruption energy equivalent to 400 million tons of dynamite (an atomic bomb every second for 9 hours)

Landslide:

- Volume: 2.8 km³
- Buried North Fork Toutle River Valley to an average depth of 45 m, maximum of 183 m
- Traveled at velocity of 113 to 241 km/hr
- Created a 250 m tsunami in Spirit Lake

Lateral blast:

- Covered 596 km²
- Temperature as high as 350° C
- At least 483 km/hr velocity

- Destroyed 390 km² of forest or 4 billion board feet of lumber (enough timber to build 300,000 homes)

Lahars (mudflows):

- Traveled 15 to 80 km/hr
- Damaged 27 bridges and 200 homes
- Caused Columbia River to shallow from 12 to 4 m, stranding ships upstream

Eruption cloud:

- Reached 24 km upward in less than 15 minutes
- Spread across US in 3 days
- Detectable ash covered 57,000 km²

Pyroclastic flows:

- Covered 16 km² up to 37 m deep
- Moved at 80 to 130 km/hr
- 700° C

Photo by Steve Austin

FIGURE 5.34. Violent pyroclastic flows surprisingly produced finely laminated beds of volcanic material. Usually geologists think of fine laminations forming in quiet, still conditions.

2. *Erosion through solid rock can happen very quickly.* Two years after the main eruption, a minor overnight eruption melted snow and ice within the crater making mud which was promptly ejected down the slopes of the volcano as a mudflow. When geologists visited the area the next morning, they were shocked to find that new, deep canyons, some hundreds of feet deep, had been cut by this single event the night before.

3. *Sometimes canyons are made before rivers.* The small stream in figure 5.35 at the bottom of the canyon did not carve this canyon. Instead,

FIGURE 5.35. Engineers' Canyon. The cliff on the left is about 30 m tall. This canyon was not cut by the stream, the canyon came first.

the canyon was quickly eroded by catastrophic processes, and now a river flows through it. Canyon formation at Mt. St. Helens can help us develop hypotheses on how other canyons formed, such as the Grand Canyon.

4. *The Yellowstone "fossil forests" were likely transported and buried.* A classic argument that has often been posed against young-Earth creation is that there are 27 consecutively buried forests in Yellowstone National Park. The argument states that standing trees in forests were buried over and over again by volcanic catastrophes. Conservatively estimating that about 1,000 years separate each layer of trees,

27,000 years must be present in the fossil tree layers, making a young-Earth creation view impossible.

Now recall the trees that washed into Spirit Lake during the 1980 eruption. Dr. Steve Austin observed that many of the trees were floating in the lake in an upright position (figure 5.36), and also sinking to the bottom in an upright position. As a result, the trees were deposited in such a way, that it looks like they had grown in place. A closer examination of the trees in Yellowstone showed that many of the roots were short and broken and the trees had no bark or branches (figure 5.37), just like the Spirit Lake logs. No fossil soil horizons could be found in the Yellowstone "forests," either. Now it appears that all of the "forests" were the result of transportation and burial, not buried sequentially over thousands of years.

FIGURE 5.36. Upright logs deposited along the north shore of Spirit Lake. The trees are slightly inclined due to manmade lowering of the lake level. The photo was taken in 1983, three years after the eruption.

FIGURE 5.37. Many of the upright trees buried in the volcanic deposits at Specimen Ridge in Yellowstone National Park lack bark, branches and have truncated roots, just like those in Spirit Lake. The trees have clearly been transported and did not grow in place.

FIGURE 5.38. As the trees floated in Spirit Lake, the bark on their trunks was rubbed off and sank to the bottom (inset).

5. *Coal can be formed by floating log mats.* Most geologists have been taught that coal is the result of large quantities of vegetation and trees buried in swamps over millions of years. In the late 1970's Dr. Steve Austin discovered serious problems with this hypothesis when he studied coal seams in Kentucky. The coal deposits lacked roots, were made of layered bark sheets, and thin coal deposits could be traced for many kilometers. Curiously, many coal deposits were also closely associated with marine deposits. From these observations Austin developed a hypothesis in which coal was formed beneath a floating log mat.

When Mt. St. Helens erupted in 1980, Spirit Lake became a laboratory to test Austin's hypothesis. As the logs floated on Spirit Lake, the bark from the trees rubbed off (figure 5.38). Austin's SCUBA work beneath the floating log mat confirmed that bark was accumulating in thick deposits on the bottom of the lake. If this material could be more deeply buried and compressed, it would turn into coal—coal that looks very much like coal deposits that we have in the sedimentary rock record that was made during Noah's Flood.

6. *Recovery of completely decimated areas can occur in short periods of time.* During the eruption, nearly 400 km² of forest was leveled by the lateral blast from the volcano. The Forest Service purposely did not replant a large area of the blast zone so natural ecological recovery could be studied. Everyone was surprised as to how quickly vegetation returned. Seeds were carried into the area by wind, birds and animals. Fish soon returned to the lakes; probably arriving to the lakes as eggs on the feet of birds. Studying ecological recovery of such an area gives us ideas of how the earth recovered after Noah's Flood.

7. *Catastrophic processes help us consider what Noah's Flood was like.* Geologists often think of the landscape forming by the normal day-to-day processes we see most often. Mt. St. Helens has shown us that landscapes don't change much until catastrophic processes occur; then change is sudden, violent, and dramatic. Eruptions, avalanches, tsunamis, pyroclastic flows, and mudflows help us appreciate (at least on a small scale) what the events of Noah's Flood must have been like.

CHAPTER 5 IN REVIEW: THE RESTLESS EARTH

5.1 Introduction

- Earthquakes and volcanoes help us to understand what goes on deep inside the earth.

5.2 Does Rock Bend or Break?

- Only soft, unlithified rock can bend. Hard rock breaks. If bending takes place at great depths and pressures, the rock will contain evidence for those depths and pressures in the form of metamorphism.
- Sedimentary rocks are deposited horizontally. Sometimes they can be folded (before the rock is lithified) into anticlines, synclines, domes, basins and a variety of other shapes.
- Hard rocks are brittle and when forces are applied to them (tension, compression, shear) they break and cause a fault. There are three basic types of faults: dip slip, strike slip and oblique slip. Dip slip faults are identified based on the relative movement of the hanging wall.

5.3 Earthquakes

- Earthquakes are measured using the Richter scale (magnitude) or the Mercalli scale (intensity).
- Seismic waves propagate from the focus of an earthquake. Body waves (P and S) travel inside of the earth and surface waves (Love and Rayleigh) travel on the surface of the earth. The surface waves cause the most structural damage during an earthquake.

5.4 The Interior of the Earth

- The solid body of the earth is divided into a crust, a mantle and a core. Geologists can only investigate the mantle and the core by indirect methods, such as seismic waves and volcanic products.

5.5 Volcanoes

- There are many types of volcanic dangers including pyroclastic flows, mudflows, poisonous volcanic gases, lava flows and landslides. Some large volcanic eruptions can significantly affect climate.
- There are three main types of conical volcanoes: cinder cones, composite cones and shield volcanoes. Each type has characteristic products, shapes, slopes and eruption style. Sometimes volcanoes collapse after an eruption forming a caldera.

5.6 Case Study: Mt. St. Helens

- Mt. St. Helens was a volcano in the Cascade Range that erupted in 1980. It has been one of the most studied volcanic eruptions in earth history. There are some important discoveries

Aa

Anticline

Asthenosphere

Basin

Body wave

Caldera

Cinder cones

Composite cones

Compressional force

Core

Crust

Dip slip fault

Dome

Elastic rebound theory

Epicenter

Fault

Focus

Fold

Footwall (footblock)

Fumarole

Hanging wall (hanging block)

Lahar

Lapilli

Law of Original Horizontality

Lithosphere

Love wave

Mantle

Mercalli scale

Moho (mohorovičić discontinuity)

Moment magnitude scale

Mudflow

Normal fault

Nuée ardente

Oblique slip fault

Pahoehoe

Primary (p) wave

Pumice

Pyroclastic material

Rayleigh wave

Reverse fault

Richter scale

Scoria

Secondary (s) wave

Seismic wave

Seismogram

Seismograph

Seismometer

Shear force

Shield volcanoes

Strike slip fault

Surface wave

Syncline

Tensional

Tephra

Thrust fault

Tsunami

Vog

Volcanic ash

Volcanic bombs

Xenolith

REVIEW QUESTIONS

1. If rock that has folds in it was bent as the result of tremendous heat and pressure, what evidence should be apparent for folding in these kinds of conditions?
2. Draw pictures of all the different types of folds.
3. Draw pictures of each of the different kinds of faults, and practice identifying them using photographs.
4. What are some of the various causes of earthquakes? What is the theory behind why they occur?
5. Are the largest earthquakes always the most devastating earthquakes to humans? What are some of the factors involved?
6. How are seismic waves used to determine the structure of the interior of the earth?
7. Describe the two different ways in which an earthquake can be measured. What are the advantages and disadvantages of each method?
8. Why do we have continents?
9. What are some observations that can be made to determine the properties of the interior of the earth?
10. What are the different kinds of products that can be ejected from volcanoes?
11. Make a chart comparing and contrasting cinder cones, composite cones and shield volcanoes.
12. Summarize the surprising things geologists found when studying the eruption of Mt. St. Helens.

FURTHER READING

Austin, S.A. Twentieth century earthquakes – confronting an urban legend. http://www.icr.org/article/twentieth-century-earthquakes-confronting-urban-le/ Are earthquakes increasing with time?

Morris, John and Austin, S.A. 2003. *Footprints in the Ash*. Master Books: Green Forest Arkansas. The story of the eruption of Mt. St. Helens and the important discoveries that have been found.

Austin, S.A. Greatest Earthquakes of the Bible. http://www.icr.org/article/greatest-earthquakes-bible/

Austin, S.A. Mt. St. Helens and catastrophism. http://www.icr.org/article/mount-st-helens-catastrophism

WEBSITES TO VISIT

http://www.usgs.gov/ United States government site for everything related to geology and natural hazards (earthquakes, volcanoes, landslides, maps, etc.).

http://www.ready.gov/earthquakes Earthquake safety tips

http://www.wbdg.org/resources/seismic_design.php How buildings are designed for earthquake prone areas

http://www.tsunami.noaa.gov/ NOAA tsunami website

http://pubs.usgs.gov/fs/2000/fs036-00/ USGS Fact sheet about Mt. St. Helens

http://earthobservatory.nasa.gov/Features/WorldOfChange/sthelens.php NASA's site discussing the recovery at Mt. St. Helens.

REFLECT ON SCRIPTURE

Genesis 7:11
Amos 1:1
Psalm 104:32
Matthew 27:51, 28:2
Acts 16:6
Revelation 6:12, 8:5, 11:13, 11:19 and 16:18

Image courtesy Marcus Ross

READING THE RECORD OF THE ROCKS

The Grand Canyon, as seen by boat along the Colorado River, provides a spectacular view of the rock units of the American Southwest. Appling the rules of geological dating helps Earth scientists determine the geological history of the area.

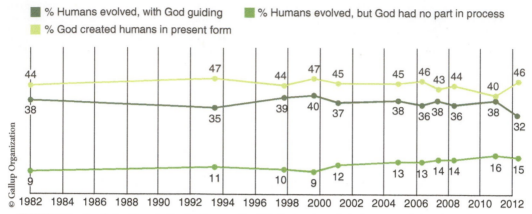

■ % Humans evolved, with God guiding   ■ % Humans evolved, but God had no part in process
■ % God created humans in present form

FIGURE 6.1. The Gallup Organization has been polling US citizens about their views on evolution since 1982. Poll question: Which of the following statements comes closest to your views on the origin and development of human beings?
1) Human beings have developed over millions of years from less advanced forms of life, but God guided this process, 2) Human beings have developed over millions of years from less advanced forms of life, but God had no part in this process, 3) God created human beings pretty much in their present form at one time within the last 10,000 years or so

The age of the Earth is a contentious issue, especially in Christian circles. Vast numbers of scientists believe that the Earth is over four billion years old, and the universe nearly fourteen billion years. Yet other scientists argue that the Earth is much, much younger, with an age of only a few thousands of years. The American public is split on this issue, and has been for quite some time (figure 6.1). But why?

6.1 A HISTORY OF EARTH HISTORY

We often view creation-evolution issues in light of contemporary controversies and arguments. But throughout the history of the Christian church, perhaps what it is surprising is that prior to the late 1700s, the voice of Christianity's greatest minds in science and theology were so *unified* in their thinking.

During the time of the early Church (from Pentecost to 325 A.D.), nearly all commentators on Genesis agreed that the days of Genesis 1 should be understood as six sequential 24-hour periods. Using the Septuagint translation of the Old Testament, most favored a date of creation at around 5,000 B.C. Notable early Church fathers who held this view included Ireaeus, Hippolytus, Lactantius, and Clement of Alexandria. Even Origen and Augustine, who held more symbolic views on the days of Genesis 1, both wrote that Earth's age was less than 6,000 years at the time of their writings. These and other Christians arrived at a young Earth by adding the ages of the men listed in the genealogies of Genesis 5 and 11 to get back to Adam. Since Adam was made on Day 6 of Creation Week, determining the age of the Earth was considered rather straightforward.

The tradition of a young Earth continued throughout the Middle Ages and Renaissance, and as scholars gained access to the Hebrew Old

Testament (the Masoretic texts) the age of Earth actually got *younger*, since it appears that the Septuagint translators inserted longer lifespans to the Biblical chronologies in Genesis 5 and 11. While many in Catholicism held to Augustine's allegorical approach to the days, reformers such as Martin Luther and John Calvin both argued that Scripture did not allow for such views, and that the days described in Genesis 1 were normal days. But again, the young age of Earth was still not questioned among Christians, even across the Catholic-Protestant divide.

The most rigorous age estimate came from James Ussher, Anglican Archbishop of Amargh (Ireland), who published a book titled *Annals of the World* in 1651. Ussher used both Biblical and extrabiblical sources to produce a comprehensive history of the world from creation to the fall of Jerusalem in 70 A.D. In so doing, he produced dates for many Biblical events, including the Exodus from Egypt, Noah's Flood, and the beginning of the Creation Week, which he placed in 4,004 B.C. Shortly after publication, Ussher's dates were incorporated into many printings of the Authorized Version (King James Version) of the Bible, which cemented Ussher's chronology in the English-speaking world and beyond.

It was shortly after Ussher's writings that one of the most influential early geologists, Nicolas Steno (figure 6.2), would lay the groundwork for modern geology. A brilliant Danish anatomist living in Florence, Italy, Steno's interest in fossils led him to formulate some of the foundational rules for mineralogy and sedimentary rock interpretation. Steno also produced the first detailed geological history of any region in the world (the rocks in Tuscany, Italy). In this work, he argued that God's actions in Creation Week and Noah's Flood were responsible for the formation of many rock layers, and that the retreating floodwaters also caused their final erosion (figure 6.3). Steno was not only the "father of geology," he was also a dedicated Christian and young-Earth creationist!

J. P. Trap 1868

FIGURE 6.2. Danish anatomist and pioneering geologist Niels Stensen (known as Nicolas Steno), 1638–1686 A.D.

Other geologists continued in this tradition, but by the late 1700s some difficult challenges to young-Earth creation and Noah's Flood were emerging. The discovery of many different types of fossils from animals and plants no longer alive today meant that extinction had occurred. Many began to think that the Earth was older than Ussher's chronology, and that Noah's Flood could not account for all of the geological features of the world. This led to the concept of **catastrophism**, an Earth history concept that held a) the world was ancient, rather than young; b) the world's geology was formed by many major destructive events (catastrophes) over time; c) following each catastrophe, God would replenish the world with new plants and animals; and d) Noah's Flood was only the most recent catastrophe, and may or may not have been global.

Many of the leading geologists of the early 1800s were catastrophists, including William Buckland and Adam Sedgwick in England and Georges Cuvier in France. Like Steno, most were Christians, and a good number were clergy. Seeking to harmonize their new

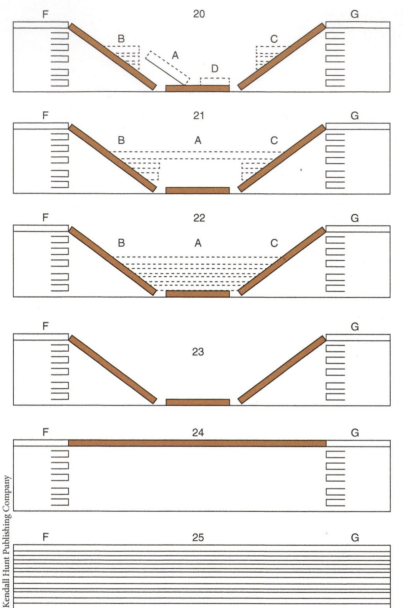

FIGURE 6.3. Steno's cross-section of the rocks at Tuscany. Reading upwards from 25 to 20, Steno believed that the lower rocks were formed horizontally and under water during Creation Week, undermined and eroded, then new horizontal rocks formed over the remaining original rocks during Noah's Flood, which were also later eroded.

© Kendall Hunt Publishing Company

ideas of an ancient Earth with Genesis resulted in two completely new interpretations of the creation account: the **Day-Age** view and **Gap Theory**. In Day-Age, each of the days of creation was considered to be some extremely long period of time, or "ages" of geology. Gap Theory tried to keep a normal six-day Creation Week by placing most of geological history in a "gap" in time between Genesis 1:1 and 1:2. These views did not come naturally from a detailed study of the Scriptures. Rather, they were attempts to harmo-nize Scripture's account of creation with the belief in an ancient Earth.

While catastrophists rose to prominence in the early 1800s, their influence in geology was short-lived. Scottish geologists James Hutton founded the idea of **uniformitarianism** in 1788, but it wasn't until the writing of *Principles of Geology* in the 1830s by English lawyer and geologist Charles Lyell that the idea would gain ground (figure 6.4). Uniformitarianism is often summed up in the phrase "the

(a)   (b)

FIGURE 6.4. a) James Hutton, the founder of uniformitarian geology; and b) Charles Lyell, whose arguments for uniformitarianism quickly dominated geological thinking.

present is the key to the past." Instead of invoking Noah's Flood or numerous catastrophes, uniformitarians insisted that the geological processes of the past operated slowly, and that only the processes and rates observed today could be applied to the geological record of the past.

Modern geologists trace their thinking back to Hutton and Lyell, though today they are more willing to consider large-scale and catastrophic events in Earth history. Most follow a modified version of uniformitarianism called **actualism** which holds that the geological processes that worked in the past were much the same as those operating today, though the scope and magnitude of events may be very different. So while modern geologists think that many geological processes operate slowly, some can be fast, catastrophic, or global (ex: dinosaur extinction from a massive asteroid or comet impact).

6.2 RULES FOR READING ROCKS

In order to determine how and when different geological events took place, geologists have developed a number of interpretive laws and rules. These rules will help us determine a geological history though what geologists call **relative dating**. This means that the methods described below help determine the sequence and order in which geological events occurred, without assigning a numerical date/age. Because relative dating relies of the physical relationship of the rocks, these methods are used by both young-Earth creation geologists and old-Earth advocates. Despite disagreements over the numerical age of the rocks (thousands vs. millions/billions of years), both groups can often agree about whether one rock was formed before or after another rock.

Based on some very common-sense principles, we start with rules that apply primarily to layers of sedimentary rocks (called **strata**) and to any of the extrusive igneous rocks (such as lava flows and volcanic ash deposits). Because both the sedimentary and extrusive igneous rocks are subject to gravity and transportation at the Earth's surface, the first three laws below apply equally.

Law of original horizontality—when sediments or extrusive igneous rocks are deposited, they are spread out in broad, flat sheets (figure 6.5). This law means that these units start off flat, and any tilting or folding of the unit happened *after* its initial formation.

Law of superposition—in an undisturbed sequence of sedimentary rocks and extrusive igneous rocks, the oldest units are found below the younger units. Much like a messy room, the oldest materials are near the bottom, and the newest materials are near the top (figure 6.6)

Law of lateral continuity—when sediments or extrusive igneous rocks are deposited, they remain the same composition in all directions until they either (a) contact an edge or wall to the depositional environment, or (b) thin out as energy levels decrease far from the source region (figure 6.7). Under the power of gravity, fluids flow and create a level area for the deposition of sediments or solidification of lava. This flow continues until energy levels change: sandy material may no longer be carried as fluid velocity decreases, and finer-grained silt or clay will be deposited further away from the sand. Alternately, the water or lava flow may encounter a barrier, such as the sides of a valley, which confines the material and stops the flow from continuing laterally.

Original horizontality, superposition, and lateral continuity were all developed by Nicolaus Steno,

FIGURE 6.5. Tibbet Knob, along the Virginia/West Virginia border. Because the sedimentary rocks exposed here were originally deposited horizontally, the tilting of the rocks came afterwards.

mentioned earlier. Steno's laws make up the basis for the discipline of *stratigraphy*, which is the study of the relationships among sedimentary rocks.

The law of lateral continuity can be combined with superposition. Developed by Johannes Walther, **Walther's Law** states: in a vertical sequence of sedimentary rocks, the types of rocks found above

FIGURE 6.6. Superposition of chalk (upper, younger) and limestone (lower, older) along the coast of Norfolk, England.

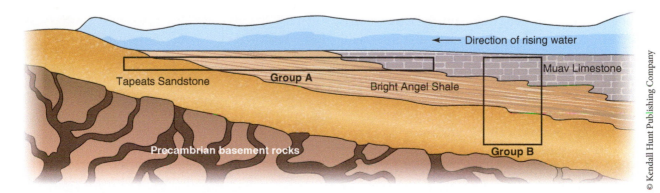

Direction of rising water

Tapeats Sandstone
Group A
Bright Angel Shale
Muav Limestone
Precambrian basement rocks
Group B

© Kendall Hunt Publishing Company

FIGURE 6.7. Lateral continuity and Walther's Law depicted in the lower Paleozoic sediments in the Grand Canyon. Lateral continuity is depicted in "Group A" rocks, where the sediment continues in a horizontal plane until conditions change. Walther's Law follows, as the laterally-adjacent environments are the sand→shale→limestone units in Group A, and the same pattern is reflected in Group B, the vertical succession of sediments.

and below each other are the same as those found adjacent sedimentary environments (figure 6.7). Walther's Law helps geologists determine shifts in depositional environments, because depositional environments are next to each other along the surface. For example, along the southeastern coast of North America, sediments transition from a sandy beach to a silty offshore region, and then a shallow-ocean limestone. If sea level changes, then these environments shift in response. For example, the sandy beach moves further inland, and the old beach (which is now in deeper water) is covered by deposits of silt.

Because of the vertical shift in geology that results from horizontal changes at the surface, Walther's Law recognizes that there should be continuity in the rock record that reflects the continuity we find in modern depositional environments. As you'll see shortly, violations of Walther's Law can help in identifying missing time in the rock record.

Rule of inclusions—when a rock includes pieces of another rock inside it, the included pieces must be older than the rock containing them (figure 6.8). This rule was anticipated by Steno regarding fossils but formally defined by Hutton, who recognized that sometimes rocks from one geological unit are found

Mike Richter/Shutterstock.com

Lagui/Shutterstock.com

(a)   (b)

FIGURE 6.8. The rule of inclusions works like gelatin with fruit in it (a). The fruit must be older than the gelatin, since the gelatin is poured around it to make the dessert. Likewise, the pebbles that make up a conglomerate rock (b) must be older than the conglomerate as a whole, since those pebbles already existed and were eroded and transported before their incorporation into the conglomerate.

CHAPTER 6: Reading the Record of the Rocks

inside another geological unit. Hutton reasoned that the smaller pieces must have existed earlier, and were included in the younger rock through erosion and redeposition for sedimentary rocks or by incorporation into an igneous magma that did not fully melt the pieces.

Principle of cross-cutting relations—geological structures or features (such as dikes, faults, or erosion surfaces) must be younger than the geological units that they affect (figure 6.9). Here is another of Steno's rules, this time dealing with instances where a rock is cut by another rock or geological feature. Common cross-cutting features include igneous intrusions and faults. If a fault is seen among a group of rocks, then the fault must be the younger feature. This is because the rock must already exist before it can be broken.

FIGURE 6.9. A fault cross-cuts and offsets a group of volcanic rocks.

Among sedimentary rocks, there are often features within or at the top of the unit. These **sedimentary structures** reflect the conditions in which the rock formed (figure 6.10). *Primary sedimentary*

(a)

(b)

(c)

(d)

FIGURE 6.10. A variety of sedimentary structures: a) graded beds, b) fossil ripple marks, c) mudcracks, and d) fossil animal tracks.

structures were made during the initial deposition of sedimentary material, and include cross-beds (formed as sand falls down the steep side of sand waves), ripple marks (wavy surfaces formed by flowing water), and graded beds (indicators of different fluid speeds). *Secondary sedimentary structures* were formed after depositions, and can include mudcracks (cracks formed as water leaves mud/clay deposits), raindrop impressions, and animal tracks and burrows.

Unconformities are surfaces within rocks that represent a break in time. In many rocks, this surface was formed by erosion that removed some of the rock record, thus leaving a gap in the time recorded by the once-present rocks. Alternatively, the surface could result from a period of low/no sedimentation (called a *hiatus*). In this case, time is not recorded because no rock was deposited.

Nearly all geology textbooks include as part of their definition that unconformities represent a "long period of time." However, this is inappropriate for the definition of an unconformity. The unconformity is simply the break in time, and the amount of time represented by that unconformity is a separate issue. In fact, there are many, many examples of unconformities that old-Earth geologists believe must represent many millions of years, yet do not appear to have taken much time to form. So once again we see that we can agree on the *pattern* (the unconformity), while disagreeing about the *process* (how much time was involved).

Three types of unconformities are generally recognized (figure 6.11):

Disconformity—an unconformity located between two roughly parallel sedimentary or extrusive igneous units. Here, rocks are deposited according to original horizontality, then erosion or non-deposition occurs, and then more rocks are deposited according to original horizontality, and parallel to the lower rocks. Disconformities may be particularly difficult to identify, unless there is a very visible erosional contact (perhaps showing scouring or

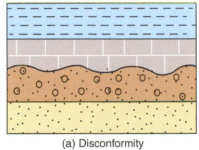
(a) Disconformity

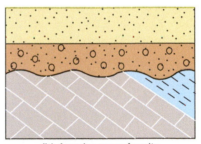
(b) Angular unconformity

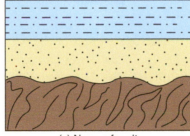
(c) Nonconformity

FIGURE 6.11. The three types of unconformities. Note that units with dashes, dots, and bricks are sedimentary rocks, wavy lines are metamorphic rocks.

Courtesy Marcus Ross

channels), or if there is an obvious violation of Walther's Law between two rock units, such as a marine limestone with fossil shells that is overlain by rounded cobbles and gravel from a mountain river or glacier. Since marine carbonates and mountain streams are not laterally adjacent environment, their superposition points to a disconformity.

Angular Unconformity—an unconformity where the sedimentary or extrusive igneous rocks above and below the unconformity are oriented at some angle to each other. These are easier to spot in the field because the rocks below the unconformity are not parallel to those above the unconformity. To form, originally horizontal strata must be deposited and later tilted through tectonics, mountain-building, or other forces. The tilted rocks are eroded, and new strata are horizontally deposited above them.

Nonconformity—an unconformity located at the contact of a sedimentary or extrusive igneous rock and an intrusive igneous or metamorphic rock. Here the unconformity is located at the contact between the sedimentary and crystalline rock (either igneous or metamorphic). To form, the crystalline rock must either be raised from depth (where it formed) to the surface then eroded and covered with new sediments, or a large magma body rises and intrudes sedimentary rocks.

6.3 FOSSILS

A **fossil** is the remains of a formerly living organism that is preserved in the geological record. They provide us with knowledge of animals, plants, and other organisms that often no longer exist. Fossils come in a variety of types (see BOX 6.1), but all of them somehow preserve the organisms' past existence, and tell us about its shape, relationship to other organisms, mode of life, and sometimes even behavior!

Paleontologists are the scientists who study fossils. This includes dinosaurs, of course, and so much more. The vast majority of fossils in the sedimentary rocks of the world are invertebrates (animals without backbones) such as arthropods, mollusks, brachiopods, corals, and many other types. Much less common are the fossils of vertebrates (animals with backbones), but they are more familiar to us: fish, amphibians, reptiles, birds, and mammals. In addition, there are many fossils of plants (including trees, ferns, and pollen) and many kinds of single-celled organisms (such as the calcite-shelled organisms, called coccoliths, which make up chalk deposits).

Most fossils were formed during Noah's Flood, as the waters of the oceans catastrophically covered the continents (figure 6.12). These waters carried many different kinds of plants and animals from different ecosystems, and deposited them in sequences on the continents. Some fossils were also made after the Flood, as smaller and more regional catastrophes buried additional plants and animals on top of the sediments deposited during the Flood. Chapter 7 provides a more detailed look at geological and life history of Earth from both the young-Earth and naturalistic evolutionary perspectives.

BOX 6.1

When you think of fossils, you often think of skeletons, skulls, teeth, and shells. These are all examples of **body fossils**, which are the now-mineralized remains of an organism's hard parts.

© Merlin74/Shutterstock.com

Courtesy Marcus Ross

Paul B. Moore/Shutterstock, Inc.

Other types of fossils include:

Impressions—when an organism dies and is buried, the body is pressed into the sediment, and this may produce an impression of the original shape, even if the organism is destroyed, as is the case with these bivalves.

© Tom Grundy/Shutterstock, Inc.

Tracks/Traces—these are the impressions left by organisms as they move. Tracks can tell us about how an animal moved, and traces can include burrows where an organism once lived. Many trace fossils were formed during Noah's Flood by animals attempting to escape the sediments being deposited, such as the amphibian or reptile that scurried across an underwater sand wave that is now the Coconino Sandstone in the Grand Canyon.

Courtesy Marcus Ross

Amber—tree sap, when hardened and fossilized, forms an orange to yellow, glassy material. Occasionally organisms, such as the insect shown here, can be trapped inside and fossilized!

Baciu/Shutterstock.com

Coprolites—sometimes even the feces of an animal can be preserved, and when it is, it can be very informative! Coprolites preserve the fossilized remains of animals' diets, and so can help us understand what (or who) they were eating.

Courtesy Department Mineral Resources

CHAPTER 6: Reading the Record of the Rocks

Petrification—this particular form of fossilization occurs when mineral-rich waters deposit microcrystalline silica (similar to quartz and opal) in the open spaces between the tissues of an organism. For example, petrified wood is wood that has all its open space filled by silica. The various colors of petrified fossils come from additional elements (such as iron) that may also be deposited during fossilization.

© Bill Florence/Shutterstock, Inc.

There are many different types of fossils and processes that form them, but they all share one thing in common: fossils must be formed *quickly*. In order to turn organic remains like bone, wood, or even their dung (!) requires very rapid burial in sediments, followed by a quick replacement of the original organic material by minerals. *Every* fossil is a testimony to rapid geological processes, because without them, there would be no fossils at all.

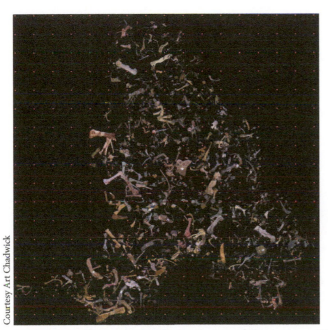

Courtesy Art Chadwick

FIGURE 6.12. Thousands of jumbled and separated dinosaur bones (high-resolution GPS map, left) were catastrophically buried in a fast-moving mudflow in the Lance Formation of eastern Wyoming (right).

While geologists may work at a single outcrop to describe the rocks exposed at that area, they are also interested in how those rocks are related to other rocks in other locations. **Correlation** is the process of linking or matching rock units over distances in which they are not seen. Correlation may be done by a number of methods, but the two basic categories are:

> *Physical correlation*—rocks are matched according to shared composition and structures. *Temporal correlation*—rocks are matched according to their time of formation. This is often accomplished by using (a) fossils, or (b) radioactive dating methods.

Physical correlation (figure 6.13) is perhaps the most straightforward method, since it focuses on the material composition of the rocks (including mineral types and sizes, physical features like cross-beds, etc.). In fact, many people have done this, even without realizing it. When driving on the highway through hilly terrain, we can often see rocks exposed along the rock cuts made for a highway. Looking to the one side, we may see a sandstone, coal seam, or igneous intrusion. And on the other side, the same rocks may be exposed in the other road cut. We intuitively connect them in our mind, physically correlating the rocks.

In temporal correlation, geologists attempt to determine if two different rocks were formed simultaneously. Often, fossils are used to temporally correlate sedimentary rocks. This is because when early scientists discovered and documented fossils, they

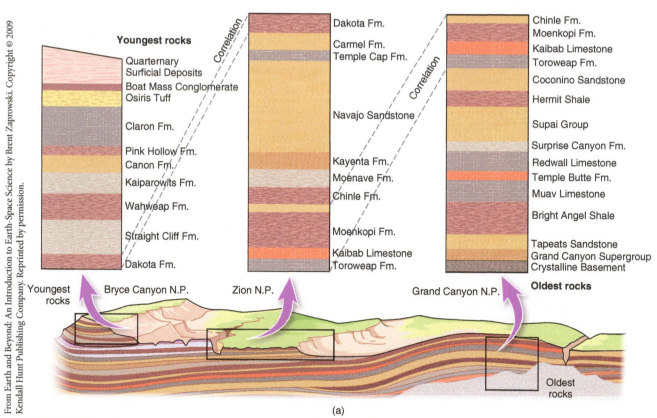

FIGURE 6.13. a) Physical correlation of rocks exposed in three national parks: Bryce Canyon, Zion, and Grand Canyon. b) Temporal correlation connects rocks by time of formation. The pattern of fossils can be the same despite differences in rock types, and is used to temporally correlate rocks.

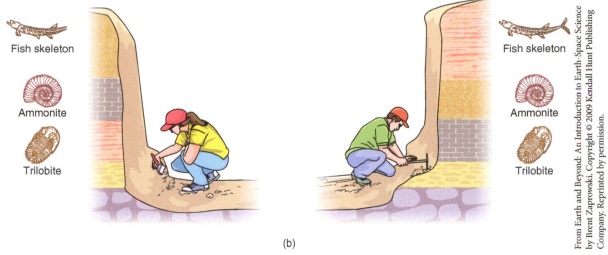

Fish skeleton

Ammonite

Trilobite

Fish skeleton

Ammonite

Trilobite

(b)

From Earth and Beyond: An Introduction to Earth-Space Science by Brent Zaprowski. Copyright © 2009 Kendall Hunt Publishing Company. Reprinted by permission.

FIGURE 6.13. *(Continued)*

realized that not all fossils were found in all rock units. The first geologist to formalize and effectively use this knowledge was the British surveyor William Smith, who also created the first large-scale geological map, showing the names and locations of rocks exposed in Great Britain (figure 6.14). In his many jobs at quarries and performing surveying work around Wales and England, Smith noticed that the fossils in the rocks he excavated remained in the same order, even if the type of rock containing them

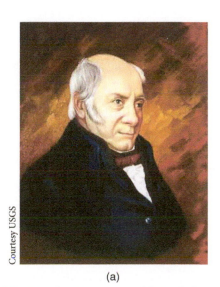

(a)

(b)

FIGURE 6.14. William Smith (a) was one of England's finest early geologists, and produced the first major geologic map of Britain (b).

differed from one place to another. From this observation he developed the **principle of faunal succession**, where fossil organisms follow one another in a definite and recognizable order within sedimentary rocks.

Geologists use the principle of faunal succession to correlate different rocks that contain the same or very similar fossils (figure 6.15). For old-Earth geologists, the assumption behind fossil correlation is based on the idea that fossils only lived for a certain period of time in Earth history, so rocks containing the same fossils must be very close to each other in time. However, if the majority of sedimentary rocks were deposited during Noah's Flood, then this assumption does not hold. Most young-Earth creationists believe that the patterns reflect the order of burial among these organisms and their ecosystems as Noah's Flood destroyed the pre-Flood world. The description of the Flood in Genesis 7 and 8 clearly indicates that the entire world was flooded and destroyed, killing off vast numbers of plants and animals (see Chapter 4). Such a destructive event needs to be considered if the true history of geology is to be understood. But in both cases, the geologist is attempting to understand the pattern of fossils in the rocks. Those are the data, and our explanations must adequately address them.

The recognition of faunal succession ultimately gave rise to the **geological column** (figure 6.15), as many rocks became correlated with one another and a large-scale pattern of fossils emerged. The many names on the geological column (such as "Paleozoic," "Mesozoic," "Permian," etc.) reflect the types of fossils that have been found in rocks from around the world. Since the geological column was built on the basis of superposition and faunal succession, it is a product of relative dating methods. Following the relative dating rules like superposition, the oldest parts of the geological column are at the bottom, and more recent events are located toward the top.

The geological column was basically completed before Charles Darwin published *On the Origin of Species* in 1859, so it was built before evolutionary theory. Many young-Earth creationists have argued that the geological column is an evolutionary idea, but this is not really the case. Like the old-Earth explanation for faunal succession, evolutionary theory attempts to explain the *pattern* reflected by the column by arguing that various life forms have evolved, lived, and died on Earth over many millions of years. So while evolutionary theory uses

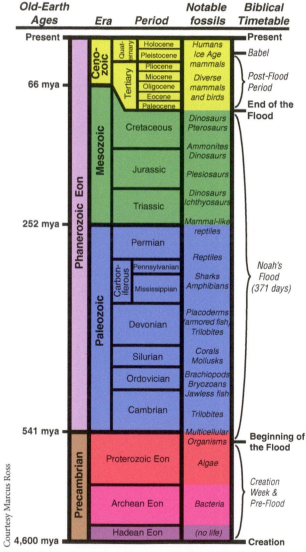

FIGURE 6.15. The geological column, with the old-Earth timescale on the left and a young-Earth timescale on the right.

Courtesy Marcus Ross

and attempts to explain the geological column, it was not responsible for building the column in the first place.

Note that there are ages (in millions of years) listed on the left-hand side of figure 6.15. These are the numerical ages used by old-Earth geologists for what they think are the ages for the various divisions within the column. These ages were introduced only after the column was established, and made more specific after the discovery of radioactivity and its use in dating rocks (discussed below). On the right-hand side of figure 6.15 is an alternate, young-Earth creation view of the column. Here we see that the rocks of the geological column may be divided into three major groups: those formed during Creation Week and prior to the Flood; rocks formed during the Flood; and rocks formed after the Flood. It is clear that young-Earth creationists believe that the Flood was the most important geological event after creation, and is responsible for most of the fossil-bearing sedimentary rocks in the world. A more detailed treatment of naturalistic evolution and young-Earth views on Earth history is given in chapter 7.

6.5 ROCKS AS CLOCKS

Determining the sequence of events in geology falls under the umbrella of *relative dating methods*, but geologists also attempt to determine the numerical ages of rocks. This is most often accomplished by the used of *radioisotope dating*. To understand how these systems are supposed to work, we must learn how radioactive decay occurs.

Atoms of the same element can have different numbers of neutrons in their nucleus. These different forms of the same element are called *isotopes*. Most isotopes are stable, retaining the same identity over time as defined by their number of protons and neutrons. Some isotopes, however, are fundamentally unstable, and over time will convert into different isotopes through **radioactive decay**. This is not a chemical reaction or process. Instead, it is a *nuclear* process, where the radioactive decay events occur in the nucleus of the atom, and are not controlled by chemical bonds or states of matter (e.g., solid vs. liquid). The element carbon has three isotopes: $^{12}_{6}C$, $^{13}_{6}C$, and $^{14}_{6}C$ (figure 6.16). Of these, carbon-12 and carbon-13 are stable, but carbon-14 is radioactive decays into nitrogen-14, or $^{14}_{7}N$.

A radioactive atom is called a **parent isotope**. When it decays, it ejects one of several particles out of its nucleus to become a new atom, called the **daughter isotope**. There are two important forms of radioactive decay, and a third event that can also change the identity of an isotope.

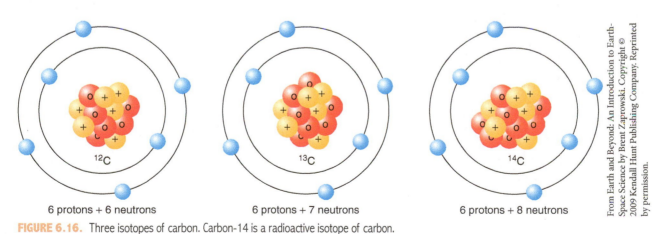

6 protons + 6 neutrons   6 protons + 7 neutrons   6 protons + 8 neutrons

FIGURE 6.16. Three isotopes of carbon. Carbon-14 is a radioactive isotope of carbon.

From Earth and Beyond: An Introduction to Earth-Space Science by Brent Zaprowski. Copyright © 2009 Kendall Hunt Publishing Company. Reprinted by permission.

6.5.1 Types of Radioactive Decay

In this section we will investigate four styles of radioactive decay. Three of them are used by geologists to determine a numerical age for rocks (those methods are discussed in the next section). The fourth actually isn't radioactive decay *per se*, but also involves a change to an atom's nucleus, and so is included here.

In **alpha decay**, the nucleus emits a cluster of two protons and two neutrons (figure 6.17), called an *alpha particle* (a helium nucleus), represented by the lower-case Greek letter, α. We can write this similar to the way we note the atomic mass and number of isotopes, like the carbon isotopes above: $^4_2\alpha$. The α particle has an atomic mass of four and atomic number by definition, so scientists do not usually write it in, but including it here helps keep track of how a parent isotope changes into its daughter isotope during α decay.

Taking a hypothetical parent (P), it undergoes radioactive decay ($\rightarrow$) to produce a daughter (D), an alpha particle (α), and other forms of energy. A basic equation can be written as follows:

$$^x_y P \rightarrow {}^{x-4}_{y-2} D + {}^4_2 \alpha + energy$$

Here the *x* and *y* values are the atomic masses and numbers of the parent isotope. Notice that, on the other side of the equation, the daughter's atomic mass and number are *x-4* and *y-2*. This is because the alpha particle ($^4_2\alpha$) left the parent isotope's nucleus, leaving behind a nucleus with fewer protons and neutrons. So by adding the atomic masses and numbers of the daughter isotope and the α particle, we arrive at the original atomic mass and number of the parent isotope.

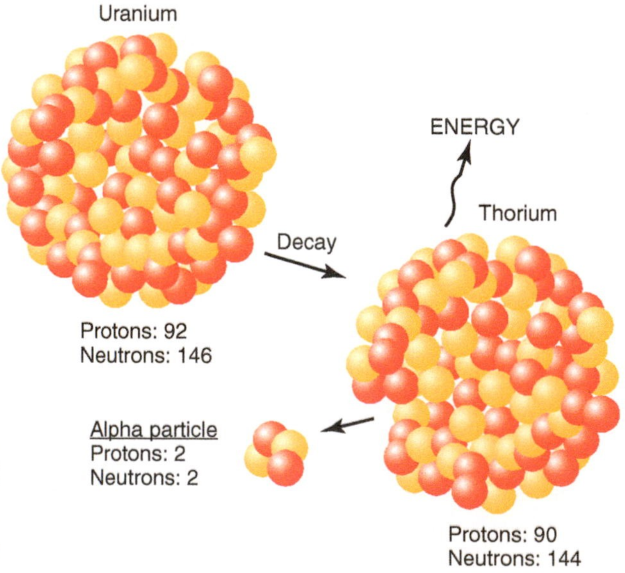

This basic equation applies to all isotopes that undergo α decay. A real-world example is seen in uranium. All atoms of uranium by definition have 92 protons. There are two isotopes with different numbers of neutrons, $^{235}_{92}U$ and $^{238}_{92}U$. Both isotopes are unstable, and $^{238}_{92}U$ is the most common. When it undergoes α decay, its decay equation is:

$$^{238}_{92} U \rightarrow {}^{234}_{90} Th + {}^4_2 \alpha + energy$$

By losing two protons and two neutrons via α decay, the uranium drops two atomic mass units and four atomic weight units to become thorium-294. The new atom of thorium is also unstable, and is the first step in a long decay series (figure 6.18) starting with uranium-238 and ending at the final, stable daughter isotope of lead-206 ($^{206}_{82}Pb$).

The second type of radioactive decay is **beta decay**. In beta decay, a neutron in the nucleus emits a negative beta particle (β^-), which is identical to an electron in size and charge (figure 6.19). This seems strange, because electrons are not found in the nucleus. The simplest way to think of this is that a neutrally charged neutron is like a positively charge proton fused together with a negatively charged electron. By emitting the nearly massless, negatively charged β^- particle, the remainder of the neutron has a positive charge, and it actually becomes a proton. This increases the atomic number of the daughter isotope, but because the *combined* number of protons and neutrons has not changed (one less neutron, but one more proton), the atomic mass remains the same. A basic equation for β^- decay can be written as:

$$^x_y P \rightarrow {}^x_{y+1} D + {}^0_{-1} \beta^- + energy$$

Uranium
Protons: 92
Neutrons: 146

Decay

ENERGY

Thorium

Alpha particle
Protons: 2
Neutrons: 2

Protons: 90
Neutrons: 144

FIGURE 6.17. Uranium-238 decays into thorium 234 by alpha decay.

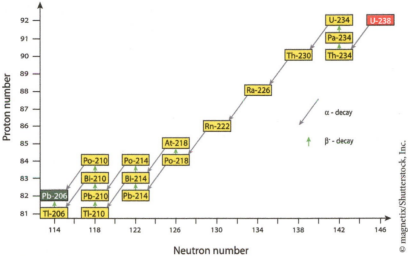

FIGURE 6.18. The decay series of uranium-238 (top right) to lead-206 (bottom left). The left-hand axis is the atomic number of the isotope, and the bottom axis is the atomic mass. α (diagonal down and to left) and β⁻ (upward) decays shown. Note that there can be variations in the decay events, but all paths ultimately end at the stable lead-206 isotope.

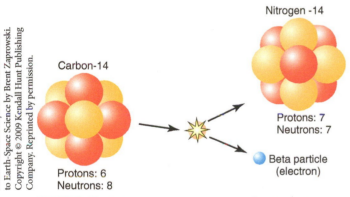

FIGURE 6.19. Carbon-14 emits a beta particle (β⁻) when it decays into nitrogen-14.

This is the type of radioactive decay that carbon-14 undergoes. Here, the radioactive isotope of carbon-14 decays into a stable daughter isotope of nitrogen-14. Unlike uranium, there is no complex decay series; a stable daughter is produced with only one β⁻ decay. The decay equation is:

$$^{14}_{6}\mathrm{C} \rightarrow {}^{14}_{7}\mathrm{N} + {}^{0}_{-1}\beta^- + \text{energy}$$

Nuclear fission is an unusual type of radioactive decay in which the parent isotope splits into two smaller daughter isotopes, often with the release of additional nuclear materials and energy. An isotope that can undergo fission is called *fissile*. The composition of the two daughter products can be variable, and here we look at the most important example of fission, that of uranium-238. As discussed earlier, U-238 usually emits an α particle as it begins its decay series to Pb-206. In very rare instances (about one time for every two million α decays), a U-238 atom will split via fission into two smaller nuclei, such as two equal-sized palladium atoms:

$$^{238}_{92}\mathrm{U} \rightarrow {}^{119}_{46}\mathrm{Pd} + {}^{119}_{46}\mathrm{Pd} + \text{energy}$$

There are a large number of other possible products, since the daughters do not have to be the same size, but in each case the combined total of protons and neutrons is conserved among the daughter products. These two smaller atoms explode out from the center of the original U-238 parent, and as they travel in opposite directions they destroy some of the crystalline structure of the mineral in which the U-238 was bound. Like shooting a small bullet through a number of large window panes, these atomic daughters create microscopic lines, called *fission tracks* in the mineral.

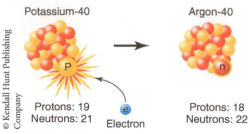

Potassium-40   Argon-40

Protons: 19   Protons: 18
Neutrons: 21   Electron   Neutrons: 22

© Kendall Hunt Publishing Company

FIGURE 6.20. Electron capture converts a proton into a neutron, changing the atomic number but not the atomic mass.

A final nuclear process is **electron capture**, which occurs when a high-speed electron from outside the atom collides with a proton in the nucleus (figure 6.20). The collision fuses the electron and proton, cancelling their charges and producing a neutron. In effect, this is the like a reversal of β^- decay above. The general decay equation is:

$$_{-1}^{0}e^- + {}_{y}^{x}P \rightarrow {}_{y-1}^{x}D + energy$$

Note that in the daughter isotope, the atomic number is reduced by one, and the atomic mass remains the same. The atomic number decreases because one proton is converted into a neutron, and because the combined total of protons and neutrons stays constant, the atomic mass remains the same between the parent and daughter isotopes. Since the electron is nearly massless, it does not contribute to the atomic mass.

Electron capture can occur in rocks and minerals where there is a high concentration of β^- emission, since β^- particles are the same as electrons (e^-). Potassium-40 ($_{19}^{40}K$) is an unstable isotope, which frequently decays by β^- emission. However, if there are many atoms of potassium-40, the emitted β^- particles can collide with other potassium-40 atoms, converting it to an argon-40 atom:

$$_{-1}^{0}e^- + {}_{19}^{40}K \rightarrow {}_{18}^{40}Ar + energy$$

6.5.2 Decay Rates and Half-Lives

As scientists have studied the radioactive decay of various elements in modern laboratories, they have discovered that radioactive materials decay at very predictable rates, at least in the present. The decay of any particular parent isotope into a daughter isotope is basically random, and it is impossible to know when any one atom will decay. However, if there are many atoms, there will always be a certain percentage that will decay over time. This is called the *rate of radioactive decay*.

The rate values are rather cumbersome to use, so scientists often convert these into a different value, called a **half-life**. The half-life is the amount of time it takes for one half of a group of parent isotopes to decay into their final, stable daughter isotopes. Figure 6.21 illustrates the concept. After each half-life, half of the parent that existed at the time decayed into daughter isotopes. So after one half-life, there is a 50% of parent and 50% of daughter. After a second half-life, half of the 50% parent decayed into daughters, resulting in 25% parent and 75% daughters, and so on.

The Greek letter lambda (λ) is the symbol for half-life. Table 6.1 lists the half-life of several important radioactive parents used in geology. The first four elements listed in table 6.1 are used for dating geological objects, such as igneous and metamorphic rocks (but not sedimentary rocks). The last element, carbon-14 (or ^{14}C) is used to date recent materials from once-living organisms. We'll return to ^{14}C dating and its impact on the age of "ancient" materials below.

When geologists attempt to numerically date a rock or mineral, they measure the ratio of parent and daughter isotopes in the sample, and plot it along a decay curve (figure 6.22). This will tell the geologist how many half-lives have transpired, and this number is multiplied by the half-life value (λ) for the parent isotope (Table 6.1):

$$Age = (\# \text{ of half-lives}) \times \lambda$$

So if a geologist measures a ratio of uranium-238 to lead-206 of 75% to 25%, then this spot on figure 6.22 represents approximately 0.5 half-lives. Thus the age of the sample is:

$$Age = 0.5 \times 4.6 \text{ billion years}$$
$$Age = 2.3 \text{ billion years}$$

So what kind of materials can be dated using radioisotopes? From table 6.1, the first five are used to

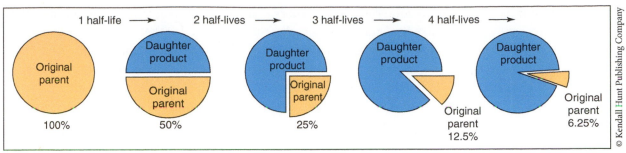

FIGURE 6.21. The change in percent of parent to daughter isotopes over four half-lives.

TABLE 6.1.
Common radioactive elements used in geology.

Parent isotope	Daughter isotope	Half life (λ)
Uranium-235	Lead-207	713 million years
Uranium-238	Lead-206	4.6 billion years
Thorium-232	Lead-308	14.1 billion years
Potassium-40	Argon-40	1.3 billion years
Rubidium-87	Strontium-87	48.8 billion years
Carbon-14	Nitrogen-14	5,730 years

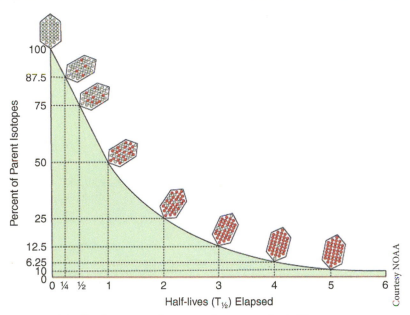

FIGURE 6.22. The decay curve of any parent isotope over five half-lives.

date igneous and metamorphic rocks (we'll consider carbon-14 dating in the next section). These rocks can be dated by radioisotopic methods because the crystals either cool from an originally molten state (igneous) or may be sufficiently heated or altered to remove previous daughter isotopes from the minerals (metamorphic). Sedimentary rocks are not dated by radioisotopes because a) the minerals of the sediment are older than the sedimentary rock (rule of inclusions), and b) sedimentary systems are very open to the loss of parent and/or daughter isotopes. As a result, attempting to date a sedimentary rock using radioisotopes is considered too prone to contamination to provide accurate dates.

So how can geologists say that the dinosaurs went extinct 66 million years ago, when the dinosaur fossils are found in sedimentary rocks? The answer comes from dating various igneous and metamorphic rocks that are in relative dating relationships to the sedimentary rocks. In the case of the dinosaurs, a layer of impact debris is found above the highest dinosaur-bearing rocks in places like Montana, North Dakota, and New Mexico. By dating the metamorphosed minerals of the impact, geologists have a date for the end of the dinosaurs, which according to superposition must be older than the debris.

6.6 YOUNG-EARTH STUDIES CHALLENGE OLD-EARTH DATES

Radioactive decay poses a challenge to those who argue that the Earth is only thousands of years old, rather than billions of years old. How do we handle this? The first is to evaluate the requirements and assumptions of radioactive dating methods, to see if any of them may be invalid. In order for a date to be assigned, three requirements must be met:

1) The number of parent and daughter isotopes has only changed by radioactive decay (that is, there has been no contamination).
2) The amount of daughter isotopes at the beginning is either zero or some amount that can be accurately determined (we must know initial conditions).
3) The rate of decay has been constant over time (uniformity of rate).

If all three requirements are satisfied, and the collection and analysis of the rock/mineral has been properly done in the field and the lab, then the date is considered accurate.

While the first two requirements may have occasional instances that are problematic, they are often satisfied. Modern equipment and collection can determine accurately the number of parent and daughter isotopes, and chemical bonding rules allow certain minerals to satisfy the requirement of known initial conditions. It is the third assumption that may be problematic. While the rate of half-lives appears stable today, there is fascinating evidence that at least some of the rates were different in the past, and that elements that should have long since decayed are still in places that, if the Earth is ancient, should not be there.

When uranium-238 decays to lead-206, in the process it releases 8 α particles (figure 6.23). Once outside the nucleus, the α particles attract two electrons to become a helium atom. Helium is a noble gas, which means it does not bond with other atoms. It also quickly escapes the minerals and rocks around it because it is extremely small. A group of young-Eaçrth creation scientists decided to look into radioactive decay. Called the **RATE team** (Radioisotopes and

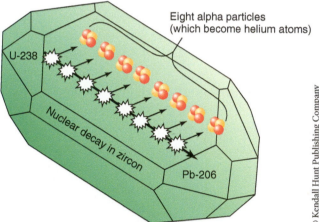

Eight alpha particles (which become helium atoms)

U-238

Nuclear decay in zircon

Pb-206

© Kendall Hunt Publishing Company

FIGURE 6.23. The RATE team determined that the escape of helium from zircon crystals could be used to date the mineral, and check if radioactive decay rates have changed over time. a) Uranium-238 decay produces 8 helium atoms before reaching the stable lead-206 isotope; b) Helium escapes the crystal at a rate determined by crystal size and temperature.

the Age of The Earth), one project looked at helium's "escape rate" from mineral crystals, and compare an age from this escape rate to the typical age derived from uranium-lead dating methods.

If we know that for every one atom of lead-206 produced by decay of uranium-238 there are also eight helium atoms, then by counting the number of lead-206 atoms in a crystal, we also know the number of helium atoms produced (8 times the number of lead-206 atoms). By measuring how fast the crystal can lose helium, the RATE team figured out a second way to date the rock, one based on the helium escape rate which can independently assess if the rate of decay has changed (figure 6.23).

Based on the amount of uranium and lead and the assumption of unchanged decay rates over time, a typical geologist would conclude that the crystals studied by RATE were 1.5 billion years old. In contrast, the RATE team found that the helium escape date was only 6,000 +/- 2,000 years! This date is compatible with a Biblically-based date for Noah's Flood. These data point to a time (likely during Noah's Flood and also Creation Week) when rates of radioactive decay were faster than they are today, because over 1.5 billion years' worth of radioactive decay actually occurred only a few thousand years ago. Maybe these rocks are not as old as most geologists believe!

Another study by the RATE team involved looking at carbon-14. Note that carbon-14 has a very short half-life of 5,730 years, far shorter than all of the other geologically important parent isotopes (table 6.1). Carbon-14 forms constantly in the upper atmosphere, as cosmic radiation converts atoms of nitrogen-14 into carbon-14. Once formed, the carbon-14 combines with oxygen to form carbon dioxide ($^{14}CO_2$). That $^{14}CO_2$ is then absorbed by plants, and used to make plant tissues, sugars, etc. By eating plants, all other organisms also get carbon-14 into their bodies (see figure 6.24). Currently, the amount of carbon-14 produced in the atmosphere is balanced nicely by

the decay rate, so all organisms maintain a small and constant amount of carbon-14 in their bodies. The daily loss of carbon-14 through decay is balanced by new carbon-14 taken in from food. Because of this dynamic connection with the atmosphere, *only organic remains* can be dated using carbon-14. But when an organism dies, it no longer replaces the decaying carbon-14 in its body with new carbon-14 from food. So when an organism dies, the ratio of carbon-14 to carbon-12 begins to change at the half-life rate. Geologists then use the ratios of carbon-14 to carbon-12 to determine the age of a sample, such as cloth, charcoal, or bone.

Since carbon-14 decays so quickly, it cannot be used to date objects older than 80,000 years, even if we grant all three radioactive dating requirements. The RATE team had a different hypothesis, though: *If most of the rocks of the geological column were formed by the Flood only a few thousand years ago (rather than over many millions of years), then there should still be carbon-14 in fossils and other organic remains buried in sedimentary deposits.*

The RATE team tested this hypothesis in two ways. First, using 10 different coal deposits from different layers of rock in different locations around North America, they discovered that all of the coal looked "young." That is, the coal seams all gave ages in thousands of years, even though the various rocks they were in are thought to be 30 million to 300 million years old by old-Earth geologists.

What these data showed is that the coal seams from many different locations all have carbon-14 in them, which they should not *IF* they were really millions of years old. Instead, the young-Earth hypothesis of carbon-14 in "ancient" rocks is confirmed (figure 6.25), and all of the coal appears to be only thousands of years old.

In addition to carbon-14 discovered in multiple coal seams, carbon-14 has been found in all other "fossil" fuels (natural gas, oil, etc.), as well as many types of fossils (including sea shells, petrified wood, and vertebrates). And these are all coming

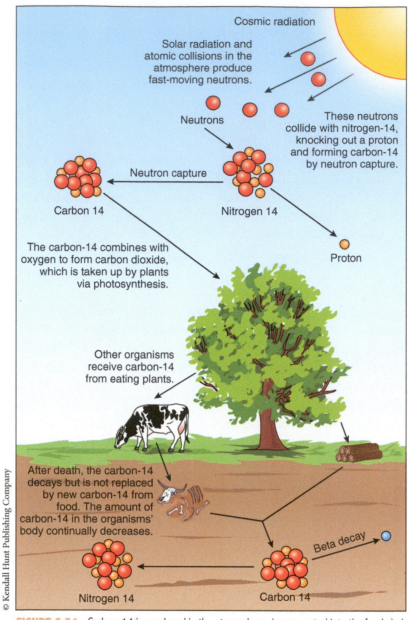

FIGURE 6.24. Carbon-14 is produced in the atmosphere, incorporated into the food chain, and is used to date material from once-living organisms.

from deposits thought to be many millions of years old, yet each with carbon-14 dates giving us ages of only thousands of years. The RATE team even discovered carbon-14 in diamonds thought to be nearly 2 *billion* years old! If these diamonds were really that old, there would be no measurable carbon-14. In other cases, using a variety of radioactive dates on *the same rock* yield huge variations in the calculated age of that rock (table 6.2). This pattern may be telling us that the old-Earth assumptions for dating rocks have serious problems. Moreover, there are additional evidences of a young Earth noted in Box 6.2. With more work, young-Earth creationists will continue to make amazing discoveries that can point us towards our great Creator!

(a)   (b)

FIGURE 6.25.

TABLE 6.2
Radioactive dates for various Grand Canyon rocks using multiple dating methods (See Snelling 2014 for details)

Rock Unit	# of samples (# of methods)	Date obtained (Ma=millions of years)	Accepted age in (Ma=millions of years)
Uinkaret lavas	13 (4)	0.01-2,600 Ma	1-4 Ma
Cardenas basalt	13 (4)	715-1,100 Ma	1,100 Ma
Diabase sills	55 (5)	656-1,375	1,070
Amphibolite	77 (4)	405-1,883	1,800

BOX 6.2   ADDITIONAL GEOLOGICAL EVIDENCES OF A YOUNG EARTH

A number of geological indicators point to a young Earth, rather than one 4.55 billion years old. See the October 2012 issue of *Answers Magazine* for more details on these and several other issues, or more detailed treatment in Snelling (2014):

Process	Maximum Age
Saltiness of the oceans	42 million years
Seafloor sedimentation rates	12 million years
Preseved biomolecules in fossils (including proteins and DNA)	10 million years
Decay of Earth's magnetic field	20,000 years

6.1 A History of Earth History

- A young-Earth view of Earth history was the only view held throughout Christianity from the Early Church period through the Reformation.
- Early geologists held to young-Earth views, but by the late 1700s, old-Earth views began to take hold with catastrophism. Attempts to harmonize the Bible with old-Earth views resulted in Day Age and Gap Theory, which were developed in the early1800s.
- Uniformitarianism is summed up as "the present is the key to the past," and has dominated geological thinking since the middle 1800s. Its modern formulation is called actualism.

6.2 Rules for Reading Rocks

- Relative dating is a collection of methods used to determining the order and sequence of events in geological history.
- The laws of original horizontality, superposition, and lateral continuity were first developed by Nicolaus Steno, and are fundamental to relative dating. Other important rules are Walther's Law, inclusions, and cross-cutting,
- Unconformities (including disconformities, angular unconformities, and nonconformities) indicate breaks in time in the rock record.

6.3 Fossils

- Paleontologists study fossils, which are the remains of once-living organisms in the rock record.

6.4 Correlation and the Geological Column

- Rocks can be linked to each other over distance or time using the tools of correlation.
- Correlating rocks using fossils led to the concept of faunal succession. Applying this and other relative dating principles resulted in the formulation of the geological column (an idealized diagram of rock relationships and relative ages).

6.5 Rocks as Clocks

- The radioactive decay of parent isotopes into daughter isotopes is used by geologists to determine the numerical ages of rocks and minerals. Alpha decay, beta decay, and nuclear fission are the three most important types of radioactive decay in geology.
- A half-life (λ) is the time it takes for one half of a radioactive sample to decay into stable daughter isotopes. Half-lives are used to calculate the ages of rocks and minerals.
- Igneous and metamorphic rocks may be dated by radioactive methods. The ages of sedimentary rocks and fossils are determined by their relative position to the dated igneous or metamorphic rocks.

6.6 Young-Earth Studies Challenge Old-Earth Dates

- Radioactive dating provides a difficult challenge to young-Earth creationism.
- The RATE team addressed a number of issues regarding radioactive dating.
- Carbon-14 is found throughout the rock and fossil record in materials that are supposed to be much older (on an old-Earth timescale) than carbon-14 can survive. The discovery of carbon-14 in diamonds and coal points to a young Earth and a global flood at the time of Noah.

KEY TERMS

Actualism

Alpha decay

Angular unconformity

Beta decay

Body fossils

Catastrophism

Correlation

Daughter isotope

Day-Age view

Disconformity

Electron capture

Fossil

Gap Theory

Geological column

Half-life

Law of lateral continuity

Law of original horizontality

Law of lateral continuity

Nonconformity

Nuclear fission

Paleontologist

Parent isotope

Principle of cross-cutting relations

Principle of faunal succession

Radioactive decay

RATE team

Relative dating

Rule of inclusions

Sedimentary structures

Strata

Unconformity

Uniformitarianism

Walther's law

1. What three rules for relative dating were discovered by Nicolaus Steno?
2. Using the rules for relative dating, determine the relative ages (oldest to youngest) for the diagram provided at the bottom of this page. Units with dashes, dots, and bricks are sedimentary rocks, "v" and "/" patterns are igneous rocks, and wavy lines are metamorphic rocks.
3. Describe in your own words the difference between physical and temporal correlation.
4. If a parent isotope with an atomic mass of 232 and atomic number of 90 decayed by alpha decay, what would be the atomic mass and number of the daughter isotope? What would be the atomic mass and number of the daughter isotope if the parent isotope underwent beta decay instead?
5. Describe how carbon-14 is formed in the atmosphere and then enters the biosphere.
6. Why does the discovery of carbon-14 in coal pose a challenge to an ancient Earth?
7. A geologist samples a mineral and finds that there is a ratio of 10% parent and 90% daughter. Assuming the system has remained closed, how many half-lives have transpired? If the half-life of the parent is 14 million years, how old is the mineral?

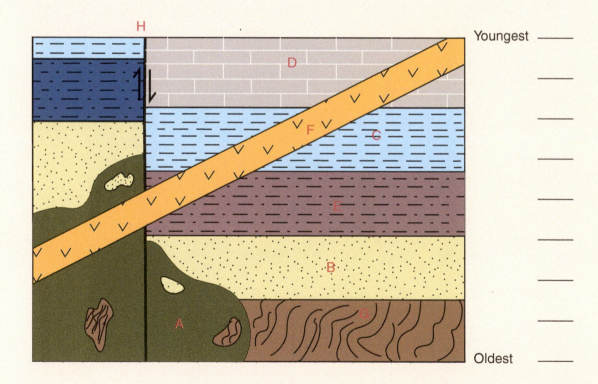

FURTHER READING

Cutler, Alan. *Seashell on the Mountaintop*. Plume Publishers. 2003.

DeYoung, Don. *Thousands . . . Not Billions*. Master Books. 2005.

McDougall, Douglas. *Nature's Clocks*. University of California Press. 2009.

Snelling, Andrew. *Earth's Catastrophic Past*, volumes I and II. Master Books. 2014.

REFLECT ON SCRIPTURE

Writing to the church in Jerusalem, Peter states "But do not forget this one thing, dear friends: With the Lord a day is like a thousand years, and a thousand years are like a day" (2 Peter 3:8). This verse has frequently been used to argue that long periods of time are allowed for the days of creation described in Genesis 1:1-2:3. Read through both II Peter 3 and the Genesis 1:1-2:3, and consider the following questions regarding the use of "day": Who/what is the subject of *day*, and are they the same? Is the timeframe past, present, or future for each case? In your opinion, does II Peter 3:8 address the timing of creation? Does Genesis 1:1-2:3?

CHAPTER 7

EARTH'S GEOLOGIC HISTORY: TWO CONTRASTING VIEWS

OUTLINE:

Catmando/shutterstock.com

An array of wonderful life forms, such as the stegosaurid dinosaur *Miragaia* and the flying pterosaur *Dorygnathus*, once filled the Earth. Earth's geological history can be understood from combining the evidence of nature with a proper reading of the Bible's treatment of creation, the Flood, and other events.

7.1 A UNIQUE HISTORY

Few topics in science are as quick to cause controversy as debates over the age of Earth or biological evolution. The topic makes headlines from court cases to school board decisions and is the subject of recurring polls and endless debate on the internet. Yet if we are to understand these issues so that we can speak meaningfully about them, we must learn *both* views fairly and accurately. Make no mistake: there is only one history to our world. Our planet is either a few thousands of years old or it is 4.6 billion years old. Perhaps both are wrong and some other model is correct, but at present these are the two main views on Earth history. As figure 7.1 illustrates, these models are not complementary; they are exclusive of each other because they seek to explain the same rocks using very different time scales and mechanisms.

All too frequently in creation/evolution discussions, the various authors spend more time tearing down and poking holes in each other's theories and models. To be honest, this has been the dominant pattern for many years in young-Earth creationism, and is less common (though still present) in naturalistic evolution. This is a shame, since this "deconstructive" approach can never produce a fully Biblical model of Earth history, and often results in hard feelings between people. Instead, here we present brief but accurate summaries of each model, and allow each to present its case without comment or rebuttal. While we advocate a young-Earth view, we want you to hear the evidence as it would be presented by an informed advocate for each side.

In both cases, we will present a history that attempts to explain the geological record as depicted in figure 7.1. This is the **geological column**, an idealized image representing the vertical (stratigraphic) relationships of rock units around the world. The geological column was formulated during the early 1800s using the relative dating principles described in chapter 6 (see section 6.4). More specifically,

the geological column was built on the basis of *patterns of fossils in stacked rock assemblages* and the comparison of fossil patterns among many different areas. The multi-million year ages on the left side of figure 7.1 were formalized in the 1900s through today, after radioactivity was discovered and applied to rocks via radioisotopic dating. The many unusual

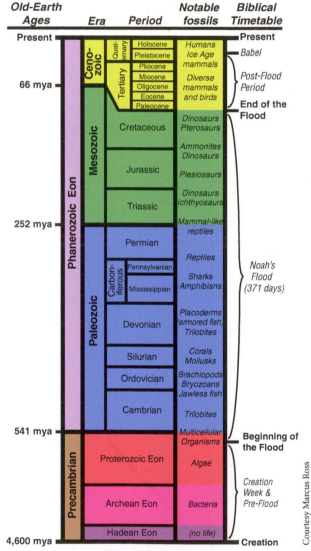

FIGURE 7.1. The geological column (center) with old-Earth and young-Earth timelines on either side.

names (such as Cambrian, Devonian, and Eocene) represent rocks with certain types of fossils or assigned certain radioisotope ages.

All old-Earth advocates (old-Earth creation, theistic evolution, and naturalistic evolution) believe that the geological column is an accurate, though idealized, representation of Earth's geology, and we believe that young-Earth creationists should also be prepared to accept its relative-dating representation as accurate (though disagreeing on the multi-billion year ages assigned to it). While there is no one location on the planet to see the column in its entirety, the *patterns* of the column are expressed in rock exposure after rock exposure around the world, and some locations display large portions of the column, lending evidence to its overall accuracy. As such, our discussions below of both naturalist evolution and young-Earth creation make use of the geological column, but provide alternatives to the timeline and mechanisms of its formation.

7.2 NATURALISTIC EVOLUTIONARY VIEW OF EARTH HISTORY

7.2.1 Early Planet Formation Through the Proterozoic

The **Precambrian** is the informal name given to the three eons prior to the Phanerozoic eon: the *Hadean*, *Archean*, and *Proterozoic* (see figure 7.1). These three eons represent the periods of Earth's initial formation, the development of oceans and continents, and the origin and evolution of early life.

Hadean Eon. This period begins at 4.6 billion years ago with the formation of Earth as a planetary body. The name Hadean derives from Hades (Greek god of the underworld), in reference to the hellish conditions of eruptions, lava oceans, and frequent impacts from asteroids and meteors (figure 7.2). At this early phase, the Earth undergoes a period of **differentiation**, where molten iron and nickel are pulled towards the Earth's center to form its core, leaving more silica-rich materials to form the mantle and Earth's surface (which did not yet have continents). Volcanic eruptions also released many gasses to produce the early Earth's atmosphere. These gasses were primarily water, ammonia, methane, and a host of others. Unlike today's atmosphere, the Hadean was low in oxygen and inhospitable to life, and many areas may have been covered by a veneer of acidic water.

Shortly after or during differentiation, Earth was struck by another early-forming planet about the size of Mars. This collision vaporized much of both planets, and blew apart a large section of Earth's mantle, which condensed under gravity to become the moon. This was followed by the **late heavy bombardment**, a period of intense meteor showers that destroyed most of the early surface rocks. The evidence for the bombardment comes primarily from lunar rock samples, and explains why no rocks on Earth have been radioactively dated past 4 billion years.

© MichaelTaylor3d/Shutterstock, Inc.

FIGURE 7.2. Earth as it might have appeared during the Hadean.

Archean Eon. Most of the oldest rocks on earth are about 4 billion years old, and this marks the beginning of the Archean Eon. By this time, the rocks of Earth's surface had cooled enough after the late heavy bombardment that they retain their radioactive daughter products and produce usable dates. Most of the geologic activity at this time continues to be volcanic, and the Earth's surface was only beginning to develop continental crust in isolated regions. By the mid- to late Archean eon, continents had formed, and plate tectonic movements had likely begun.

Oceans formed 3.5 billion years ago or earlier, which allowed life to evolve from previously non-living chemical interactions, though how this came about is uncertain. Various chemical indicators point to life's origin at 3.5 billion years ago, or even to 3.7 billion. Life most likely arose near underwater hydrothermal vents where high temperatures, low-oxygen conditions, and nutrients fostered life's origin. At about 3.5 billion years ago, the first fossil evidence of single-celled bacteria and/or archaea is seen in the rocks, as well as the presence of mounded layers of microbes and sediment called **stromatolites** (figure 7.3).

Proterozoic Eon. The Proterozoic Eon began 2.5 billion years ago and continued until the beginning of the Phanerozoic Eon at 541 million years ago. This is a long and transitional period of time for Earth and its living inhabitants. Continued volcanic degassing contributed large amounts of water vapor and other gasses to the atmosphere, whose composition was low in oxygen at the start of the Proterozoic. This is evidenced by many **banded iron formations** (figure 7.4), which are layered and iron-rich sedimentary deposits. However, over time oxygen levels increase, and the ozone layer develops in the stratosphere, though at a fraction of today's levels.

Twice during the middle- to late Proterozoic the Earth became completely glaciated, with ice sheets covering the planet from pole to pole, or nearly so. This period is frequently referred to as **Snowball Earth**, since the earth would look like an ice-covered ball from space. These conditions formed during the breakup of the first major supercontinent, called **Rodinia**, which existed from about 1.1 billion years ago until 750 million years ago.

FIGURE 7.3. Modern stromatolites from Shark Bay, Australia. Inset: possible stromatolite layers from the 3.5 billion year old Pilbara chert, Australia.

CHAPTER 7: Earth's Geologic History: Two Contrasting Views

FIGURE 7.4. Banded iron formations may indicate low-oxygen conditions where iron did not rust. The red layers are jasper and the black are magnetite.

FIGURE 7.5. The fossils of the Ediacaran Period are often preserved as impressions on sandstones, such as this *Dickensonia*. Many may have been solitary, filter-feeding animals, but there are still many questions about what they were and how they are related to living organisms.

The best early fossil evidence for eukaryotes (organisms with a cell nucleus) dates from 1.65 billion years ago, and possible remains are known from 2.1 billion year old rocks. Starting about 600 million years ago and after the end of Snowball Earth, the first complex, multicellular organisms appear in the fossil record. The **Ediacaran fauna** consists of a variety of unusual organisms often believed to be, or be relatives of, animals. Yet many of these strange creatures were two-dimensional, frond-like organisms which may have absorbed nutrients directly from seawater (figure 7.5). Also found among the Ediacaran fauna are the first true animal fossils, including sponges, worm tracks, and fossils that may be early mollusks and arthropods.

7.2.2 The Paleozoic Era

CONTINENTS AND CLIMATE The Paleozoic Era is the first of three eras that are collectively known as the Phanerozoic Eon (figure 7.1). Phanerozoic means "visible animals," since the sedimentary rocks of this eon are known to produce the numerous, varied, and sometimes familiar fossils, such as trilobites, ammonites, bivalves, dinosaurs, and mammals. The Paleozoic Era is divided into six periods

(seven in North America, where the Carboniferous is divided into the Mississippian and Pennsylvanian periods).

Following the breakup of Rodinia, the early Paleozoic saw the many components of future North America, Europe, and Asia located at the equator, while **Gondwana** (composed of South America, Africa, India, Antarctica, Australia, and some northern continental fragments) lay mostly in the southern hemisphere with significant portions still covered by glaciers. Active subduction along the eastern margin of North America (which was actually facing south at the time) formed a volcanic island arc that eventually merged with North America, forming early components of the proto-Appalachian mountains during the Taconic orogeny (a mountain-building event). The largest mountain-building events occurred as the northern continents assembled into **Laurasia** and collided with Gondwana to form **Pangaea** (figure 7.6). The collision formed the majority of the Appalachians, the Atlas Mountains (Morocco), the Scottish highlands, and the Scandanavian Mountains during the Alleghehanian orogeny.

Temperatures rose and sea levels were high during the early Paleozoic as ocean ridges produced

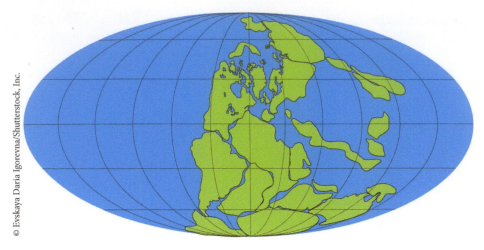

© Evskaya Daria Igorevna/Shutterstock, Inc.

FIGURE 7.6. The supercontinent Pangaea as configured during the Permian Period.

significant amounts of new, buoyant ocean crust and continental ice sheets melted. Large **epicontinental seas** covered continental interiors and margins with vast regions of shallow oceanic waters, often filled with limestone and abundant fossil life (figure 7.7). These conditions reversed in the middle and late Ordovician (coinciding with a significant extinction), but sea levels and temperatures rose once again during the Silurian and Devonian.

The later Paleozoic saw a return to cooler global temperatures, particularly at the poles. Nearer to the equator, lush tropical and temperate swamps created enormous quantities of peat that later became important coal beds. Indeed, the Carboniferous gets its names from the many coal beds found in Europe, and the equivalent Mississippian and Pennsylvanian periods

of North America are also rich in coal. As Pangaea assembled in the Permian, the geography of that supercontinent kept much of the interior land far from the oceans, destroying the Carboniferous swamps and resulting in vast desert regions now recognized by iron-rich **red beds** found on many continents.

LIFE IN THE PALEOZOIC As mentioned above, the Paleozoic Era is the first of three eras that are collectively known as the Phanerozoic Eon. Crossing from the Proterozoic Eon into the Paleozoic Era we are met with the **Cambrian explosion** (figure 7.8), a geologically sudden appearance of a wide variety of very different types of fossils than what is seen in the rocks below. Diverse and complex arthropods, mollusks, echinoderms, and primitive chordates (jawless fish)

© Alena Hovorkova/Shutterstock, Inc.

FIGURE 7.7. Shallow epicontinental seas covered over large regions of continents during the early Paleozoic and hosted abundant, if unusual, marine life.

along with unusual forms unlike any seen alive today all make their first appearance in the fossil record, and they are very different from the earlier Ediacarans, which largely go extinct before the Cambrian explosion.

After the Cambrian explosion, marine life diversifies and evolves with the origin of jawed fish and a wide diversity of invertebrate fossil groups. By the end of the Devonian, a variety of freshwater fish have evolved into the earliest amphibians (figure 7.9). Other important early life forms on land include unusual plants and early scorpions, centipedes, and insects during the Carboniferous Period. These rocks are believed to have formed in semitropical bogs and marshes, where low-oxygen sedimentary conditions produced peat that later formed extensive coal beds.

By upper Carboniferous time (the Pennsylvanian Period in North America), amphibians were the dominant terrestrial vertebrates, and the very first reptiles had evolved. The reptiles quickly expanded during the Permian, as their ability to live and lay eggs away from the water opened up many new ecological niches,

especially as Pangaea formed and the Carboniferous swamps drained to form vast large arid regions. Carboniferous and Permian insects reached giant proportions possibly due to higher oxygen concentrations in the atmosphere. For example, a dragonfly-like insect named *Meganeuropsis* (figure 7.10) had a wingspan of 71 cm (28 in), and *Arthropleura* was a millipede genus that grew to 2.8 m long (9 ft)!

At the close of the Permian Period, the largest extinction in Earth history destroyed the vast majority of life. About 90% of marine species go extinct at this time, including many types of corals, brachiopods, and the last of the trilobites. On land, 70% of terrestrial species go extinct, including large numbers of amphibians and reptiles leaving only a few groups alive that will be the ancestors of dinosaurs, pterosaurs, lizards, birds, and mammals. The cause of the extinction is a matter of considerable interest and debate among geologists and paleontologists. Currently, many experts believe that a significant portion of the deep ocean was anoxic, and that a

Courtesy Marcus Ross

FIGURE 7.8. A variety of Cambrian explosion fossils including worms, arthropods, brachiopods, and others from the Burgess Shale of British Columbia, Canada.

CHAPTER 7: Earth's Geologic History: Two Contrasting Views

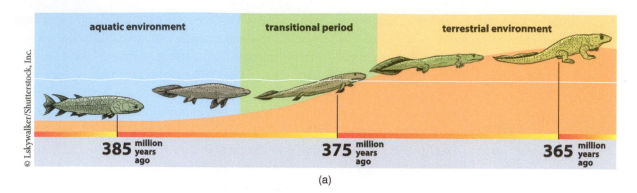

(a)

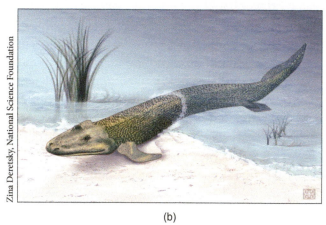

(b)

FIGURE 7.9. The evolution of fish to amphibians (a), and the fossil fish *Tiktaalik rosea*, which has a head and limbs similar to early amphibians.

sudden rise of these waters up to the surface killed off many species, leading to a broad ecological collapse. Other factors, including massive volcanic flows in Siberia and climate changes may also have played major roles in this extinction event.

FIGURE 7.10. The largest flying insect of all time was the Permian giant, *Meganeuropsis, measuring over 2 feet wide.*

7.2.3 The Mesozoic Era

CONTINENTS AND CLIMATE The Mesozoic Era is the time of "middle animals," and is divided into three periods: the Triassic, Jurassic, and Cretaceous. The Mesozoic begins with Pangaea fully assembled, and the initial separation begins in the late Triassic as Laurasia splits from Gondwana. Laurasia was the northern portion of Pangaea and consists of North America, Europe, and Asia while Gondwana is composed of South America, Africa, Australia, Antarctica, and India. Recall that much of Wegener's arguments for continental drift (chapter 4) centered on shared features and fossils for the continents that once were united as Gondwana.

In the late Jurassic, the early Atlantic Ocean develops, separating North America from Europe and Asia and ending the Laurasian continent. Gondwana persists longer, as the southern Atlantic opened much

FIGURE 7.11. The white to yellow layers of bentonite are deposits of volcanic ash falls in the rocks of Capitol Reef National Park in Utah.

later during the Cretaceous. Even as the southern Atlantic separates South America from Africa, the components of Gondwana remain connected, such that South America connected to Antarctica, which then connected to Africa and Australia in a chain-like series. India was separated early and began its journey northward towards Asia as an isolated continent.

Extremely long subduction zones occupied the western margins of North and South America and southern Asia during the Mesozoic. These subduction zones helped to produce enormous amounts of volcanism during this time. Large emplacements of granites and other intrusive igneous rocks formed within the continents, and the violent eruptions produced numerous ash layers called **bentonites** (figure 7.11) that are preserved within the sedimentary record. These granites and bentonites are frequently used for radioactive dating, providing numerical dates that can be connected to the relative dating methods of stratigraphy.

Throughout the Mesozoic, the climate became warmer and wetter, with increasingly high sea levels. The sea level was primarily influenced by the increase in mid-ocean ridge volcanism as the Atlantic Ocean formed. The water pushed upward as the new ocean crust advanced inland into the continents, and in the Cretaceous these produced large epicontinental seas that once again covered vast continental regions with ocean water. Only in the late Cretaceous do sea levels start to drop and the epicontinental seas begin draining off of the continents as the Rocky Mountains began to form during the onset of the *Laramide orogeny*.

During the Jurassic and especially in the Cretaceous, global temperatures were much warmer than today, with temperate regions that extended far towards the poles. This is evidenced by dinosaur communities in both northern Alaska and Australia, which was closer to the South Pole during the Cretaceous than it is today. High carbon dioxide levels from large amounts of volcanism contributed to this greenhouse world.

LIFE IN THE MESOZOIC Known as the "age of reptiles," Mesozoic animal communities are dominated by a variety of reptile groups, especially the **archosaurs**: a group that survived the end-Permian extinction and includes crocodilians, dinosaurs, and pterosaurs (figure 7.12).

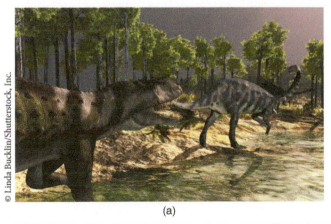

(a)

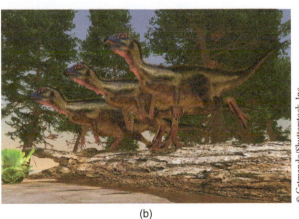

(b)

FIGURE 7.12. Whether small or large, archosaurs were the dominant land animal group of the Mesozoic. (a) *Prestosuchus* (a crocodylian) threatens a *Massospondylus* (long-necked prosauropod dinosaur; (b) the 2.3 m (7.5 ft) plant-eating dinosaur *Hypsilophodon*.

FIGURE 7.13. *Archaeopteryx*, the first bird in the geologic record, has a mix of dinosaur and bird traits, including feathers, teeth, a tail, and clawed wings.

Several different types of dinosaurs first appear in the late Triassic, and later become the dominant terrestrial vertebrate group of the Mesozoic. They include many herbivorous and carnivorous groups, such as the long-necked sauropods, duck-billed hadrosaurs, horned ceratopsians, and many different and fearsome theropods. The skies were ruled by the flying pterosaurs, but beginning in the late Jurassic, the first birds (figure 7.13) evolved from a branch of the theropods known as maniraptorans (a group that includes the well-known

Velociraptor). Primitive mammals rather unlike those of today survived mostly underfoot of the ruling reptiles. Ferns and gymnosperm plants such as cycads and diverse pine-like trees dominate much of the Mesozoic flora, though a major radiation of angiosperms (flowering plants) occurred during the Cretaceous.

In the oceans, non-archosaur reptiles topped the food chain. Fish-shaped ichthyosaurs (figure 7.14) were common in the Triassic and Jurassic oceans but go extinct by the middle Cretaceous. The Jurassic and Cretaceous plesiosaurs all had four equal-sized flippers, and include both long- and short-necked groups that ate small and large prey, respectively. In the late Cretaceous, up to 18-m (59 feet) aquatic lizards called mosasaurs dominated the seas. Notable invertebrate groups of the Mesozoic include the coiled-shelled and squid-like ammonites, various bivalve mollusks, and many kinds of crustaceans.

Like the Paleozoic, the end of the Mesozoic is defined by a major extinction, known as the Cretaceous-Tertiary (or K-T) extinction (figure 7.15). At the end of the Cretaceous, a huge meteor or comet approximately 10 km (6 miles) wide struck the Yucatan Peninsula of Mexico, vaporizing that region and sending immense volumes of dust and debris into the atmosphere. Wildfires were rampant as the superheated rocks rained down over many parts of North America. Dinosaurs, pterosaurs, and many marine reptiles and

FIGURE 7.14. Jurassic seas: a school of *Ichthyosaurus* try to escape the short-necked plesiosaur *Liopleurodon*.

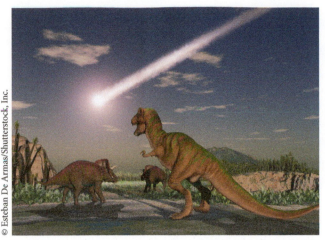

© Esteban De Armas/Shutterstock, Inc.

FIGURE 7.15. Dinosaurs and many other Mesozoic animals died in a major extinction at the end of the Cretaceous Period.

invertebrates died out as the photosynthesis-based food chains on land and water were decimated by years of ash-darkened skies. Survivors were frequently organisms based in river, lake, or deep-water environments that relied on *detritus* (dead organic materials) as the primary food source. Early modern mammals, birds, amphibians, crocodiles, shellfish, and crustaceans survived the extinction to become important components of the Cenozoic ecosystems.

7.2.4 The Cenozoic Era

Continents and Climate The Cenozoic ("recent life") is the last and shortest of the three eras of the Phanerozoic Eon, and it is divided into two periods consisting of seven epochs (see figure 7.1). During the Cenozoic the continents continued their migration towards our modern configuration. The Laramide orogeny continued across the K-T boundary, building the Rockies throughout the Paleocene and into the Eocene. North and South America continued migrating westward with extensive subduction zones at their westward margins, producing large volcanic mountain chains such as the Andes. In North America, this subduction zone was overrun by the continent and replaced by a transform boundary (the San Andreas Fault and many associated faults), resulting in subduction-based

volcanism only in the Cascade Mountains of the Pacific northwest.

On the other side of the globe, India began its collision with Asia to form the Himalayas at about the mid-Eocene, an orogeny that continues to this day. At about the same time, Africa begins a collision with Europe that produces the Alps. During the Pleistocene, the Isthmus of Panama formed, connecting North and South America for the first time since the separation of Laurentia from Gondwana during the Triassic. This profoundly affected ocean currents and climate by sealing off the equatorial connection between the Atlantic and Pacific Oceans, and allowed many animals to migrate between the two continents.

Following the K-T extinction, the world remained warm and wet through much of the Paleogene and Eocene stages. During the Oligocene, though, temperatures began to fall and produced the first sustained ice sheets in Antarctica since the Triassic Period. Temperatures continued to drop throughout the Miocene and Pliocene, resulting in widespread grasslands, particularly in continental interiors. Temperatures had dropped so low that during the Pleistocene, large continental ice sheets covered vast regions of North America, Europe, and Asia during the **Ice Age**. A series of advances and retreats of the glaciers points to a complex history, with the last large melt occurring about 11,000 years ago, just before the Holocene stage, which is defined at 10,000 years B.C.

Life in the Cenozoic The Cenozoic Era is characterized by the most familiar fauna of the three Phanerozoic eons. The loss of dinosaurs, pterosaurs, and marine reptiles at the K-T extinction paved the way for birds, mammals, modern crocodilians, and lizards to take their place as the dominant vertebrates of the Cenozoic. Combined with numerous flowering plants and pollenating insects, the Cenozoic ecosystems are much more familiar looking, even if the particular families, genera, and species are different than those of the modern world.

Without dinosaurs to dominate the land, mammals rapidly filed the vacant ecological space during

FIGURE 7.16. At 4.8 m tall (16 ft), the Oligocene *Paraceratherium* was the largest land mammal of all time.

the early Cenozoic, and their prominence throughout this era is why it is often referred to as the "Age of Mammals." Early modern mammals were still generally smaller than their modern counterparts, but during the Eocene the first appearances of hooved mammals, various carnivores, rodents, insectivores, and other forms had all taken place. This "Eocene explosion" saw the first appearance of many different modern and extinct mammal orders. As temperatures began to cool after a peak in the Eocene, many mammals grew in size (figure 7.16), ultimately reflected in extremely large forms of elephants, rhinoceroses, camels, horses, and even a 2 m-long giant beaver as the Ice Age commenced. Connections between North and South America in the late Pliocene, and also between North America and Asia during the Pleistocene Ice Age facilitated an extensive migration of animals among these continents.

Primate-like mammals are first recognized in the fossil record after the K-T extinction, and the earliest true primates to become the ancestors of modern lemurs, monkeys, and apes are known from the lower Eocene. The oldest ape fossils are found in late Oligocene deposits in Tanzania. About 4 million years ago in the middle Pliocene, the first upright-walking ape (*Australopithecus*) is found in eastern Africa. Members of our own genus (*Homo*) are known as early as 2.4 million years ago in the late Pliocene. From that point, a variety of human species, including *Homo erectus*, *H. heidelbergensis*, *H. neandertalensis*, the Denisovans, and *H. sapiens* evolved and spread from Africa across the globe during the Pliocene and Pleistocene (figure 7.17). All but our own species (*H. sapiens*) are now extinct, though there is genetic evidence of several groups (particularly the Neandertals and Denisovans) in modern human DNA.

7.3 YOUNG-EARTH CREATION VIEWS OF EARTH HISTORY

7.3.1 Creation Week

Genesis 1:1-2:3 describes the creation of the world as happening in an orderly fashion over a 6-day period, followed by a day of rest. The creation happens according to commands made by God ("and God said,

'Let there be . . .' "), and His creation responds immediately to fulfill its Maker's desires (". . . and it was so"). Table 7.1 summarizes the events of creation week:

Table 7.1 organizes the days of creation into parallel sets of three days each, followed by an unparalleled Day 7. Following the rules of Hebrew

TABLE 7.1.

Day of creation	Events	Day of creation	Events
Day 1	Creation of the heavens and the earth in unformed fashion. Creation of light and its separation from darkness	Day 4	Creation of sun, moon, and stars to serve as time keepers.
Day 2	Formation of atmosphere from separation of "waters above" from "waters below".	Day 5	Creation of marine animals and flying animals.
Day 3	Formation of land and oceans. Creation of plant life.	Day 6	Creation of land animals and humans (God's image-bearers).
Day 7 God rests from His work of creation			

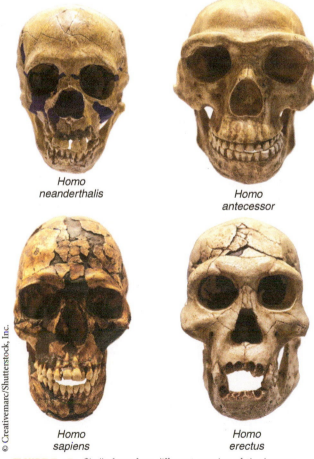

Homo
neanderthalis

Homo
antecessor

Homo
sapiens

Homo
erectus

© Creativemarc/Shutterstock, Inc.

FIGURE 7.17. Skulls from four different species of the human genus *Homo*.

grammar and literary style, these days can only consist of seven sequential days just as we experience them (24-hour periods of time). But the structure of the text does more than tell us about what God created each day. Genesis 1:1-2:3 is also a structured by the author to help Israel understand both God's sequence of creation (space is first created and partitioned, then filled with objects and living things in an orderly and logical fashion) as well as draw attention to the unparalleled day of rest, which is later used as a justification for the Sabbath day in the Ten Commandments (see Exodus 20:11). In creating the world, God purposefully chose a timeframe so that He could set an example for us. There are also many important theological points and claims made in these passages relating to God's uniqueness (there are no other gods but the LORD), His omnipotence (all-powerful nature), His sovereign ownership of all creation, and His focused creation of mankind as His image-bearers. The text is amazingly rich in both style and substance.

Each day of creation is important and necessary to the next and to understanding Earth history. Geologically, one of the most important days is Day 3, where God said "Let the waters under the heavens be gathered together into one place, and let the

FIGURE 7.18. During the Great Upheaval on Day 3 of Creation Week, God raised the land out of the ocean.

dry land appear" (Gen. 1:11). Creation geologists refer to this as the **Great Upheaval** period of creation (figure 7.18), where God raises the continents out of the water-covered world to separate and establish the land and the seas. Such immense energies from tectonics, earthquakes, and erosion likely produced many of the deep Hadean, Archaean and Proterozoic igneous, metamorphic, and sedimentary rocks

(which lack any evidence of animal fossils, since animals would not be created until days 5 and 6). Since the land and oceans are separated, young-Earth creationists maintain that the created world was divided into continental and oceanic crust, as well as having the crust separated from the mantle and core deeper below. Once the creation was declared "very good" on Day 6, it seems unlikely that many violent geological catastrophes such as volcanic eruptions, earthquakes, and tsunamis would exist, at least prior to the Fall. At the end of Creation Week, the Earth may have been structured something like Rodinia (figure 7.19).

Day 4 recounts the creation of the sun, moon, and stars, yet in a seemingly strange fashion. The sun and moon are not mentioned by those titles, but rather as "greater" and "lesser" lights, whose function is to govern the day/night cycle first established by the creation of light and its separation from darkness on Day 1 (this means there was some other source of light for Earth before the sun). Why not use "sun" and "moon"? The most likely reason is that the ancient Hebrew words for sun and moon are the same in the Canaanite language (much like Spanish and Italian share many words from their parent language, Latin), where "sun" and "moon" meant both the *objects* as

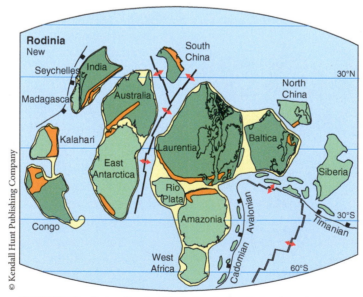

FIGURE 7.19. During Creation Week, the continents may have been arranged similar to this reconstruction of Rodinia.

well as the sun and moon *deities*. Lest the Israelites misunderstand and think that God created other gods to rule the day or night (and then worship them), God describes the sun and moon only by their *function* and relation to humans. When creating the stars, there are only two Hebrew words: "stars also." This is amazingly brief, given God's discussions with Abraham and others referring to the stars as innumerable!

Days 3, 5, and 6 document the creation of plants, aquatic life, flying animals, and the many types of terrestrial animals. Since at the end of Day 6 God had finished creating all that he would create (calling it "very good" and resting on Day 7), young-Earth creationists expect that all the different "kinds" of plants and animals that have ever lived were created and living at the same time, though not in all of the same areas (figure 7.20). What do we mean by "kinds," as the Bible describes them? The term does not mark a particular level of taxonomy (the study of naming and classifying organisms), however all creationists believe that the "kind" is not the same as *species*. Following a number of studies, it seems likely that the "kind" is closer to the taxonomic rank of *family* among vertebrates, at least. This means that all living and fossil members of the cat or dog "kinds" (biological families *Felidae* and *Canidae*) are variations of the original cat and dog "kinds" whose ancestry goes back to Creation Week. Further, we see

that in their initial created state, the many "kinds" of land-dwelling animals were intended to be herbivores, rather than carnivores (Gen. 1:28-30).

The largest focus in the text is on the creation of humans on Day 6. All of creation to this point has actually been made *for* humanity, not as something that can or should exist apart from us. When God says "Let us make man in our image, after our likeness" (Gen. 1:26), it is not that we take on the physical form of God (for He is spirit). Among the peoples of the ancient near east, the "image" of a king or ruler (such as the stone image of Ramesses the Great in Egypt) was an extension of that king's presence. In the same way, being made in God's image places humans as the extension of God on Earth. We are His representatives, which is why Adam and Eve were given dominion and authority over all of the creation which God had made.

7.3.2 From the Fall to Noah

Once the fruit of the Tree of the Knowledge of Good and Evil was eaten (Gen. 3:6-7), sin entered the world and was swiftly followed by God's judgment on all parties. Not only were Adam, Eve, and the serpent punished, but the ground itself was cursed (Gen. 3:17-19). The serpent's curse was "above all the livestock and all the beast of the field," indicating

FIGURE 7.20. By the end of Day 6, God had finished creating the world and all the kinds of living organisms.

that the curse had profound effects across the entire realm of creation, not just these three participants. After the Fall (figure 7.21), we find that sin pervades humanity quickly with murder and strife.

A set of genealogies carries the narrative quickly through the history between Adam and Noah in Genesis 5. These genealogies appear to be complete father-son pairs, and adding the ages from the text indicates over 1,600 years between the Fall and the Flood. By Noah's time, the world has fallen so far into sin that God decides to destroy humanity and animal life because "every intention of the thoughts of [man's] heart was only evil continually" (Gen. 6:5).

7.3.3 Noah's Flood

After the great upheaval, no event in history had such profound effects on the planet as Noah's Flood, which was essentially an undoing of the creation and a new start. The text of Genesis 6-9 focuses on Noah and his family before, during, and after the Flood, with the Flood account described in Genesis 7 and 8. More than some boat filled with barnyard animals, the ark was the largest wooden ship

FIGURE 7.22. Noah's ark was a huge wooden vessel constructed to save Noah's family and representatives of the land-dwelling animal "kinds" from the Flood's destruction.

in human history (figure 7.22), measuring somewhere between 135–155 m long (450–510 ft), 23–25 m wide (75–83 ft), and 14–15 m tall (45–50 ft). Chapter 4 introduced the idea of **catastrophic plate tectonics**, and the role that tectonics played as God's mechanism for causing the Flood. What follows here is a chronology of geological events as the Flood progressed.

INITIAL DESTRUCTION OF THE PRE-FLOOD WORLD
One of the most universal geological features is the **Great Unconformity** (figure 7.23). Recall from chapter 6 that an unconformity is a surface between rocks that indicates a period of erosion or non-deposition.

FIGURE 7.21. The expulsion of Adam and Eve from the Garden of Eden (by Gustave Dore).

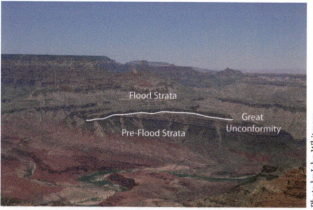

Flood Strata

Pre-Flood Strata

Great Unconformity

FIGURE 7.23. The Great Unconformity exposed in the Grand Canyon, Arizona. This is an angular unconformity (see chapter 6) between pre-Flood Precambrian sedimentary rocks and the flat-lying, early Flood Cambrian sedimentary rocks.

The Great Unconformity is an erosional unconformity between deep Proterozoic rocks and either uppermost Proterozoic or the lowest Phanerozoic (Cambrian) rocks. It is called the Great Unconformity because of its vast extent: in North America it can be seen from California to New York and from British Columbia to Georgia. It is also seen on every other continent on Earth, where it separates either crystalline rocks (igneous and metamorphic) or fossil-poor sedimentary rocks below the Great Unconformity from the fossil-rich sediments above. Young-Earth creationists believe that the Great Unconformity represents the initial destruction and erosion of the pre-Flood world, and the fossil-rich sediments above it are the first ecosystems deposited by the Flood waters.

SEQUENCES OF ROCKS AND FOSSILS Above the great unconformity in nearly every location studied is a pattern: sandstones topped by shales topped by limestones. Geologists call these regular pattern of stacked geological formations **sequences** (figure 7.24), and young-Earth creationists recognize them as representing different speeds, depths of water, and different sources of sediment and organisms from various times and places during the Flood. These sequences are often repeated many times over, and continual tectonic upheaval during the Flood created unending tsunamis, underwater mudslides, and other catastrophes that deposited immense volumes of rock onto the continents. Indeed, the fact that the continents have much *larger* thicknesses of ocean sediments on top of them than do the oceans themselves is one of the many good evidences for a global Flood.

Not only are the sequences found in varied geographic locations, many of the geological units themselves cover vast areas of a continent. Whether the rocks have the same names in different locations or not, a broad sheet of sandstone lies above the Great Unconformity across nearly the entire North American continent.

At the base of the first sequence lies the first fossil evidence of major animal life. Either Ediacaran forms or, more frequently, animals from the Cambrian

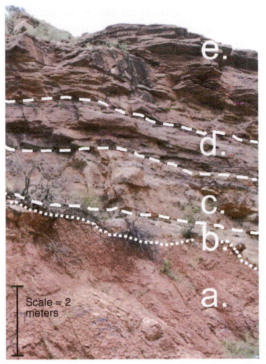

Courtesy Steve Austin

FIGURE 7.24. The Great Unconformity (b) separates the Pikes Peak Granite (a) from a sequence of Cambrian through Ordovician sandstone (c), shale (d), and limestone (e) in Manitou Springs, Colorado.

explosion are found in huge numbers in the rocks immediately above the Great Unconformity. This first occurrence in the fossil record resulted from the raging waters churning up the pre-Flood seafloor sediments and ecosystems and driving them up and onto the continents. The deposition of the Cambrian explosion marks the first of many continental-scale and ecosystem-wide destructions as the waters of the Flood rose to cover the whole Earth.

Higher in the geologic record, sandstones like the Jurassic dinosaur-bearing Morrison Formation extend from Utah to Saskatchewan, and the Cretaceous Pierre Shale extended from the Gulf of Mexico to the Arctic Ocean, dividing North America in half during the end period of the Flood. The mechanism of catastrophic plate tectonics helps to explain why so much oceanic material can be forced onto the continents, and how the many terrestrial ecosystems ended up washed into the same depositional areas.

There is also a trend in the distribution of fossils. As seen in figure 7.1, the dominant forms of life found lower in the geological column are marine animals, and especially deep/bottom-dwelling forms such as trilobites, brachiopods, bryozoans and mollusks. The Cambrian explosion (figure 7.8) represents the burial of complete and complex ecosystems, rather than the first appearance or evolution of life. Higher up in the column, more mobile aquatic organisms such as sharks and bony fish are encountered, and still higher more terrestrial organisms including amphibian, reptiles, some mammals, and birds. Young-Earth creationists believe that the overall organization of fossils in the column reflects a combination of pre-Flood habitat zones (which may have been strongly separated by their distance from the ocean, elevation, or community composition) as well as the processes that destroyed, transported, and buried them during the Flood. Rather than a chaotic mix, the fossil record is very orderly, and any understanding of the Flood must account for this.

In this order are some very interesting patterns that make sense from a young-Earth creation perspective. One is that fossil trackways (called **ichnofossils**) of animals are frequently found below their bone and shell body fossils. This seems strange, because the tracks are made in soft mud and sand, and are easily destroyed under modern conditions. Naturalistic evolution predicts the opposite: hard shells and bones that fossilize more easily should be found below the delicate and temporary tracks. Once we think about the Flood, however, this strange pattern makes sense: animals run over the top of wet sediment as they try to escape the Flood, only to be caught and deposited in higher sediments. There was not much time between the two events.

MOVEMENT OF THE CONTINENTS AND OCEANS As discussed in chapter 4, catastrophic plate tectonics proposes a complete replacement of the pre-Flood ocean crust with new material. As this occurs, the position of the continents shifts rapidly, moving at rates of a few meters per second! Based on paleomagnetic data, at the time of the beginning of the Flood, the continents were probably in a configuration similar to Rodinia. As "the fountains of the great deep burst forth," the pre-Flood continent (or continents) were violently broken and began their "continental sprint."

As waters sliced across the continents forming the Great Unconformity and then piling on oceanic sediments on land, the continents later collided violently with each other to form Pangaea. The collision of Africa, North America, and Europe provided the energy to form an immense mountain belt across all three continents, now represented by the Atlas, Appalachian, and Scandanavian mountians. These mountains are united by the types and relative ages of the rocks that comprise them: early Paleozoic sedimentary rocks and deeper Proterozoic rocks.

Pangaea existed as a supercontinent only during the Flood, and even then for a brief period of time (perhaps a month or two while under the Flood waters). A new fissure opens up between the northern and southern portions of Pangaea (called Laurentia and Gondwana) while another then splits them along a north-south trending ridge. Once again, runaway subduction on the western margins of North and South America and the eastern margin of Asia pull the continents away from each other, and a massive outpouring of basalt forms new ocean basins, such as the Indian and the Atlantic. As this occurs any remaining terrestrial animals (such as the dinosaurs; figure 7.25) are buried in Mesozoic sediments on these continents (figure 7.26). We know that their deposition occurs after Pangaea's breakup because there are no Jurassic or higher rock units that are continuous between North America and Europe or South America and Africa, while identical and previously-deposited Paleozoic sediments are seen across the Atlantic.

As the Flood neared its end, several of the world's largest mountain ranges were formed.

Courtesy National Park Service

FIGURE 7.25. Many fossil sites are "bone bed" like this one at Dinosaur National Monument. The jumbled bones and partial skeletons are evidence of the world-wide flood at the time of Noah.

The Himalayas, Rocky, and Andes mountains are all made of younger material than mountains like the Appalachians and the Atlas (figure 7.27). They seemed to be formed as the last major push of catastrophic plate tectonics as the continents began to slow down and the floodwaters began to subside (see chapter 4 for more details). Large-scale draining of the continents empties huge amounts of sediment first towards the oceans, and later into localized depressions called **basins**.

© Byron W.Moore/Shutterstock, Inc.

FIGURE 7.26. Dinosaurs and other very mobile animals fled the advancing floodwaters, resulting in stratigraphically higher fossils than many shelly ocean creatures.

© Galyna Andrushko/Shutterstock, Inc.

FIGURE 7.27. The towering Himalaya Mountains were formed during the Flood by catastrophic plate tectonics. Cretaceous seashell fossils are found at Mt. Everest's peak, indicating that the mountain was formed after ocean sediments were deposited under water and was later lifted as India collided into Asia.

7.3.4 The Post-Flood to Today

Somewhere at or above the Cretaceous-Tertiary (K-T) boundary marks the end of the Flood. Precisely where is a point of active debate among young-Earth creationists, who often favor a position either at/near the K-T boundary or higher, such as the Eocene or right before the Ice Age. These positions each have their strengths and weaknesses, but it appears that when we consider a large variety of different types of evidence (including fossils, volumes of sediment, post-Flood erosional power, sediment restricted to local/regional basins, and the location of current landforms), a K-T boundary fits the most data (see Whitmore and Garner's paper in the Further Reading section at the end of the chapter). We will assume this boundary for the discussion below.

When the floodwaters finally receded from the Earth, Noah, his family, and the animals on the Ark exited into a new and strange world. The animals then went out ahead of people, who did not disperse

FIGURE 7.28. A variety of animals including the rhino *Teleoceras* (1, 3) and horse *Cormohipparion* (2) were buried in a post-Flood catastrophe at Ashfall Fossil Beds.

until after the Tower of Babel (see Gen. 11:1-9). This explains why human fossils appear much later in the post-Flood fossil record, and perhaps also why they are so varied. Most creationists believe that many of the species within the genus *Homo* (figure 7.17) are actually post-Flood people groups, not non-human primates. Local and regional catastrophes continued to play an important role in the post-Flood world (see Chapter 8 for more discussion). Volcanic eruptions, for example, were commonplace in western North America, and were responsible for producing many amazing fossil sites, such as Ashfall Fossil Beds in Nebraska (figure 7.28). Sites like this document the many changes that occurred during the post-Flood period as animals spread out to the far reaches of the world, and continually diversified into many new species within their kinds.

During the post-Flood period, the continents continued moving towards their present-day locations. Unlike the speeds seen during the Flood, the continents continually slowed during the post-Flood period. The most recent geological event was the Ice Age. As warm ocean waters evaporated in the centuries after the Flood, large snow accumulations created continental ice sheets over large sections of North America and northern Europe and extensive mountain glaciers in many regions of the world (see figure 10.4). As water from the oceans was stored in continental ice sheets, global sea levels dropped, a land bridge formed between Siberia and Alaska, allowing the migration of large numbers of animals (figure 7.29) and the first people to follow them into North and South America.

FIGURE 7.29. Wooly mammoths crossed from Asia into North America over the Bering Strait, which was dry land during the Ice Age. Some fell through frozen ponds and are discovered today in blocks of ice in the permafrost.

7.4 CHALLENGES AND OPPORTUNITIES

The naturalistic evolutionary and young-Earth creationist accounts presented above represent the two most common views on Earth's history and development over time. As such, they are both valuable to know and understand. Each has strengths and weaknesses, but in the end there is only one true history of the planet.

To be completely honest, the naturalistic evolutionary view is more detailed and thoroughly established than the young-Earth view. This is because over the past two centuries, countless numbers of Earth scientists have worked within this model to expand, refine, and adjust it to the many discoveries made over time. This model is robust, makes testable predictions, and has been very successful. As such, it deserves respect and consideration, even from those who do not agree with it. A major drawback to the naturalistic model is the lack of recognition that divine or other intelligent activity can be considered a part of Earth and life history. We find the philosophical walls erected by naturalism to be too strict. After all, if God does exist, and He did act in history, such actions may be discernable through observation and study, even without the testimony of Scripture (cf. Romans 1:20). To rule out God's possible action within the world based on a philosophical preference applies an unneeded set of shackles upon science, which we would do without.

The young-Earth creation view has had far less work done to establish a coherent model than the naturalistic view, but in the recent past more creationists have applied their knowledge and skill to this problem, with limited but sometimes significant success. Discoveries in plate tectonic and radioisotopic dating (see chapters 4 and 6) have encouraged deeper scientific investigation into various aspects of Noah's Flood and the creation of Earth. Other work on determining the Biblical "kinds," modeling of the Ice Age, astronomical models, and Biblical studies of the relevant passages of Scripture continue to refine the young-Earth model. We believe that over time this model will continue to mature. Our hope and prayer is that in presenting the model that we do have (even briefly), we may encourage and inspire future Earth and space scientists to lend their skills in producing a robust history of our Lord's world, bringing glory to Him in the process.

7.1 A Unique History

- There are two primary views on Earth history: naturalistic evolution and young-Earth creation. Both attempt to explain the same data, but do so according to very different timeframes (billions vs. thousands of years) and mechanisms (only natural processes, or natural *and* supernatural processes).
- The geological column was first formulated in the 1800s, and is an idealized, composite diagram showing the relationship of rock units found throughout the world. It is based on patterns of fossils in stacked sets of sedimentary rocks.

7.2 Naturalistic Evolutionary View of Earth History

- The Precambrian is an informal name for the first three eons of geologic history: the Hadean, Archean, and Proterozoic. It spans from 4.6 billion years ago to 541 million years ago.
- The Hadean Eon is the period of early Earth development prior to a stabilized, solid crust and liquid-water oceans. The Hadean Eon includes a major impact with another planet (resulting in the formation of the moon) and closes with the Late Heavy Bombardment.
- The Archean Eon produced the Earth's oldest rocks, oceans, and first fossils.
- The Proterozoic Eon is a time of significant change, with increasing oxygen, globe-covering glaciers, the first supercontinents, and the first complex fossil organisms (the Ediacaran biota).
- The Phanerozoic Eon is divided into the Paleozoic, Mesozoic, and Cenozoic eras. It lasts from 541 million years ago to today.
- Continents were mostly separated at the beginning of the Paleozoic, but at the end, the supercontinent Pangea had formed. Life in the Paleozoic (sometimes called the "age of fish") is marked at its beginning by the Cambrian explosion, and at the end by the end-Permian mass extinction (where 90% of all marine species go extinct).
- The Mesozoic Era begins with Pangaea assembled, but the supercontinent begins to split and the continents start migrating towards present-day locations. The highest sea levels are reached during the Cretaceous Period. Life is dominated on land, sea, and air by large reptiles such as dinosaurs, ichthyosaurs, pterosaurs. There is a mass extinction at the end of the Cretaceous caused by a meteor impact, resulting in the extinction of many groups, including dinosaurs.

- The Cenozoic begins with overall warm temperatures globally, but cools over time to produce the Ice Age, which ended only 11,000 years ago. Major mountain chains such as the Rockies, Alps, and Himalayas are produced during the Cenozoic. Mammals become the dominant land animals, and flowering plants and grasses become important plant groups.

7.3 Young-Earth Creation Views of Earth History

- Genesis 1:1-2:3 describes the creation of the world as happening in an orderly fashion over a 6-day period, followed by a day of rest. The "days" are understood as 24-hour periods of time, and the nature of the text indicates that this is a historical narrative passage (documenting history), rather than poetic or allegorical. The greatest emphasis is seen in Day 4 (heavenly bodies) and Day 6 (land animals and humans). Much geological work was accomplished in the Creation Week.
- The widespread and deep levels of sin among mankind resulted in God's judgment of the world by sending a global flood at the time of Noah. Representative pairs of all land-dwelling, air-breathing animal "kinds", along with Noah and his family were saved through the Flood on board the ark.
- The Great Unconformity generally marks the beginning of the Flood, with Flood-derived rocks found above it. These rocks are found arranged in a series of sequences. In each sequence course-grained sedimentary rocks transition vertically into more fine-grained rocks.
- The Cambrian explosion reflects the initial destruction of the Flood. Whole marine ecosystems were destroyed and driven up onto the continents, where they became part of the first sedimentary rocks containing abundant animal fossils.
- The pattern of fossils in the rock record (especially the Paleozoic and Mesozoic rocks) reflects a combination of pre-Flood habitat zones as well as the processes that destroyed, transported, and buried them during the Flood.
- The catastrophic movements of the Earth's plates was the driving geological mechanism behind the Flood. The supercontinent Pangaea was produced and separated during the Flood.
- The post-Flood period is represented by Cenozoic rocks. The animal and plant fossils of the Cenozoic represent the recolonization of the world after the Flood. The Ice Age occurred late in the post-Flood period.

7.4 Challenges and Opportunities

- The old-Earth and evolutionary perspectives have a more thorough and detailed model of Earth history than young-Earth creationism at this time.
- Though young-Earth creationists have more frequently approached these issues as anti-evolution or anti-old-Earth, growing numbers are seeking to understand Earth history from a distinctly young-Earth perspective. Encouraging results from plate tectonics, radioactive dating, Ice Age modelling, and other areas point towards a greater ability to refine and strengthen this model in the future.

KEY TERMS

Archosaurs

Banded iron formations

Basin

Bentonite

Cambrian explosion

Catastrophic plate tectonics

Differentiation

Ediacaran fauna

Epicontinental sea

Geological column

Gondwana

Great Unconformity

Great Upheaval

Ice Age

Ichnofossil

Late heavy bombardment

Laurasia

Pangaea

Precambrian

Red beds

Rodinia

Sequence

Snowball Earth

Stromatolites

FURTHER READING

Naturalistic Evolution

Benton, Michael. *Vertebrate Paleontology*, 4th edition. Wiley-Blackwell. 2014

Levin, Harold. *The Earth Through Time*, 10th edition. Wiley Publishers. 2013.

Young-Earth Creation

Brand, Leonard. *Faith, Reason, and Earth History*, 2nd edition. Andrews University Press. 2009.

Snelling, Andrew. *Earth's Catastrophic Past*, volumes I and II. Master Books. 2014.

Wise, Kurt. *Faith, Form, and Time*. B&H Books. 2002.

Whitmore, John, and Paul Garner. Using Suites of Criteria to Recognize Pre-Flood, Flood, and Post-Flood Strata in the Rock Record with Application to Wyoming (USA). *Proceedings of the Sixth International Conference on Creationism*. 2008.

Whitmore, J. H. 2008. Continuing Catastrophes. *Answers Magazine* 3(4): 70-72.

≈USGS
science for a changing world   preliminary interpretation by R.J.Haugerud

0   0.25   0.5   1 MILES

D
B
C   C
D   D
D
D
C   A
B   2014 Oso slide
A
C   D
D   C   B
D   C   B

AP Photo/Ted S. Warren

CHAPTER 8

SOILS, WEATHERING AND MASS WASTING

On Saturday morning, March 22, 2014 a mountainside in Oso, Washington failed and slid across the North Fork Stillaguamish River. It buried 30 houses and forty-four people were killed. Some people died as they were driving along State Route 530. The landslide traveled about 1.6 km from beginning to end. It covered the distance in about a minute! This chapter will cover all aspects of weathering including soil and soil production, the processes that cause weathering and breakdown of rock, and mass wasting. The Oso landslide is only one type of many different kinds of mass wasting processes that can have a sudden and devastating impact on humans. Inset: A USGS map showing previous landslides in various colors where "A's" are the youngest and "D's" are the oldest slides.

There are an important number of physical and chemical processes that happen in the environment which help sustain biological life of all kinds. Without these processes, life on Earth would be impossible. God has designed a remarkable number of processes to help Earth sustain itself, even in a cursed world. For example, there are processes in plants that convert the inorganic carbon from the air into organic carbon that is found in plants (on which all other life depends). Bacteria in the soil break the very strong triple covalent bonds between the nitrogen atoms in N_2 gas to make nitrogen more accessible to plants; a very important nutrient. On the physical side of things, various acids found in the soil can alter things like feldspar minerals to make the all-important clay minerals. Most clay minerals are ionically charged, allowing them to hold various nutrients like Ca^{+2} and even water. In this chapter we will examine soils and the weathering processes that help make them. We will also examine what happens when erosion happens on large scales—with often devastating consequences on mankind. Landslides, avalanches and rock falls are all the result of weathering on large scales.

Weathering is the process by which larger rocks get broken down into smaller pieces. A rock can be weathered by mechanical means, chemical means, or both. Weathering is the breakdown of the rock into small pieces and **erosion** is removal of the weathered products. Sometimes these two processes happen hand-in-hand and they are difficult to distinguish, but there is a technical difference. Weathering usually happens slowly to produce things like soils involving some type of chemistry. In this chapter we will examine various types of soils, various chemical and physical weathering processes, and large scale erosion and mass wasting events, such as avalanches, landslides, and rockfalls. All of these things can have important economic implications. In 2014, a large landslide originated on a highly weathered slope and killed 44 in the small town of Oso, Washington (see opening photo). Some areas like the San Joaquin Valley or the Mississippi River floodplain have rich soils, but also formidable geological challenges when utilizing the soils. The San Joaquin Valley is very dry and needs to be heavily irrigated, which causes nutrients to be lost from the soils and severe issues with adequate groundwater. The Mississippi River floodplain usually doesn't have to worry about drought conditions, but sometimes it catastrophically floods and ruins the crops (but also enriches the soils).

One of the things that we are reminded of in Scripture is that Earth is wearing out like a garment (Ps. 102:25-27; Is 51:6). One of the ways we can think about passages like this is that Earth is groaning, waiting for the curse to be over (Romans 8:22); we see evidence of that decay all around us (e.g., weathering and erosion). Notice the contrast in the passage from Psalms and Isaiah. Although Earth is wearing out, and by implication constantly changing; God does not change. He is unchangeable. We can always count on Him to always be the same. God is not subject to the curse or decay processes that are so familiar to us.

Psalm 102:25-27 25"Of old You founded the earth, And the heavens are the work of Your hands. 26"Even they will perish, but You endure; And all of them will wear out like a garment; Like clothing You will change them and they will be changed. 27"But You are the same, And Your years will not come to an end.

Isaiah 51:6 "Lift up your eyes to the sky, Then look to the earth beneath; or the sky will vanish like smoke, And the earth will wear out like a garment And its inhabitants will die in like manner; But My salvation will be forever, And My righteousness will not wane.

CHAPTER 8: Soils, Weathering and Mass Wasting

Soil is a mixture of mineral and organic matter. The mineral matter can consist of various sizes of rocks (from fine particles to boulders) and chemically weathered products of rocks (clay). The organic matter contains various decay products of plants. Good soils have a balanced mixture of all of these substances. If there is not enough organic material, the soil will probably not be suitable to grow a variety of plants. If there is not enough clay, the soil will not hold nutrients and water very well. If there is too much clay, the soil will not drain properly and probably will be difficult to use for farming.

The amount of rainfall and the temperature of a climate have enormous impacts on soil. In desert climates, soils are usually non-existent or very thin. Water plays a very important role in the weathering process because of the various acids that can be produced from rainwater. If the climate is warm and there is plenty of water, weathering of the host bedrock normally takes place very quickly and a thick soil will be produced. A warm climate is important because this speeds the rate of chemical weathering reactions. Because winds generally blow east-to-west in the Hawaiian Islands, the eastern sides of the islands are in a tropical rainforest type environment. As a result, thick soils are produced, often many meters thick (figure 8.1a). However, on the western sides of the islands it is very dry, and as a result, soils are very thin and sparse (figure 8.1b).

Over time, if an area remains undisturbed by geological and human processes and if there is sufficient precipitation, thick soils will develop. If one were to dig down into the ground, all the way to bedrock, they would find that a typical soil would be distributed into layers called **soil horizons** (figure 8.2). The horizons usually have indistinct contacts between them, grading from one layer into another; and will have varying thickness, mostly dependent on climate. The **"O" horizon** or organic zone contains dead and decaying plant and animal remains. It will usually have masses of roots penetrating into the deeper layers. It will normally be dark in color (brown) because of the large amount of **humus**, or organic material that it contains. Below the "O" is the **"A" horizon** which contains a mixture of the organic material above and some mineral matter. Mixing can occur between the two layers due to burrowing animals and root growth. It will usually be slightly darker than the "O" horizon, and sometimes black. This is usually the most productive and nutrient rich portion of any soil horizon. The thickness of these two layers will depend on how vigorous plant growth is, and how much of the organic material is buried over time.

(a) (b)

FIGURE 8.1. The eastern sides (a) of the Hawaiian Islands get considerably more precipitation than the western sides do (b), and as a result thicker and redder soils develop.

SOIL LAYERS

FIGURE 8.2. A typical profile of soil layers or "horizons."

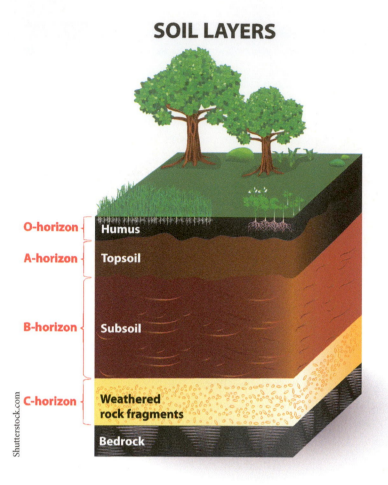

O-horizon — **Humus**

A-horizon — **Topsoil**

B-horizon — **Subsoil**

C-horizon — **Weathered rock fragments**

Bedrock

Shutterstock.com

Loam is a soil that contains approximately equal amounts of clay, silt and sand. It is desirable for the "A" horizon to have this consistency. Clay particles help hold moisture and various charged particles (like potassium, K^+) that are vital for plant growth. Many clay minerals are negatively charged, so they hold onto the positive sides of water molecules and various positively charged ions quite nicely. This is beneficial for plants; roots can access the water and various nutrients that they need as they interact with clay minerals. However, too much clay is detrimental because it would not allow excess water to drain from the soil. Sand and silt are vital for creating pore spaces for drainage and not allowing the soil to become too compact.

Some geographic areas have an **"E" horizon**, or zone of leaching below the "A" horizon. It will generally be lighter in color than the overlying horizons. Leaching happens in climates where there is a lot of rainfall and significant downward movement of water toward the bedrock. The leached zone is poor in nutrients and is usually rich in iron. Iron does not dissolve very well in water, and is usually left behind, making some soils quite red or orange. The **"B" horizon**, is the "subsoil" or the "zone of accumulation" which consists of a variety of substances (organics, clay, small pieces of bedrock), most of which have come from above. Next is the **"C" horizon**, or the parent material. Its composition is highly dependent on the type of bedrock below and usually consists of broken pieces of the underlying bedrock. Keep in mind that this is a very general description of soil horizons and this will vary from area to area as climate changes. Soil scientists subdivide all of the horizons mentioned above. Figure 8.3 shows several different soils and their respective horizons, and figure 8.4 shows a map of the major soil types in the United States.

Forest litter

Mixture of humus and minerals

Slit, loam soil

Clay, soil, minerals

Parent material

O
A
E

B

C

Temperate deciduous soil

Thick, acidic organic debris

Light colored, acidic

Humus, iron, aluminium

Parent material

O

E

B

C

Coniferous forest soil

Thick, alkaline, dark, very rich in humus

Clay and calcium compounds

Parent material

A

B

C

Grassland soil

Negligible, since organic matter is decomposed and recycled quickly

Acidic, light-colored

Iron and aluminium compounds mixed with clay

Parent material

O
A

B

C

Tropical rain forest soil

Thin, humus-mineral mixture

Thick, dry, containing variable accumulations of clay, calcium cabonate, soluble salts

Parent material

A

B

C

Desert soil

© Kendall Hunt Publishing Company

FIGURE 8.3. Several different soils and their respective horizons.

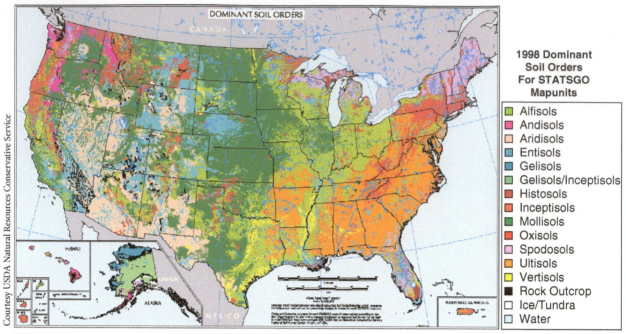

FIGURE 8.4. Map of the United States showing the extent of the 12 recognized soil orders.

The map legend reads:

1998 Dominant Soil Orders For STATSGO Mapunits

- Alfisols
- Andisols
- Aridisols
- Entisols
- Gelisols
- Gelisols/Inceptisols
- Histosols
- Inceptisols
- Mollisols
- Oxisols
- Spodosols
- Ultisols
- Vertisols
- Rock Outcrop
- Ice/Tundra
- Water

Courtesy USDA Natural Resources Conservative Service

8.3 WEATHERING AND EROSIONAL PROCESSES

8.3.1 Chemical Weathering

Chemical weathering occurs when a rock disintegrates due to chemical reactions that happen within the environment. One of the more common chemical means of weathering is solution by carbonic acid. Carbonic acid is made by very common environmental ingredients: water and carbon dioxide. Any time these two ingredients are present, a small amount acid can form. Note that the reaction is reversible, so it can go in the other direction too. Carbonic acid can dissolve calcite, the mineral that makes up limestone (figures 8.5, 8.6). When this happens, hydrogen ions are produced, which

The Formation of Carbonic Acid

Water + Carbon Dioxide $\leftrightarrow$ Carbonic Acid
$H_2O + CO_2 \leftrightarrow H_2CO_3$

The Solution of Calcite by Carbonic Acid

$CaCO_3 + H_2CO_3 \leftrightarrow Ca^{+2} + 2H^+ + 2CO_3^{-2}$
Calcite + Carbonic Acid $\leftrightarrow$ Calcium + Hydrogen + Bicarbonate

Kayo/shutterstock.com

FIGURE 8.5. Limestones are particularly susceptible to chemical weathering, able to be dissolved by various acids. In this case the limestone has been dissolved to create a large sinkhole that penetrates deep into the earth. The Macocha Abyss is a sinkhole in the Moravian Karst cave system of the former Czech Republic near the town of Blansko.

Photo by John Whitmore

FIGURE 8.6. Carbonic Acid can dissolve limestone and dolomite and is responsible for the cavities and holes that have developed on the face of the Cedarville Dolomite in Cedarville, Ohio.

float around in the water, making it acidic. The reaction of carbonic acid with calcite is fairly simple. Carbonic acid can also react with other minerals to make various other products. For example, the reaction of carbonic acid with K-feldspar is an important process in the production of clay minerals.

Rocks can rust. If iron is present in the rocks, it can combine with oxygen to make iron oxide, or rust. This chemical reaction can also help alter the rocks original chemistry. There are several different kinds of iron oxide. Hematite (rust, Fe_3O_4) is one of the more common ones and makes the rocks have a rusty brown color. Limonite is another common iron oxide ($HFeO_2$) and it gives rocks a yellowish

color. It actually doesn't take much iron oxide (< 1%) to stain a rock red or yellow (figure 8.7).

If you live in the northeastern part of the United States, you know the consequences winter road salt can have on your car. Solutions of salt water can weather rocks as well, through a combination of both chemical and mechanical weathering. In coastal areas, salty seawater splashes onto rocks. Saltwater can get into cracks and crevices and begin a chemical weathering process. However, when the sun comes out, the water evaporates allowing tiny salt crystals to grow. Much like frost wedging (discussed below) the growing salt crystals can wedge the rock apart, making room for other types of weathering to occur.

The Rusting of Rocks

Iron + Oxygen → Rust (Hematite)
$4Fe + 3O2 → Fe_3O_4$

FIGURE 8.7. The Redwall Limestone of the Grand Canyon is the large red cliff in the middle of the photo. Its natural color is gray, but it is stained red from the red rock layers above it. The red color is due to the iron in the rocks rusting, making hematite.

FIGURE 8.8. Cold temperatures in the winter causes liquid water below road surface to freeze, causing the road to heave. This makes the road uneven, causing it to crack and making it even more susceptible to other types of damage. It is important to install good drainage below road surfaces to minimize the effects of frost heaving.

8.3.2 Mechanical Weathering Processes

Mechanical weathering is the breakdown of rocks by physical or mechanical processes. If you took a hammer and hit a rock, breaking it into many pieces, that would be mechanical weathering. There are many different ways mechanical weathering can happen in nature. Water is fairly unique among chemical compounds. Most things contract or shrink when they get cold; water does this, but only until 4° C. When the temperature falls below this, it begins to expand until it finally freezes at 0° C. Ice floats because frozen water is less dense than liquid water. If you place a water bottle in the freezer and don't allow room for expansion of the freezing water, the bottle will burst, making a mess. A similar process can happen if water gets into the cracks of rock and then proceeds to freeze. This is known as **frost wedging**. A related process is called **frost heaving**. This can happen beneath a roadway in the winter or under the foundation of your house, see an example in figure 8.8. The freezing water can lift (and crack) the road, causing damage that must be repaired the following summer. This is why places like Michigan, Ohio, and New York do so much road repair every summer. If a

small amount of water gets under the foundation of your house, it can lift the house slightly and cause the frustrating experience of not being able to close (or open) a door on a cold winter day. When roads and houses are built, especially in cold, frost-prone areas, good drainage is paramount to avoid these problems.

In areas with considerable moisture and exposed bedrock, plants of all types will grow on the rocks.

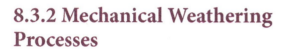

Types of Mechanical (or Physical) Weathering Processes:

- Frost wedging
- Frost heaving
- Thermal expansion/contraction
- Root wedging
- Unloading (pressure release)
- Water abrasion
- Hydraulic action
- Cavitation
- Wind abrasion
- Fire
- Groundwater sapping

This can include microscopic algae, moss, larger plants, and trees. When the plants die, their decay will create various acids that can help decompose the rock chemically. However, roots can provide an avenue of mechanical weathering called **root wedging** (figure 8.9). As the roots force open even small fractures, this leads the way to other weathering processes, such as frost wedging in the winter.

Rocks and other things (like glacial ice) exert a tremendous amount of downward force on a landscape. If a large amount of rock is removed suddenly by erosion, or if a glacier melts and retreats, the force being exerted on the landscape no longer exists. In response, the landscape will respond by expanding upward as a result of the lost material (figure 8.10). This is called **isostatic rebound** and can have mechanical consequences on the remaining rock. Rock has incredible compressive strength; in other words you can put an incredible amount of downward force on it and nothing much will happen. However, rock has extremely poor tensile

FIGURE 8.9. Both large and small roots from the tree and the moss can get between small fractures in the rock, enlarging them and leading to other types of mechanical and chemical weathering.

strength; in other words if you pull on it, it will fracture. When the weight of a landscape is unloaded, or the pressure of the overlying rock is released, the rocks respond by slightly expanding. The expansion that takes place has the same effect as pulling the rock apart. As a result, the rocks fracture, parallel to

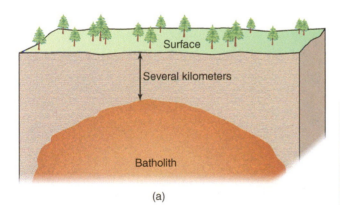

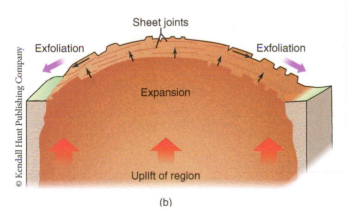

FIGURE 8.10. Rocks can expand when a large mass is deeply buried, and then the material from above it is removed (unloading). Rocks do not expand very well, and usually crack, creating sheet joints. The weakened rock can abruptly break and fall down the mountain, creating dangerous rock slides. Yosemite National Park, California.

the surface. This process of mechanical weathering is called **pressure release** or **unloading**.

There are several ways in which a stream can mechanically weather a rock. These processes will be discussed in more detail in Chapter 9, so they will only be introduced here. When you smooth a piece of wood with sandpaper, you are using **abrasion** to remove the rough spots. Rocks that tumble around in stream beds are smoothed through the process of abrasion. Sand that is carried along in rivers can also help to round the corners of angular rocks. **Hydraulic action** is the breaking rocks apart by the sheer force of moving water. Fast moving water can get into a crack or a crevice and break a rock or cliff face apart. If you have ever taken your car to the car wash and use the pressure washer to get mud and dirt off your car, instead of the foaming brush, you are using hydraulic action to clean your car. **Cavitation** is a special, but very powerful type of mechanical weathering process. It occurs when very fast moving water flows over small depressions in

bedrock. When this happens, small vacuum bubbles form and then implode (not explode). This acts like a mini sledge hammer pounding on the rock.

Several other types of mechanical weathering include abrasion by wind and **groundwater sapping**. Wind is actually not as important of a weathering process as you might imagine. Wind can pick up small sand grains and forcefully blow them against a rock or other object. If a glass bottle is left out in the sand dunes, the glass will soon become frosted due to the sand impacts on the glass. Sometimes rocks can be smoothed and shaped by wind abrasion, but it generally is incapable of removing a large amount of material. Groundwater sapping is a process that occurs as groundwater pushes from the inside-out of a rock face. It will typically happen in an area where a spring is flowing out of a rock (figure 8.11). As water pushes from the inside-out on the rock face, typically large pieces of rock will weaken and fall to the base of the cliff. Over time, this process will create a large arched alcove (figure 8.12).

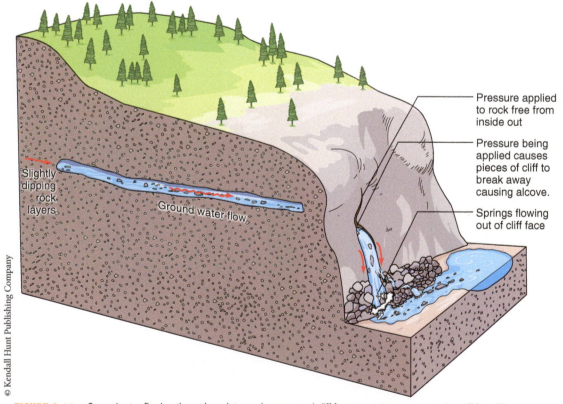

Pressure applied to rock free from inside out

Pressure being applied causes pieces of cliff to break away causing alcove.

Springs flowing out of cliff face

Slightly dipping rock layers

Ground water flow

© Kendall Hunt Publishing Company

FIGURE 8.11. Groundwater flowing through rock toward an exposed cliff face can put pressure on the cliff face. The pressure exerted by the flowing groundwater can cause pieces of the cliff face to calve off and create an alcove on the cliff.

CHAPTER 8: Soils, Weathering and Mass Wasting

Photo by John Whitmore

FIGURE 8.12. Groundwater sapping can ultimately produce a large arched alcove. In this photo from Zion National Park white stains from groundwater flowing out of the rock can be seen inside the alcove.

Courtesy NASA

FIGURE 8.13. International Space Station photograph of Zion National Park, Utah, showing large parallel joints that cut across the landscape. Canyons are beginning to form in some of the joints.

8.3.3 Special Types of Weathering

There are a few special types of weathering that can be seen over and over again if you travel and look at rocks much. One of the things that you should notice is that when rocks have faults, fractures, or joints in them, these places are more subject to weathering than rock without fractures (figure 8.13). Water, roots, carbonic acid, salt, rivers, and many other things can get into the cracks and widen them. When rock is fractured, it has been broken into smaller pieces and these pieces are easier to move by erosion processes. Slot canyons sometimes form due to a river that is being controlled by a fracture or joint (figure 8.14). **Spheroidal weathering** is when rocks (typically granites) become weathered into rounded shapes because at least two sets of perpendicular joints crisscross the granite (figure 8.15). Mechanical and chemical weathering attacks the joints and widens them turning cubes into spheres. **Differential weathering** is when one rock type weathers faster than another. For example, a shale usually weathers faster than an adjacent sandstone or limestone, often causing the more resistant layer to stand out in comparison to the less resistant one (figure 8.16). Limestones weather differently depending on the type of environment. In wet tropical areas, limestones weather very quickly (due to carbonic acid) while limestones usually are the hardest and most resistant rock units in arid regions.

Rolf_52/shutterstock.com

FIGURE 8.14. A slot canyon is a deep and narrow canyon that is often excavated along a joint. They are deadly places to be during a flash flood. This is a feature known as "the Bear" in Antelope Canyon near Page, Arizona.

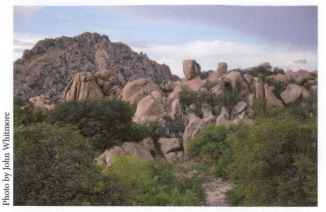

Photo by John Whitmore

FIGURE 8.15. Spheroidal weathering in granite near Dragoon, Arizona. You can see the weathering in the foreground and on the mountainside in the background. Note the joint pattern on the mountainside.

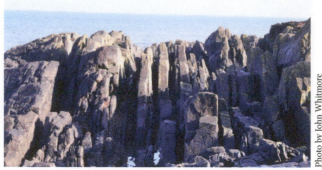

Photo by John Whitmore

FIGURE 8.16. Pictured here are nearly vertical sandstone and shale beds at Siccar Point, Scotland. The softer shale beds weather faster than the more resistant sandstones creating the parallel ridges of sandstone. This is an example of differential weathering, when one rock layer weathers faster than another.

8.4 MASS WASTING

Recall that there is a difference between weathering and erosion, although sometimes it is hard to tell where one stops and the other begins. Weathering is usually responsible to some degree in preparing rocks and soils for erosion. **Mass wasting** (or mass movement) is when large quantities of rock or soil move downhill under the influence of gravity, like landslides, rock falls, and slumps. Mass movements can be spectacular and rapid, or they can be slow processes that occur over periods of years and decades.

Mass wasting events are important to study because of the disastrous effects that they can have on humans. Some of these events are quite unexpected and can destroy lives, homes and roads within seconds. On March 22, 2014 a large part of a hillside slipped downslope in Oso, Washington (see opening photo). The landslide blocked a river, blocked a road and buried a housing development, taking 44 lives. In 1983, a slow moving landslide in Thistle, Utah, blocked a road, railroad trunk line and a river. People had time to escape the slow moving slide, but the slide blocked the river, the town of Thistle, and a railroad switching yard (figure 8.17).

This became the most expensive landslide ever in the United States, costing more than 200 million dollars. Geologists can help predict where things like mudflows, landslides, and rock falls will occur so the damage to roads, homes, and human lives can be minimized.

Courtesy USGS

FIGURE 8.17. A giant earthflow happened in the spring of 1983 near Thistle, Utah. The slide blocked the Spanish Fork River (creating the lake, and flooding the town of Thistle) and covered a major highway and railroad line. At the time, it cost $200 million in damage, making it the most expensive landslide in US history.

When a large amount of material moves down a slope, there are several factors that often contribute to the movement. Usually two or three of these factors are present when a mass movement occurs. Areas that have failures often have great relief. **Relief** is the distance between the highest and lowest point in a landscape. The greater the relief, the greater the likelihood a failure will occur. **Slope** (the angle of the land surface) is another factor that is critical. The higher the angle, the more likely a slope is to fail. Whether a slope has vegetation on it or not can make a difference too. Roots from trees and grass can help hold a hillside together and keep it from failing; but sometimes even slopes with vegetation will fail. Most mass wasting events happen after prolonged rain events or after large winter snowmelt. Water can seep into hillsides and is often the key additional factor that initiates a mass wasting event. It is "the straw that breaks the camel's back." Water adds weight to a hillside. It also gets into the pore spaces and has a buoyancy effect, causing particles to loose frictional contact with one another. Thus, water reduces friction which ultimately holds everything in place. Earthquake or seismic activity is often a trigger which can initiate a large slide or rock fall. Often slopes are "prepared" ahead of time by chemical and physical processes that deteriorate rock and soil making it more susceptible to eventual failure. Weathering processes can considerably weaken a slope so that when a heavy rain or an earthquake comes along it is more apt to fail. The nature of the rock in a hillside can also play a role. If the rock is poorly lithified or doesn't hold together well in the first place, it will be more apt to fail. Also, bedrock that is highly jointed, fractured or faulted in certain directions may be more apt to fail.

8.4.1 Falls

There are four general types of mass wasting events: falls, slides, slow flows, and rapid flows. **Rockfalls** and **soilfalls** happen when material is undercut, and a large section of the overlying material drops through the air (figure 8.18). As it lands it may create a large pile of debris at the base of a cliff (called **talus**). While on a trip to northern Alaska to collect dinosaur bones, Dr. John Whitmore and his team encountered some spectacular rockfalls coming from a steep cliff. They were rafting down the Colville River when they heard some big "booms" up ahead around a bend in the river. As they rounded the bend in the river, much to their horror, large sections of the cliff were dropping into the river ahead

Factors that often contribute to mass movements:

- Relief
- Slope angle
- Lack of vegetation
- Water
- Seismic activity
- Amount of previous weathering
- Character of the bedrock

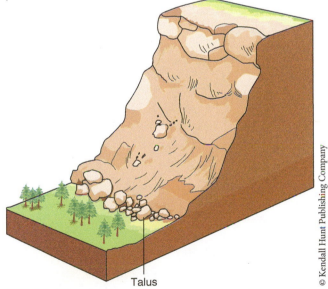

Talus

FIGURE 8.18. A rockfall happens when material free falls from a cliff face. The fallen material usually accumulates at the base of a cliff as a pile of talus.

of them. The strongest current of the river was pulling them directly into the path of the falling rocks, some as big as automobiles! They had to paddle like mad, but eventually were able to get to the other side of the river and watch the falling rocks from a safe location (figure 8.19). A combination of weak rocks, undercutting by the river, little vegetation, and meltwater from the permafrost on top of the cliff all contributed to the rockfalls along the Colville River that day. You can read more about this story in *The Great Alaskan Dinosaur Adventure.*

8.4.2 Slides

Slides (or **landslides**) is a general term used to describe rock or soil sliding down a slope. A playground slide is a good analogy for how geological slides work. In order for a child to slide down, the surface must be more or less frictionless and have a steep enough slope for a full ride. Bigger children have more momentum and travel further down the slide than smaller children. And if the slide is wet, the ride is even faster. All these parameters apply to landslides as well.

Rock slides can be a single block of material that slides down a hillside or they can be composed

Types of Mass Movements		
Type of material →	Rock	Soil or mixture of rock and soil
1. Falls	Rockfalls	Soilfall
2. Slides (landslides)	Rock slump Rock slide	Soil slump
3. Slow Flows	Rock creep Talus creep	Soil creep Solifluction
4. Rapid Flows	Rock avalanche Snow avalanche Debris avalanche Turbidity currents Hyper-concentrated subaqueous flows	Earth flow Mudflow Debris flow

of innumerable particles that slide all at once. Single units that slide are sometimes called block slides. Slides generally move along a flat planar surface (figure 8.20) whereas **slumps** move along a curved surface or scoop-like surface (figure 8.21). Both soil and rock can slide or slump, but slumps are more common in soils. Think of how you might slump while sitting in a chair. When you do this, it is kind of a backwards rotation. Geological slumps have the same behavior. After a slide or a slump a large **scarp** or cliff face is generally left on the hillside. In the case of a soil slump, sometimes several curved scarps will be present afterwards. As is common in other kinds of mass wasting events, slides will often

FIGURE 8.19. Rockfalls along the Colville River of the North Slope of Alaska. The cliff is about 50 m high and the splash is about 5 m high.

Photo by John Whitmore

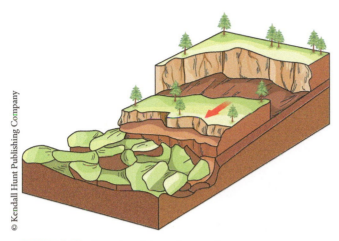

FIGURE 8.20. Slides travel down planar surfaces.

transition into other types of mass movements. Depending on how much water is present in the slide material, earthflows and debris flows are common, especially at the front edge of the slide (figure 8.22).

Late in the evening of August 17, 1959, a 7.3 magnitude earthquake occurred in the vicinity of Yellowstone National Park at Hebgen Lake, Montana. The earthquake caused a large section of rock, estimated at 28-33 million cubic meters, to break away from a mountainside in the Madison River canyon, and slide to the bottom of the valley (figure 8.23). The slide dammed up the river and buried a campground with 26 people in it to depths up to 120 m. The slide created a new lake (appropriately named "Earthquake Lake") which filled to a depth of 60 m in just a few weeks.

On Saturday morning, March 22, 2014, a mountainside slid across the North Fork Stillaguamish River and buried 30 houses near Oso, Washington (Introductory photo). Forty-four people were killed in their homes and as they were driving along State Route 530, suddenly engulfed by the mud and debris. Greater than normal precipitation amounts during the months of March and February probably contributed to the failure of the glacial sediments. The slide started on a steep bank that was about 180 m high just above the river. The size of this slide was about 8 million cubic meters, and traveled over

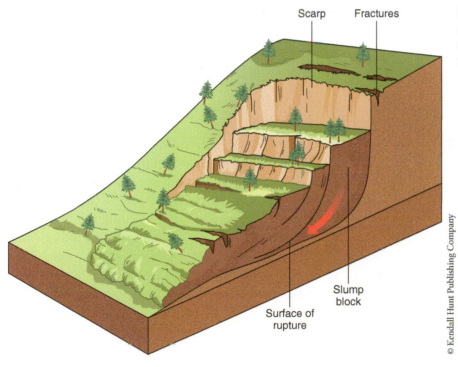

FIGURE 8.21. Slumps fail along a curved surface. Notice the crescent-shaped scarp that is left behind after the movement has happened.

Scarp　　Fractures

Surface of rupture

Slump block

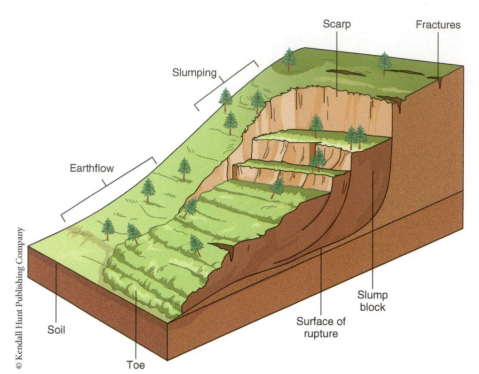

Scarp   Fractures

Slumping

Earthflow

Soil

Toe

Surface of rupture

Slump block

© Kendall Hunt Publishing Company

FIGURE 8.22. It is often difficult to identify a mass wasting event as a "single" process. More often than not, several processes are involved. At the top of this hillside, "slump" was the dominant process. At the bottom of the hill, there was so much water that "earthflow" was the dominant process. Notice the "toe" left by the earthflow.

1 km from the top of its slope to the toe. The event temporarily dammed the river, causing upstream flooding and further property damage. The housing development was built in an area of known landslide activity (see introductory photo inset), but no one expected the kilometer-long run out of the catastrophic failure.

Photo by John Whitmore

FIGURE 8.23. "Earthquake Lake" along the Madison River in Montana. The landslide originated from the slope in the center background, blocking the river in the center-right of this photograph.

8.4.3 Slow Flows

Slow flows can take days, weeks, months or years to materialize. While they do not cause immediate danger to human life, they still can cause considerable financial damage. All of the risk factors that lead to other mass movements still apply here (slope, relief, water, etc.), but usually friction is greater so things don't move as fast. Some rocks are pliable and can slightly bend over time, a phenomenon known as **rock creep** (figure 8.24). Poorly consolidated rocks are more likely to undergo this type of movement. **Soil creep** is more common than rock creep (figure 8.25). In various types of creep, particles closer to the surface will move faster than particles deeper in the ground. Usually the only parts that move are within a meter of the surface. Signs that a hillside is creeping are bent tree trunks (figure 8.26), cockeyed telephone poles or fence posts, tilted gravestones, sections of roads that are out of place and unusual mounds of dirt at the bottoms of slopes. If purchasing hillside property, especially if you plan

FIGURE 8.24. Fractured rocks slowly moving downhill in Marathon, Texas. Notice that the rocks closer to the surface are "bent" more than the deeper rocks. The rocks are not actually bent; they are fractured and the broken pieces are moving.

Courtesy USGS

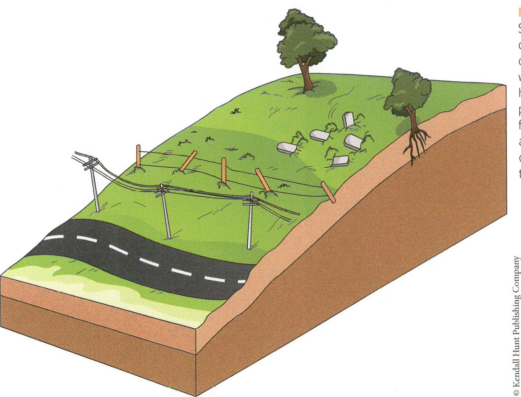

FIGURE 8.25.

Soil creep causes downslope disturbance of the soil and objects within the soil. Notice how trees, telephone poles, tomb stones, fences, roads, etc., are all effected by the slow downward movement of the soil.

© Kendall Hunt Publishing Company

FIGURE 8.26. As soil slowly slides downhill, tree roots go along with it sometimes causing the trunk to bend near is base. The root mass is sliding, but the tree trunk bends to continue to grow upright.

to build a home, make sure the slope does not have any signs of creep. Over time building foundations will surely crack and fail.

Solifluction is a special process that happens in arctic areas where there is **permafrost**. In the artic summer, the sun can thaw the soil to a depth of a meter or so. The soil below that remains frozen, hence the name "permafrost." The ground above the permanently frozen soil becomes like mush—thick and soupy. The liquid water cannot move downward because the ice below prevents its flow. If the soupy mush is on any kind of slope it will flow downhill. Some of the towns in northern Alaska are built in areas of permafrost where bedrock is not close enough to the surface for a firm foundation. When buildings are built in places like this, usually pylons are sunk into the permafrost and the structure is then built, perched on the pylons. This way, if thawed soil moves as a result of solifluction, it will move under the structure and not damage it, and heat from the building will not melt the permafrost.

8.4.4 Rapid Flows

Rapid flows move large amounts of rock and other debris very quickly. There is often a tumbling motion within the flow, especially near the front of the flow. Types of rapid flows that most people are familiar with are snow avalanches, but there are other types as well. **Rock avalanches** involve large amounts of rock tumbling down a mountain side and **debris avalanches** involve a mixture of rock and other debris (snow, trees, etc.). In 1970, a 7.8-magnitude quake struck Peru. The earthquake initiated a snow and rock avalanche beginning high on the slopes of Mt. Huascaran which came down the mountain at speeds of 210–335 km/hr. As it traveled it continued to pick up glacial debris and it was estimated that the final volume was 50–100 *million* cubic meters. Altogether the avalanche traveled about 18 km, decimating the towns of Yungay and Ranrahirca. About 20,000 people were buried alive. It was not the first time that something like this had happened in the area; in 1962, a similar event killed 4,000 people.

Rapid flows can also happen on the bottoms of lakes and oceans. These are generally referred to as **density currents** (or **turbidity currents**) because they are denser than the water through which they are moving, and are thick slurries that can be generated by an earthquake which in turn causes large amounts of mud and sand to flow downslope. An earthquake in eastern Canada in 1929 caused a large underwater landslide. As the density current swept across the ocean floor, it successively broke all the submarine telegraph cables in its path (figure 8.27). This was long before satellite and radio communication between continents, so it disrupted communication between Europe and North America for some time to follow.

The Redwall Limestone contains evidence of a large density current. A 2-meter thick bed stretching from the Grand Canyon to Las Vegas, Nevada has over 6 billion shelled nautiloids buried within it (figure 8.28). Nautiloids are squids that had ice cream cone-shaped shells. It is assumed they were

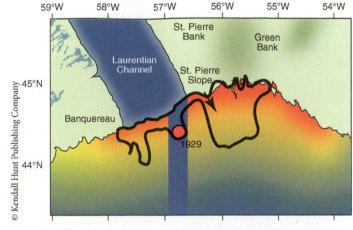

FIGURE 8.27. The 1929 Grand Banks earthquake in Nova Scotia, Canada caused an enormous part of the ocean floor to fail and slide into deeper water (the entire area outlined in black). The event generated a tsunami and underwater density current, which damaged numerous submarine transcontinental cables that were the only means of rapid communication between Europe and North America at the time.

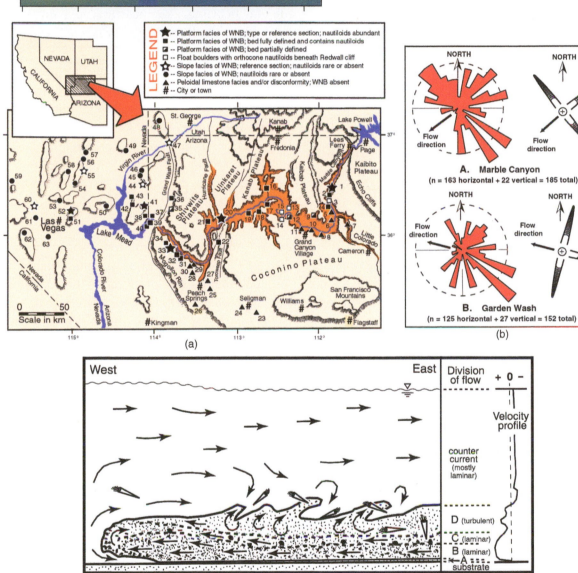

Courtesy Steve Austin

FIGURE 8.28. At the bottom of the Redwall Limestone a 2 m thick layer of limestone with an estimated 6 billion nautiliods in it has been found (a). Steve Austin mapped and measured the orientation of the nautiloids at many locations throughout the southwestern United States (map and rose diagrams). The nautiloids have a preferred direction of orientation (b), which means they were buried by a moving current. Austin's hypothesis is that living nautiloids were overcome and trapped within a large density current that flowed catastrophically along the ocean bottom during the Flood (c).

CHAPTER 8: Soils, Weathering and Mass Wasting

very capable swimmers, just like today's squid. As the flow swept across the ocean bottom during Noah's Flood, billions of these living squid were caught up and buried by the flow. The entire bed, from one end to the other, was probably deposited within hours.

Earth flows occur when soils and other unconsolidated materials become extremely water saturated and begin to slowly flow downhill. The damage done to Thistle, Utah, was from an earthflow. **Mudflows** (called lahars in volcanic settings) happen anytime large quantities of dry soil or volcanic ash are suddenly mixed with large volumes of water (making up 10–60% of total volume). This makes them extremely dense and capable of floating things like huge boulders and houses. They can move very swiftly, much like a flash flood, but with the consistency of wet cement. Mudflows can be extremely destructive, causing vast amounts of erosion within minutes.

Debris flows usually have smaller quantities of water (about 5–30%) compared to mudflows. There is no sharp distinction between the two types of flows, other than that mudflows tend flow more like fluids. As debris flows begin to tumble downslope, water and rock gets entrained, and the momentum can continue to carry the material into rivers further downstream. In the Grand Canyon, debris flows are the major process that delivers sediment from the side canyons into the Colorado River, and all of the major rapids in the Grand Canyon are the result of boulders that have been brought in by debris flows. On July 14, 2012, a debris flow swept down National Canyon into Grand Canyon, bringing boulders the sizes of automobiles (figure 8.29). The reason that the Colorado River cannot carry the boulders further downstream is because river water is not dense enough to move the debris. Mud and debris flows are much more dense than water, so things like boulders actually can float in the flows! When the boulders reach the Colorado River, they sink and sit in the rapids, staying there for many decades.

FIGURE 8.29. A debris flow "floated" these large boulders down National Canyon and deposited them at its confluence with the Colorado River in the Grand Canyon (the Colorado is just behind the pile of boulders). John Whitmore is standing on the pile of boulders, some of which are the size of small cars.

8.4.5 Mediation of Mass Movements

Is it possible to prevent mass wasting events? Civil engineers and geomorphologists (geologists who specialize in the shape of the land) study landscapes and hill slopes and try to predict where mass movements may occur and how they might affect mankind. If it appears that slope failure in a particular area is imminent building is curtailed in the danger zone. Some mass movements can be prevented, or at least slowed down (under normal conditions) if some of the issues that commonly cause mass wasting are addressed. For example, if a slope is steep and has high relief, terraces may be cut into the hillside to prevent any failures from moving further down hillsides (figure 8.30), better drainage may be installed, or vegetation may be planted. If rocks are steeply dipping along a highway and there is a chance of a rock slide, the rocks might be bolted down to keep them in place (figure 8.31). In California, sometimes hillsides are covered in large plastic sheets to prevent water from seeping into the dry soils during the rainy season. This helps prevent mudflow and earth flow on unstable slopes. In other cases, water

FIGURE 8.30. Terraces cut into hillsides can serve two purposes. First, they provide flat areas for farming. In this case, they are growing rice. Second, if a small mass movement begins on the hillside, it might be stopped as it reaches a terrace below. It looks as if a small slump may have occurred in the top center of the photo, just above the upper terrace.

FIGURE 8.31. Rock bolts can be placed in the ceilings of mine shafts (b) or drilled into cliff faces to hold unstable rocks in place.

drainage pipes are installed in problematic slopes in order to drain the water out quickly. Remember all the factors that contribute to mass movements? Mass movements can be curtailed if one or more of these factors can be lessened or eliminated. Since many mass movements occur soon after the addition of water, often one of the best things to do is to either prevent the water from getting into a hillside or to help it be removed quickly when it arrives. Good drainage is often the key that will prevent most types of failures. Some things are very difficult to prevent or are unpredictable. For example an earthquake can't be stopped, but if a potential slide or fall is identified that will move during an earthquake, appropriate adjustments to lessen human impact can be made.

8.5 MASS WASTING FOLLOWING NOAH'S FLOOD

During Noah's Flood the Bible tells us the continents were completely submerged with water. Beneath the waters of the Flood, vast layers of sediment were being deposited, many of which still cover the continents today. Following the Flood, the continents and mountains rose up and the ocean floors began to subside (Psalm 104:8). The immediate post-Flood time was likely a time of extreme landscape instability.

Think about all the factors that contribute to mass wasting events. All of these factors would have been present following the Flood. Sediments deposited during the Flood and then lifted out of the Flood waters as the continents rose, would have been full of water. Mountain building would have been accompanied with seismic activity and caused great increases in slopes and in relief. Slopes would have been bare of any substantial type of vegetation. Freshly made rock layers would not have been fully hardened and could have failed much more easily. It is likely that massive storms swept across Earth's surface in post-Flood times dumping incredible amounts of rain and snow. The precipitation would have increased already rapid erosion rates considering all of the other factors present.

comparison to the events that must have happened in the immediate post-Flood times. Imagine mountains sliding, huge canyons being cut by mudflows, water suddenly draining from large lakes, giant slumps and slides on unstable hill slopes, and earthquakes causing frequent failures of steep slopes. One example is Heart Mountain that we believe slid at least 50 km (from west to east) as massive volcanism was happening in the Yellowstone area just after the Flood (figure 8.32). In Chapter 5, you will find other examples of gigantic volcanic landslides and other examples of mountains that have slid great distances.

At the end of the Flood, continents were still not in their present locations (see chapters 4 and 7); additional continental drift would have continually caused earthquakes and shifting climatic patterns all of which can cause additional erosion and mass wasting. As we have considered large modern mass wasting events in this chapter, they pale in

Photo by John Whitmore

FIGURE 8.32. Heart Mountain near Cody, Wyoming. It is thought this mountain (and some others around it) slid at least 50 km from west to east as the Yellowstone area was experiencing massive volcanism after the Flood. We think many mass movements like this happened in post-Flood times, radically altering the landscape.

8.1 Introduction

- Weathering is the disintegration and decomposition of rock, erosion occurs when the material is picked up and carried.

8.2 Soil

- Soil is a mixture of mineral and organic matter. Loam has nearly equal amounts of clay, silt and sand and is the most desirable soil type for crops. Precipitation, temperature, and rock type are the biggest factors in which types of soils will develop.
- Soil is typically stratified into horizons (O, E, A, B, C). Soil horizons will vary from place to place to place and are classified into 12 different soil orders.

8.3 Weathering and Erosional Processes

- Rocks disintegrate via mechanical and chemical processes. Chemical weathering is primarily by carbonic acid and is important in producing soil.
- Mechanical (or physical) weathering processes are numerous and include frost wedging, frost heaving, thermal expansion and contraction, root wedging, unloading, abrasion (water and wind), hydraulic action, cavitation, fire and groundwater sapping.
- Some types of rock weather faster or differentially compared to other rock types. Joints and fractures make good conduits for the weathering process to begin and can lead to spheroidal weathering.

8.4 Mass Wasting

- Mass wasting is when large quantities of material move downhill under the influence of gravity. It can happen quickly or slowly. Types include falls, slides, slumps, creep, solifluction, avalanche, mud flows, debris flows, earth flows, and various types of subaqueous flows.
- Factors that contribute to mass movements are relief, slope angle, water, vegetation (or lack of), seismic activity and the character of the bedrock. Mass movements can be prevented (or mediated) by controlling or limiting these factors.

"A" horizon

"B" horizon

"C" horizon

"E" horizon

"O" horizon

Abrasion

Cavitation

Chemical weathering

Debris avalanche

Debris flow

Density current

Differential weathering

Earth flow

Erosion

Frost heaving

Frost wedging

Groundwater sapping

Humus

Hydraulic action

Isostatic rebound

Landslide

Loam

Mass wasting

Mechanical weathering

Mudflow

Permafrost

Pressure release

Relief

Rock avalanche

Rock creep

Rock slide

Rockfall

Root wedging

Scarp

Slide

Slope

Slump

Soil

Soil creep

Soil horizon

Soilfall

Solifluction

Spheroidal weathering

Talus

Talus creep

Turbidity current

Unloading

Weathering

REVIEW QUESTIONS

1. What are the two main kinds of weathering?
2. What is the difference between weathering and erosion?
3. What is soil and why does it vary from place to place?
4. What are some different ways to classify soils?
5. Can you identify pictures of mechanical weathering if shown pictures (like on a test)?
6. List and describe the various types of mechanical weathering processes?
7. Be able to identify special types of weathering (joints, spheroidal, differential) from photos.
8. What are the factors that cause mass wasting? Which ones are more important? If you wanted to prevent a potential mass wasting event, how would you go about doing it?

9. Be able to identify various mass wasting types from photos.
10. Describe several different circumstances in which mudflows can occur.
11. There were many historical mass wasting events described in this chapter. What are some of the things they have in common?
12. Why was the post-Flood time ideal for mass wasting processes?
13. What are some measures that could be taken to prevent or reduce the consequences to humans regarding mass wasting processes?

FURTHER READING

- Davis, B., Liston, M. and Whitmore, J. 1998. *The Great Alaskan Dinosaur Adventure*. Green Forest, Arkansas: Master Books.
- Whitmore, J.H. 2013. The potential for and implications of widespread post-Flood erosion and mass wasting processes. In *Proceedings of the Seventh International Conference on Creationism*, M. Horstemeyer, ed. Pittsburgh, Pennsylvania: Creation Science Fellowship.
- Austin, S.A. 2003. Nautiloid mass kill and burial event, Redwall Limestone (Lower Mississippian), Grand Canyon region, Arizona and Nevada. In *Proceedings of the Fifth International Conference on Creationism*, R.L. Ivey, Jr., ed., pp. 55–99. Pittsburgh, Pennsylvania: Creation Science Fellowship.
- Pierce, W.G. 1960. The "break-away" point on the Heart Mountain detachment fault in northwestern Wyoming. U.S. Geol. Survey Professional Paper, v. 400B, p. 236–237.

WEBSITES TO VISIT

- http://geology.com/usgs/landslides/ Landslide hazard information at geology.com.
- http://landslides.usgs.gov/ United States Geological Survey landslide hazard website.
- http://www.pbs.org/wgbh/nova/earth/killer-landslide.html A NOVA movie titled "Killer Landslides." It talks in detail about the Oso, Washington, slide.
- http://www.nrcs.usda.gov/wps/portal/nrcs/detail/soils/edu/?cid=nrcs142p2_053588 Soil orders and pictures of soil profiles (USDA)
- http://www.cals.uidaho.edu/soilorders/orders.htm The twelve soil orders (University of Idaho).
- http://geology.com/records/largest-landslide.shtml Heart Mountain slide.
- http://media.wr.usgs.gov/movies/index.html?id=debris_flow_dynamics Debris flow movie at USGS.

REFLECT ON SCRIPTURE

- Genesis 8
- Job 9:5,6; 14:18,19
- Psalm 102:25-27; 104 (especially verse 8)
- Isaiah 51:6

CHAPTER 9

STREAMS AND GROUNDWATER

OUTLINE:

This chapter will discuss rivers, streams and groundwater. Water is the single most important natural resource that we have. All life depends on it. One of the most fascinating subjects in this chapter is hot groundwater which includes geysers and hot springs. This is the Strokkur Geyser in Iceland.

kavram/Shutterstock.com

231

Of all the surface water on earth, ocean water makes up 96.5% of the total. Only 2.5% of the earth's water is fresh and most of that is tied up in glacial ice in Greenland and Antarctica. Most of the liquid fresh water occurs underground and is the major source of drinking water around the world. The fresh water bodies that we are most familiar with (lakes and rivers) only makes up about 0.3% of Earth's surface water! Figure 9.1 shows the distribution of water on the surface of the earth.

Have you ever thought about the importance of water in Scripture? Jesus told the woman at the well in John 4, that he could provide her with living water so that the woman would never thirst again. Peter says in II Peter 3:5 that the earth was formed out of water and through water. In Genesis 2 we find a river flowing out of the Garden of Eden and again in Revelation 22 we find the river of the water of life flowing directly from the throne of God. In 1 Corinthians 10, Paul says that the water that flowed out of the rock in Exodus 17 and then again in Numbers 20 was a spiritual drink for the Children of Israel (and Paul identifies Christ as the Rock). Of course baptism is a symbol of Christ's death and resurrection, a practice of Christians to demonstrate that they are sincere followers of Christ. The act of going under water represents the cleansing of sin by Christ's blood. The world was "baptized" and cleansed by water during the Flood of Noah's day (I Peter 3:20–21).

Read the passage in Genesis 2:10–14 that describes the river flowing out of the Garden of Eden.

Genesis 2:10–14

10 A river flowed out of Eden to water the garden, and there it divided and became four rivers.
11 The name of the first is the Pishon. It is the one that flowed around the whole land of Havilah, where there is gold.
12 And the gold of that land is good; bdellium and onyx stone are there.
13 The name of the second river is the Gihon. It is the one that flowed around the whole land of Cush.
14 And the name of the third river is the Tigris, which flows east of Assyria. And the fourth river is the Euphrates.

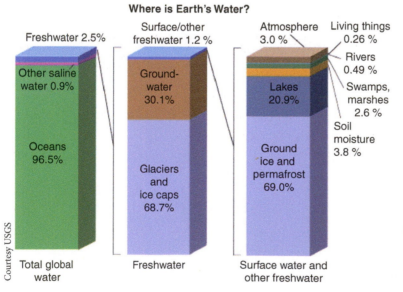

Where is Earth's Water?

Freshwater 2.5%
Other saline water 0.9%
Oceans 96.5%
Total global water

Surface/other freshwater 1.2 %
Ground-water 30.1%
Glaciers and ice caps 68.7%
Freshwater

Atmosphere 3.0 %
Living things 0.26 %
Rivers 0.49 %
Swamps, marshes 2.6 %
Soil moisture 3.8 %
Lakes 20.9%
Ground ice and permafrost 69.0%
Surface water and other freshwater

Courtesy USGS

FIGURE 9.1. The distribution of water on Earth. Humans depend on freshwater every day. However, only 2.5% of Earth's water is fresh, and most of that is inaccessible because it is tied up in glacial ice.

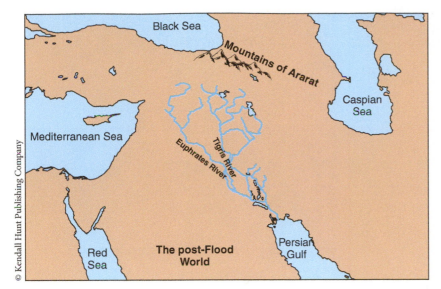

FIGURE 9.2. The modern Tigris and Euphrates Rivers flow through Iraq. These cannot be the same rivers described in Genesis 2 because the Bible says the Tigris and Euphrates were two of four rivers that split from a larger river that comes out of the Garden of Eden, while today they join downstream before entering the Persian Gulf. The original geography no longer exists today because it was destroyed during Noah's Flood.

Do you notice anything strange about this river? First, you might recognize the names of two rivers, the Tigris and Euphrates. Some believe that the location of the Garden of Eden might be able to be located because of this information. However, note that the river which leaves Eden *splits* into four rivers: the Gihon, Pishon, Tigris and Euphrates. Rivers don't split to form other rivers today, in fact they do just the opposite, they join together to form larger rivers. The Tigris and Euphrates Rivers in Genesis 2 are probably not the same rivers flowing in Iraq today, because at the beginning these rivers all had the *same* headwaters, whereas now their sources are separate (figure 9.2). What happened?

The entire Earth's surface went through a radical transformation during Noah's Flood. The geography completely changed; the post-Flood geography was nothing like the pre-Flood geography. As Noah's descendants began to explore and settle in the new world, they likely renamed new places in memory of old places. When settlers came to the Americas, they did similar things, naming the new towns and places, such as Plymouth, Dover, and New England.

9.2 THE HYDROLOGIC CYCLE

The book of Ecclesiastes begins with describing three endless cycles that we should learn about in any Earth science course: the day/night solar cycle (v. 5), the circular nature of winds in the atmosphere (v. 6) and the hydrologic cycle (v. 7). This description of the **hydrologic cycle** (or water cycle, figure 9.3) is poetic, but it outlines the essential nature of what happens to water in this cycle: the ocean never fills up with water because somehow, the water gets back to where it started from. Water evaporates from the ocean as water vapor. In the sky the vapor condenses into liquid or frozen droplets (depending on temperature) to form clouds. Once large enough, the drops fall from the sky back to earth as precipitation. Most water that falls onto the continents becomes part of **groundwater** (water found below the surface of the earth), while some water runs off the surface and makes it to rivers or lakes, and these waters return to the ocean, continuing the cycle.

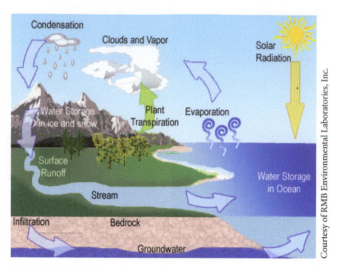

FIGURE 9.3. The hydrologic cycle.

Courtesy of RMB Environmental Laboratories, Inc.

9.3 THE WORK OF STREAMS

9.3.1 Stream Erosion

Streams do several different kinds of work: erosion, transportation, and deposition. In a Chapter 8 we discussed the difference between weathering and **erosion**. Sometimes they are hard to distinguish from one another, but *weathering* is the breakdown of rock into smaller pieces while *erosion* is the removal of the rock from its original location. Streams can be responsible for quite a few different types of erosion. **Abrasion** is similar to the action of sandpaper and can happen as rocks and sediments rub against each other. **Hydraulic action** is erosion by the sheer force of moving water. If you use a pressure washer to clean a building or spray mud off your car, you are

using hydraulic action. In some circumstances rivers can remove rock in this way. Another way in which rivers can remove rock is by **solution**, or dissolving the rock. This is particularly important in humid tropical areas when rivers flow across limestones or volcanic rocks. The minerals in these kinds of rocks are particularly unstable and decompose as warm water continually flows across them. **Cavitation** is a special kind of erosion that takes place when water flows rapidly across a depression and leaves a vacuum bubble in the hole. Vacuum bubbles do not explode, instead they collapse (or "implode") and create a powerful sledge-hammer effect on whatever medium the water is flowing across.

The Glen Canyon Dam forms Lake Powell along the Colorado River. In 1983, warm weather and rapid snow melt caused the water level in the reservoir to rise very rapidly, to near the top of the dam. Engineers decided to increase the flow of water through cement-lined spillway tunnels so the dam would not be overtopped. These 12-meter tunnels bypassing the dam had been drilled through sandstone bedrock and covered with steel-reinforced concrete. Several days after water had begun to be released

Streams Do Three Kinds of Work:

• Erosion of rock and sediment
• Transportation of eroded products
• Deposition of transported products

through the tunnels, engineers began to feel intense vibrations coming from the dam, and workers saw discolored water and large pieces of rock coming from one of the tunnels. Immediately the flow of water was shut off so the spillway tunnel could be investigated. Surprisingly, a hole $10 \times 12 \times 46$ *meters* was found in the damaged tunnel, probably carved out in just *minutes* by cavitation (figure 9.4). What a powerful process!

9.3.2 Stream Transportation

Weathered and eroded materials transported by streams are called the **load** (Figure 9.5). Depending on the size of the material, it can be transported along the bottom of the stream as **bed load** or carried in the water column as **suspended load**. Larger objects like rocks and gravel on the bottom of the stream can be transported by rolling, sliding or bouncing down the stream bed, while finer material is carried along by the water's flow. Sand is usually too heavy to be carried directly by suspension, and tends to bounce along the stream bed. Sometimes a sand grain gets picked up in the current, carried a short ways and then propelled to the bottom of the stream. As the sand grain hits, it usually causes several other sand grains to jump up into the current,

FIGURE 9.4. Large hole cut by cavitation damage in a spillway tunnel in Glen Canyon Dam in 1983.

Courtesy of U.S. Bureau of Reclamation

and this process it is referred to as **saltation**. Rivers also carry a **dissolved load** which consists of various chemicals, salts, and pollutants that are dissolved in the water.

The ability of a stream to carry a load is called its **competence**. The larger the load, the higher the stream's competence, and a stream's competence is affected by numerous factors like water velocity, gradient (river slope), channel width, and water volume.

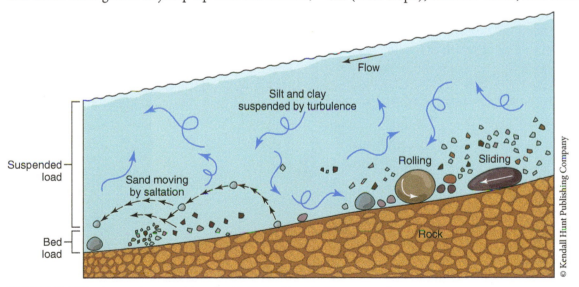

FIGURE 9.5. Different ways that material can be transported in a stream.

© Kendall Hunt Publishing Company

Rivers typically have episodes of much greater competence during flooding which is when rivers accomplish most of their work of erosion and transportation; very little work is done during the non-flooding periods.

9.3.3 Stream Deposition

Transportation in streams happens because competence increases and causes various particles in the stream bottom to be picked up and moved. Likewise, every rock on a stream bottom is there because of a drop in competence. If you look in the bottom of any stream bed you might find rocks, sand or mud. Deposits in the bottom of a stream bed are called **channel deposits** and they will always form as a result of a drop in competence. Even the large boulder in the stream bed in figure 9.6 is a channel deposit. Sometimes sand and gravel bars can be made alongside or even within stream channels. These kinds of deposits are often well-sorted because as the competence drops, all the same-sized particles drop out at once. So, typically we will find sand bars (with not much gravel) or gravel bars (with not much sand).

As water spills out of its channel onto the **floodplain** during periods of flooding, **natural levees** (figure 9.7) can be made alongside the stream. The levees help confine the flow during flooding. When a river spills out of its channel, the water suddenly shallows causing a drop in competence and natural levee formation.

In places like Death Valley, flash floods come racing down steep and narrow mountain canyons with tremendous competence. However, when the rivers approach the valley floors, the gradient of the stream suddenly shallows, causing the water to slow down and the competence to drop. As a result an apron of gravels ends up getting deposited at the mouths of the canyons; deposits are referred to as **alluvial fans** (figure 9.8).

When rivers enter into the ocean or a lake, they no longer can flow downhill and are said to have

FIGURE 9.6. A large boulder deposited in a stream bed during a period of great flow. Anza Borrego Desert, California.

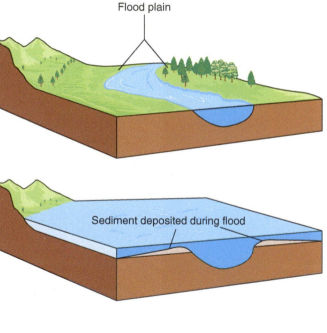

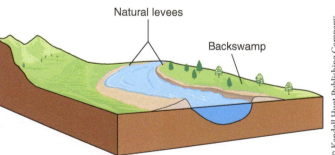

FIGURE 9.7. Natural levees form as a steam spills out of its banks during flood stage.

Photo by John Whitmore

© Kendall Hunt Publishing Company

CHAPTER 9: Streams and Groundwater

FIGURE 9.8. Three small alluvial fans at the base of a mountain in the Anza Boreggo Desert of California.

reached **base level**. As a result, the river can no longer carry its load and the sediments are dumped to form a **delta** (figure 9.9). The word "delta" is derived from the Greek letter shaped like a triangle, and was first applied to the Nile River. Not all deltas are triangle-shaped, though, and delta shape is controlled by factors like the amount of sediment being delivered, currents, and tides. If you look closely at the images of the river deltas, you will see that the main river splits apart into many different

FIGURE 9.9. The Yukon River of Alaska as it enters into the Pacific Ocean. Note the many distributaries branch off from the main river channel.

channels as it approaches the sea. Upstream, smaller **tributaries** came together to make the larger river. On the delta, the large stream splits into smaller river channels called **distributaries**.

9.4 FLOODS

By far, floods are the most numerous and devastating type of natural disaster. More people lose their lives as the result of flooding than earthquakes, tornados, landslides, or hurricanes. There are many floods that happen every year with devastating consequences. We will look at three famous examples of floods: the 1931 Chinese floods, the 1993 Mississippi River flood and the Ice-Age Missoula flood. Of course, all of these examples are local/regional floods, and pale in comparison to Noah's Flood described in Genesis 6–9. But by studying these events we gain understanding and appreciation for what Noah's Flood may have been like.

Perhaps the modern world's greatest natural disaster of all time was the central China floods that began in July of 1931. The area experienced several years of drought that was followed by a winter of heavy snows. During the summer of 1931, the area was hit by at least seven tropical storms that led to massive flooding along the Yangtze, Hwang, Hwai, and Huai Rivers. Many of the dikes and reservoirs along the rivers failed, causing months-long, meters deep standing water in many farming communities, ruining the crops. It is impossible to know the death toll, but it is estimated nearly 150,000 died by drowning and 3.5 million died from starvation and disease in the subsequent months. It is reported that during

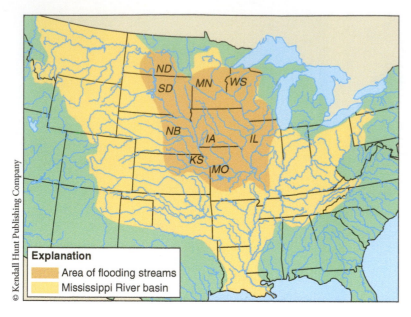

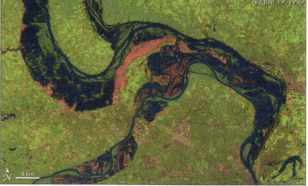

FIGURE 9.10. The area of the Mississippi River drainage basin that experienced many severe thunderstorms during the summer of 1993 and contributed to the historic flooding of the river.

the height of the flood, the Yangtze River averaged more than 60 km wide over a 1,400 km stretch along it course! The famous American pilot, Charles Lindbergh, took photos of the city of Gaoyou being flooded in September of 1931.

In 1993, there were a record number of thunderstorms that pummeled the Midwestern portion of the United States. The water in this area is funneled to the Ohio, Missouri, and Mississippi Rivers, which drain a large portion of the country (figure 9.10). The floods began in April and lasted through October becoming one of the most costly natural disasters in the United States causing approximately 15 billion dollars in damages. Heavy snowfall and steady spring rains caused the soil in the area to become saturated with water, causing additional precipitation during the summer to quickly runoff. Satellite photos of the confluence of the Missouri and Mississippi Rivers in St. Louis show just how extensive the flooding was (figure 9.11). Many of the dikes and levees that had been built to hold in the rivers during flood stage failed, and some areas remained flooded even six months afterward.

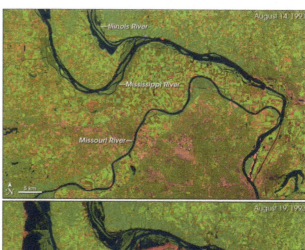

FIGURE 9.11. Images of the confluence of the Mississippi and Missouri Rivers in the St. Louis area before and during the flooding of 1993.

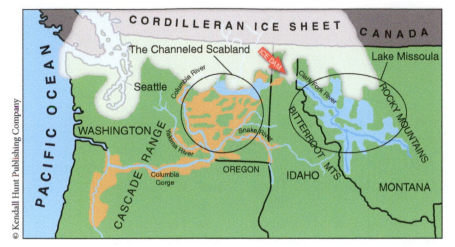

FIGURE 9.12. A continental glacier moving south from Canada blocked the Clark Fork River in Idaho and created an ice dam which caused the formation of Lake Missoula in Montana. When the ice dam failed, the water rushed down the formerly blocked valley and rapidly eroded the "Channeled Scabland."

During the Ice Age, glaciers advanced from western Canada southward into the states of Washington, Idaho, and Montana. As the glaciers marched forward, one eventually blocked the drainage of the Clark Fork River (figure 9.12) in what is now the state of Idaho. Water began to back up behind the natural dam and eventually formed glacial Lake Missoula, and enormous lake the size of Lake Erie and Lake Ontario *combined*. In time, it is believed the ice dam began to leak, weakening the ice until it failed, catastrophically, releasing all of Lake Missoula's water, which carved a magnif-icent landscape in Washington called the Channeled Scabland (look for it on Google Earth®). It is estimated that the water flowed 100 kph as it raced across the state of Washington towards the Columbia River and eventually to the Pacific Ocean. It is thought that the water took about two weeks to drain. J. Harland Bretz was a geologist who began to work all this out in the 1920's, but it took him a while to put together the whole story. At first, no one believed him because the story was too fantastic and sounded too much like the biblical Flood.

Bretz's Evidence for the Missoula Flood:

1) Dry waterfalls with deep plunge pools at their bases (figure 9.13).
2) Giant potholes in the valley bottoms carved into solid bedrock (figure 9.14).
3) Dry hanging valleys above the main valley floors with no evidence of glacial origin.
4) A giant underwater debris delta in the Pacific Ocean at the mouth of the Columbia River.
5) Giant gravel bars, some 120 meters high
6) Giant ripple marks made out of gravel and boulders, 4—9 meters high (figure 9.15).
7) Ancient shoreline marks on hillsides in places like Missoula, Montana.
8) Lake sediments in the valley floors in places like Missoula, Montana.
9) A very weird drainage pattern of anastomosing streams in the eastern part of Washington (figure 9.16).
10) Large erratic boulders that littered the landscape

FIGURE 9.13. Dry Falls, near Coulee City, Washington. During the Missoula Flood, floodwater rushed over the 100 m high cliff in the background. The falls were five times the width of Niagara Falls and the water was estimated to be about 60 m deep and flowing at a rate of 65 km/hour. As the water cascaded over the cliff it cut giant plunge pools (now filled with water) at the cliff's base. It is recommended you look at this area with Google Earth®.

FIGURE 9.14. A large pothole about 70 m across excavated during the Missoula flood.

FIGURE 9.15. Giant ripple marks made near Camas, Montana, by deep fast moving water from the collapse of the ice dam that held back Lake Missoula.

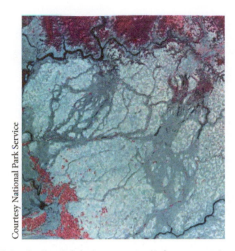

FIGURE 9.16. A weird drainage pattern of anastomosing streams created in eastern Washington by the catastrophic drainage of Lake Missoula.

As Bretz began his study of the Channeled Scabland in the state of Washington, he came across many interesting and associated features (see box). No single hypothesis of which he was aware could explain the features. He couldn't think of a way to explain the features by normal stream activity or by glacial activity. This led him to think of another radical hypothesis that might explain the data. He began publishing his ideas about a large flood carving the landscape and as an explanation for all the odd features in the 1920's. As the years passed, he continued to find more data to support his radical theory, which wasn't widely accepted until he published a final paper on the topic in 1969.

9.5 TYPES OF STREAMS

There are many types of streams, and finding them is easy using Google Maps® or Google Earth®. One of the easiest types of streams to identify is the **meandering stream** (figure 9.17). These streams snake back and forth across a generally flat valley bottom, making giant "S" curves. These rivers often change course, especially during times of flooding. Meandering rivers actively erode on the outsides of the meander bends, forming a **cut bank**; and actively deposit on the shallow inside portion of a meander

FIGURE 9.17. A whole series of meander scars and oxbow lakes on the Rio Negro in Patagonia, Argentina. Inset: A point bar is the inside of a meander loop (top side of river in this picture) and usually consists of sand or gravel deposits.

FIGURE 9.18. The "Goosenecks" of the San Juan River, Utah. This is an example of entrenched meanders. This would be a good site to examine on Google Earth®.

bend called a **point bar** (figure 9.17 inset). At the cut bank, the water is deeper and has greater competence than the shallow water on the inside of the meander bend. Sometimes two cut banks will erode towards each other and actually join. When they join, the river takes a short cut through the new channel and the old mender is typically abandoned. The old meander can still contain water and form a horseshoe-shaped lake called an **oxbow lake**. Over time, the oxbow lake can fill with debris forming a **meander scar**. Figure 9.17 shows a series of meander scars, oxbow lakes and point bars.

If an entire landscape is lifted up under a meandering stream, or if the stream is able to cut quickly down into the bedrock below its channel, it can become a meandering river in a canyon. When this happens, the meanders are referred to as **entrenched meanders** (figure 9.18).

Braided streams (figure 9.19) get their name from the numerous crisscrossing channels that resemble a hair braid. They are often found in mountain valleys where high gradient rivers deliver great quantities of sand and gravel to the valley. During spring snowmelt and flooding, the river is often a single channel and is able to carry the increased sedimentary load. However, as summer approaches, water levels decrease and the river channel is filled a huge load of un-transported sand

and gravel. As this happens, the river splits into a number of more shallow braided channels among the low points of the gavel.

Disappearing streams are streams that "disappear" for one reason or another. This can happen in a number of different ways. One of the most common ways is that the stream disappears into a sinkhole and begins to flow underground, usually within a cave. Streams can also disappear in the desert, usually due to evaporation of the water and the water

FIGURE 9.19. A braided stream system (green) in Madagascar.

Photo by John Whitmore

FIGURE 9.20. A misfit stream is a small stream that is in an exceptionally large valley or canyon. The Colorado River in the Grand Canyon could be considered a "misfit" stream.

Steve Byland/Shutterstock.com

FIGURE 9.21. There are several terraces along the Snake River at the base of the Teton Mountains in Wyoming. The lowest floodplain is next to the river. Two other floodplains can be seen in the grasslands above the river.

soaking into the desert sands and gravels. The stream really doesn't disappear of course; it just ceases to be seen above ground and instead becomes part of the groundwater system.

When we think of a "misfit" we usually think of something that isn't quite up to par. The vast majority of rivers, all around the world are **misfit streams**. This means that most rivers are in very large valleys compared to the relatively small amount of water that they carry (figure 9.20). In most cases, rivers are probably misfit because they had orders of magnitude more flow in the past than they currently have flowing in them today. In areas that have been glaciated, this might be explained do to rapid melting of glacial ice or a megaflood, as in the case of the Missoula flood. It may be that some areas simply had much greater amounts of precipitation in the past.

A floodplain is a flat area onto which a river spills during flood stage. Larger rivers meander on at least a small floodplain; some rivers, like the Mississippi, have enormous floodplains many tens of kilometers wide. Floodplains usually are composed of thick deposits of sand and gravel through which the river meanders. Occasionally an entire landscape is lifted up from below and the river begins to cut downward into its channel, abandoning its former floodplain. When the landscape becomes stable again, the river cuts a new floodplain at the lower level, leaving a bench on the valley side. This process can happen several times creating a whole series of steps or terraces along the valley wall. One of the most famous (and scenic) sets of **river terraces** are along the Snake River on the east side of the Teton Mountains in Wyoming (figure 9.21).

9.6 THE ROLE OF GROUNDWATER IN THE HYDROLOGIC CYCLE

Groundwater is simply water that is below the Earth's surface. If you have ever dug a shallow hole, sometimes you will notice the bottom of the hole will fill with water as you dig. This is groundwater. Some groundwater is close to the surface, and

some can be found deep underground. Although groundwater only accounts for 0.7% of Earth's surface water (see figure 9.1), it is a very important part. Most water that humans consume comes from wells that pump groundwater out of the Earth. You have probably used groundwater already today. Water gets into the ground from precipitation that seeps into rocks and soils and becomes part of the groundwater (see figure 9.3). Only a small portion of precipitation finds its way into rivers and streams, and in fact, most rivers and streams get their water from groundwater!

9.6.1 The Water Table

As groundwater seeps through rock and soil, it flows downward because of gravity. At first, the water seeps through cavities that are full of air. This area is called the **unsaturated zone** (sometimes referred to as the **zone of aeration**). Eventually the water reaches a place where there is no more air space and the ground is completely full of water. At this, point the downward moving water has reached the **water table** (figure 9.22). If you have ever looked down into an old-fashioned well, the water level you see is the water table. Just above the water table is a zone called the **capillary fringe**. Have you ever placed the corner of a napkin in a small puddle of water and watched the water magically get drawn upwards into the napkin, seemingly defying gravity? The same thing happens just above the water table. Water gets drawn upwards just slightly above the water table, by capillary action, hence the name.

Because the term "table" is used to describe the water table, many people have the connotation

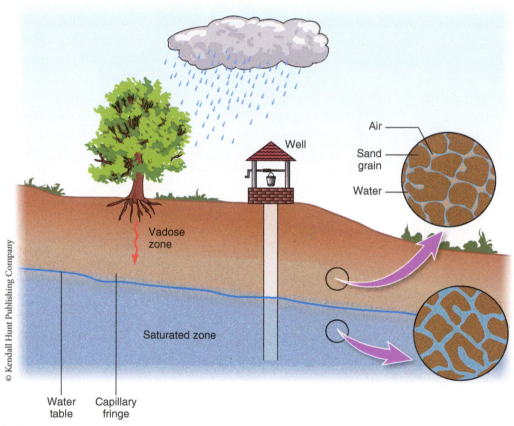

FIGURE 9.22. The water table is the boundary between the zone of aeration (unsaturated or vadose zone) and the zone of saturation. Note that the water table is not a flat line, but follows surface topography.

that it is flat. It is not. It often mirrors the same contours that are on the surface of Earth. If there is a hill, the water table is higher. If there is a valley, the water table is lower. Because there are high and low places in the water table, water will flow (under the influence of gravity and pressure) from high areas to low areas. During times of drought, the water table becomes lower because the groundwater leaks out into rivers and streams In fact, you can think of lakes and streams as places where the land surface is so low that it gives you a "window" to see the groundwater.

9.6.2 Wells and Springs

Most people who live in rural locations have a well somewhere on their property as a water source. If groundwater is particularly abundant, even small municipalities will service their citizens from a local well field (a collection of closely spaced wells). Larger municipalities will usually treat water from a reservoir or river as their water source. In developed countries, a hole is usually drilled down through the soil into bedrock to obtain groundwater. In many countries, wells are still dug by hand, and these wells are typically shallow (< 15 m in depth) and usually cannot penetrate bedrock. Probably billions of people around the world access their water from shallow hand-dug wells.

When a well is drilled, the goal is to drill the hole into an **aquifer**, far below the water table (figure 9.23). An aquifer is a layer of rock, sand or gravel through which water can easily flow. It is necessary to drill the well into an aquifer so that when water is removed, it can quickly be replaced by new water. The ability of water to flow through the aquifer is referred to as **permeability**. Well drillers seek aquifers that have good permeability and good **porosity**. Porosity is the amount of empty space in an aquifer. Sandstones and gravels tend to be the best kinds of aquifers because they have both good porosity and permeability. Look at the sandstone in figure 9.24. The blue color is the empty space between all the sand grains, or the porosity. If these empty spaces are well-connected, the rock will have good permeability as well. Limestone can

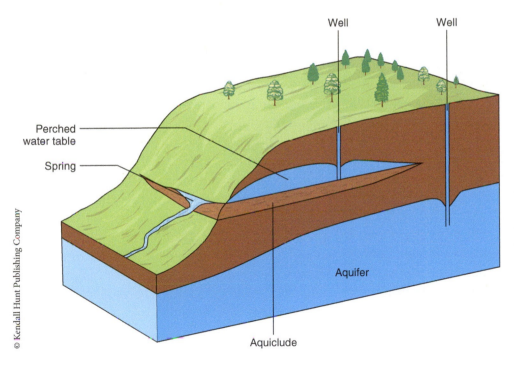

Well   Well

Perched
water table

Spring

Aquifer

Aquiclude

FIGURE 9.23. The goal of a water well is to drill the well into an aquifer below the water table. Sometimes water tables can be "perched" if a layer of clay or shale (an aquiclude) prevents the water from moving deeper into the earth. Sandstones are the best aquifers.

100 μm

FIGURE 9.24. This is a microscopic view of the Coconino Sandstone. The picture is about 1.5 mm wide. The blue color is porosity, or empty space within the sandstone. This is where water could be stored if it were present.

be a good aquifer, particularly if the well encounters fractures where water flows easily along the cracks in the rock. An **aquiclude** is a rock or substance that does not have good permeability. A layer of clay or shale usually acts as an aquiclude because it is composed of particles that are very tiny (< 1/256 mm) and permeability is very low.

Sometimes there is more than one water table. A **perched water table** (figure 9.23) is one that rests higher than the main water table. It is formed by a layer of clay, shale, or other impermeable rock layer that is local in occurrence, and traps water above it. If the perched water table holds enough water, wells can be drilled into them.

Springs are places where water naturally flows out of the Earth, either continually or intermittently. There are several different ways in which springs can form (figure 9.25). Some springs are simply caused by a high water table that forces water above the surface of the ground, while others form from a perched water table that intersects a hillside, causing it to flow out and down the hill. Joints and faults in bedrock usually make good conduits in which water can flow. If the cracks intersect a hillside, springs will

usually be found in the same place. Carbonate rocks like limestone can develop a whole network of solution cavities and tunnels, and some are large enough to form caves. Sometimes the flow is so great that it could be considered an underground river. If the river then comes out of a hillside, it is called a "giant" spring.

Calcium carbonate is fairly soluble and is typically one of the main minerals found in spring water. **Spring deposits** are made when minerals such as calcium carbonate are deposited as a spring leaks out of a rock. In Cedarville, Ohio, groundwater flows through dolomite and limestone, dissolving some of the minerals as it flows. The groundwater is prevented from flowing deeper in the ground because of a shale unit that acts as an aquiclude, forcing the water to flow horizontally. The water comes out of a cliff face at the contact between the carbonate rocks and the shale (figure 9.26) making spring deposits wherever the water flows. If spring deposits are full of small holes, the carbonate rock is called **tufa**. If the deposit is more solid and lacks holes, it is called **travertine**.

Sometimes water can flow out of the ground under natural pressure. If it flows out of a well in this manner, it is called an **artesian well** (figure 9.27). Circumstances can also occur when the water naturally flows out of the ground; in this case it is called an **artesian spring**. In both of these cases, water flows out of the well or spring because the water table is higher than the outlet. Some spring water is very good to drink, and can be used without treatment. However, extreme caution should be used. Not all spring water is good to drink. Some springs contain various parasites like *Giardia* that can give you a nasty case of diarrhea and cause further health problems later. Other spring water can have various minerals or natural pollutants that would not be good to consume. Many of the natural springs in the Yellowstone National Park area smell quite bad because the water is enriched in sulfur.

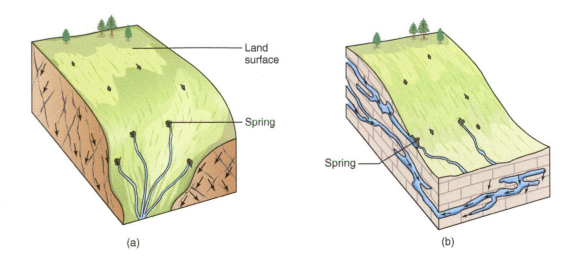

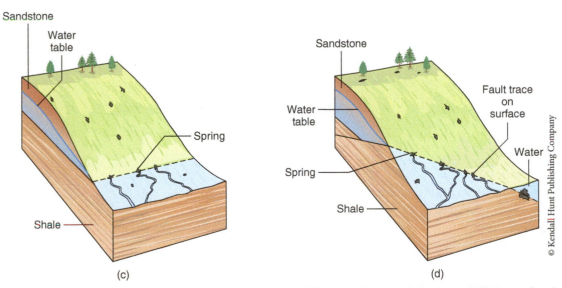

FIGURE 9.25. Several different ways in which springs can form. (a) Water can flow through fractures. (b) Water can flow through solution cavities in limestone. (c) Water can be prevented from flowing deeper by coming into contact with an aquiclude, such as a layer of shale. (d) Water can flow along a fault.

FIGURE 9.26. A limestone spring deposit in Cedarville, Ohio. Groundwater easily flows downward through the fractured dolomite, but as it flows deeper it encounters the shale which acts as an aquiclude. Since the water can no long go deeper, it flows horizontally and out of the cliff face making the moss covered spring deposit.

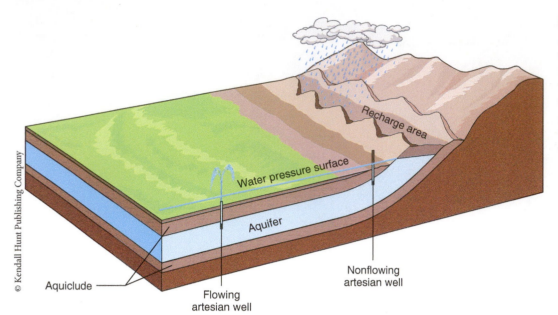

© Kendall Hunt Publishing Company

FIGURE 9.27.
Artesian wells form when an aquifer is confined between two aquicludes and the water pressure surface is higher than the top of the well head.

Recharge area

Water pressure surface

Aquifer

Nonflowing artesian well

Aquiclude

Flowing artesian well

9.7 GROUNDWATER USE AND PROBLEMS

9.7.1 Home Uses

Perhaps you live in a home that has a well and/or septic system or perhaps you will buy one someday. Figure 9.28 shows a sketch of a typical well and septic system on a rural property. The well is a deep hole (typically greater than 30 m deep) and about 15 cm in diameter. A pump is located near the bottom of the hole and forces water up the hole to a water storage tank, which is pressurized (using a thick rubber balloon inside) to push the water through the pipes and out of the faucet. As the pump fills the tank with water, the pressure becomes greater and greater on the balloon and eventually a point is reached where a pressure switch shuts the pump off. As you use water, the switch is turned back on to pump in more water.

When a home is located far from a sewer system, all drains in the house lead to a septic tank where sludge, toilet paper and garbage that makes its way down the drain is decomposed. You can even buy special enzymes at the hardware store to flush down the toilet to help decomposition take place. Most septic systems also have a second tank usually with an overflow that goes out to a drain field. This will usually be an area in your yard that is green, moist and damp (because it has been well-fertilized!). In a properly-functioning septic system, the water released in the drain field is ready to go back into the aquifer, and the water's journey through the rock and soil will further filter the water, making it drinkable again. Regular maintenance is critical to keeping a septic tank working efficiently. Proper maintenance includes periodic pumping of the septic tank. There are local companies that will come with a tank truck and pump your system. Another bothersome issue is a high water table. After periods of heavy rain or rapid snow melt, the water table may rise above the level of the septic system, preventing normal flow. When this happens, your toilets will back up and your drains won't work, and all you can do is wait for the water table to return to normal! Buying a home that is on slightly elevated terrain is usually well-advised so this problem can be avoided.

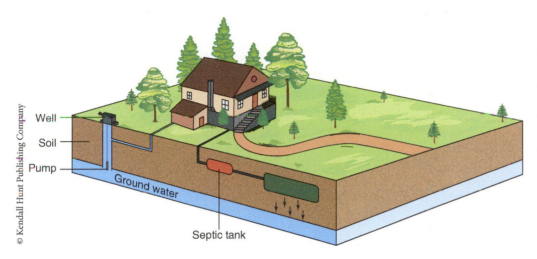

© Kendall Hunt Publishing Company

FIGURE 9.28. A typical rural house has both a well and a septic tank, usually in close proximity to each other. It is important that both are working correctly to avoid contamination of drinking water.

Well

Soil

Pump

Ground water

Septic tank

9.7.2 Large-Scale Use

When a well is pumped, the level of the water table around the well lowers, creating a **cone of depression** (figure 9.29). This cone forms because water cannot move through the aquifer to refill the well hole as fast as the water is being pumped out. The larger the well, and the more water that is pumped out, the bigger the cone of depression. Normally this is not a problem unless the well belongs to a large farm (where enormous quantities of water is used for irrigation) or a municipality. In these cases the cone of depression can become so large, that it lowers the water table below other local wells, causing these wells to become dry. Managing the sizes of the cones of depression is usually advisable so the cones do not fall below the bottoms of other wells.

Landfills can pose significant challenges, since groundwater flows both to and away from them. If the landfill contains toxic chemicals and leaks, it is nearly impossible to clean the groundwater. In the days before federal regulations, any old hole was filled with debris. It didn't matter if it was garbage, demolition debris or industrial waste; it all went to the landfill, unregulated. This caused enormous problems later, because rain water filters down through the landfill and causes any chemicals present to become part of the groundwater. For this reason, landfills are now carefully regulated. New landfills are lined with clay, shale or some other impermeable

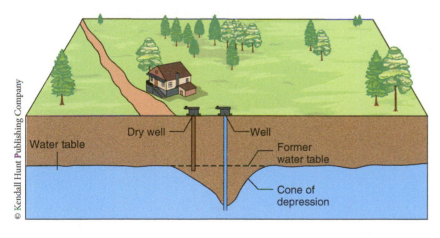

© Kendall Hunt Publishing Company

Water table

Dry well — Well

Former water table

Cone of depression

FIGURE 9.29. A cone of depression develops around a well that is being pumped because water from the water table cannot flow back into the well as fast as it is pumped out. Large wells can have a cone of depression so large that it causes smaller wells to go dry.

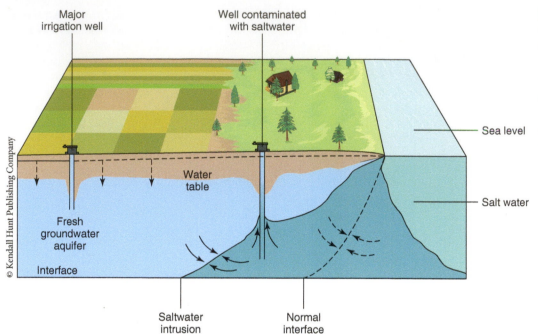

Major irrigation well

Well contaminated with saltwater

Sea level

Salt water

Water table

Fresh groundwater aquifer

Interface

© Kendall Hunt Publishing Company

Saltwater intrusion

Normal interface

substance to prevent hazardous materials from being introduced to the groundwater. However, in lining the landfill, an artificial perched water table is created, which contains toxic water. Modern landfills have wells that are pumped so the toxic water can be treated and released back into the environment safely.

Big cities that are close to the ocean usually have groundwater problems because of their close vicinity to saltwater. Not all groundwater is fresh; close to the ocean is a saltwater water table (figure 9.30). The two types of water usually do not mix easily and the freshwater rests on top of the saltwater because freshwater is less dense than the saltwater. When wells are placed, caution needs to be exercised not to pump the wells too heavily. As normal, a cone of depression will form on top of the freshwater table. However, if too much pumping occurs, saltwater will be drawn upwards into the well, forming a **cone of ascension**, and causing the well to produce saltwater instead of freshwater. Like polluted water from a landfill, this is also a difficult problem to correct and requires pumping large quantities of

fresh water back into the well to force the saltwater table back down.

Some areas pump much more groundwater than is replaced by rainfall every year. This happens in places like Las Vegas, Los Angeles and Mexico City. These large cities all occur in desert environments and receive very little rainfall. As a result the water table in these areas continues to fall every year. Some areas have exceptionally rich soils, but are also located in arid lands. The San Joaquin Valley in California (figure 9.31a) provides a huge quantity of nuts, olives and vegetables for much of the United States. In order to grow all of the crops, the farmers need to pump large quantities of groundwater that is not replaced by annual rainfall. As a result, the ground in the area is subsiding (sinking) at an alarming rate (figure 9.31b). So much water has been pumped out that the soil particles are compacting which is causing the ground surface to sink. Mexico City (the largest city in the world) is also having trouble with subsidence due to too much pumping. The subsidence often occurs unequally, causing structures and buildings to tilt and become unstable.

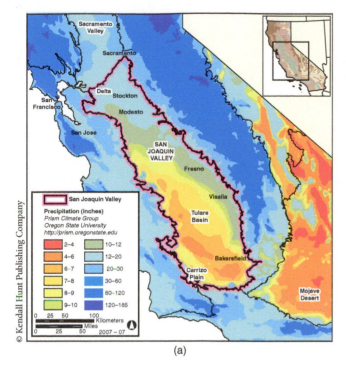

Courtesy USGS

FIGURE 9.31. The San Joaquin Valley, receives very little rainfall (a). The soils are rich, but they do not receive adequate amounts of rainfall for most types of agriculture, so they must irrigate using groundwater. This has caused significant subsidence, as shown by the telephone pole markings between 1925 and 1977 (b). As of 2015 California is experiencing a severe prolonged drought in this area, which is continuing to put stress on already scarce water resources.

Groundwater management is one of the primary areas in which geologists are employed. Water is one of our most important natural resources and clean water is perhaps the greatest priority for any home, village, town, or city. Without clean water, it is impossible to live in that place for very long.

This is an area of great opportunity for Christian geologists to have influence in our world. Access to clean groundwater is a life and death situation for many small villages in third world countries; countries that are often closed to traditional missionary work.

9.8 CAVES AND KARST FEATURES

Caves are large underground openings that sometimes are in the form of tunnels, rooms, or crawlspaces. They come in many different varieties and can have beautiful and intriguing cave formations. Most caves are in some type of carbonate bedrock (limestone, dolomite, marble). That is because this type of rock is susceptible to solution by various types of acids. It was thought that carbonic acid produced at the surface (as water migrated through soils) dissolved the limestone to form caves from the top-down. However, this type of thinking is changing and now it is believed that many caves may have formed by some very different processes

(consult the readings by renowned cave geologist, Dr. Emil Silvestru). Many caves have interesting erosional features that seem to point to various fast-acting, catastrophic erosion processes. Silvestru believes that many caves formed late in the Flood or early in the post-Flood period as a result of acidic water (sulfuric acid) rising up from depth, not coming down through soil and rock. Most caves become larger the deeper you go (like Mammoth Caves in Kentucky or Carlsbad Caverns in New Mexico) because the acids were more concentrated at depth and were diluted by surface waters as they rose.

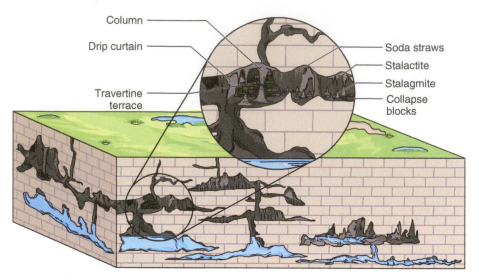

Column
Drip curtain
Travertine terrace
Soda straws
Stalactite
Stalagmite
Collapse blocks

FIGURE 9.32. The variety of different types of cave formations is astounding. The most common are stalactites, stalagmites, soda straws, and columns.

Once the cave forms (i.e., there is some empty space within the rock), the cave can begin to be filled with various **speleothems** (precipitated cave formations) if the water table is located below the cave. When water from the surface drips down inside the cave, it has small amounts of calcium carbonate ($CaCO_3$) dissolved in it. The dripping water can deposit the calcium carbonate in the form of travertine in various ways (figure 9.32). The limestone can come out of solution by temperature changes or by loss of carbon dioxide in the dripping process. Whether this process is slow or fast greatly depends on how much and how fast water drips in the cave. Cave-like formations can often be found on the undersides of bridges and monuments that haven't taken long at all to form.

The walls of the cave can be coated with **flowstone**. Icicle-shaped rocks hanging from the ceiling of the cave are called **stalactites**. Sometimes stalactites are hollow inside, and are called **soda straws**. Water can drip from stalactites and soda straws down to the cave floor. Here, a cone-shaped deposit can grow from the floor of the cave called a **stalagmite**. Over time, a stalactite and a stalagmite can join to form a **column**. Most cave formations are vertical, but if there is a draft in the cave, some cave formations seem to almost defy gravity. These are called **helictites** (figure 9.33). Some cave formations are rare and poorly understood.

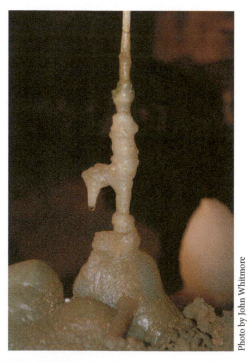

Photo by John Whitmore

FIGURE 9.33. A helicite called "the town pump" at Ohio Caverns near West Liberty, Ohio.

FIGURE 9.34. Shields are relatively rare cave formations, but occur in abundance in Lehman Caves in Great Basin National Park in Nevada.

FIGURE 9.35. In the vicinity of Winter Haven in central Florida, hundreds of sink holes are filled with water and form round lakes. Occasionally new sink holes develop and swallow houses and roads.

For example, Lehman Caverns in Great Basin National Park (Nevada) has an extraordinary number of shields (figure 9.34); as to why they are common in this cave and rare in others, no one knows. Most cave formations are made of travertine limestone, but sometimes other minerals can be involved too. Ohio Caverns in West Liberty, Ohio, has some unusual and very colorful iron carbonate cave formations.

Because caves are large underground cavities, sometimes their roofs collapse, making holes at the surface of the Earth. These holes are called **sinkholes**. Sinkholes can form suddenly, or it can be a slow process that happens over time. If the water table is close to the surface, the sinkhole will fill with water and form a round lake (figure 9.35). Sometimes a river will flow into a sinkhole and form a disappearing stream.

Landscapes that are characterized by caves, sinkholes and disappearing streams are referred to as areas of "**karst topography**" (figure 9.36). In the United States, the two most famous areas of karst topography are south central Kentucky and central Florida. Both of these areas have carbonate bedrock just below the surface that can easily dissolve in the presence of acid. In 2013, a sinkhole formed suddenly below a house in Seffner, Florida, killing one of the occupants as part of the house crumbled into the hole as he slept. In 2014, a sinkhole formed directly below the National Corvette

Museum in Bowling Green, Kentucky (figure 9.37). Eight priceless cars fell into the sinkhole. It took several months to recover the cars, some of which were damaged beyond repair.

9.9 HOT GROUNDWATER

One of the most amazing experiences that you should experience in your life is a visit to Yellowstone National Park. It is especially famous for its display of hot groundwater features which number in the thousands. There are four types of **thermal features**: **hot springs**, **mud pots**, **fumaroles** and **geysers**. Why does Yellowstone have so many thermal features? Most of the park rests in a giant volcanic caldera and there is still hot rock and magma close to the surface. Sometime shortly after the Flood, the giant Yellowstone volcano blew itself apart, sending ash over half of the United States. So much

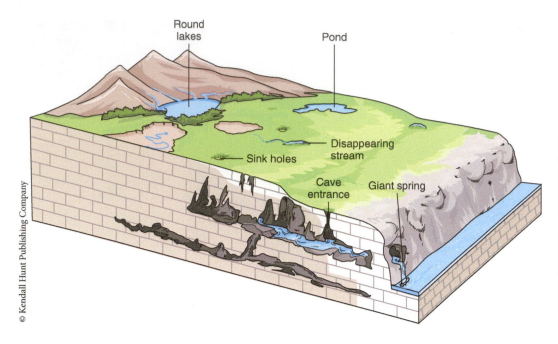

Round lakes

Pond

Sink holes

Disappearing stream

Cave entrance

Giant spring

© Kendall Hunt Publishing Company

FIGURE 9.36. Areas of karst topography have limestone below the surface that has dissolved and is filled with caves. The surface is characterized by disappearing streams, sink holes, round lakes and cave entrances.

material was erupted from the magma chamber, that the top of the giant volcano collapsed into the underground void after the eruption. Many fractures formed in the brittle rock during the collapse which provide conduits for the downward movement of water. As the water travels downward, it is heated and rises back up to the surface (figure 9.38).

The hot springs (figure 9.39) in Yellowstone are very hot! Some of the water even boils as it comes out of the ground. It's amazing, but specially designed bacteria actually thrive in the hot water. Different kinds of bacteria make different colors. The hottest water is blue, followed by yellows and browns. Many of the hot springs emit clear water, but occasionally the water rises up through volcanic ash which weathers to clay. As the hot water rises, it becomes mixed with the clay and makes mud. The thick pasty mud rises to the surface and forms a mud pot (figure 9.40). Mud pots are often display many pastel colors and are fascinating to watch because blobs of mud often shoot unexpectedly up into the air.

Fumaroles occur when hot gases escape from the ground. Fumaroles often have a rich sulfur smell accompanied with a hissing sound, kind of like an open steam pipe. Geysers form as fountains of steam

National Corvette Museum/Rex Features

FIGURE 9.37. In 2014 a large sinkhole developed in the floor of the National Corvette Museum in Kentucky, swallowing eight cars. Some of the cars are now displayed as part of a sinkhole damage exhibit.

and water shoot up into the air. Some geysers erupt continuously, others intermittently (sometimes years between eruptions) and others are more or less regular, like Old Faithful (figure 9.41). It is thought that geysers form because cool groundwater fills cavities at depth. The water is then heated to the boiling point, flashes to steam, which forces the mix upward through open holes or pipes forming the geyser.

CHAPTER 9: Streams and Groundwater

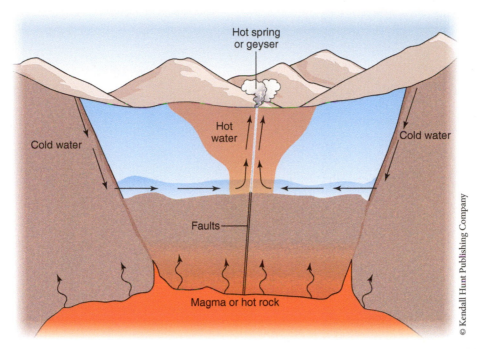

FIGURE 9.38. Hot springs and other thermal features form as cold water is circulated deep underground, is heated, and then is forced back to the surface.

FIGURE 9.39. The Grand Prismatic Hot Spring in Yellowstone National Park in Wyoming. Each color is due to a different type of bacteria, which each live in a different water temperature, blue being the hottest.

FIGURE 9.40. Fountain Paint Pot, Yellowstone National Park, Wyoming. Hot groundwater rises to the surface through volcanic ash layers making the bubbling mud.

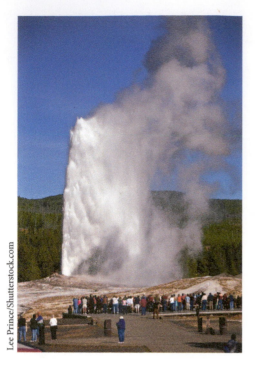

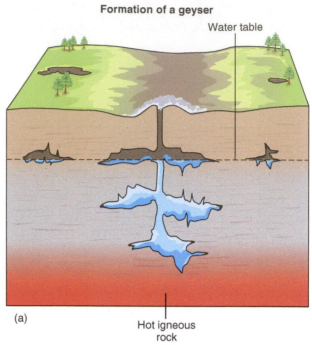

Formation of a geyser

(a)

(b)

FIGURE 9.41. Geysers like Old Faithful in Yellowstone National Park (a) are formed when groundwater fills empty cavities, gets heated and then flashes to steam (b).

9.2 The Hydrologic Cycle

- Water is perhaps the single most important natural resource that we have. It is circulated via the hydrologic cycle.

9.3 The Work of Streams

- Weathering is breaking larger rocks into smaller pieces, while erosion involves picking up and transporting those pieces. Streams can erode by the processes of abrasion, hydraulic action, cavitation and solution by various acids.
- Material is transported in a stream in the bed load, the suspended load and the dissolved load. The ability of a stream to transport a certain sized particle is called competence.
- Streams deposit their load when water velocity and/or the volume of the stream are reduced. Deposits can be in the form of bars, channel deposits, alluvial fans, deltas and natural levees.

9.4 Floods

- One of the most devastating natural disasters in terms of lives lost is floods. It is important for us to understand these processes so we can build in appropriate areas and evacuate during periods of high risk. The study of large floods like the Missoula Flood might help us understand how the Grand Canyon might have been formed or processes that might have occurred during Noah's Flood.

9.5 Types of Streams

- The two major types of streams are meandering and braided. Each have characteristic features such as point bars, meanders, and cutbanks (meandering streams) or gravel bars and crisscrossing channels (braided streams)
- Changes in the stream's behavior can make stream terraces. A floodplain is the relatively flat area nearby a stream that is periodically flooded.

9.6 The Role of Groundwater in the Hydrologic Cycle

- Groundwater is where most humans get their drinking water.

9.7 Groundwater Use and Problems

- The water table is the top of an underground reservoir of water. Wells are dug or drilled into the water table to access clean drinking water. A cone of depression forms around a well as the water table drops due to pumping. It is important to understand how wells and septic tanks work, because you may have property with these utilities on them someday.

- An aquifer is layer of rock that supplies drinking water. Sandstones are the best aquifers because they are both porous and permeable. Aquicludes prevent the flow of water; shales or clay make good aquicludes.
- Groundwater can sometimes be polluted by landfills or septic tanks that are not working properly. Sometimes so much water can be drawn out of the ground that the surface of the ground sinks. Many desert communities exact far more water out of the ground than is replaced by precipitation.

9.8 Caves and Karst Features

- Carbonate rocks are susceptible to cave formation because the calcite they are composed of can dissolve in the presence of various acids (to make caves). Caves can be filled with a variety of different types of cave formations. The ceilings of caves can collapse to make sinkholes on the earth's surface.

9.9 Hot Groundwater

- Groundwater can be heated to form hot springs, geysers, mud pots and fumaroles. This features are often found either in mountainous or volcanic settings in places where groundwater can circulate downward and become heated by hot rock.

KEY TERMS

Abrasion	Dissolved load
Alluvial fans	Distributaries
Aquiclude	Entrenched meander
Aquifer	Erosion
Artesian spring	Floodplain
Artesian well	Flowstone
Base level	Fumaroles
Bed load	Geyser
Braided stream	Groundwater
Capillary fringe	Helictite
Cavitation	Hot springs
Channel deposits	Hydraulic action
Column	Hydrologic cycle
Competence	Karst topography
Cone of ascension	Load
Cone of depression	Meander scar
Cut bank	Meandering stream
Delta	Misfit stream
Disappearing stream	Mud pots

Natural levees	Spring deposit
Oxbow lake	Stalactite
Perched water table	Stalagmite
Permeability	Suspended load
Point bar	Thermal features
Porosity	Travertine
River terrace	Tributaries
Saltation	Tufa
Sinkhole	Unsaturated zone
Soda straw	Water table
Solution	Weathering
Speleothem	Zone of aeration
Spring	

REVIEW QUESTIONS

1. Why are rivers and groundwater such an important topic to study? List four or five reasons.
2. Be able to sketch the hydrologic cycle and all of its components.
3. What are some mechanisms by which streams can erode?
4. What are some different mechanisms by which streams can transport material?
5. Describe at least six types of stream deposits. In each case, why was the particular deposit made?
6. Why is it important to study floods?
7. What is the water table? How does water get to it?
8. Be able to define terms like *aquifer, aquiclude, permeability* and *porosity*. Why are these things important when considering well placement?
9. Explain the layout of a home that uses a well and septic tank. Why is this important for you to understand?
10. What are some of the potential problems that can be encountered with getting fresh groundwater? What are some of the potential opportunities we have as Christians in this area?
11. Be able to identify various types of cave formations when shown images of them.
12. Be able to identify various types of thermal features when shown images of them.

FURTHER READING

Austin, S.A. (ed). 1994. *Grand Canyon Monument to Catastrophe.* El Cajon, California: Institute for Creation Research.

Bretz, J.H. 1969. The Lake Missoula Floods and the Channeled Scabland. *The Journal of Geology*, v. 77(5), p. 505–543.

Silvestru, E. 2003. A hydrothermal model of rapid post-Flood karsting. In *Proceedings of the Fifth International Conference on Creationism*, R.L. Ivey, Jr., ed., pp. 233–241. Pittsburgh, Pennsylvania: Creation Science Fellowship.

Austin, S.A. 1984. Rapid erosion at Mount St. Helens. *Origins*, v. 11(2), p. 90–98.

http://www.pbs.org/wgbh/nova/megaflood/ NOVA video on the *Mystery of the Megaflood* (Missoula Flood)

http://www.nps.gov/yell/index.htm Yellowstone National Park

http://www.water4lifeministry.org/index.php Water 4 Life Ministry

http://pubs.usgs.gov/circ/2004/circ1254/pdf/circ1254.pdf (USGS) The world's largest floods, past and present

http://www.corvettemuseum.org/ National Corvette Museum

REFLECT ON SCRIPTURE

- Ecclesiastes 1:5–7
- Genesis 2:10–14
- Revelation 22:1–2; River of Life
- Exodus 17, Numbers 20, I Corinthians 10:1–4; passages of water coming out of rock

CHAPTER 10

GLACIERS AND DESERTS

OUTLINE:

Denis Burdin/Shutterstock.com

Sand dunes at sunset in the Sahara Desert, Libya. Glaciers and deserts are two of the world's most extreme climates and yet they cover a significant portion of our planet and contain some of Earth's most breathtaking landscapes. The geology of these areas is important to understand especially as it relates to discussions of climate change.

Glaciers and deserts represent two of the most extreme environments on the surface of Earth. Together they cover about 38% of Earth's land surface, and these two climate extremes have continually changed their coverage since the end of the Flood. For example, there is fossil evidence that islands in far northern Canada were warm enough to sustain large trees immediately following the Flood. As global temperatures continued to drop, glaciers began to form in the deep continental interior of Canada, Antarctica, Greenland, and Asia as well as many high mountain ranges all over the world. The subsequent melting of many of these glaciers have left telltale evidence in the landforms they produced and left behind. One of the goals of this chapter is to help you realize that climates have changed radically and quickly in post-Flood times, largely without man's influence.

Likewise, deserts have developed, migrated, and disappeared since the time of the Flood. While much of the southwestern part of the United States is desert today, during the Ice Age this area was well-watered, with freshwater lakes present in areas like Death Valley. In contrast, places like Nebraska were desert areas during that same period of time. Today's grass-covered prairies of Nebraska are actually underlain by sand dunes—called the Nebraska Sandhills. It is the largest pile of sand in the western hemisphere and is now well-watered enough to sustain grass. Climate changes caused most of these glaciers to retreat and deserts coincided with the time the Native Americans first migrated here from Asia, and the climate continued to change after that, and continues to change today.

10.2 GLACIERS AND THE ICE AGE

10.2.1 What Is a Glacier?

Glaciers are large bodies of ice that move under the influence of gravity. They can be so large as to cover large islands and continents (Greenland and Antarctica) or they can be rather small and take the place of flowing water in a valley. Glacial ice typically moves slowly, but glacial ice has tremendous *competence*, or the ability to carry enormous amounts of load in comparison to the same volume of water in a stream. Glacial ice forms when more snowfall than can melt over a summer builds up year after year. As a deep snowpack forms, the weight of the overlying snow begins to compact snow at the bottom, squeezing the snow into **granular ice** (small pellets of ice). The granular ice further compresses into **firn** (larger pellets of ice), and the firn compacts into solid glacial ice at the bottom of the snowpack.

The largest and thickest types of glaciers are **continental glaciers** or **ice sheets** (figure 10.1).

Courtesy NASA

FIGURE 10.1. Almost all of Antarctica is covered by a glacial ice sheet, with ice nearly 5,000 meters thick in some areas.

They are immense, with surface areas in excess of 50,000 km². Only two of these are left today: Greenland and Antarctica. The thickness of this ice is almost unbelievable: nearly 4,000 m deep in Greenland 5,000 m in Antarctica, and both were probably thicker during the Ice Age! The areas that once had vast thicknesses of glacial ice covering them caused the rock underneath of them to be pushed down. Now that the glaciers have melted, places like Maine and Norway are rising in response to the missing ice. This process is called **isostatic rebound** and causes lakes and seas (like the Great Lakes or Hudson Bay) to become shallower over time if they fill one of these glacial-made depressions, or causes rocky coastlines to rise out of the ocean faster than erosion can wear them down.

During the Ice Age that came after Noah's Flood, many continental glaciers existed, including ones that covered Canada, the northern United States, northern Europe and parts of Asia. **Ice caps** (figure 10.2) are large, but cover smaller areas (< 50,000 km²) than continental glaciers. Ice caps generally occur at high elevations in mountainous areas and may cover or exist in between many mountains. Imagine how a bird nest sits down in the intersection of several branches of a tree. Some ice caps are similar in that they can be nestled down in a group of several surrounding mountain peaks. **Valley glaciers** (sometimes also called mountain or alpine glaciers) occupy the same places rivers would if temperatures were warm enough to sustain liquid water (figure 10.3). In other words if the valley glacier were not present, a river would be in its place. Ice caps generally feed several valley glaciers, but valley glaciers can also be independent of ice caps.

10.2.2 What Caused the Ice Age?

It is no longer debated whether the Ice Age was a real event or not. Both old-Earth and young-Earth scientists believe vast glaciers once covered areas like Canada and Europe (figure 10.4). Vast amounts of evidence suggest that the Great Lakes and the Finger Lakes of New York were carved by glaciers; that glaciers carried rocks from Canada to places like Ohio and Indiana and that long ridges of unsorted rocks and debris mark the ancient edges of where glaciers stood. The question is how it all got started, and why it ultimately ended.

FIGURE 10.2. An ice cap is a large glacier that might feed into a number of smaller glaciers around its perimeter. This is the Heiltskuk icefield in British Columbia.

Courtesy USGS

FIGURE 10.3. A valley glacier descends down a mountain in a place that would normally be occupied by a river if the temperatures were warmer.

Narongsak Nagadhana/Shutterstock.com

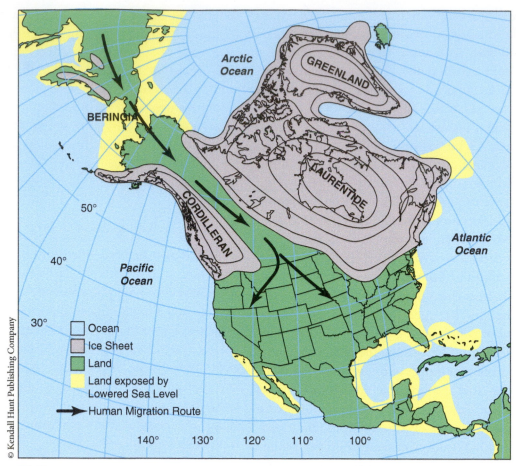

FIGURE 10.4. Humans and animals were able to easily migrate during the Ice Age because of lower sea levels and land bridges that connected most of the continents. The map shows ice-free paths through Asia, Alaska and Canada.

Because of the volcanic and tectonic processes that happened during Noah's Flood (see Chapter 4), the oceans became quite warm. As newly-formed ocean crust cooled, its heat would have been released into the oceans causing them to warm. There is also evidence in post-Flood rocks that explosive volcanic activity contributed large amounts volcanic ash and gases to the atmosphere. When volcanic products are added to the atmosphere, it causes cooling because the dust and gas reflects some of the incoming solar radiation back to space. For example, as a result of the volcanic products added to the atmosphere during the 1991 Pinatubo eruption in the Philippines, global temperatures dropped about 0.5 °C the following year. It appears that this combination of warm oceans and volcanic aerosols (gases) caused the Ice Age.

It's easy to understand how volcanic aerosols and ash could cause global cooling, but it seems counterintuitive that warm oceans would be necessary for glacial ice formation. The reason is that warm post-Flood oceans allowed for great amounts

> Genesis 9:1 And God blessed Noah and his sons and said to them, "Be fruitful and multiply and fill the earth."

© Kendall Hunt Publishing Company

of evaporation. Volcanic aerosols would cause cooling of continental interiors, especially far away from oceans. Temperatures did not have to be bitterly cold; they simply had to be below freezing to allow snow to accumulate. As more snow fell than was melted over the post-Flood summers, glacial ice began forming. In North America, it is thought that the greatest thickness of glacial ice developed in the Hudson Bay area and spread out in all directions. These conditions caused glaciers to cover parts of Europe and Asia, and all of Antarctica.

It is still an area of active scientific research, but we believe the Ice Age probably began several hundred years or so after the Flood. We base this on different kinds of rocks that we find at different locations within the rock record. For example we find evidence of very warm post-Flood climates in areas like Wyoming (the Green River Formation) and in northern Canada (Axel Heiberg Island). Warm climates in the immediate post-Flood era would have allowed animals and plants to migrate quickly to fill the Earth, just as God commanded them to do in Genesis 9. Although there is evidence of animals spreading and filling the Earth quickly following the Flood, man did not follow suit. Instead, we find in Genesis 11 that man decided to "all come together" and build a large city at Babel. God came down, and forced them to spread by changing their languages. We think this probably happened at least several centuries after the Flood because widespread human remains don't occur in the fossil record until the Ice Age deposits in the Pleistocene. Thus the events at the Tower of Babel probably happened just before the Ice Age began (see chapter 7).

The Pleistocene (see figure 7.1) is a conventional name for the recent period of time that was characterized by extensive glacial ice. There is biblical evidence for the Ice Age in the ancient book of Job. From various textural references in the book of Job and comparison with names in the book of Genesis; it is thought that Job lived at about the same time of Abraham, probably several hundred years or more

References in Job to Snow, Ice and Cold

37:6 For to the snow he says, 'Fall on the earth,' likewise to the downpour, his mighty downpour.

38:22 "Have you entered the storehouses of the snow, or have you seen the storehouses of the hail,

37:9–10 From its chamber comes the whirlwind, and cold from the scattering winds.

By the breath of God ice is given, and the broad waters are frozen fast.

6:16 which are dark with ice, and where the snow hides itself.

38:29–30 From whose womb did the ice come forth, and who has given birth to the frost of heaven?

The waters become hard like stone, and the face of the deep is frozen.

after the Flood. The book of Job has more references to snow, ice and cold than any other book in the Bible, suggesting the Ice Age was probably happening during this time.

Since the time of the Ice Age, a significant amount of glacial ice has already melted. Consider that places like Ohio, Wisconsin, and New York all had glacial ice. As the ice melted, it turned into water, flowed down rivers into the oceans, and caused sea level to rise. Conversely, as glaciers formed they did so by keeping precipitation (snow) on the continents, and there was a significant drop in sea level. This drop in sea level during the peak of the Ice Age allowed animals and people to migrate from places like Asia to North America following Noah's Flood. Low sea levels caused land bridges in places like the Bering Strait that connects Alaska with Asia.

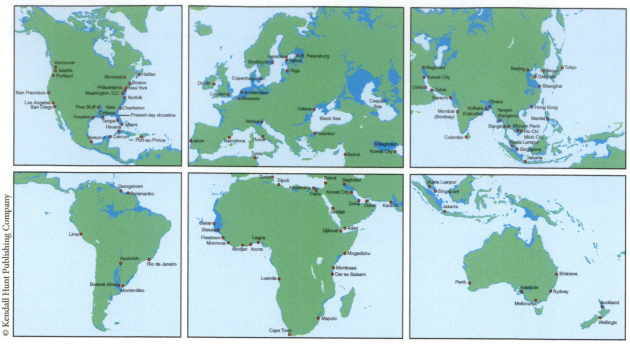

FIGURE 10.5. New continental outlines if all of the glacial ice were to melt. Note the effect it would have on the East and Gulf coasts of the United States (especially Florida), the Amazon region of South America, and even southeastern Europe.

During the Ice Age, many of the world's continental shelves would have been exposed, creating a very different continental outline than what we have today. The warm waters from the nearby oceans kept the coastal areas ice-free, allowing migration further south into North America. As glaciers retreated to today's locations, world sea levels rose on the order of about 125 m. If the remaining glacial ice that we have in the world today were to melt, sea level would probably rise an additional 60 m or so, creating a very different outline for the United States. Much of Florida, Louisiana, Mississippi and most of the United States' large cities would be under water if all of the remaining glacial ice were to melt (figure 10.5).

10.2.3 Glacial Movement

Glaciers move under the influence of gravity from areas of higher elevations to areas of lower elevations. Most glaciers move rather slowly—a meter per day or less. Ice first accumulates in areas of higher elevations (where it is colder, and snow does not melt) or in areas of great snowfall. This is called the **zone of accumulation** (figure 10.6). As the glacier flows downhill to lower (and warmer) elevations, the ice reaches a place where it begins to melt, or waste away. This area is called the **zone of wastage**. The **firn limit** separates the zone of accumulation from the zone of wastage. The firn limit will migrate seasonally; during the summers it will be at higher elevations and during the winter months it will be further down the glacier at lower elevations.

The leading edge of a glacier is called the **terminus**. The terminus can do one of three things over time (figure 10.6): it can advance, retreat, or remain in one location (equilibrium). In the case of an **advancing glacier**, the glacial ice is moving forward faster than it is melting; hence the terminus will push further down the hill. In the case of a

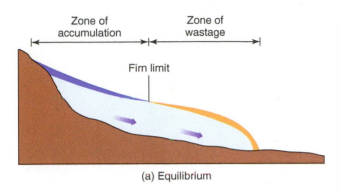

(a) Equilibrium

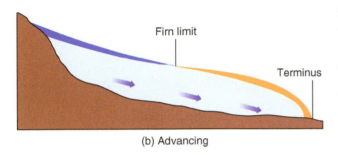

(b) Advancing

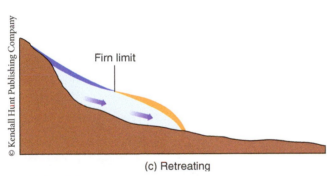

© Kendall Hunt Publishing Company

(c) Retreating

FIGURE 10.6. Glaciers move downhill under the force of gravity (shown by the forward arrows of movement in all three cases). A. If glaciers move forward as fast as they melt, the terminus does not move and the glacier is in equilibrium. B. The terminus of a glacier can advance if forward movement is faster than the rate of melting. C. The terminus of the ice will retreat if melting is occurring faster than the forward movement of the ice.

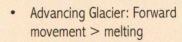

The Movement of a Glacier:

- Advancing Glacier: Forward movement > melting

- Equilibrium: Forward movement = melting

- Retreating Glacier: melting > forward movement

faster than forward movement, and the terminus of the ice "retreats" uphill.

Although much has been made by global warming advocates about the retreat of many glaciers, glacial retreat is not a new phenomenon. Keep in mind that glacial ice used to be much more extensive and most ice has retreated to near its current positions well before the Industrial Revolution, which is often marked as the beginning of global warming. In reality, an incredible amount of warming has occurred since the end of the Ice Age and only the most recent glacial retreats may be linked to mankind's activities.

Recall also that advance and retreat is affected not only by temperature, but also by the amount of snow that falls. Increased amounts of snowfall causes glacial advance, and vice versa. Immediately after the Flood there was a critical time when ocean temperatures were warm enough to produce huge amounts of precipitation and yet cool enough to not overly warm the continental interiors. As ocean temperatures dropped, a time was reached when the oceans were no longer warm enough to produce enough precipitation (snowfall) to sustain Ice Age glacial growth. When this point was reached, the glaciers began their retreat because of decreased precipitation. Higher temperatures probably play a role in today's glacial retreats, but they were not the primary cause for the end of the Ice Age.

glacier in **equilibrium**, the ice is melting just as fast as it is moving forward, so the terminus remains in one location. In a **retreating glacier** the ice is still moving forward by gravity, but melting is happening

10.2.4 Landforms Made by Continental Glaciation

When glaciers extend over any area and then retreat, they leave characteristic evidence of their presence. As glaciers move, they can "pluck" large rocks from their bases, which can become entrapped in the glacial ice. When the ice melts, the rocks are released from the ice and are usually too large to be transported by the less competent **outwash streams**. The ice-transported rocks are called **glacial erratics** and are often easy to identify because they do not match the local bedrock where they have been transported (figure 10.7). For example, states like Illinois, Indiana, Ohio, and the lower peninsula of Michigan have sedimentary bedrock at the surface; but often boulders and cobbles of igneous and metamorphic rocks can be found scattered at the surface. These rocks were transported from Canada and then dumped as the glaciers melted. As glaciers scrape against bedrock, rocks embedded in the glacial ice leave scratches and **glacial striations** in the bedrock (figure 10.8). These features can be used to discern the direction of glacial movement long after the ice has melted.

Because glaciers have tremendous competence, they can carry practically any size of rock that

FIGURE 10.8. Glaciers often have rocks embedded in them that will scratch the surface of any bedrock across which they flow. These are known as glacial striations and can help geologists determine flow directions of ancient glaciers.

becomes entombed in the ice. As a result, glacial deposits are usually a mixture of poorly sorted boulders, gravels, sands and muds. A general term for these poorly sorted deposits is **glacial till or glacial moraine** (figure 10.9). On the other hand, outwash streams usually only pick up and carry only the smallest particles (mud, sand and some gravel) left by the glacier. When outwash stream competence drops, and a deposit is made by the stream, it tends to be better sorted than the glacial deposits (see Chapter 9 for a discussion on competence). Hence, ice makes

FIGURE 10.7. Glacial erratics are rocks that are transported by glaciers, often over great distances. This glacial erratic near West Liberty, Ohio, was transported from somewhere in Canada.

FIGURE 10.9. Glacial till (sometimes called glacial moraine) consists of ice-deposited material of all different sizes. Notice the wide variance of material from sand to boulders.

poorly sorted deposits and water tends to make better sorted deposits.

A glacial deposit made at the terminus of a glacier when the glacier is in equilibrium is called and **end moraine** (figure 10.10). The end moraine that marks the furthest advance of the glacial ice is called the **terminal moraine** (figure 10.10). For example, in Ohio during the ice age glaciers advanced over the northwest half of the state. Thus, the terminal moraines can be found along a line approximately between the cities of Cincinnati, Columbus, and Youngstown (figure 10.11). A flatter deposit of till, called a **ground moraine**, forms as the glacier quickly retreats. A whole host of different kinds of glacial features and deposits (besides moraines) can form during continental glaciation). Sometimes rivers or small streams can form on top of the glacial ice. They can flow into a hole in the ice, filling it up with gravel. When the ice melts, the gravel-filled hole is left behind as a gravel hill called a **kame**. Rivers can also flow inside or just below the glacial ice making deposits in their channels. When the ice

melts, a ridge of sinuous sand and gravel deposits is left which is called an **esker** (figure 10.12). Kames, eskers and outwash deposits often form important sand and gravel resources that are easily quarried.

Long, cigar-shaped hills parallel to ice flow, are called **drumlins** (figure 10.13). Drumlins are rather mysterious features and their formation is not completely understood because they form beneath glacial ice. It is thought they probably form either as a result of moving ice shaping the hills, or by catastrophic flooding beneath the ice depositing, or even eroding the hills. Areas affected by continental glaciation often are covered with small lakes. The lakes can form in a number of different ways, but one common way is that a block of ice gets left surrounded by moraine or till, forming the lake depression as the ice melts. These are commonly referred to as **kettle lakes** (figure 10.14). Most of Minnesota's "10,000 lakes" are kettle lakes. These are common features in areas like the Midwestern United States that were covered by continental ice sheets.

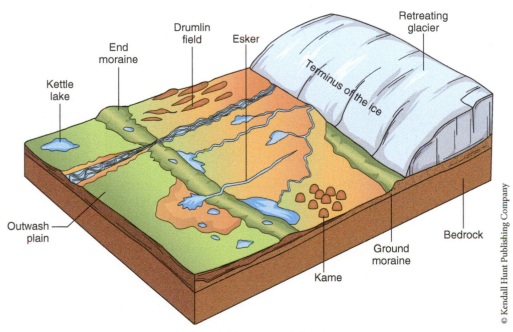

© Kendall Hunt Publishing Company

FIGURE 10.10. Various features and landforms left by continental glaciation.

GLACIAL MAP OF OHIO

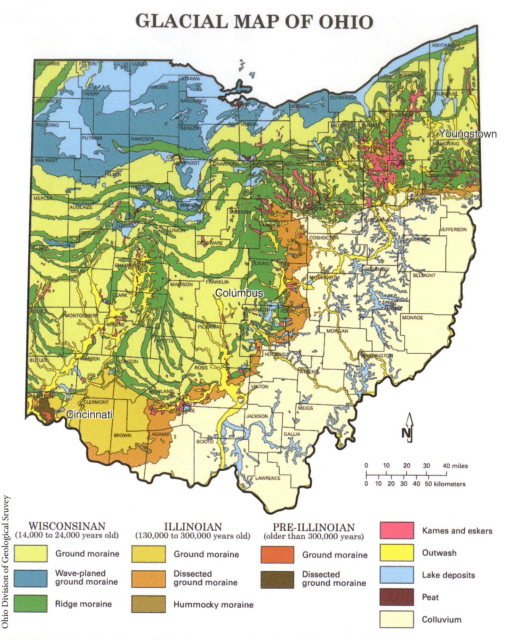

Ohio Division of Geological Sruvey

WISCONSINAN (14,000 to 24,000 years old)	ILLINOIAN (130,000 to 300,000 years old)	PRE-ILLINOIAN (older than 300,000 years)	
Ground moraine	Ground moraine	Ground moraine	Kames and eskers
Wave-planed ground moraine	Dissected ground moraine	Dissected ground moraine	Outwash
Ridge moraine	Hummocky moraine		Lake deposits
			Peat
			Colluvium

FIGURE 10.11. A glacial map of Ohio. The unglaciated portion is the light tan color in the southeastern part of the state. Terminal moraines roughly follow orange belt just beside it. Note: age designations are from an old-Earth view. A young-Earth view holds that the Ice Age happened shortly after the Flood, only thousands of years ago.

10.2.5 Landforms Made by Alpine Glaciation

As in continental glaciation, alpine glaciers produce end moraines, terminal moraines, and ground moraine. However, two other types of moraines can also be found. **Lateral moraines** (figure 10.15) form along the edges of the alpine glacier where it comes into contact with the steep valley wall. **Medial moraines** form when two valleys come together, hence two lateral moraines will join, making a moraine in the center of the glacier. When the

FIGURE 10.12. An esker in northern Manitoba.

FIGURE 10.13. A drumlin field north of Milwaukee, Wisconsin. The large square road patterns are one square mile each.

FIGURE 10.14. Dozens of small lakes often dot areas covered by glacial till. Many of these are kettle lakes, formed when a block of ice gets left behind by a retreating glacier which leaves a depression in the moraine. This is an area in Siberia.

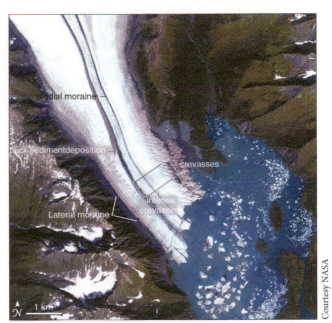

FIGURE 10.15. Bear Glacier, Gulf of Alaska. In this photo, the dark lines in the center of the glacier are medial moraines, and the dark edge of the glacier is a lateral moraine. Notice the crevasses and the ice bergs.

glacier melts, these moraines will be left as ridges of glacial till along the sides or centers of the valleys.

Mountainous areas that have experienced alpine glaciation are some of the most picturesque areas in the world. This type of glaciation causes a distinctive topography to develop. Valley glaciers often cut very deep and wide valleys causing them to have "U"-shaped profiles instead of "V"-shaped profiles often characteristic of river erosion (figure 10.16). Alpine glaciation also causes very rough, jagged and sharp features compared to more smooth profiles caused by river erosion. Mountain peaks often develop sharp tops called **horns** because of opposing glaciers cutting into the mountain side. Deep bowl-shaped amphitheaters called **cirques**) are often cut by small glaciers near

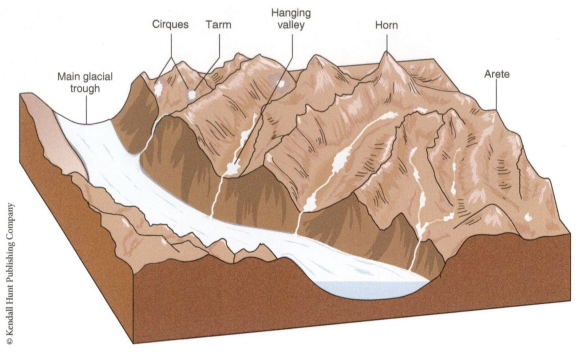

Cirques   Tarm   Hanging valley   Horn   Arete

Main glacial trough

FIGURE 10.16. Alpine glaciation causes a mountainous landscape to become sharp and angular.

the mountain peaks. If two cirques are next to each other, a sharp knife-edge ridge called an **arête**) will develop between the two cirques. Arêtes will often extend down the mountain from horns. Often times a cirque will have a deep depression at its bottom that will fill with water post-glaciation. The small lake is called a **tarn**. Because the main glacier often has much more ice (and more competence) than tributary glaciers, it often cuts valleys that are much deeper than the tributary glaciers. Hence, the side valleys are left hanging high above the main glacial trough when all the ice melts. These **hanging valleys** (figure 10.17) often have spectacular water falls, sometimes referred to as **bridal veil falls**.

Sometimes glaciers end in the ocean. Before the glacier enters the water, it is capable of cutting deep valleys. Since ice is lighter than water, a glacier is buoyed upwards as it enters the sea and no longer can cut downward into bedrock. When these glaciers retreat, they leave long water-filled valleys with shallow entrances to the ocean, called

fjords. Hence, fjords often are very deep channels, but have shallow entrances. **Icebergs** form when glaciers calve (pieces break and fall) into the ocean (figure 10.18).

FIGURE 10.17. Perhaps one of the world's most famous glacial landscape and hanging valley is found in Yosemite National Park in California. One of the main glacial troughs makes up the valley floor. Another smaller glacier came in from the right, but was unable to cut to the same depth as the glacier in the main valley. After the ice melted a waterfall (a bridal veil falls) marks the location of the "hanging valley."

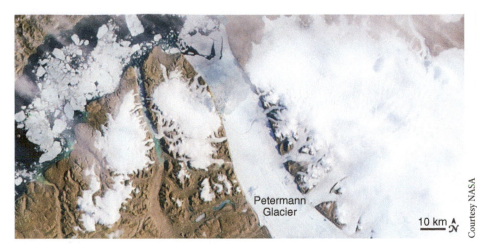

Petermann Glacier

10 km N

Courtesy NASA

10.3 DESERTS

10.3.1 Why Do Deserts Form?

Geologists love to study in deserts not only because of the fantastic scenery, but also because of well-exposed rocks which are not covered by soil or vegetation. Many famous national parks occur in desert and semi-desert areas, such as the Grand Canyon, Death Valley, and Zion National Parks in the United States. By definition, deserts are areas that receive less than 25 cm of rain per year (about 10 inches). Many deserts receive much less rain than this. Because of the lack of water, many plants and animals are specially adapted to conserve water. For example, few desert plants have leaves; they perform photosynthesis in their stems or in their spines. Desert plants that do have leaves often have small, waxy leaves that prevent water loss. Plants are often widely spaced and small compared to forest plants, which allows them to conserve water (figure 10.19).

Desert and semi-desert areas cover about 18% of Earth's land surface, making them one of the most important landscapes on Earth (figure 10.20). Most deserts are hot, but technically the continent of Antarctica is world's largest desert because most of it receives less than 25 cm of precipitation (water equivalent) per year. Deserts usually form due to one or more of the following factors:

1) *Lack of rain due to dominant high pressure systems* (figure 10.21). The Earth's large-scale atmospheric system results in rising air at the equator eventually sinking back down at about 30° north or south latitude and forming surface high pressure systems (see chapter 15). As the air sinks it becomes warmer and expands, resulting in very little precipitation. If you look carefully at figure 10.20 most of the world's

Rigucci/Shutterstock.com

FIGURE 10.19. Desert plants are often widely spaced and have small thick waxy leaves (if they have leaves at all) in order to conserve water.

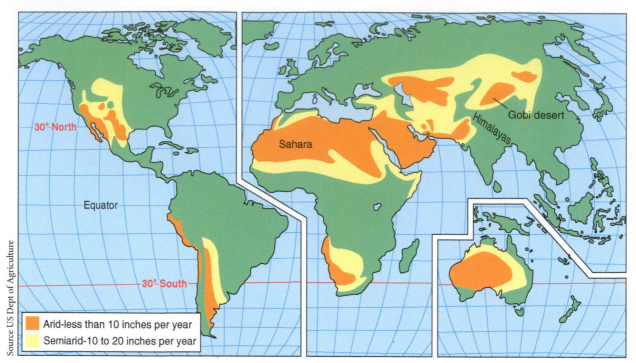

Source US Dept of Agriculture

Legend:
- Arid-less than 10 inches per year
- Semiarid-10 to 20 inches per year

FIGURE 10.20. World Distribution of Nonpolar Deserts.

deserts occur at or near 30° latitude both in the northern and southern hemispheres.

2) *Distance from an ocean.* Many of Earth's clouds form over the ocean and then move over

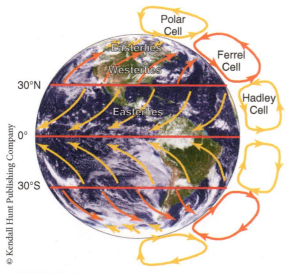

FIGURE 10.21. Many deserts occur in areas of high pressure. Because of atmospheric circulation patterns, air sinks at about 30° in latitude creating high pressure and clear skies.

continents where they drop their rain. Further inland the air has exhausted its moisture, and places like central Australia and Mongolia are too far from oceans receive much precipitation.

3) *Rain shadow effect.* When moist air rises into the atmosphere it cools and the moisture it contains condenses to form precipitation. Sometimes air can be forced over a mountain range or a high plateau (figure 10.22). Moisture is often dumped on the windward side of the slope, before the air can make it completely over the topographic high. When the air gets to the other side of the mountain, it sinks and warms, allowing very few clouds to form. Hence, the lee side of the mountain gets very little precipitation. This is the cause for the deserts in Nevada; moisture blocked by the Sierra Nevada Mountains which run along the California-Nevada border. The Cascade Mountains in Washington and Oregon block moisture from reaching the eastern parts of these states, which are semi-deserts.

© Kendall Hunt Publishing Company

CHAPTER 10: Glaciers and Deserts

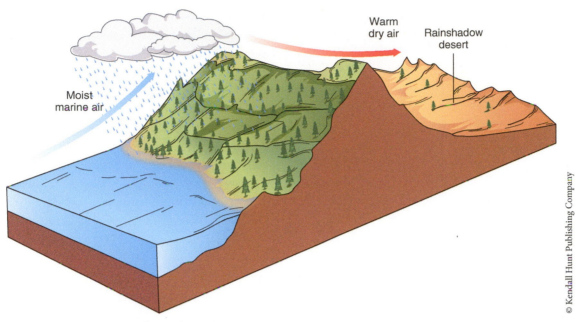

Moist marine air

Warm dry air

Rainshadow desert

© Kendall Hunt Publishing Company

FIGURE 10.22. Development of a rainshadow desert.

4) *Cold ocean currents.* Compare the distribution of deserts in figure 10.20 with the surface ocean currents in figure 12.1. Notice from these maps deserts are located in relationship to cold ocean currents on the western sides of continents in both the northern and southern hemispheres. This is because cold water evaporates at much slower rates than warm water, forming fewer clouds and much less precipitation.

5) *A combination of these factors.* Most often multiple factors combine to form deserts. The Atacama Desert in Chile is the driest place in the world (figure 10.23), and some areas in this desert that have never had recorded precipitation. Several factors created these harsh conditions: it is in the vicinity of 30° of latitude (high pressure), the waters of the western side of South America are some of the coldest in the world, and the prevailing wind patterns move east-to-west (taking evaporating moisture from the sea away from the desert). It is also in a rain shadow behind the immense Andes Mountains, which prevents moist air from Brazil from reaching this region.

Deserts Can Form Due to:

1. Lack of rain due to high pressure (30° of latitude)
2. Far away from an ocean
3. Rain shadow effect
4. Cold ocean currents
5. Combination of factors

Richard Nowitz/National Geographic Cretive

FIGURE 10.23. The Atacama Desert is one of the most severe deserts in the world.

10.3.2 Desert Landforms

Contrary to popular belief, deserts are not all covered with sand dunes. In fact, only a small percentage of deserts contain dunes, perhaps 10%. Most deserts are rocky areas consisting of bare bedrock surfaces and mixed sand and gravels. The mixture of sand and gravels that cover most desert floors is called **desert pavement** (figure 10.24). Also contrary to popular belief, wind is not the major erosional factor that shapes deserts. Surprisingly, the most important erosional agent is water! Because deserts do not receive very much rainfall, plants are sparse and there are not many roots to hold the soils and desert pavement in place. Significant rainfall may only occur once or twice within a decade or longer; but when it happens rainfall is quickly funneled to dry river beds (called **wadis**) and causes flash flooding, mud flows and debris flows (see more on these in Chapter 8). These processes can move significant amounts of material, cut deep canyons and cause radical changes to a landscape within a few hours. In deserts (and

My brothers have acted deceitfully like a wadi, like the torrents of wadis which vanish... (Job 6:15, NASB)

FIGURE 10.24. Most deserts are covered by loose rock and gravel called "desert pavement."

Photo by John Whitmore

probably many other areas too) landscape change happens quickly, and then the landscape sits dormant until the next major event. In deserts we can judge when landscape change happened by examining such things as **desert varnish**, the size of canyons and the location of various mass flow deposits.

In North America, the **Basin and Range Province** of the western United States is a classic area where many desert landforms can be studied. This area receives sparse rainfall due to being in the rain shadow of the Sierra Nevada Mountains on the California/Nevada border, being located in an area of high pressure (close to 30°) and low air moisture from the cold water of the coast of California. The Europeans who first explored this area were Spanish, and hence many of the landforms have names derived from that language.

Plateaus, **mesas** and **buttes** (pronounced "byoots") are landforms that are often capped by a resistant layer of rock. In the Basin and Range province, limestone frequently makes up the resistant layer that caps the top of flat-topped landforms, because the low precipitation rates do not chemically erode the limestone's calcite minerals (Chapter 8). A plateau is a very large flat landform that is bounded by cliffs (figure 10.25). Often plateaus are so large you won't be able to see the whole landform, except with the aid of something like Google Earth®. There is no "official" size designation of how large plateaus can be. The Kaibab Plateau in the area of the Grand Canyon is capped by the Kaibab Limestone; it covers the general area of north-central Arizona. Some plateaus are very large and are flat only in a general sense. The Colorado Plateau is a large area covering the Four Corners area of Utah, Colorado, New Mexico, and Arizona. It is called the Colorado

See if you can find these the Colorado Plateau and the Tibetan Plateau on Google Earth®.

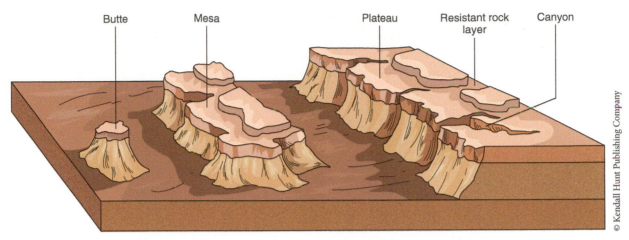

Butte   Mesa   Plateau   Resistant rock layer   Canyon

© Kendall Hunt Publishing Company

FIGURE 10.25. Plateaus, mesas, and buttes all have flat tops. Note the size differences between the three landforms.

Plateau because of the presence of the Colorado River which flows across it. Occasional small mountain ranges can be contained within these larger plateaus. The Tibetan Plateau, north of the Indian subcontinent, is another example of a large plateau that has many other features contained within it.

The word *mesa* comes from the Spanish word meaning table. It is much smaller than a plateau and the table-like nature can often be recognized from a distance (figure 10.26). Like plateaus, there is no "official" size for how big or small a mesa can be, but the flat table-like nature should be clearly seen by an observer on the ground. A butte (figure 10.27) is smaller than a mesa and typically has a pinnacle-like shape. Usually it has a resistant rock layer at the top of the landform, but occasionally it may be missing.

Alluvial fans (figure 10.28 and also figure 9.8) are desert and semi-desert landforms that form near the bases of mountain ranges. Rivers in mountain drainages often have steep gradients and therefore

Photo by John Whitmore

FIGURE 10.26. A mesa is a flat-topped "table land." These two mesas are capped by resistant lava flows in the northwest corner of New Mexico.

Francesco R. Iacomino/Shutterstock.com

FIGURE 10.27. Some of the most famous buttes in the world are those in Monument Valley, Arizona. In this case, the buttes and mesas are capped by the resistant DeChelly Sandstone.

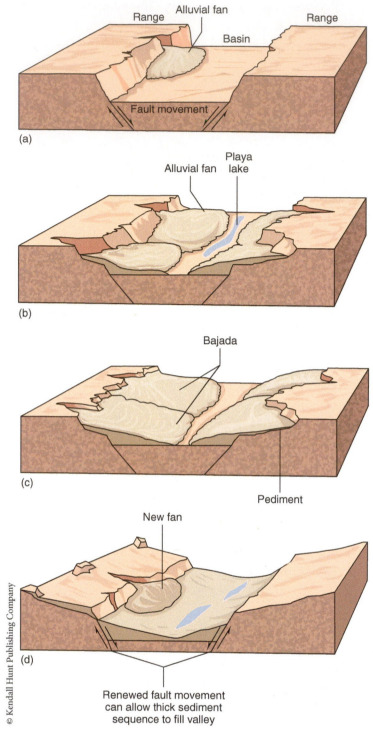

FIGURE 10.28. A block diagram showing some of the different types of deposits in the Basin and Range province of the southwestern United States.

CHAPTER 10: Glaciers and Deserts

high competence to carry a large load. Water is often funneled over a large area into a narrow drainage in mountain valleys. When the rivers get to the base of the mountains, the slopes become lower and the streams dumps large volumes of material near the mountain base as the water slows down. The deposit ends up in a fan shape, radiating outward from where the river exits the mountains. If a mountain front is particularly steep, or a number of drainages come out along the base of a mountain front, sometimes the alluvial fans overlap with one another forming a **bajada**.

In the Basin and Range Province, faults are often located along the bases of mountains and the alluvial fans and bajadas are often deposited directly on top of the fault scarp. In between mountain ranges, sometimes a shallow lake, or **playa** forms that has no external drainage out of the mountains. Playas only rarely have water on them; in fact roads can be built across them. When they do get water, the water is very shallow and often becomes salty as the lake evaporates. Often the sediments in playas are mixtures of mud, silt, and various salts. Sometimes great accumulations of salt can form making salt flats. When playas and streams in the desert dry up, the mud in them shrinks causing **mud cracks** to form (figure 10.29). Mud cracks often have polygonal shapes which can occur as various sizes (centimeters to meters in size). Sometimes the polygons can be so large, they can be observed from the air. If a road is built across the playa, the mud cracks can actually break the road apart. When the cracks are this large, we call them **playa cracks**.

FIGURE 10.29. Racetrack Playa in Death Valley has many rocks in the middle of it—with skid mark behind them. Occasionally the playa is covered with shallow water that freezes during the winter. Rocks that have rolled out onto the playa from the nearby mountain get trapped in the ice. Then, wind blows the ice sheet across the lake and the rocks leave a skid mark in the mud.

Sometimes mountains have an erosional ramp that approaches them known as a **pediment** (figure 10.30). The surface of the pediment is usually covered with desert pavement. Many times large boulders can be found on pediments that indicate some type of mass flow processes have operated on them in the past. Pediments are often dissected by modern streams, so it appears to be evident that modern streams did not cause the erosional ramp. The pediment must have been formed by some other process not related to modern streams, but their origins have not been satisfactorily explained in either old-Earth or young-Earth models.

Pediment Surface

FIGURE 10.30. A pediment surface in central Nevada. Pediments are erosional surfaces that form ramps at the bases of mountain ranges.

Wind does cause some erosion in deserts, but it is generally a much smaller contributor to landscape formation than water. Sand and silt picked up by the wind can polish and smooth rocks by abrasion making a **ventifact**. Generally particles of sand do most of the work, but they are only transported a meter or two above the desert surface. A **yardang** (figure 10.31) is a larger rock or outcrop that might look like an overturned sinking ship, with one end (the end facing the predominant wind direction) polished by the wind.

In deserts with internal drainages, the erosional remnants of mountains often fill the valleys. As a result, mountains end up burying themselves in their own erosional products eventually leaving only a remnant of the mountain protruding out of the desert floor. The feature is called an **inselberg** meaning "island mountain." Ayer's Rock or Uluru, located in Australia is probably the most famous example of such a feature (figure 10.32).

10.3.3 Sand Dunes

Wind-blown material in deserts is common because of the lack of vegetation which holds soil in place. Clay, silt, sand, and sometimes even large material can be transported across the desert by wind. Clay and silt particles are often swept out of the desert, but the coarser material can accumulate in piles called dunes. Sand dunes often form against some type of barrier, accumulate in a low depression or on the lee side (back side) of a hill or mountain where wind velocities are not as great. Dunes can be formed from sand or silt (windblown silt dunes are referred to as **loess**). Sand dunes don't necessarily have to form in deserts. They can also form in coastal areas next to the ocean or a large lake with strong onshore winds, such as the sand dunes of North Carolina's Outer Banks (where the Wright Brothers successfully flew the first airplane).

Figure 10.33 shows a profile of a typical sand dune. Notice the wind direction and the slope angle of the *stoss* and *lee slopes* of the dune. The stoss slope or windward side of the dune faces the wind and has a slope of 10 degrees or so. The lee slope or slip face of the dune is downwind and usually has a slope of about 32 degrees because this is the **angle of repose**, or the angle at which sand grains become unstable. In other words, if the slope is steeper than 32 degrees, the sand will tumble down until the pile has a slope of 32 degrees or less. Sand grains get blown or more commonly bounced up the stoss slope of the dune and over the **crest** to the lee slope. Notice that sand dunes have angled layers within them called **cross-beds**. Because sand from the stoss slope is carried over to the lee slope, sand dunes will migrate over

FIGURE 10.31. A yardang has been shaped by blowing sand and resembles an overturned ship hull. This one is from Xinjiang, China.

FIGURE 10.32. Ayers Rock or Uluru, Australia, is the most famous inselberg.

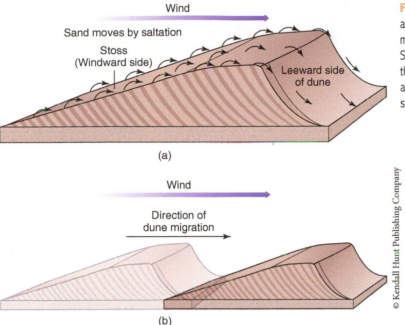

Wind

Sand moves by saltation

Stoss
(Windward side)

Leeward side
of dune

(a)

Wind

Direction of
dune migration

(b)

© Kendall Hunt Publishing Company

FIGURE 10.33. A sand dune has a windward side and a steeper leeward side. Over time, the dune will migrate as sand gets blown up and over the crest. Sand avalanches down the leeward slope to form the internal cross-beds. These cross-beds are often about 32°, which is the angle of repose for dry sand.

time moving with the direction of the prevailing wind (figure 10.34).

Wind is a much more efficient sorting agent than water; therefore wind-deposited sandstones tend to have better sorted sand grains than sand deposited by water. About 90% of the sand grains in most deserts tend to be made out of the mineral quartz because it is abundant and fairly resistant to both chemical and mechanical weathering. Because

sand grains tend to get bounced around by the wind, the corners of larger sand grains tend to get broken off during grain-to-grain collisions and the grains become relatively round. Quartz is relatively hard (7.0 on Mohs scale of hardness), so it lasts fairly well in the dune setting. Other common, but softer minerals like feldspar (hardness of 6.0) and mica (hardness of 2.0-3.0) disappear quickly in the dune environment because of grain-to-grain collisions.

Photo by John Whitmore

FIGURE 10.34. An old road next to the telephone poles is being overcome by small barchan sand dunes in the Salton Sea area of California.

Sand dunes come in a variety of different shapes and sizes; all of which are dependent on availability of sand, wind strength, wind direction, various obstacles, and some other factors (figure 10.35). **Barchan dunes** are the most basic type of sand dune. They are horseshoe-shaped with the tips (or horns) of the dune pointing downwind. They are

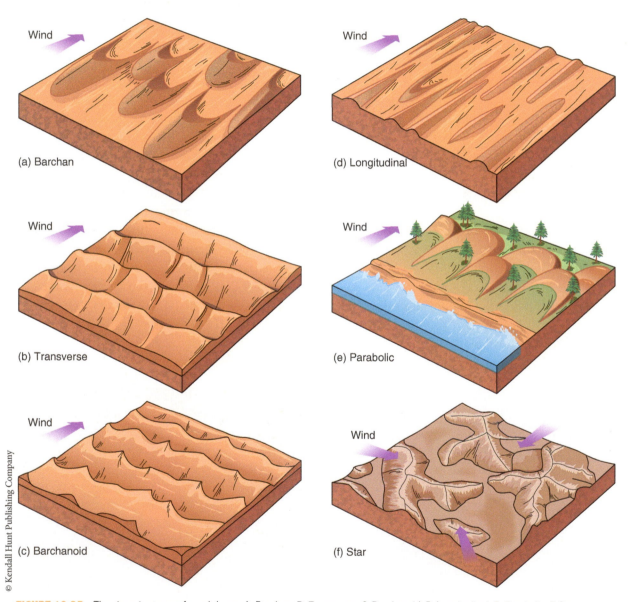

(a) Barchan

(b) Transverse

(c) Barchanoid

(d) Longitudinal

(e) Parabolic

(f) Star

FIGURE 10.35. The six major types of sand dunes. A. Barchan, B. Transverse, C. Barchanoid, D. Longitudinal, E. Parabolic, F. Star.

© Kendall Hunt Publishing Company

small in height compared to other types of dunes and usually form on flat desert plains where sand is rather sparse. They migrate over time and can encroach upon various manmade structures. A classic field of barchan dunes can be found on the southwest side of the Salton Sea in California and along the Pan American Highway northwest of Vitor, Peru.

Sometimes the crests of dunes orient themselves perpendicular to the prevailing wind direction, forming long rows of parallel dune crests. Sand is generally abundant, but wind velocities are comparatively weak. These are known as **transverse dunes** (figure 10.36). The Nebraska Sandhills is the largest accumulation of sand in the western hemisphere (now all covered with grass) and contains many examples of transverse dunes. **Barchanoid ridge** dunes appear to be kind of a mix between barchans and transverse dunes. The Little Sahara Dunes in Utah have some nice

examples of these. **Linear** or longitudinal **dunes** are among the highest dunes in the world and occur in some of the world's greatest deserts. Their crests are parallel to the wind instead of perpendicular to it as in transverse dunes. They form as the result of strong cross winds that push the sand into long linear ridges. The dunes can be as much as 200 m high and 100 km long! The Simpson Desert of Australia has some great examples of these. **Star dunes** (figure 10.37) are shaped much like their name. They have at least three ridges (slip faces) that extend outward from a central peak. They result from constantly changing wind directions, strong winds, and plentiful sand. Star dunes do not migrate, instead they grow in height. Great Sand Dunes National Park in Colorado has classic examples of these dune types. **Parabolic dunes** have a similar horseshoe shape as barchans dunes except the horns of the dune are facing into the wind instead of pointing in the direction the wind is blowing. These types of dunes are very common in coastal settings with strong onshore winds as in the Sleeping Bear Dunes along the east coast of Lake Michigan. These types of dunes also do not migrate because they are partially anchored by vegetation. They can occur in areas that receive abundant precipitation.

Courtesy NASA

FIGURE 10.36. Transverse dunes, White Sands, New Mexico.

Courtesy NASA

FIGURE 10.37. Star dunes in the Sahara Desert, Algeria.

The typical interpretation for sandstones with large and "steep" cross-beds is that they were deposited in a desert setting sometime in the ancient past. Examples include the Coconino Sandstone of Grand Canyon National Park in Arizona (figure 10.38) and the Navajo Sandstone of Zion National park in Utah. The interpretation of these sandstones as desert deposits has caused some difficulties for young-Earth creationists because we wouldn't expect to find desert deposits in the midst of a global flood! However, a closer examination of the Coconino Sandstone by Dr. John Whitmore and his colleagues has shown some unexpected features suggesting that it was deposited underwater, not in a desert.

1) *Cross-beds in the Coconino Sandstone do not match desert dunes.* After measuring hundreds of cross-bed dips, they found the average dip was only about 20°, not the 32° expected from desert sand dunes.

2) *Sand grains are neither well-rounded nor well-sorted.* Microscope analysis showed that the sand grains are not well rounded or well-sorted,

as would be expected if the Coconino was a desert deposit. Instead, the grains are sub angular and only moderately sorted (figure 10.39).

3) *Sand injectites point to watery formation.* Large sand filled cracks below the Coconino (figure 3.25) have usually been interpreted as arid mud cracks filled by dry sand. However, the cracks do not have features common to mud cracks, and instead it has been shown that the Coconino sands were injected downward by highly pressurized water.

4) *Well-preserved fossil tracks.* The vertebrate footprints in the Coconino Sandstone (figure 10.40) are best explained as being made underwater. Dr. Leonard Brand showed through a series of observations and experiments the Coconino tracks are very similar to underwater salamander tracks that he produced in a water filled aquarium in the laboratory.

5) *Folds in the sand.* Large folds in the Coconino were discovered by Guy Forsythe in Sedona, Arizona (figure 10.41). These folds are *parabolic*

FIGURE 10.38. One of the many places the Coconino Sandstone can be found is in the Grand Canyon. It is characterized by angled cross-beds.

100 μm

FIGURE 10.39. Microscopic view of the Coconino Sandstone from Hance Trail in the Grand Canyon. Note that the sand grains are quite angular and they are not very well sorted. This is typical of the Coconino.

recumbent folds which form from fast moving water overturning the cross beds shortly after they were deposited.

All of these features and others suggest the Coconino was deposited in a marine setting, not in a desert. We also now know that there are sand dunes which form on the ocean floor; they are referred to as **sand waves**. The most famous examples can be found at the entrance to San Francisco Bay in California (figure 10.42) and in Long Island Sound in New York. In both situations, strong ocean currents provide the energy to make these underwater "dunes". It appears there may have been similar processes (albeit on a much larger scale) that were occurring during Noah's Flood.

Photo by John Whitmore

FIGURE 10.41. Large parabolic recumbent folds occur in the Coconino Sandstone in Sedona, Arizona. These folds were made by rapidly flowing underwater currents.

Photo by John Whitmore

FIGURE 10.40. A large footprint and many smaller footprints that can be found on the sloping Coconino cross-beds along Hermit Trail in the Grand Canyon. All of the tracks go in one direction, up the slopes of the cross beds.

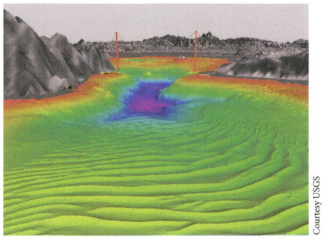

Courtesy USGS

FIGURE 10.42. Large sand waves, very similar to desert sand dunes, on the ocean floor at the entrance to San Francisco Bay, Califronia.

10.1 Introduction

- Climate change has been an ongoing phenomenon in earth history since the time of the Flood; most caused by non-human factors.

10.2 Glaciers and the Ice Age

- Glaciers are large bodies of moving ice. They come in three varieties: ice sheets, ice caps and valley glaciers.
- The Ice Age was a real period in earth history, probably beginning several hundred years after Noah's Flood. It was likely caused by a combination of factors including exceptionally warm ocean water, volcanic aerosols, and shifting continents and ocean currents. The Ice Age ended because of the cooling down of the ocean basins.
- Glaciers form when more snow falls than can melt over a season. Glaciers can advance, retreat or remain in equilibrium.
- A number of features can be produced by continental glaciation including moraines, outwash plains, kames, kettles, drumlins, and eskers.
- Alpine glaciers cause mountains landscapes to become very rugged and jagged producing features like U-shaped valleys, horns, cirques, arêtes and hanging valleys.

10.3 Deserts

- Deserts form because of persistent high pressure (30° latitude), being far away from an ocean, rain shadows, cold ocean currents, or a combination of these factors.
- Desert varnish is a coating often found on desert rocks made by bacteria. Its presence is important because it shows that many landscapes are relics; they have not changed for very long periods of time.
- There are a number of important desert landforms including plateaus, mesas, buttes, playas, alluvial fans, bajadas, wadis, desert pavement, pediments, ventifacts, yardangs, inselbergs, mud cracks, and playa cracks. Water is the primary shaping agent in a desert that is responsible for most of these landforms.
- Sand dunes migrate as wind carries sand up the stoss slope and as sand avalanches down the lee slope. There are six major kinds of dunes: barchan, transverse, barchanoid, longitudinal, star and parabolic.

10.4 Are There Fossilized Sand Dunes?

- Some have suggested the Coconino Sandstone is a wind-blown desert sand dune deposit. However, it contains poorly sorted angular sand grains, mica, footprints suggestive of underwater formation and a host of other features suggesting it is a subaqueous sand wave deposit.

KEY TERMS

Advancing glacier

Alluvial fan

Angle of repose

Arête

Bajada

Barchan dune

Barchanoid ridge dune

Basin and range province

Bridal veil falls

Butte

Cirque

Continental glacier

Crest

Cross-bed

Desert pavement

Drumlin

End moraine

Equilibrium (of a glacier)

Esker

Firn

Firn limit

Fjord

Glacial erratic

Glacial moraine

Glacial striations

Glacial till

Glacier

Granular ice

Ground moraine

Hanging valley

Horn

Ice cap

Ice sheet

Iceberg

Inselberg

Isostatic rebound

Kame

Kettle lake

Lateral moraine

Linear dune

Loess

Medial moraine

Mesa

Mud crack

Outwash deposits

Outwash streams

Parabolic dune

Pediment

Plateau

Playa

Playa crack

Retreating glacier

Sand injectites

Sand waves

Star dune

Tarn

Terminal moraine

Terminus

Transverse dune

Valley glacier

Ventifact

Wadi

Yardang

Zone of accumulation

Zone of wastage

REVIEW QUESTIONS

1. What are some things that you could look for in a particular area to find out if the area had been glaciated or not?
2. Briefly summarize a hypothesis for how the Ice Age began and ended from a biblical perspective.
3. How did the Ice Age help humans, animals, and plants repopulate the Earth after the Flood?
4. How is it possible for a glacier to move forward and yet have the terminus retreat at the same time?
5. Be able to identify various glacial features from photographs.
6. What is the difference in sediment deposited by glacial ice compared to that deposited by a river?
7. What is the most important erosional agent in a desert (and why)?
8. What are the five factors that can form a desert? Provide examples of each.
9. Be able to identify various desert features from photographs (including sand dunes).
10. What are some of the evidences that the Coconino Sandstone was formed underwater and not in a desert?

FURTHER READING

Gollmer, Steven. Initial conditions for a post-Flood rapid ice age. In *Proceedings of the Seventh International Conference on Creationism*, M. Horstemeyer, ed. Creation Science Fellowship, Pittsburgh, Pennsylvania.

Oard, Michael. 2004. *Frozen in time: The wooly mammoth, the ice age and the Bible.* Master Books: Green Forest, Arkansas.

Patrick, K. 2010. Geomorphology of Uluru, Australia. *Answers Research Journal*, v. 3, p. 107–118.

Whitmore, J. H. 2006. The Green River Formation: A large post-Flood lake system: *Journal of Creation*, v. 20, no. 1, p. 55–63.

Whitmore, J.H., Forsythe, G., and Garner, P.A. 2015. Intraformational parabolic recumbent folds in the Coconino Sandstone (Permian) and two other formations in Sedona, Arizona (USA). *Answers Research Journal*, v. 8, p. 21–40.

Whitmore, J.H., Strom, R., Cheung, S. and Garner, P.A. 2014. Petrology of the Coconino Sandstone. *Answers Research Journal*, v. 7, p. 499–532.

WEBSITES TO VISIT

http://sand.xboltz.net/index.html Sand Dunes of the Southwest (United States).
http://nsidc.org/cryosphere/glaciers/ National Snow and Ice Data Center.

Genesis 9 & 10. Consider the post-Flood climate and the conditions under which people repopulated Earth after the Flood.

Look up passages in Job that refer to snow, ice, cold, etc.

Kobby Dagan/Shutterstock, Inc.

CHAPTER 11

EARTH RESOURCES: PROVISIONS FROM GOD'S CREATION

OUTLINE:

This picture of a summer night at Times Square, New York City, reminds us of the vast energy and material resources used to power modern society.

People use energy. This energy comes in many forms, from the food that we eat to the fuels that power our cars. The human body uses the equivalent of 100 watts of energy just to move, breathe, and live each day. Each one of us is like a 100-watt light bulb turned on 24 hours a day, every day of our lives.

When Adam and Eve were first created, they were given roles and tasks to perform. In Genesis 1:28, God spoke to them saying "Be fruitful and increase in number; fill the earth and subdue it. Rule over the fish in the sea and the birds in the sky and over every living creature that moves on the ground." To further instruct them, "the LORD God took the man and put him in the Garden of Eden to work it and take care of it" (Gen. 2:15). These verses help to define the **dominion mandate** of mankind. From the beginning, the expectation was that mankind would serve as a ruler over creation, as a representative of God on Earth (this is in part what being "made in the image of God" means). Man is to rule over all creation, and to do so in a way that cares for God's world, recognizing that it is His creation, not ours.

From shortly after the creation of Adam and Eve, we see their descendants using energy. Tubal-Cain, mentioned in Genesis 4:22, made implements (tools) of metal. This required him to find metal-rich rocks and heat them with fire to extract the metal, then use still more energy to craft and shape the tools. God made these material resources available from the beginning of Creation, with the intention that we would use them as part of our role to act as stewards of this world. So from the time before Noah's Flood and then afterward, humans used available materials for energy sources. Plus, the Flood's profound effect on Earth's surface (see Chapter 4) had a dramatic impact on the location and availability of many mineral resources, both in destroying some and creating many others, particularly our fossil fuels.

The Dominion Mandate

The care and overseeing of creation was commanded from the very beginning, and is theologically referred to as the dominion mandate. It is derived from various parts of scripture, but most notably from two passages in Genesis. In these passages, we see our role as caretakers, stewards, and managers of God's creation.

Gen. 1:28 (ESV)

And God blessed them. And God said to them, "Be fruitful and multiply and fill the earth and subdue it, and have dominion over the fish of the sea and over the birds of the heavens and over every living thing that moves on the earth."

Gen. 2:15 (ESV)

The Lord God took the man and put him in the garden of Eden to work it and keep it.

Over most of human history, the main sources of energy to sustain people have been human effort, domestic animal effort, fire, and wind power. Beginning with the Industrial Revolution in the late 1700s, drastic changes in energy sources (first water, then later coal) and technology (both agricultural and manufacturing) allowed people to exert far less personal energy in order to accomplish far greater amounts of work, and the trend continues to this day. Along with the use of these new energy resources and others added more recently (such as petroleum and nuclear power) came an increase in the number of people that could be supported on Earth, which also increases our material consumption as standards of living continue to rise.

Today, the average U.S. citizen uses an enormous amount of materials each year to maintain a very comfortable standard of living. While the human body uses 100 watts of energy on average throughout the day (figure 11.1), per capita energy use for each American is *10,000* watts. This is a 100-fold increase in our energy use prior to the Industrial Revolution, and is equivalent to each person having about 100 "energy servants" working for them around the clock! Obviously these energy servants are not people; they are many energy **resources** utilized in homes, businesses, and industries all over the country. A *resource* is any material that can be used by people, and here we are thinking primarily of Earth materials and energy resources. Table 11.1 summarizes the very large use of energy and rock/mineral resources by each U.S. citizen in 2014 alone.

In reviewing this table, it is clear that the average U.S. citizen uses *a lot* of resources: nearly

FIGURE 11.1. The human body generates approximately 100 W of energy when doing normal activities and a bit more during exercise

17,500 kilograms (about 38,600 pounds) per person, per year. But these resources aren't stockpiled in your back yard; they are in the buildings, foundations, pipelines, bridges, automobiles, airplanes, roads, televisions, cell phones, windows, and everything else you can think of around you. As they say in the mining industry, "If you can't grow it, you have to mine it."

While the above discussion of resources refers to the earth and energy materials used by people each year, the term *resource* has a different meaning in economics when we think about what resources are (or may be) available to us. As defined above, a *resource* is any material that can be used by people. In terms of economic forecasting a resource represents the highest possible amount of that material that may available in an area. A **reserve** is a resource that has been determined to exist or has a good likelihood of existing in the resource area, and a **proved reserve** is a reserve that is known to exist and can be recovered economically. Each term becomes more narrowly defined, and the potential recovery value gets smaller. For example, let's say that a state geological survey estimates that a shale deposit contains a large amount of natural gas (methane), and reports the estimated volume of gas at 1 trillion cubic feet. This is the resource value. An energy company specializing in natural gas then estimates that the amount of gas that they could possibly extract is 100 billion cubic feet (the reserve value), and they expected to actually recover 10 billion cubic feet (the proved reserves). Proved reserve values can change with the price of the material: if the price goes up, less-rich or technically difficult and more expensive deposits may become economically recovered, and the proved reserve value increases. Likewise, if the price goes down, companies will shift to fewer and more economical deposits and the proved reserve value decreases. When world leaders, corporations, and reporters discuss the world's natural resources, it is helpful to keep these terms in mind.

TABLE 11.1.

2014 U.S. per capita consumption of mineral and energy resources (kg/person)

Rock and Mineral Resources	
Bauxite (aluminum)	33
Cement	260
Clay	70
Copper	5
Iron	142
Lead	5
Phosphate rocks	100
Salt	158
Sand, gravel, and stone	6846
Other rocks and minerals	317
Energy Resources	
Coal	2650
Natural gas	3920
Petroleum	2967
Uranium	0.1
TOTAL	**17,474 kilograms**

2014 data amended from National Mining Association report at http://www.nma.org/pdf/m_consumption.pdf.

11.3 ENERGY RESOURCES: FOSSIL FUELS

The majority of energy used in the world comes from **fossil fuels**. These resources are ultimately the products of photosynthetic organisms that have been buried in sedimentary rocks and altered into carbon-rich materials. Three sources are the most important to our society: coal, petroleum, and natural gas. Figure 11.2 illustrates the range of U.S. energy resources by type and by sectors of use. Fossil fuels account for *86% of all energy sources used*. Should the U.S. seek to significantly reduce carbon dioxide emissions, it will need to dramatically alter the current mix of energy sources that currently (and for the foreseeable future) power our society. Coal, petroleum, and natural gas (most of which

formed during Noah's Flood) dominate the world's energy resources because of the enormous amount of energy stored in the chemical bonds of their molecules, which is released by combustion.

11.3.1 Coal

Coal is a brittle, carbon-rich sedimentary deposit that can be burned (figure 11.3). It is composed of compressed and altered plant materials. Because it is a very abundant fuel source found on every continent and is inexpensive to mine, coal has been the largest electricity-generating fuel throughout history, though in the U.S it is now rivalled by natural

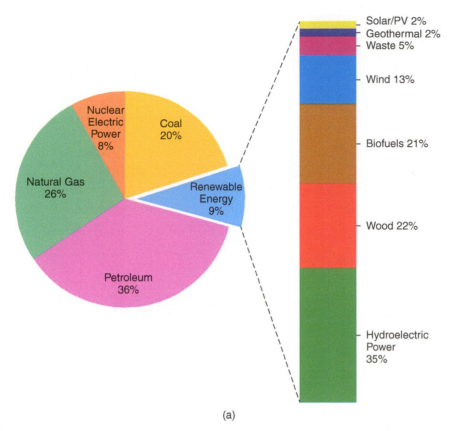

(a)

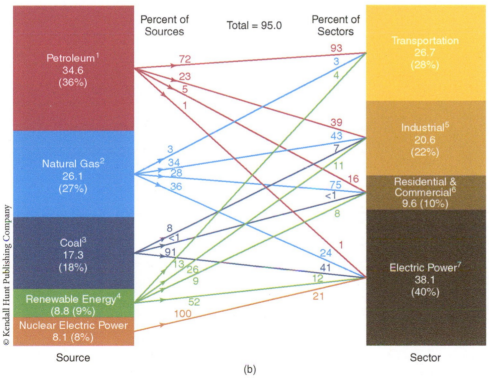

(b)

FIGURE 11.2. (a) Energy resource use in the U.S. (2011) by type and (b) energy sources and sectors of use (b). Note that in (a) the renewable resources are listed by percent of all renewables, which together are only 9% of all energy sources. So hydropower's 35% of renewable energy represents 3.15% of all U.S. energy.

indykb/Shutterstock.com

FIGURE 11.3. Coal is a sedimentary deposit composed of plant material, often buried between layers of sandstone or shale (pictured here).

gas. Coal is subdivided into a number of types, or *ranks*, based on the degree of alteration and carbon content (figure 11.4). "Low-rank coal" includes lignite and sub-bituminous coal. These types of coal are rather soft, dark brown to black, dusty, and have a higher moisture content and lower carbon content compared to higher grades of coal. Lignite and sub-bituminous coal are used almost exclusively for generating electricity. Bituminous and anthracite are considered high-rank coal, and are more solid, less dusty, and have low moisture content and high carbon content. They burn hotter than low-rank coal, and are often used in cement production and

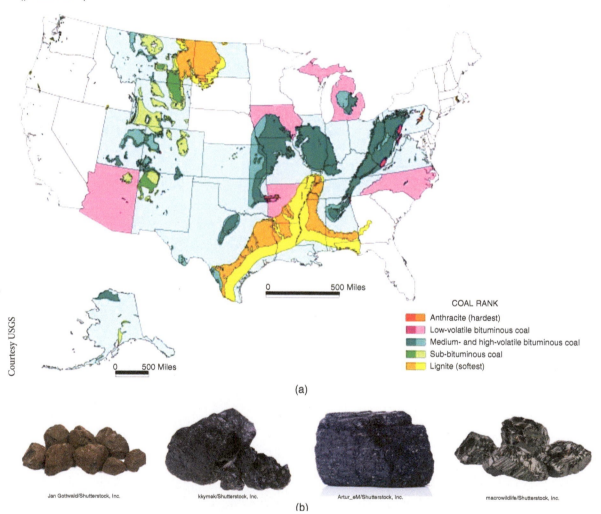

COAL RANK
- Anthracite (hardest)
- Low-volatile bituminous coal
- Medium- and high-volatile bituminous coal
- Sub-bituminous coal
- Lignite (softest)

Courtesy USGS

(a)

Jan Gottwald/Shutterstock, Inc.   kkymek/Shutterstock, Inc.   Artur_eM/Shutterstock, Inc.   macrowildlife/Shutterstock, Inc.

(b)

FIGURE 11.4. (a) Distribution of coal by rank in the U.S. Colors indicate surface (darker) and subsurface (lighter) coal reserves. (b) Ranks of coal. From left to right: lignite, sub-bituminous, bituminous, and anthracite. Carbon content increases with higher rank.

CHAPTER 11: Earth Resources: Provisions from God's Creation

metallurgy. In electricity production, low-rank coal is used to provide **base load power**: the minimum level of electricity that is needed day or night. Coal-fired power plants run 24 hours a day, every day, to efficiently cover the basic electricity needs of the world.

Most old-Earth geologists believe that coal is formed by low-oxygen conditions in ancient swamps. Accumulating and rotting plant matter is later buried, and over time compressed and heated to form coal. Dr. Kurt Wise developed a model in which the deeper coal layers of the Mississippian and Pennsylvanian periods were formed by vast pre-Flood floating forest systems that were ripped apart and buried during Noah's Flood. This floating forest model has successfully described a number of deposits in Pennsylvania, Kentucky, and West Virginia.

Coal often contains trace amounts of other materials, including mercury, sulfur, arsenic, uranium, and thorium. Each of these pose health risks to humans and the environment, and the burning of coal releases some of this material into the atmosphere. However, other types of byproducts (particularly sulfur emissions, carbon monoxide, and particulate matter) are efficiently trapped at the site, greatly reducing pollution in countries with laws requiring treatment. In other countries, pollution levels can reach dangerous levels (figure 11.5). After combustion, large amounts of **coal ash** remain, which is often mixed into concrete as a filler.

The mining of coal can present additional environmental impacts. **Underground mining** is the traditional form of mining coal. It is still an important method, but is rarer now due to safety and better technology. Today, about 60% of all coal is mined via **strip mining** (figure 11.6), the dominant mining method in the western U.S. There, several meters of overlying soil and rock are removed to access a coal seam typically 15–30 meters (50–100 feet) thick. U.S. laws require that strip mines restore the region to its pre-mined state, and the mines do so by removing topsoil at the front of the mine, then filling in the back side with aggregate and dirt, then placing the topsoil on top. As a result, a strip mine's quarry literally moves through the landscape. In the eastern U.S., **mountaintop removal** is sometimes the favored mining method. Large sections of a coal-bearing mountain, as much as 300 meters thick (1,000 feet) are removed by blasting. The coal is harvested and the remaining rocks are deposited in a nearby mountain valley. This method has come under significant pressure from environmental groups, as the

FIGURE 11.5. Severe pollution from coal-fired power plants obscures the sun in Beijing, 2014.

FIGURE 11.6. A large strip mine. For scale, the dump trucks are able to haul 80 tons of mined material.

scar left on the mountain and alterations to valley/stream environments are quite significant.

11.3.2 Petroleum

The single largest source of energy used in the U.S. is **petroleum**, and it dominates the transportation sector, making up 93% of all transportation fuels (figure 11.2). Petroleum is a type of hydrocarbon, an organic compound composed of chains of carbon bonded to hydrogen atoms, and is often referred to as **crude oil**. Unlike coal, petroleum is formed by a variety of microscopic marine organisms collectively called *plankton* and *algae*. As the plankton are buried in sediments (primarily during Noah's Flood), heat and pressure reorganized their cellular materials into liquid hydrocarbons. To become oil, the hydrocarbons must reach temperatures within the "oil window" of 90–160 °C.

Lower temperatures produce kerogen (a thick, waxy substance), while higher temperatures produce natural gas or even graphite. If the rock that the organisms were deposited in is porous, the hydrocarbons may then migrate to other rocks and be trapped in high concentrations forming a petroleum reserve.

Petroleum reserves can be either conventional or unconventional, depending on the type of technology required to extract the crude oil. Conventional wells extract crude oil from a petroleum reserve with a vertical well. The crude oil is under pressure, and it will naturally rise up the well. However, because the oil is viscous (thick and resistant to flow), a pump is usually needed to draw the oil fully up to the surface. There are a number of different types of geological structures (called traps) that concentrate petroleum into a reserve. One such trap, called an anticline trap, is illustrated in figure 11.7. Geologists identify

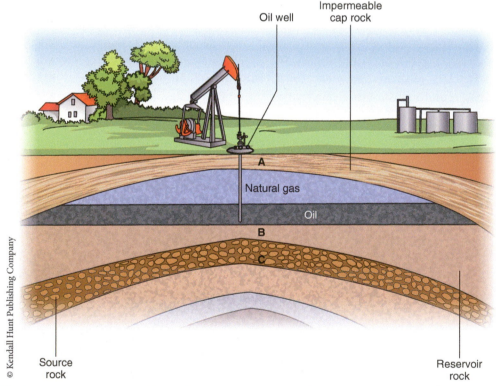

Oil well

Impermeable cap rock

A

Natural gas

Oil

B

C

© Kendall Hunt Publishing Company

Source rock

Reservoir rock

FIGURE 11.7. An anticline petroleum trap. Crude oil rises towards the surface from the source rock (c) into the reservoir rock (b), but is trapped by an impermeable cap rock (a). A conventional well is drilled to extract the oil.

CHAPTER 11: Earth Resources: Provisions from God's Creation

potential petroleum reserves through a combination of surface geology and the use of seismic reflection surveys (which simulate earthquake waves to "see" into deep geologic units that may contain oil).

Most oil fields of the world have been accessed by conventional wells. Wells that recover crude oil from extremely deep sedimentary systems or low-porosity rocks are called unconventional wells. Beginning in the late 1990s, a pair of pre-existing technologies were successfully combined to allow oil companies to extract oil from "tight" deposits (shales and other low-porosity rock). The two technologies were directional drilling and hydraulic fracturing, often together referred to as "fracking" (figure 11.8). In directional drilling, a vertical drill is sent into the rocks. As it approaches the petroleum reserve rock (usually a mile or more below the surface), the drill is turned so that it can follow along inside of the sedimentary rock, often for hundreds to thousands of meters. Instead of the well sticking vertically into the reserve like a straw into a cup (figure 11.7), the well is now oriented along the orientation of the rock formation for very long distances, increasing the amount of oil that can be recovered. Moreover, from a single vertical well at the surface, many horizontal "spokes" can be made in all directions into the reservoir rock.

But the rock's low porosity still impedes the extraction of oil. This is where hydraulic fracturing comes in. This technique injects sand, water, and lubricants at very high pressures to create fractures within the deep rock. The mix is then removed from the well, and at the surface pumps work to extract the crude oil. The combination of these two technologies has opened up vast new areas of formerly inaccessible rocks, greatly increasing the proved oil reserves of the U.S. while at the same time reducing the number of surface wells that need to be drilled.

Both conventional and unconventional wells can negatively affect their surrounding environments. Small and localized spills and contamination are the most frequent problems, and developed nations tend to have very strict laws requiring cleanup and

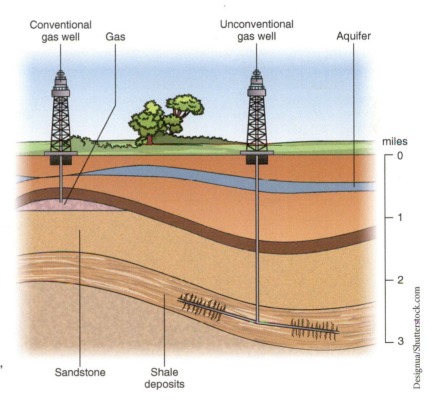

FIGURE 11.8. Directional drilling and hydraulic fracturing (right) have been combined to access large resources of oil and natural gas from "tight" shale deposits.

Designua/Shutterstock.com

FIGURE 11.9. The *Deepwater Horizon* was an ultradeep oil platform capable of drilling several miles deep to reach petroleum reserves. In 2010, an explosion resulted in its destruction and the release of over 3 billion barrels of oil into the Gulf of Mexico.

disposal should a spill occur. Far less frequent but more attention-getting are the large spills or explosions that can send millions of barrels of crude into the ocean, such as the 2010 explosion and sinking of the *Deepwater Horizon* drilling rig in the Gulf of Mexico (figure 11.9) or continued environmental damage in Nigeria due to aged infrastructure and rampant sabotage and violence.

Once extracted, the crude oil must be **refined** into usable products. The refining process is accomplished at large facilities (figure 11.10) that separate out the components of the crude oil by temperature differences (a process called *fractional distillation*). Refined products include not only transportation fuels such as gasoline, diesel, and jet fuel, but also home heating oil, asphalt, and lubricants. Though plastics are not made from crude oil itself, they are made from other materials that are frequently recovered from oil extraction, and therefore owe much of their existence and low costs to the oil industry. The oil industry's supply chain is perhaps the most efficient of any industry in the world. Consider this: despite all the effort that goes into finding, recovering, transporting, refining, and finally delivering a gallon of gasoline, that gallon of gasoline is less expensive than a gallon of locally produced milk!

11.3.3 Natural Gas

Natural gas is the simple hydrocarbon methane, composed of one carbon atom covalently bonded to four hydrogen atoms. Like petroleum, natural gas is

FIGURE 11.10. An oil refinery is a giant chemical factory dedicated to separating crude oil into various useful fuels, lubricants, and other substances.

formed from the breakdown of plankton and algae at moderate geological temperatures. Natural gas begins to form in the "oil window," and continues to form up to 225°C. Above this temperature, the hydrogen-carbon bonds break, releasing hydrogen gas and forming graphite (a non-fuel carbon mineral). It is often found in association with petroleum, such as above the crude oil in figure 11.7. This is because, as a gas, methane is less dense than the oil, but is also trapped by the overlying impermeable rock. In other cases, natural gas exists alone as a trapped resource, either in structural traps similar to petroleum, or in impermeable, organic-rich shales.

Natural gas is a very important energy resource, and has becoming more important in recent years, particularly in the U.S. In industry, natural gas is used as a base material for the formation of many types of chemicals. Some vehicles (such as city busses) use it as a transportation fuel, and there is considerable interest in using compressed natural gas as a competitor to diesel fuel, which is both more expensive and more polluting. Natural gas is probably most familiar to us in its use for home space and water heating, cooking, or for Bunsen burners in science labs (figure 11.11). Natural gas is actually odorless; the rotten-egg smell we associate with it is a harmless chemical added to alert us if there is a gas leak.

The most important role for natural gas, however, is in electricity production. Natural gas is becoming a more common source of baseload power, but it also excels at power plants that can be quickly adjusted to serve the shifting power needs during the day, such as when warm weather results in widespread air conditioning use in the afternoon. These **peaker plants** compliment the baseload power and are needed to avoid blackouts and brownouts. Over the past decade, though, a significant drop in the price of natural gas in North America, combined with new regulations on carbon emissions, has made natural gas-fired electricity generation more popular and economical than even low-cost coal. In addition to the price drop of natural gas, this fuel source emits roughly

FIGURE 11.11. Natural gas is frequently used as a fuel for cooking, water heating, and home space heating.

half the carbon dioxide compared to coal, and none of the pollutants, ash, or other wastes associated with coal. This makes natural gas a cleaner and low(er) carbon emitter for our future energy needs.

Why did the price of natural gas drop? The answer lies in the same unconventional drilling technology that has rejuvenated the oil industry in the U.S.: horizontal drilling and hydraulic fracturing. The combination of these technologies has opened up immense geological units that are rich in natural gas but previously untapped due to low-porosity sedimentary rocks, such as the Marcellus Shale of Pennsylvania (figure 11.12). This one geological formation is estimated to hold up to 100 years' worth of current U.S. natural gas needs, and additional

FIGURE 11.12. The Marcellus Shale in Pennsylvania is a large natural gas resource made possible by hydraulic fracturing technologies.

Source: U.S. Energy Information Administration based on data from various published studies. Canada and Mexico plays from ARI.
Updated: May 9, 2011

FIGURE 11.13. Current and prospective "tight" shale plays for oil and gas in North America.

formations, such as Texas' Eagle Ford Shale and the Niobrara Chalk of the Denver Basin may hold similar, or even greater promise of large natural gas and oil reserves (figure 11.13).

11.4 ENERGY RESOURCES: NUCLEAR POWER

Nuclear power accounts for roughly 8% of all energy use in the U.S, and all of it is dedicated to electricity production. In a nuclear reactor (figure 11.14), **fission** (the splitting of a large atomic nucleus into two smaller nuclei) is used to generate heat. The heat from fission boils water to produce steam, which then flows through a series of tubes. The heat from these tubes is used to boil water in a second set of tubes, which then makes more steam that then turns a turbine to produce electricity. Various reactor designs use this basic setup to ensure that any radiation within the reactor core does not come in contact with the outside water sources that are heated to turn the turbines.

FIGURE 11.14. The nuclear power station at San Onofre, California, which closed in 2013. The large dome structures are the containment units for the nuclear reactors.

Fission is accomplished in a nuclear reactor by colliding a neutron with an atom of uranium-235 (figure 11.15) in **fuel rods** located in the reactor core. The U-235 to splits into two smaller nuclei of barium and krypton, and also produces three more neutrons plus other forms of energy, such as heat (used to boil water in the reactor) and electromagnetic radiation:

$$_0^1 n + {}_{92}^{235}U \rightarrow {}_{56}^{142}Ba + {}_{36}^{91}Kr + 3{}_0^1 n + \text{energy}$$

The three additional neutrons produced from fission are moving fast enough to collide with other nearby atoms of U-235, continually creating more fission events and causing a **chain reaction**. To prevent the exponential increase in fission events, **control rods** are placed among the fuel rods to absorb

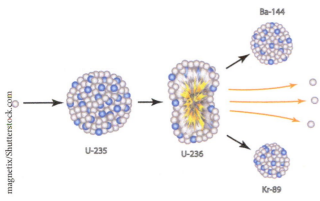

FIGURE 11.15. The fission reaction of uranium-235.

two of the three neutrons produced during fission. Water is continuously circulated and heat transferred to the second set of tubes to keep the temperature inside the reactor stable.

In nature, 99.3% of all uranium is U-238, a non-fissile form of uranium that cannot be used in reactors. Only 0.7% of is U-235. To be suitable for use in reactors, the amount of U-235 must be increased about 5% by one of several types of **enrichment** processes. Enrichment is difficult and expensive, but it is essential to produce the proper type of fuel. It is also necessary to make nuclear weapons, but weapons-grade U-235 must be enriched at much higher levels than nuclear fuel, and is an extremely difficult process to master. Because nuclear fuel rods are only enriched to 5% U-235, they cannot explode like a nuclear bomb. However, if the chain reaction is not controlled properly and the water constantly cooled, the reactor can suffer a **meltdown**, in which the temperatures rise too high and damage the building and/or cause other materials to explode, contaminating the surrounding area with radioactive debris.

Despite frequent fears of meltdown, the safety record of the nuclear industry is actually the best among all major energy sources. With 435 nuclear power plants currently active worldwide, only three significant events have occurred in the past 60 years. So far only one of those has resulted in direct human casualties. The first event was in 1979, when a partial meltdown occurred at of one of the two reactors at the Three Mile Island facility in Pennsylvania (figure 11.16a). In that case, damage was done to the containment unit, and a very small amount of radiation (equivalent to 1/100 of an X-ray per person) was released into the area around the reactor. No casualties or verified long-term health issues resulted from the accident and the remaining reactor is still operational.

In 1986, reactor 4 in Chernobyl, Ukraine (former Soviet Union; figure 11.16b) suffered a sudden power spike during a test, which resulted in a significant

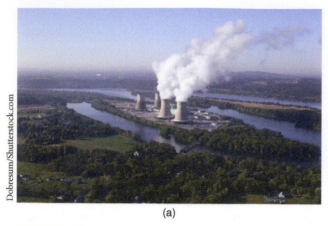

(a)   (b)

FIGURE 11.16. The partial meltdown at Three Mile Island (a) in 1979 was ultimately minor, and the station continues to operate today. The disaster at Chernobyl (b) in 1986 was the worst nuclear accident to date.

meltdown and a pair of large explosions. The explosions killed two workers and sent radioactive material widely around the area. Another 28 workers died from radiation poisoning within the next four months. Hundreds of thousands were evacuated and thyroid cancer rates spiked among children in the years following (though nearly all were successfully treated). Chernobyl remains the worst nuclear catastrophe in history.

The most recent meltdown occurred in 2011 at the Fukushima-Daiichi plant in northern Japan. The reactor suffered no damage when a magnitude 9.0 earthquake struck offshore, but the quake produced a massive 15-meter high tsunami, which breached the 10-meter tall sea wall and disabled the backup generators that would cool the reactor during an emergency. Four reactors melted down, and a large explosion spread nuclear material into the area (but primarily the ocean). Over 19,000 Japanese citizens died from the earthquake and tsunami, but amazingly no deaths occurred at Fukushima despite the multitude of disasters, and no instances of radiation poisoning in the population have emerged. The site is currently closed as efforts to contain radioactive materials continue.

These three examples illustrate the risks, but also the overall safety, of nuclear power. Many more instances of environmental and human casualties could be made for other types of energy, including renewables. One final potential hazard for nuclear power remains: **spent nuclear fuel**. This is the residual, radioactive material that remains after a fuel rod has been exhausted. Many countries send their spent material to a facility in France capable of recycling much of the fuel. In the U.S., laws preventing spent fuel recycling result in temporary storage in specialized pools or sealed containers at several locations around the country. The final storage location for this material is uncertain, since Congress de-funded the previously-approved disposal site of Yucca Mountain in 2011.

Despite high costs and uncertainties over spent fuel, nuclear power remains the only large-scale source of base-load power from carbon-free sources. Its fuel is inexpensive and the amount of hazardous waste produced is very small, but site-specific building requirements and stringent regulations result in very high costs. Countries that seek to reduce the amount of carbon dioxide produced from power plants can look to nuclear power as a clean and reliable alternative to fossil fuels. France provides an example of the success of nuclear energy, where 80% of the country's electricity is generated by nuclear power plants.

CHAPTER 11: Earth Resources: Provisions from God's Creation

Fossil fuels and nuclear power together provide approximately 91% of all the energy sources used in the U.S. The remaining 9% comes from a variety of generally carbon-free or carbon-neutral sources (figure 11.2), dominated by water and biomass. Several of these (such as solar and wind power) show significant promise for expansion in the future, while others (hydroelectric power and biomass) have little growth potential. None of these energy resources is currently able to provide the steady and reliable baseload electricity at the scale of coal and nuclear power, and only biofuels contribute to transportation, though in very small quantities.

11.5.1 Solar

Solar power is the most abundant renewable energy resource available. The sun's radiation primarily comes in visible and ultraviolet wavelengths (see chapter 13). Once we factor in the spherical shape of the planet, losses through the atmosphere, and the changing angle of the sun over the course of the day, each square meter of land on the planet receives more than 200 watts of solar energy per hour continuously (averaged from about 1,000 watts at noon, and zero at night). That's enough to run nearly four 60-Watt light bulbs on each square meter on Earth, and just one square kilometer would power 4 *million* 60-watt bulbs, day and night! The energy is free each day the sun is shining, and solar energy is harnessed for two primary purposes: heating and generating electricity.

Solar heating has been used for millennia, and modern construction designers have much to learn from those in the past who adapted their homes to hot and cold climates without the aid of modern heating and cooling systems. **Passive solar heating** refers to designs that purposefully employ sunlight to heat a building. Such systems often use south-facing windows and overhangs to allow sunlight into the home during the winter, but shade from the sun during the summer. They may also use particular materials inside the building that efficiently store and release thermal energy, such as dark-colored stone flooring and interior walls. **Active solar heating** captures insolation and then moves the thermal energy using pumps or fans. A common example of this is a solar water heater (figure 11.17), which sits outside or on top of a building, absorbing sunlight into a vacuum tube filled with antifreeze. The antifreeze is circulated and heat exchanged with water in the home's hot water heater.

Generating electricity from solar energy is accomplished in two very different ways. Most familiar to us are ground- or roof-mounted **solar photovoltaic panels** (or solar panels, figure 11.18), which are large, dark-colored modules made from sets of silicon wafers. These wafers are semiconductors that generate an electrical current when struck by photons of light. Typical residential solar panels

FIGURE 11.17. A vacuum-tube solar water heater.

Richard Waters/Shutterstock.com

FIGURE 11.18. Workers install solar photovoltaic panels on the roof of a home.

FIGURE 11.19. The Ivanpah power plant in California is the largest solar concentrator in the world. It uses hundreds of mirrors to heat a three towers, each containing molten salt. The tower is glowing from the mirrors' reflected sunlight and the heat it generates.

can convert 15% or more of the insolation into electricity. Since their first entrance into the consumer market in 1977 prices for solar panels have dropped dramatically. In the past decade, this trend has accelerated because of strong competition and the large-scale production of panels from China.

The second method of electricity generation is **concentrated solar power**. A variety of methods are used, but all employ mirrors to focus solar energy to a particular area to produce heat. That heat boils water to produce steam, which then turns a turbine to produce electricity. Concentrated solar power can use parabola-shaped mirrors focusing on a water-filled metal pipe (creating steam directly), or may employ hundreds of sun-tracking mirrors to concentrate sunlight on a tower filled with molten salt, which then transfers its thermal energy to water to make steam (figure 11.19). These systems are very efficient when the sun is shining, but are rendered inoperable under cloudy conditions and at night.

Solar power is so abundant that many countries could replace all of their existing electricity production with just solar power in a fairly small area of land. For example, *all* the U.S. electricity needs could be provided by covering an area of 22,000 km² (8,500 mi²) with solar panels. This is a lot of land, but it represents just 7.5% of the size of Arizona, a prime location for solar power. A project like this is not feasible (due to transmission, weather, security, and

many other factors), but it still helps us understand that solar power has the capacity to scale up significantly into a substantial component of electricity production from its currently tiny 2% of renewable energy, and a mere 0.2% of all U.S. energy!

11.5.2 Wind

About 1% of all solar energy captured by Earth is converted into wind, and the use of wind for various forms of power has a very long history, beginning with the first wind-powered sailing ship in ancient Egypt. Windmills were invented at various times and locations independently to crush grain into flour or to pump water up from the ground. The most important role for wind now is in electricity generation, made possible by **wind turbines** (figure 11.20). A wind turbine generates electricity when the wind rotates the large blades. This rotation spins a magnet within a coil of copper, which creates an electrical current. Most wind turbines are built on land in places where the wind is fairly constant or at least predictable, such as in the Great Plains and along mountain tops. Others are built offshore in bays and other shallow ocean waters. While the land-based turbines are less expensive to build and maintain, the offshore turbines generate more power more

FIGURE 11.20. Large wind turbines like these in Minnesota provide a significant fraction of electricity generation in the Great Plains.

consistently. The U.S. has excellent wind resources, but the areas that are best-suited for wind power are often far from major population center, requiring large (and expensive) transmission infrastructure.

11.5.3 Geothermal Energy

Below our feet lies a vast reservoir of **geothermal energy**, heat from the rocks of the planet. The deeper into the crust, the higher the temperature gets, at an average rate of 25°C for each kilometer down. This is known as the *geothermal gradient*. While some form of geothermal energy can be used nearly everywhere, it is most cost-effective in areas where the geothermal gradient is significantly higher than the average, such as near volcanic or other tectonically active regions. In these areas the hot rocks below the surface may be injected with water to produce steam. The pressurized steam is recaptured and used to turn turbines and generate electricity (figure 11.21) There are no other emissions produced, making geothermal an ultra-clean energy source.

In areas where the rocks are cooler, the near-surface soil may be used as a source for heat. The temperature of the soil and surface rocks is more a function of annual temperatures than the heat from the Earth, and in these low-temperature cases a **ground-source heat pump** uses the constant temperature of the soil as a place to exchange thermal energy with the building interior. During warm summer months, thermal energy from the building is transferred to the ground, and during the cold winter months, thermal energy from the ground is transferred to the building. Ground-source heat pumps are much more efficient than standard heat pumps and air conditioning systems, which use the air to transfer thermal energy.

11.5.4 Hydroelectricity

Hydroelectric power plants are the oldest and most mature of the carbon-free energy sources around the world. Dams hold back water and then release it either over the top of the dam or (more often) through sluices or tubes below the water line (figure 11.22). The water flows by force of gravity and turns a turbine to produce electricity. Hydroelectric plants are the most efficient of all the electrical generating systems, with efficiencies between 80–90%. Even the best coal, natural gas, and nuclear plants can only reach 45%, half that of hydropower.

If hydropower is so efficient and carbon-free, why not use it for all our electrical needs? The answer is that most areas suited for hydropower have already been developed, and there are few large-scale locations left (there is good potential for many smaller sites, but their cumulative effect is still small). As U.S. energy use increases into the future, hydroelectric power will play a smaller role, since most of the new generation will come from natural gas, coal, wind, and solar.

FIGURE 11.21. A geothermal electrical power plant in Iceland. The white smoke is steam.

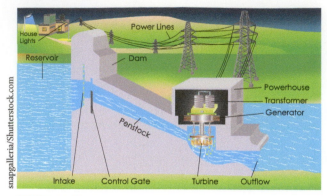

FIGURE 11.22. A cross-section diagram showing the main components of a typical hydroelectric power plant.

FIGURE 11.23. The pumped storage facility at Smith Mountain Lake in Virginia was the first of its kind in the eastern U.S., completed in 1963.

One application in which hydropower excels is **pumped storage**, where a hydroelectric power plant acts like a giant battery (figure 11.23). During the day, water is released from the dam to match fluctuating demand, like the natural gas "peaker plants" discussed earlier. At night, baseload power from coal or nuclear plants "recharge" the water behind the dam by drawing the water backwards, up through the same turbines the water came through during the day with very high efficiency (70–80%).

11.5.5 Biomass and Biofuels

After human muscle, **biomass** is the oldest energy source. The living world is anchored and supported by photosynthesis, in which plants, algae, and other organisms use the energy from sunlight into power growth and reproduction. Several of these primary producers are used for energy. Trees, corn husks, and fast-growing grasses may all be used as fuel sources (wood pellet stoves are a good example).

Other types of biomass come from other organisms that have consumed them. Animal dung, for example, is still an important fuel in many areas of Africa, South America, and Asia.

Biofuels are usually liquids (especially oils) that are produced by organisms with the intention of including them in transportation fuels. Ethanol is a common biofuel and gasoline additive produced from corn in North America and sugarcane in Brazil. Biodiesel is produced from algae and other microbial organisms and may be used in diesel vehicles either directly or blended with petroleum-based diesel fuel. Among the biofuels, only sugarcane-derived ethanol has proven economical, and only then when gasoline prices are high. Corn-based ethanol requires subsidies to be competitive, and biodiesel is still in development stage. None of the biofuels are currently capable of replacing even a small fraction of the petroleum-based transportation fuels currently used in the U.S. or worldwide.

11.6 LOOKING FORWARD: RESOURCES AND CLIMATE

Our discussion of the source and use of various resources leads us to some forward-looking questions: what are the amounts available to us, and what are the consequences of their use? These resources are obviously extremely useful, if not vital to our current standards of living, and so provide enormous benefits to us. But few activities are purely helpful or harmful to society.

CHAPTER 11: Earth Resources: Provisions from God's Creation

11.6.1 Resources

The rate at which developed nations, and the world as a whole, utilize natural resources continues to increase. The US per capita materials use in Table 11.1 is larger this year than it was last year. While many of these resources are abundant (we will never run out of granite!), others are very rare, and none are infinite. Economists, industry experts, and governments are all interested in knowing just how much is available, and how long it might be until we need to look for other resources as substitutes. These estimates can fluctuate, as they depend on the shifting values of *reserves* and *proved reserves*, which change with the price of these commodities. Furthermore, estimates for how long these resources can last are heavily influenced by price changes, capacity to substitute for other materials, new discoveries, political instability, and many other factors. Often such forecasts dramatically underestimate what is available.

In the oil industry, geologist M. King Hubbert was the first to calculate and correctly predict when the U.S. would reach peak oil production among conventional oil wells in the "Lower 48" states. Hubbert had recognized that oil well production followed a normal curve (bell curve) from discovery to closure, with the peak production at the midpoint of the curve. The term "Hubbert's Peak" or "peak oil" commonly refers to the point in time where one-half of the globally available oil reserves has been recovered. Globally, conventional oil field production appears to have reached, or will reach, peak sometime this decade. Advances in unconventional oil (including ultradeep sources, advanced recovery techniques, and shale oil accessed by hydraulic fracturing) have greatly increased the total reserves available (figure 11.24). These stretch Hubbert's Peak somewhere into the future, perhaps to 2050, which is not as far into the future as it may seem. The peak for oil is likely sooner than either natural gas or coal, which are each at least 100–150 years into the future. However, the fact that oil is the source for nearly all transportation fuels makes finding alternative fuel sources quite important.

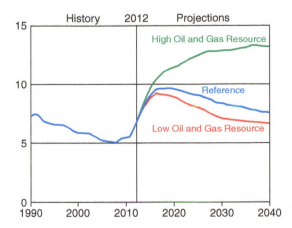

FIGURE 11.24. Historic and projected production of petroleum and related products under three oil-price scenarios.

11.6.2 Climate

In Chapter 13 we discuss the issues of greenhouse gasses, climate change, and global warming. As shown in figure 11.25, atmospheric levels of CO_2 have been rising since their first direct measurements starting in 1958. Measurements of CO_2 from ice cores indicate that the pre-industrial concentration was approximately 280 parts per million (ppm), and current levels are now near 400ppm. Because CO_2 is a greenhouse gas, increasing its atmospheric concentration increases the atmospheric effect (see chapter 13), resulting in global warming.

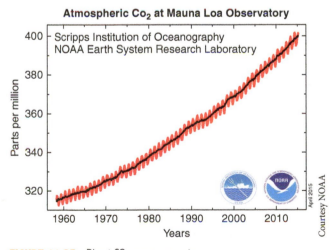

FIGURE 11.25. Direct CO_2 measurements.

One question is whether human production and combustion of fossil fuels is the main source of the increase in CO_2 levels. The answer appears to be yes. By studying the carbon in the atmosphere, scientists have discovered two confirming evidences: first, calculating the amount of CO_2 created by burning fossil fuels accounts for nearly all the CO_2 in the atmosphere above pre-industrial levels; and second, most of the new carbon is carbon-12, rather than other carbon isotopes. This is because the carbon from coal, petroleum, and natural gas has higher amounts of carbon-12 and lower amounts of carbon-14 in it than the natural atmospheric concentration.

While determining that the current increase in CO_2 is directly related to carbon emissions from fossil fuels is fairly straightforward, it raises far more difficult questions. First is whether the trend of increasing global temperatures will continue to follow the trend of increasing CO_2. As discussed in chapter 13, the modeling can be influenced by various assumptions (which can introduce biases), but they also have proven useful and the scientists who construct them are careful about their work. Further, energy use transitions (especially the current shift from coal to natural gas, and an increase in renewables) is helping reduce CO_2 emissions in some countries (see figure 11.26).

Let's assume that the models are generally correct and we should anticipate increases in both global temperature and sea level, along with other changes in climate. It is prudent, after all, to plan for the worst while hoping for the best. A second question, then, is whether the changes are a net positive or net negative for both people and the world over which we are charged to care. No doubt there will be both good and bad outcomes, such as more arable land in currently marginal climates (good), or increased displacement of people by rising sea level or diseases (bad). A third question is what can, or even should, be done next? Can CO_2 levels be reduced? This would take a combined and monumental effort at immense capital expense to convert

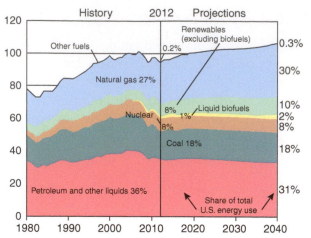

FIGURE 11.26. Past and projected energy source distribution for the United States.

the energy infrastructure of all the developed nations of the world, and the effect would not be seen for many decades or even centuries. Recall that at the moment, fossil fuels make up 83% of all energy resources, and do so at very low costs. Renewable energies simply do not have the scale or pricing to compete against coal, natural gas, and petroleum at this time, though costs for solar are falling rapidly. Quite possibly many countries (including your own) may not cooperate for fear of damaging their economies. In a globally connected world, any competitive advantage is useful, especially low-cost energy and resources.

Within the Christian community there are a number of voices on these issues. In the U.S., the *Evangelical Environmental Network* and the *Cornwall Alliance* (see websites in the Further Reading section at the end of the chapter) are two organizations that seek to encourage good stewardship of our world and care for the poor. These organizations differ on how to go about it: the *EEN* favors strict carbon and other pollution reductions while the *CA* prefers projects to raise living standards and combat poverty. Whatever the approach chosen, it is Christians who can bring an understanding of mankind as God's chosen stewards of the world, encouraging others to care and keep it as commanded in Genesis 1 and 2.

11.1 Caretakers of Creation

- The commands to care and oversee the world given Adam and Eve in Genesis 1:28 and 2:15 form the basis of the dominion mandate for mankind.
- The availability of high-quality energy sources (including coal, petroleum, and nuclear power) allows for great increases in work efficiency and the large human population on Earth today.

11.2 Resource Use in the Modern World

- Modern societies use a tremendous amount of natural resources.
- Resources, reserves, and proved reserves are terms for describing the total, potentially recoverable, and likely recoverable materials from the Earth.
- Prices of resources are determined by a variety of factors, which can change significantly based on the resources availability, technological advances, and other factors.

11.3 Energy Resources: Fossil Fuels

- The majority of energy use in developed countries comes from fossil fuels (coal, petroleum, and natural gas).
- Coal is the most common fuel for generating electricity. It is especially important in maintaining base load power levels.
- Petroleum, which is refined from crude oil, is the dominant fuel for transportation around the world.
- Natural gas has a variety of uses: chemical manufacturing, electricity generation, and home heating. It is particularly important in meeting short-term changes in electricity use.

11.4 Energy Resources: Nuclear Power

- Nuclear power uses the fission of enriched uranium to produce a chain reaction. This generates heat to boil water, turning turbines to produce electricity.
- Nuclear power has the best safety record of all power generation methods, but three notable nuclear accidents in the U.S., Ukraine, and Japan are reminders that all forms of energy generation have risks.
- One of the largest concerns about nuclear power is the question of what to do with spent nuclear fuel.

11.5 Energy Resources: Renewables

- Renewable energy sources make up about 9% of U.S. energy production, and is dominated by hydro-electricity and biomass.
- Solar power has the largest potential for energy generation, including electricity via photovoltaic (PV) and concentrated solar, or space and water heating via passive and active solar thermal methods.
- Wind can turn wind turbines to generate electricity, and this technology is currently the largest carbon-free method after nuclear and hydroelectric power.
- Geothermal energy uses the Earth's heat from the crust to provide steam for electricity generation or heat for space heating. Ground-source heat pumps use the soil or bedrock for a heat exchanger.
- Hydroelectric power is the most mature carbon-free energy source. Utilizing dams, hydroelectric plants use flowing water to turn turbines. Some dams can be used for pumped storage, in which the system acts like a giant battery.
- Biomass uses plant materials (like wood) for heating fuel. Biofuels are liquids (such as ethanol and bio-diesel) that can be used in combustion engines.

11.6 Looking Forward: Resources and Climate

- All non-renewable resources are finite by definition, but some are more limited than others.
- M. King Hubbert accurately predicted the peak production of conventional oil in the U.S., and "peak oil" refers to the point in time where one-half of globally available oil reserves been recovered.
- Carbon dioxide is a byproduct of fossil fuel combustion, and its concentration in the atmosphere has been rising steadily over the past century.
- If global temperatures continue to increase, there will be both positive and negative effects, but it will be very difficult to transition to a low-carbon energy profile.
- There are a variety of views among Christians as to how big the problem is, and the type of responses that should be explored.

KEY TERMS

Active solar heating	Control rod
Base load power	Crude oil
Biofuel	Dominion mandate
Biomass	Enrichment
Chain reaction	Fission
Coal	Fossil fuels
Coal ash	Fuel rods
Concentrated solar power	Geothermal energy

Ground-source heat pump	Pumped storage
Hydroelectric power	Refine
Meltdown	Reserve
Mountaintop removal	Resource
Natural gas	Solar photovoltaic panels
Passive solar heating	Spent nuclear fuel
Peaker plant	Strip mining
Petroleum	Underground mining
Proved reserve	Wind turbine

REVIEW QUESTIONS

1. Review the list of resources in Table 11.1. Thinking of where you live, what are some places where rock and mineral resources are currently being used?
2. Which of the rock and mineral resources is largest? Why do you think that is?
3. What percentage of total energy use in the United States is derived from fossil fuels?
4. Make a table on your own with three columns: electricity, transportation, and space (home) heating. Under each column, list the energy resources that are used for each of these. Looking at your list, which column has the most diverse energy sources? Which is the most restricted? Which resource(s) are can be used for the widest variety of needs?
5. What are the ranks of coal, and for what purpose are they used?
6. Describe in your own words the difference between a conventional oil/gas well and a hydraulically fractured well.
7. Coal, oil, and natural gas are ultimately made from photosynthetic organisms. What types form each?
8. What are some of the reasons why natural gas is a cleaner electricity source than coal?
9. What is the type of fuel used in a nuclear power plant? What must be done to naturally occurring uranium to make it into a suitable fuel?
10. Which of the renewable energy resources have the greatest potential for total energy production? What are some reasons why?
11. What are some of the factors that currently limit the amount of renewable energy resources compared to fossil fuels? How might this change?

FURTHER READING

American Association of Petroleum Geologists: http://www.aapg.org/

Cornwall Alliance: http://www.cornwallalliance.org/

Evangelical Environmental Network: http://www.creationcare.org/

MacKay, D. *Sustainable energy—Without the hot air.* UIT Cambridge, Ltd. 2009. Available free online at: www.withouthotair.com

Snelling, Andrew. The origin of oil. Answers Magazine, Jan.-March., 2007.

Switch Energy Project: http://switchenergyproject.com/

The Rocky Mountain Institute: www.rmi.org

U.S. Energy Information Administration: http://www.eia.gov/

Wise, Kurt. Sinking a floating forest. *Answers Magazine,* Oct.-Dec., 2008.

CHAPTER 11: Earth Resources: Provisions from God's Creation

In the Gospel of Matthew, Jesus presents a parable of wise and foolish virgins (Matthew 25:1-13). In the parable, the virgins are preparing for a wedding (they are bridesmaids), and of the ten, five prepared well, taking lamps and oil, while five of them took lamps but forgot their oil. When the bridegroom shows up to start the wedding party, the forgetful five were out buying oil, and missed the call to the wedding. The foolish virgins forgot to plan ahead for their future, and were left out because of their lack of foresight. As you think about energy resources, ask yourself, "What future lies ahead, and how might we best prepare ourselves for it?"

CHAPTER 12

OCEANS AND COASTAL SYSTEMS

OUTLINE:

Richart777/Shutterstock.com

For centuries, sailors aboard ships like these relied on the wind and surface currents to travel great distances across the world's oceans.

Ocean water is constantly flowing within each ocean basin and from one basin to the other. We refer to these masses of flowing water as **currents**. Several different forces act to produce currents, including winds, density differences between different masses of water, and tides. Together, these forces help to diffuse gasses into the oceans, redistribute nutrients, and moderate both ocean and world temperatures.

12.1.1 Surface Currents

Throughout the year, relatively stable high pressure systems produce persistent winds that transfer energy into the ocean, producing **surface currents** (figure 12.1). These "rivers in the ocean" flow at a velocity of 0.5-2m/s (between 2 and 5 km/hr; figure 12.2),

and play a vital role in distributing heat energy within the oceans and the world as a whole. Because they are formed by winds, the surface currents shown in figure 12.1 match well with the map of major wind belts of the world in figure 15.6. In particular, the subtropical high pressure systems located at approximately 30° north and south of the equator (like the Azores/Bermuda High in the North Atlantic) are key sources of winds for surface currents. About 5° north or south of the equator, these systems produce the east-to-west blowing trade winds, while around 30-60° north or south latitude they produce the west-to-east blowing westerlies. Within large ocean basins, these high pressure systems produce several related surface currents connected together in large, circular-flowing systems called **gyres** (from the Greek word *gyros*, meaning

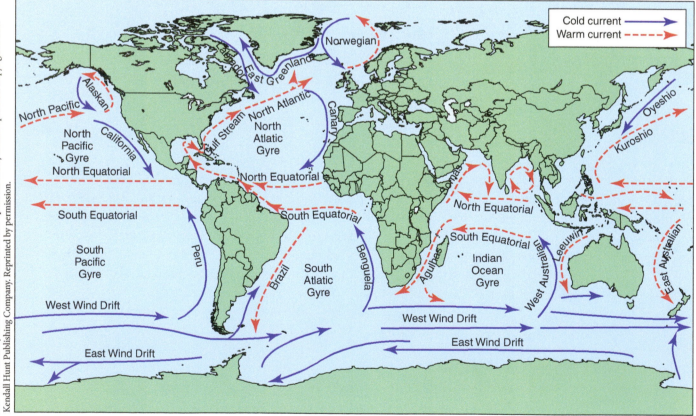

From Earth and Beyond: An Introduction to Earth-Space Science by Brent Zaprowski. Copyright © 2009 Kendall Hunt Publishing Company. Reprinted by permission.

FIGURE 12.1. Major currents and gyres of the oceans.

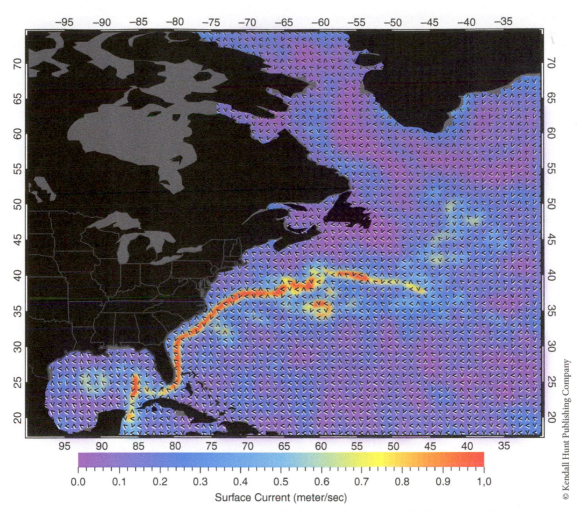

FIGURE 12.2. A velocity map of ocean currents in the northeast Atlantic Ocean. The fast-flowing Gulf Stream is easily seen in red.

"circle"). Looking closer at figure 12.1, you can see that the surface currents of the gyres distribute warm equatorial waters (red arrows) towards the poles, and bring cold polar waters (blue arrows) back to the equator.

Gyres are circular in nature because of the **Coriolis Effect**. If the Earth did not rotate on its axis, surface winds would travel in straight-line paths based on the location of high and low pressure regions. Without rotation, there would be no trade winds or westerlies, and the subtropical highs would produce "northerlies" and "southerlies." But as the Earth rotates under the winds, the winds appear to be deflected as they blow. For example, if a plane left the North Pole heading due south, its flight path would appear to be deflected because North America rotated underneath of it (figure 12.3). This relative deflection results in changes in wind direction over the ocean. In the northern hemisphere, winds are deflected to the right of their direction of motion, resulting in the clockwise rotation of the North Atlantic and North Pacific gyres. In the southern hemisphere, winds are deflected to the left, and the gyres flow counter-clockwise, as seen in the South Atlantic, South Pacific, and Indian Ocean gyres.

The air flowing from the subtropical highs that form gyres also results in two notable areas with very light winds, resulting in no or very slow currents. The center of the gyre (or mid-gyre) is characterized by sinking air in the middle of the subtropical high,

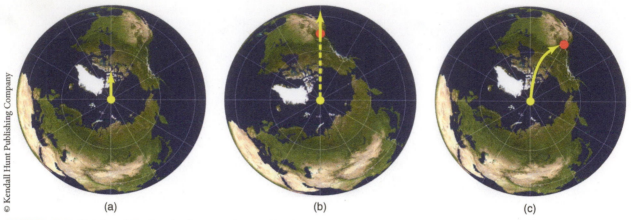

FIGURE 12.3. Rotation of Earth makes it appear that an object in straight-line motion is actually veering to the right in the Northern Hemisphere. This is called the Coriolis Effect.

resulting in clear skies and very little wind. Nearest the equator is a region of calm water called the **doldrums**. Here, the trade winds to the north and south of the equator meet at the Intertropical Convergence Zone, or ITCZ. The air from the trade winds rises at the equator, producing only variable winds that do not create sustained currents. Both of midgyre and doldrums were feared and despised by mariners on sailing ships, since storms that blew ships into the these regions could leave the crew stranded out in the open ocean for days or even weeks until they slowly migrated back to windier areas.

Thus far, we have discussed how winds produce currents that mix water horizontally along the ocean surface. At the coasts, wind direction also plays an important role in mixing water vertically. In places where winds force large amounts of water toward the coast, the excess waters pile up and are forced deeper, bringing warm waters to the cooler, deeper ocean. This process is called **downwelling**. In other areas where winds blow along or away from the coast, waters are pushed away from the coast, creating a deficit of water that is replaced by rising cold waters in a process called **upwelling** (figure 12.4). The deep waters are nutrient-rich because organic material has sunk over time and accumulated at depth. By returning these nutrients to the surface, seasonal upwelling regions are some of the most biologically productive zones on Earth, including many of the most important ocean fisheries. On the other hand, if the upwelling areas are oxygen-depleted, they can result in a dead zone, where very little aquatic life can live while such conditions dominate.

12.1.2 Thermohaline Circulation

Surface currents, as their name indicates, are features of the ocean's surface and affect rather shallow waters.

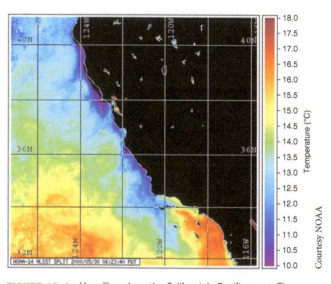

FIGURE 12.4. Upwelling along the California's Pacific coast. The darker blue colors indicate upwelling cold waters. Note how far the blue trails extend out from the coast, indicating very strong currents.

The deep ocean is unaffected by surface currents, yet there are other currents moving water, driven by density rather than wind. The density of ocean water is a function of both its temperature and its salinity (salt content). Like other materials, water initially becomes denser as temperature decreases. Unlike most materials, water's greatest density is reached prior to its freezing point (maximum density is at 4°C in fresh water), and its density then *decreases* as the temperature approaches and passes the freezing point. Ice floats because it is less dense than the water it is in.

Salinity's relationship to water density is much more straightforward. Higher volumes of dissolved salts increase the density of the water. Combined, the density-driven changes produce a **thermohaline** (temperature and salinity) **circulation** system in the ocean. In the polar regions of the North Atlantic and Arctic oceans, the formation of sea ice results in the formation of very cold and very salty water (sea ice is made primarily of fresh water that left its salt in the ocean water below). This water becomes very dense, and sinks down towards the ocean floor (figure 12.5). From there, it travels south, exiting the Atlantic near the tip of South America, where it will continue its journey. As the water warms, it rises in either the Indian or Pacific oceans, where it travels below the surface until it makes its long, slow loop back to the North Atlantic. The amount of time it takes for water to fully travel through this system is about 1,000 years, much slower than the velocity of surface currents.

12.1.3 Tidal Currents

Discussed more fully below, the gravitational pull of the sun and moon draw water vertically to produce the tides, and these tides also flow horizontally towards or away from the land. This horizontal component to the tides produces **tidal currents** (figure 12.6). Unlike surface and deep-water currents, tidal currents do not follow more-or-less linear paths for long periods of time. Instead, they switch back and forth as the tide rises

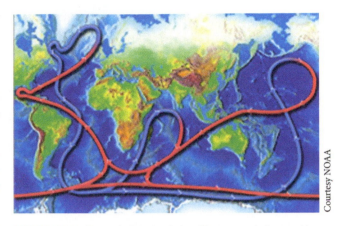

FIGURE 12.5. Thermohaline circulation. Blue arrows indicate cold, deep waters while red arrows indicate warmer, shallower waters.

Courtesy NOAA

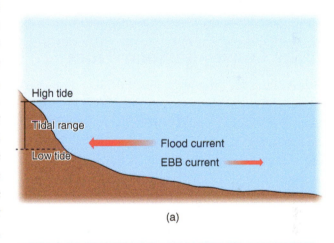

(a)

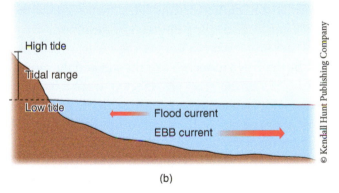

(b)

© Kendall Hunt Publishing Company

FIGURE 12.6. Tidal currents. An advancing flood current dominates as the tide rises (a) and a receding ebb current dominates as the tide falls (b).

and falls (this is called a rectilinear path). Tidal currents flowing from the ocean towards the shore are called flood currents, while those flowing from the shore out to the ocean are called ebb currents. Both flood and ebb currents operate at the same time, but switch back and forth in strength as the tide rises (stronger flood currents) and falls (stronger ebb currents). Tidal currents have their greatest effect in bays, inlets, and estuaries, where their velocities increase as water is concentrated into a restricted area. The most dramatic flood currents are **tidal bores** (figure 12.7), which create a wall of rushing water as it flows into a bay that is large enough to surf or kayak!

FIGURE 12.7. A bore tide advances in the Qiantang River in eastern China. Ocean water from the East China Sea advances into the bay of the river, which narrows to funnel the incoming waters into a tidal bore.

12.2 TIDES

12.2.1 Lunar and Solar Tides

Tides are the daily rhythmic rise and fall of the ocean's surface elevation. They occur primarily because of the gravitational interactions between the Earth with both the moon and the sun. Though the sun is by far the largest object whose gravity affects Earth, it is very far away, resulting in a small but measurable effect on the tides. The moon is much smaller, but it is very close to Earth and its gravitational interaction is much more prominent. We'll first consider the effect of the moon's gravitational pull on Earth's tides, then see how the sun's gravity alters the pattern set by the moon.

Because water is a fluid, it can be deformed easily by the gravitational pull of the sun and the moon. The rocks of the crust are affected as well, but this cannot be detected by people without the aid of precise instruments. As the moon orbits Earth, it draws the Earth's water toward itself, creating a lunar **tidal bulge** in the waters facing the moon. One might think that there should be just one high tide each day at any location, when that location is facing the moon as Earth rotates on its axis. But in reality there are *two* lunar tidal bulges: one facing the moon, and one on the opposite side of Earth. Why?

The answer lies in in the difference in distance the two sides of Earth are with respect to the moon, and the fact that the gravitational force of an object decreases quickly as you move away from that object. With this in mind, we can describe the effect of the moon's gravity at different locations in terms of vectors (figure 12.8). The average distance from Earth to the moon is about 61 times the radius of the Earth. So the water on Earth closest to the moon is 61 Earth-radii away, and is being pulled on strongly by the moon's gravity (the long arrow in figure 12.8a) and forming a tidal bulge. On the opposite side of Earth the water is 63 Earth radii away (Earth's diameter = 2 radii). Since it is further from the moon, it is affected less by its gravity (the short arrow in figure 12.8a). Sitting in between these is the Earth itself, whose core is located at a central 62 Earth radii from the moon, and not deforming noticeably because it is solid. It is being pulled on by the moon's gravity at an intermediate strength (the medium arrow in 12.8a).

Using vectors to illustrate the relative difference in gravitational strength, and using Earth's center as a reference point (figure 12.8b), we can illustrate how both tidal bulges form from the moon's gravitational

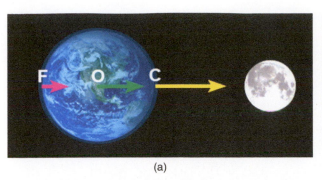
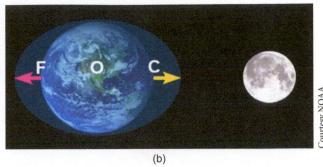

Courtesy NOAA

(a)   (b)

FIGURE 12.8. Two tidal bulges form from the moon's gravitational pull on Earth, which decrease as the distance from the moon increases. (a) Vectors showing the strength and absolute direction of movement due to the moon's gravitational pull on the near, center, and far side of Earth. (b) Vectors referenced to Earth's center, showing relative tidal direction. F = far side of Earth from moon; O = center of Earth; C = close side of Earth from moon.

pull. The tidal bulge closest to the moon is a result of the strongest gravitational forces pulling the water towards the moon faster than its pull on the Earth or the water on the far side. The water on the far side forms a bulge as well, because there the force of the moon's gravity is less than both the water closest to the moon and the Earth. In effect, the water on the far side of Earth forms a bulge because it is being left behind. The lunar tidal bulges appear to move in opposite directions to us here on Earth, though in reality they are moving in the same direction (toward the moon), but at different rates.

As the Earth rotates on its axis, most coastal areas travel through both of the two tidal bulges, and their sizes and central locations depend on where the moon is during its orbit around Earth. When the coastal area passes into the peak of the tidal bulge, it experiences **high tide**. When it is perpendicular to the tidal bulges, it experiences **low tide**. The difference in elevation between high tide and low tide at the shore is called the **tidal range**.

The sun also produces two tidal bulges on Earth, because of the same gravitational and distance-related factors as the moon. However, the solar tidal bulges are much smaller than the lunar tidal bulges. As the moon orbits Earth, the position of its two tidal bulges changes with respect to the two solar tidal bulges. The interplay of these two sets of tidal bulges results in the monthly **tidal cycle**. About twice each month the sun, moon, and Earth are aligned, producing either

a full or new moon. The lunar and solar tidal bulges are thus aligned and their elevations combine, resulting in a **spring tide** (figure 12.9a), which is expressed as the largest tidal range of the month with the most pronounced high and low tides. Spring tides were not named after the season (after all, they happen twice each month), but for the unusually high tides where the water seemed to "spring" up towards the land. In contrast, when the moon is oriented perpendicular to the direction of the Earth-sun line, the lunar tidal bulges are likewise oriented perpendicular to the solar tidal bulges, and a **neap tide** (figure 12.9b) occurs as the bulges pull water in different directions from each other. Neap tides are expressed by the smallest tidal range of the month and resulting in only modest high and low tides. Like spring tides, neap tides occur roughly twice each month, and occur with the first and third quarter moon (half-moons).

12.2.2 Tidal Patterns

A tidal monitor is used to measure the tides, and tide predictions are made in the U.S. by the National Oceanic and Atmospheric Administration (NOAA). In the past, tidal monitoring was done with mechanical floats and recorders. Now, advanced electronic and acoustic methods more accurately record changes in the ocean's elevation from tides, as well as other marine and atmospheric data, which can be accessed in real-time from the NOAA website (figure 12.10).

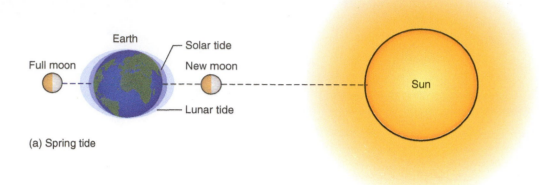

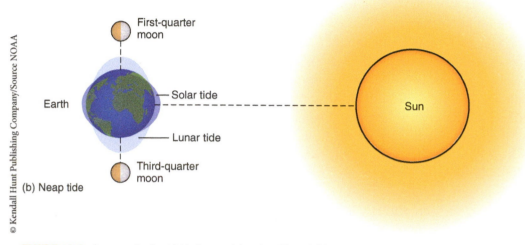

© Kendall Hunt Publishing Company/Source NOAA

FIGURE 12.9. Lunar and solar tidal bulges at (a) spring tide and (b) neap tide.

At any particular shoreline, a variety of factors will contribute to how the daily and monthly tidal patterns are expressed locally. Overall, however, there are three basic types of daily tidal patterns expressed at the shorelines of the world (figure 12.11). A **diurnal** tidal pattern features a single high tide and single low tide each day. Why just one, instead of the expected two? In places like the Gulf of Mexico, where diurnal patterns dominate, the opening to the Atlantic Ocean is partially blocked by Florida, the Yucatan Peninsula of Mexico, and the Caribbean islands. This prevents the Gulf waters from gathering with the Atlantic waters to form a second tidal bulge when on the far side of the Earth. A **Semi-diurnal** tidal pattern features two similar-magnitude high and low tides each day. This type of tidal pattern is seen across much of the Atlantic coast of North and South America, and is the most common throughout the world. A **mixed tidal pattern** occurs when there are two high and low tides each day, but the two highs are not of the same magnitude, nor are the two lows. Rather, each day there is a pronounced high followed by a pronounced low, then a modest high followed by a modest low, finally rising to a pronounced high and restarting the cycle. The Pacific coast of North America is characterized by mixed tidal patterns.

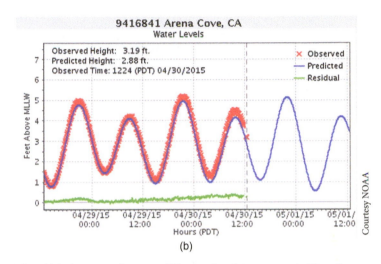

9416841 Arena Cove, CA
Water Levels

Observed Height: 3.19 ft.
Predicted Height: 2.88 ft.
Observed Time: 1224 (PDT) 04/30/2015

× Observed
— Predicted
— Residual

Courtesy NOAA

(a)   (b)

FIGURE 12.10. A) A small tide monitoring station attached to a pier, with its instrument box open. Tide elevation data are recorded from the white tube on the left. B) Tidal data recorded (red) and predicted (blue) for the station at Arena Cove, California on April 30, 2015. Compare this to the mixed tidal pattern in figure 12.11.

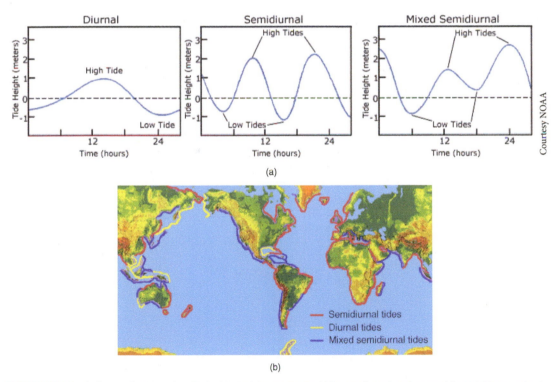

(a)

(b)

FIGURE 12.11. A) Types of tidal cycles. B) A global tidal pattern map. Yellow = diurnal; red = semidiurnal; blue = mixed.

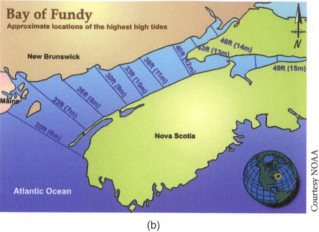

(a)

(b)

FIGURE 12.12. A boat lies on the tidal flat at low tide in a harbor along the Bay of Fundy in Parrsboroa, Nova Scotia. When high tide returns, the boat will float to near the top of the pier. Inset: Map of tidal ranges in the Bay of Fundy.

In most areas, the tidal range is a few meters. Oceanographers define coasts as *micro-tidal* if the tidal range is <2m, *meso-tidal* for 2-4m, and *macro-tidal* for >4m. In general, the US Atlantic coast is microtidal, while the Pacific coast is mesotidal. A few areas, however show extremely large tidal ranges. The most extreme example is in the Bay of Fundy in Nova Scotia (figure 12.12), which can see tidal ranges of up to 17m (56 feet) in a single day!

12.3 WAVES: FORMATION AND TRAVEL

12.3.1 The Making of a Wave

The waves that we enjoy at the beach often formed hundreds, or even thousands of miles away from the shore. A **wave** is an expression of energy passing through the water, produced by oscillating (circular-moving) water particles. Waves traveling through the open ocean are characterized by a particular set of features, illustrated in figure 12.13. The highest point of the wave is called the **wave crest**, and the lowest point between two crests

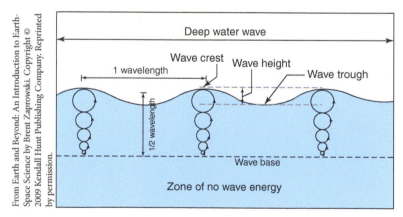

FIGURE 12.13. Basic components of an open-ocean wave.

is called the **wave trough**. The difference in elevation between them is the **wave height**. The distance between one wave crest and another is the **wavelength**. Perhaps surprisingly, the water particles within the wave are not moving along with the wave. Instead, they move in circular paths (orbits) as the wave passes through the water. This is why a boat will bob up and down in the waves without moving very far: the boat is riding the same circular paths as the water. Wave energy is strongest at the surface of the ocean, and decreases with depth, because friction reduces the energy transferred by the circulating water particles interacting through the water column. At a depth equal to one-half of the wavelength we encounter the **wave base**, and the water below it is no longer turbulent.

Ocean waves are formed by friction that builds from wind travelling across the air/ocean surface. The speed, height, and wavelength of a wave depend on three factors: wind speed, wind duration, and **fetch** (the distance over which the wind blows). Obviously a faster wind speed can impart more energy into the ocean to make waves than a slower wind speed can, if all other factors are equal. But the duration of the wind over a patch of the ocean is extremely important: a 60 km/h gust of wind lasting 30 seconds will not produce a wave nearly as big as a 10 km/h wind lasting two hours. Likewise with fetch: a storm that blows strong winds over a short distance to shore (say, 15 km) will not create the same set of waves that the same storm system can

make if it travels over 300 km, constantly imparting more energy into the ocean. This is a major factor for why the Pacific Ocean (the largest ocean basin) has larger waves than other oceans, seas, and lakes.

12.3.2 Wave Interactions at the Shore

When an open-ocean wave approaches the shore, it is altered by the shallowing depths of the coast. The first way is in the development of surf. Recall that wave base is the depth at which the wave's energy no longer affects the water below: water below wave base is tranquil, and water above wave base is turbulent. As the wave comes closer to shore, the sea floor rises through the wave base, and into the turbulent zone above. At first the effect is minimal; sediment on the sea floor is stirred up a bit. But as the wave comes closer to shore and the sea floor rises closer to the ocean surface, the circular oscillations of the water particles in the wave are stretched into more elliptical shapes, causing the wavelength to shorten and the wave height to increase. Often you can see this at the beach: there will be an area where the waves seem to suddenly grow taller than those behind them and further out from shore. Eventually, the orbits become so elliptical and the wave is so steep that water molecules are unable to complete their oscillation. Instead, water falls under the force of gravity and the wave crest collapses (or *breaks*) in the **surf zone** (figure 12.14). Similar to wave

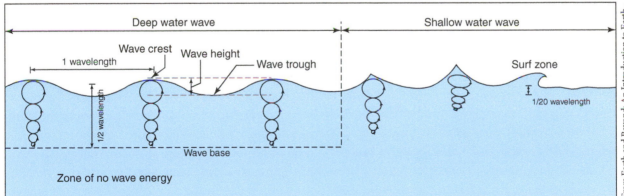

FIGURE 12.14. As an open-ocean wave approaches shore, the shallow sea floor rises above wave base, causing the wave to alter in shape as it approaches shore. Wavelength shortens and water particle orbits become more elliptical until the wave breaks in the surf zone.

FIGURE 12.15. A surfer riding the crest of a breaking wave in the surf zone. Note that the surfer is riding the wave parallel to the shoreline.

FIGURE 12.16. Straws in a glass of water appear shifted due to refraction.

base, the surf zone is also mathematically related to wavelength: the surf zone is always found where water depth is 1/20th of the wavelength. This means that as someone is surfing (figure 12.15), they are not trying to ride the wave *towards* the shoreline. Rather, they are trying to ride *parallel* to the shoreline, along the surface expression of the surf zone at a depth of 1/20th of the wavelength.

A second way that waves are altered as they approach shore is by **wave refraction**, which is the bending of ocean waves to become more parallel to the shoreline. Since an ocean wave is an expression of energy (including a wavelength, wave height, etc.), it can be bent, similar to how other wavelengths of energy can be bent (figure 12.16). Most waves do not often approach a shoreline directly; they frequently approach from some angle. As the wave approaches the shoreline, friction with the sea floor affects the part of the wave closest to the shore, slowing it down while the rest of the wave continues at its original velocity. As the wave continues to move closer to the shore and more parts of the wave are slowed by friction, the whole wave becomes oriented more parallel to the shoreline. This is true both for simple, linear beaches as well as complex, irregular beaches (figure 12.17). On irregular beaches, refraction of waves causes wave energy to be concentrated on headlands (exposures of the land's bedrock), wearing them down. In protected beaches and coves, the wave energy is lessened, allowing these areas to accumulate sediment. The combined actions work to smooth out irregular shorelines into elongated, linear shorelines. However, competing processes such as tectonics and other changes in land elevation can operate faster than the erosional and depositional forces at the shore, and can maintain or alter the existing shorelines in other ways.

Though the waves refract, they still tend to strike the shoreline at a slight angle, which is usually controlled by seasonal wind direction. This difference in angle between the incoming wave and the shoreline results in **longshore transport**, an overall direction for the erosion, transport, and deposition of sediments on and near the shore (figure 12.18). Longshore transport has two primary components. *Beach drift* is the zig-zag movement of sand caused by the waves lapping up at the angle to the shore and pushing sediment up along the beach, followed by their return back to the ocean perpendicular to the shoreline. This moves the sand right at the ocean/land interface. *Longshore current* is responsible for much more sediment transport, where it is found as a corkscrew-like flow of water parallel to the shoreline in the surf zone. The large amounts of energy released by breaking waves churn up vast amounts of sediment and transport them in the water

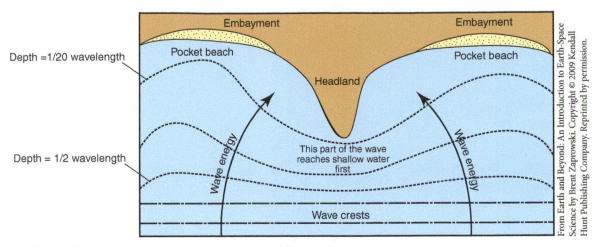

From Earth and Beyond: An Introduction to Earth-Space Science by Brent Zaprowski. Copyright © 2009 Kendall Hunt Publishing Company. Reprinted by permission.

FIGURE 12.17. Waves refract (bend) to become more parallel to shoreline.

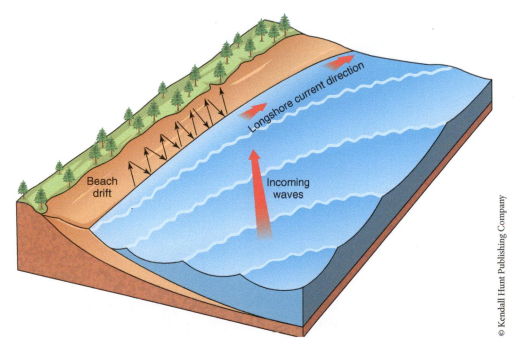

© Kendall Hunt Publishing Company

FIGURE 12.18. Components of longshore transport. Beach drift moves sand at the ocean/land interface in zig-zag patterns while longshore current transports large amounts of sediment in the surf zone.

just in front of the shoreline. If you've ever "wiped out" while surfing, you have felt the power of the waves as you were tossed and spun under the breakers as just another bit of sediment, along with all the sand that ended up in your swimsuit!

Because waves never stop impacting the shore, at times the water nearest the shoreline begins to pile up, its return to the ocean slowed by the continued influx of new waves. These waters may pool together and create a fast-moving return flow called a **rip current** (figure 12.19). The strength of the rip current varies with the strength of the incoming waves and shoreline geometry, and they can be extremely powerful and dangerous (even deadly). Fortunately, the rip current is narrow, and if caught in it, swimming sideways (not against it!) is the fastest escape, since rip currents are usually narrow (a few tens of feet across) and frequently dissipate just beyond the surf zone.

Courtesy NOAA

FIGURE 12.19. Formation of a rip current. (1) Incoming ocean waves; (2) Lateral movement of water along the shoreline; (3) Rip current return flow; (4) Dissipation.

Courtesy NOAA

(a)

© het Mitchell/Shutterstock.

(b)

FIGURE 12.20. (a) A rip current at the beach disturbs incoming ocean waves. (b) A sign warns of rip currents at a beach.

Strong rip currents can be identified from the beach and avoided because the rip current's flow can be seen as a linear disturbance affecting incoming waves (figure 12.20).

12.4 COASTAL AREAS AND LANDFORMS

In the above section, we focused primarily on the features and movements of the water, and on its ability to transport materials. But we also need to understand the land and underwater features that make the shore such a dynamic interface of the hydrosphere, geosphere, and atmosphere. Figure 12.21 illustrates the components of the area we variously call the shore, beach, or coast. This region and its components are defined by the areas that are (and are not) affected by ocean waves as they approach and impact the land.

12.4.1 The Shore Region

We begin with the **shoreline**, which is the point of intersection of the ocean and the land. The shoreline migrates towards and away from the land daily because of the tides (discussed above), so we recognize a high tide shoreline and a low tide shoreline. These two shorelines bracket the area known as the **foreshore** (also called the *intertidal zone*), and the sand in this area is usually compacted and moist. Further inland from the foreshore is the area known

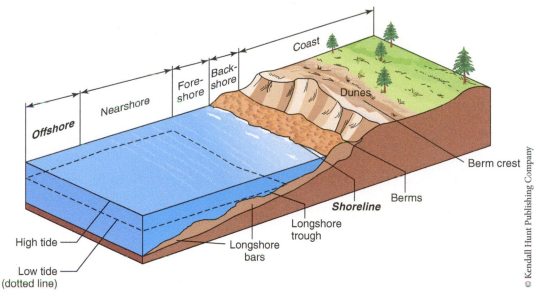

FIGURE 12.21. A cross-sectional profile of the shore region.

as the **backshore**. This region is not affected by the tides, but it is affected and defined by large storm activity. Its sands are dry and loose, and this is where most people put their beach towel when they visit the shore. Within the backshore may be a **berm**, which is a steepened ridge of sand followed by a sandy platform or terrace. The berm's face was formed by recent storms, and shows how far inland the waves were able to erode beach sand. If storms have been infrequent, winds may smooth out the beach sands, resulting in no berm at all. Taken together, the foreshore and the backshore are referred to as the **shore**, where all features are affected by wave action.

Further inland, a series of large sandy mounds, called **dunes** represent the first structures that are no longer formed by water, but rather wind. The dunes mark the beginning of the **coast**: a relatively flat region that continues inland until there is a significant change in terrestrial topography. If the land of the coast is low-lying and frequently flooded at high tide, but drained at low tide, it will often support a variety of salt-tolerant grasses, trees, or shrubs plants as part of a **coastal wetland**. The plants of the coastal wetlands anchor the sandy or muddy sediment, both protecting from erosion as well as often facilitating

sediment accumulation. Grassy wetlands often dominate temperate regions, while mangrove swamps (whose roots prop up the trees above the water line) are frequently seen in the tropics (figure 12.22). Wetlands host a rich diversity of plant and animal life, and are frequently breeding grounds for a multitude of bird and fish species.

Back at the foreshore, we turn our attention towards the water. Continuing from the low tide shoreline, we enter the **nearshore** region. This is the area of the water that is permanently submerged and whose bottom sediments are affected by wave activity. Frequently there is an elevated sand bar within the nearshore region. The sand bar often migrates onto the coast during the summer, when low-energy, net-depositional waves dominate, and is moved into the nearshore region during the winter, which is characterized by high-energy, net-erosional waves. Further out to sea, the ocean sediments drop below *storm wave base* (the depth of reach of the largest storms in the area, which is ½ of the wavelength of the storm waves, as discussed above). This deeper region of quiet bottom waters is called the **offshore** region, and extends for the remainder of the continental shelf.

(a)   (b)

FIGURE 12.22. Wetlands can include grassy salt marshes (a) and mangrove swamps (b).

12.4.2 Erosional and Depositional Landforms

This simplified description of the shore region is expressed in a wide array of variations in the real world. While we may often think of and greatly enjoy long stretches of white sand beaches, most beach areas are small and isolated, filled not with soft, uniform sands but with coarse sands, gravels, or cobbles. Many shores are characterized by steep cliffs, while others are muddy. Why such variation?

The answer lies in a combination of factors, but much of it is a result of sediment input. Large rivers that empty into the ocean provide sand to make and replenish beaches. In the northeastern United States, for example, there are few large rivers, so the shores of the New England states are erosional. Beaches form in isolated coves and bays where low-energy waves erode out gravel deposits left over from the Ice Age. Beginning in southern New Jersey and extending to Florida, the southeastern beaches are fed by larger rivers, and benefit from being downstream of generally north-to-south flowing longshore current.

Where wave-generated erosion dominates, a variety of physical features may be seen. Commonly erosional shorelines will feature **wave-cut cliffs**, a steep wall of rock exposing the headland. The cliffs are formed as strong wave actions pound and undermine

the headland resulting in rock falls and large collapse events. At the bottom of the cliff is often a relatively flat foreshore area, called a **wave-cut platform**. If tectonic or other factors raise the land upward, this set of wave-cut cliff and platform is lifted out of the reach of waves, and forms a **marine terrace**. Below and further seaward of the marine terrace, a new wave-cut cliff and platform are scoured by the waves (figure 12.23).

Where limestone or dolostone forms the bedrock of the headland, chemical as well as mechanical erosion can produce very interesting structures. Limestone and dolostone can both be dissolved by weak acids, forming cave and cavern systems (see Chapter 9). When exposed to the impact of waves,

Mike Brake/Shutterstock.com

FIGURE 12.23. This golf course is built on the generally level surface of a marine terrace.

CHAPTER 12: Oceans and Coastal Systems

FIGURE 12.24. The sea arch (right) and sea stack (left) shown here are made of eroded limestone.

the limestone may form a **sea arch** (figure 12.24), a remnant of a cave composed of a headland with rock bridge connection out to a column of limestone sitting in the ocean. When the arch collapses, the isolated column becomes a **sea stack**.

In areas where deposition dominates, such as in protected bays, beaches are typically deep and sandy or muddy (depending on the sediment available). Out in more wide-open regions, longshore transport moves sediment to areas where it is ultimately deposited at the end of the shore. A **spit** is a linear deposit of sand extending from the beach out into a bay or open water (figure 12.24). Since the spit grows from the beach outward, you can determine the direction of longshore transport by looking at the direction that the spit extends from the beach. If a spit extends far enough across a bay so that it seals the bay off from ocean waves, it is then called a **baymouth bar** (figure 12.25). Both spits and baymouth bars extend in the same direction as the beach. In contrast, a **tombolo** is a ridge of sand that develops perpendicular to the beach because an object (such as a large rock or a man-made breakwater) blocks incoming waves, or because converging longshore currents combine their sediment loads (figure 12.29). The reduced energy levels behind the object allow sediment to build up, and may eventually connect to the object. A **tied island** (figure 12.26 is a large rock or small island that is connected to the shore by a tombolo.

Barrier islands are also important depositional features. These narrow (1-5km) but long (tens of km), sand-dominated islands parallel the mainland coast and are separated from it by a lagoon. The highest elevation on the island is found at the dunes, which may reach up to 10m. Over 2,100 barrier islands are estimated to exist, protecting roughly 10% of continental shorelines. In the U.S., 25% of the

(a)

(b)

FIGURE 12.25. Two depositional shoreline features. a) the Provincetown Spit at Cape Cod, Massachusetts; b) Looking north from Redwoods National Park, California, as the Klamath River enters into the Pacific Ocean. A large baymouth bar has formed at the mouth of the river. There is a small opening between the river and the ocean at the far end of the bar.

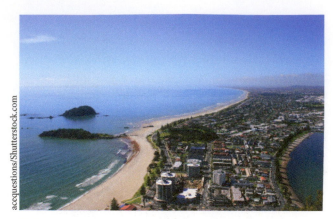

acequestions/Shutterstock.com

FIGURE 12.26. A tombolo extends from the beach to connect Moturiki Isand to Tauranga Beach in New Zealand.

coast is protected by barrier islands and large spits, with most located along the Atlantic and Gulf coasts. Barrier islands are typically found along low-lying coasts with small daily tidal ranges (less than two meters). The lagoons are typically more muddy than sandy, due to the low-energy setting behind the protection of the barrier islands, and may host extensive tidal flats of exposed sediment during low tides. The quiet waters of the lagoons permit easy travel by boat compared to the choppy waters of the ocean. You can use Google Earth™ to find barrier islands along the coasts of North Carolina and Texas.

12.4.3 Subemergent and Emergent Coastlines

The coastal regions of the world are wildly diverse, involving different types of rocks and sediments, orientations, features, wave intensities, storm frequencies, and tidal cycles. One helpful way of organizing and characterizing so many variations is to look at the changes of elevation between the land and the sea. Areas where land is rising up out of the ocean, or where deposition of new materials builds more land areas, are called **emergent coastlines**. Those areas where the sea level is rising and/or erosion removes material from the land are called **submergent coastlines**. Each type of coastline is characterized by certain types of features.

Emergent coasts form under two very different settings. In the first setting, a significant elevation rise has lifted the land up and out of the ocean. This may occur because tectonic activity has lifted the land, or the land may rise up out of the ocean when a large weight has been removed from the land, such as the melting of a glacial ice sheet. Once the weight of the glacier is removed by melting, the continent readjusts to the lower pressure by rising upward, like removing your finger from an ice cube in a glass of water. This is the situation seen in the coasts of Maine and Scandanavia. In both of these cases, the *relative* (local comparison) sea level has dropped, regardless if *eustatic* (global) sea level has risen, fallen, or remained the same.

The second setting for emergent coasts is seen in locations with high rates of sediment influx from nearby rivers. Here the advance of the land is more horizontal than vertical. The continued influx of sediment is seen in growing deltas and shorelines, such as that seen in the advancing delta of the Ganges and Brahmaputra rivers along the India-Bangladesh border. Fed by runoff from the Himalaya Mountains, the Ganges and Brahmaputra rivers form an immense and fertile plain along the India-Bangladesh border (figure 12.27). The eastern portion of the delta has grown at a rate of 7 km² per year over the past 225 years

In contrast to the emergent coasts, submergent coasts occur where ocean waters are invading the land. In contrast to the eastern Ganges-Brahmaputra delta, the western side of the delta is submergent. It no longer receives sediment from the rivers, and the area's loose sediments are gradually settling and compacting, resulting in a land loss of 1.9 km² per year. In these areas, salty ocean water may migrate upstream into rivers, making them brackish and causing changes to local plant life (such as a shift to salt-tolerant species).

When the ice sheets melted at the end of the Ice Age, eustatic sea level increased dramatically. Rivers that once flowed far out past the present-day coastlines were filled by the advancing water, forming **estuaries**. The Chesapeake Bay (figure 12.28) is one of the best examples of a former river valley now

India Bangladesh India

Brahmaputra

Ganges

Ganges-Brahmaputra Delta

Burma

Courtesy NASA

(a)

Courtesy NASA

FIGURE 12.28. The Chesapeake Bay is a classic example of an estuary, or drowned river valley.

Daniel J. Rao/Shutterstock

(b)

FIGURE 12.27. (a) The Ganges-Brahmaputra river delta along the India-Bengladesh border. (b) The broad, flat floodplain is home to much agriculture, particularly fishing, tea, and rice (shown here).

Simon Dannhauer/Shutterstock.com

FIGURE 12.29. Sognefjord, a fjord in Norway.

"drowned" by ocean waters, and displays a dendritic coastline derived from the former river drainage pattern. Estuaries are often biologically diverse, as they host a wide variety of habitats including freshwater river systems, wetlands and marshes, tidal flats, and brackish water communities. In Scandanavia (among other places), deep, U-shaped valleys formed by flowing glaciers were likewise filled by rising seas, forming a **fjord**: an elongated, narrow, salt-water filled valley flanked by vertical rock cliffs (figure 12.29).

Today, 90% of the people on Earth live within 10km (6 miles) of a major body of water. Starting in the early 20th century, many people began visiting coastal areas for recreation. Over the past 100 years, countries all around the world have seen tremendous population increases in coastal areas, along with ever-expanding cities and the construction of homes, hotels, and businesses. As with any human activity, our knowledge and understanding of the coastal region has grown over time as we have recorded events and experiences, but we tend to learn about the risks and hazards of an area only after we've been there for some time. Such is the case with the coast.

12.5.1 Living with a River of Sand

We tend to think of the shore as a fairly stable area. After all, you may go to the same beach year after year. But the sand that makes up that beach isn't the same as when you last saw it. Coastal geologists often refer to the beach as "a river of sand," because they recognize that winds, waves, and storms make the beach an ever-shifting, dynamic area. Recall the earlier discussion of longshore transport, which is the combination of beach drift and the longshore current. Incoming waves erode, transport, and deposit sediment along the beach, forming structures like spits and baymouth bars. Depending on where a home, hotel, or other man-made structure is located, this could be a problem. In response, people have developed a number of strategies to combat, control, or adjust to the river of sand.

A **groin** (figure 12.30a) is a hard stabilization structure built perpendicular to the shoreline and downcurrent of the direction of longshore transport. Groins are typically built using concrete or large rocks, and their purpose is to capture sand that is eroding/migrating from the beach. The sand trapped by the groin typically fills in as a wedge-shaped accumulation. But because the sand is trapped, the beach downcurrent of the groin is eroded because its incoming waves

no longer carry much sediment. This may cause the property owner downcurrent to build a second groin to retain sand for their beach, and the result is a series of groins down the beach. In the end, groins to not stop erosion or longshore transport, they only modify their actions. Due to their many drawbacks, many coastal states have stopped allowing the construction of groins.

Jetties (figure 12.30b) are paired shore-perpendicular structures build from the same kinds of materials as groins. Like groins, their purpose is to trap sediment. But unlike groins, jetties are built to keep sand *out* of an area, typically a bay or harbor. As longshore transport builds spits and baymouth bars, the people with boats still want a way to get in and out. The jetties are built to trap the moving sand, preventing it from sealing off the harbor. Like groins, the beach downcurrent of the jetties is sediment-starved, and erodes. The channel in the middle of the jetty may also fill with sediment during storms or via tidal currents, and must occasionally be dredged to remove the material.

In some cases where erosion rates are high, a **sea wall** (figure 12.30c) is constructed of large rocks or concrete, located directly on the land where the danger of erosion and collapse is high. The sea wall protects the land from the high-energy waves, but those same waves tend to rapidly erode the sediment at the base of the sea wall. In the worst scenarios, the sea wall is undermined and collapses into the ocean. A **breakwater** (figure 12.30d) is a hard stabilization structure built parallel to the shoreline, often to create an artificial harbor. The breakwater absorbs the force of incoming waves, making the waters behind it calm and safe for boat moorings. On the beach, the lower wave velocities result in an increase in deposition rates, causing a lobe of sand to grow out towards the sea wall (similar to a tombolo/tied island).

Some communities experiencing beach erosion opt for an alternative to hard stabilization structures: **beach nourishment**. Beach nourishment is the process of transporting sand to the beach, either by

(a)

(b)

(c)

(d)

FIGURE 12.30. A variety of hard stabilization structures used to control the movement of sand along the shore: (a) groin, (b) jetties; (c) sea wall, and (d) breakwater. Note the north-to-south longshore transport direction in (b), which has piled sand behind the northern jetty.

dredging and pumping nearshore sand deposits on to the beach or hauling by truck from nearby terrestrial sources. The sand is spread across the beach by heavy construction equipment. The results can be quite dramatic (figure 12.31), though the costs are often very high and the effect is quite temporary, since the forces of erosion do not stop because new sand has been added to the beach. Over time, the community will require additional and ongoing nourishment projects.

12.5.2 Rising Seas and Sinking Land

While beach erosion can be troublesome to coastal communities, they tend to be local concerns. In contrast, rising sea levels and submergent coastlines affect more areas and pose much more difficult (and expensive) challenges. On a local and regional scale, like the Louisiana coast and the western part of the Brahmaputra-Ganges delta of Bangladesh discussed above (figure 12.27), the soft, wet sediments deposited by rivers settle and compact over time. Without the input of new sediments to maintain or increase the land's elevation, these coastal regions gradually sink (geologists call this process **subsidence**). As they do, ocean waters and saline groundwater invade the land, reducing its areal extent, drowning wetlands and agricultural areas, and disrupting fresh groundwater sources. Louisiana has seen nearly 5,000 km² (1,900 mi²) of land lost to subsidence since 1932 (figure 12.32). In cases such as these, human activity is not a major factor;

FIGURE 12.31. Before and after photos of a 2012 beach nourishment project at Wallops Beach, VA.

natural changes in river and sediment patterns are the dominant causes for regional subsidence.

On a global scale, sea levels have been rising for over a century (figure 12.33). Data from tidal gauges indicate that the rate of sea level rise is likely faster now (3.2 mm/yr) than the average rate of the 20th century (1.7 mm/yr). This average rate is not seen in all regions of the ocean. Because warmer waters are less dense than colder waters, sea level rise is more pronounced in tropical regions, and less significant near the poles (figure 12.34). The rate of sea level rise is not large in an absolute sense, but it has been continuous, increasing, and cumulative over the

FIGURE 12.32. Land lost to subsidence in Louisiana from 1932 to 2011.

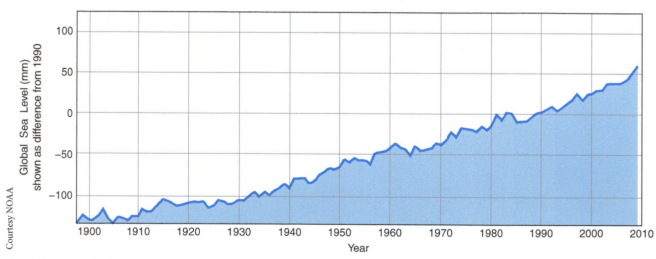

FIGURE 12.33. Sea level has risen about 18 cm (8 inches) from 1900–2009.

CHAPTER 12: Oceans and Coastal Systems

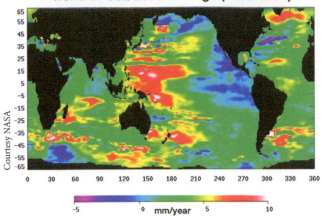

Trend of Sea Level Change (1993–2008)

Courtesy NASA

FIGURE 12.34. Changes in sea surface elevation from 1993–2008.

past century, and it is the cumulative nature of the change makes the impact of rising seas a topic of concern.

The computer models used by climate scientists at the National Oceanic and Atmospheric Administration (figure 12.35) forecast that sea level will rise between 0.2-2.0m (8 inches to 6.5 feet) by 2100, with a median range of 0.5-1.2m (1.6-4 feet). Low-lying coastal communities are more susceptible to changes in sea level than are those with more steep slopes, because a small vertical increase will migrate further inland in an area with a shallow slope to the land compared to

areas with steeper slopes. So as sea level rises, low-lying coastal areas may see large swaths of land overtaken by the ocean. This not only reduces land area, but also means that major storms like hurricanes and typhoons could impact regions that are currently safe from the destructive power of their storm surge (see chapter 15).

Why is global sea level rising? In contrast to the subsidence issues discussed above, most scientists believe that humans have played a major role in global sea level rise, and they note that the recent increase in the rate of sea level rise in particular appears to be a signal of human activity. As carbon dioxide formed by the combustion of fossil fuels enters the atmosphere and traps heat (see chapter 13), a large amount of that heat is absorbed by the uppermost ocean water. As the temperature of the water rises, its density decreases slightly, causing thermal expansion and resulting in higher sea levels. In addition to thermal expansion, the warming of the atmosphere has contributed to the melting of mountain glaciers and continental ice sheets (the glaciers of Greenland and Antarctica). Taken together, thermal expansion and increased glacial runoff accounts for 75% of the measured rise in sea level. This current rise is small compared to the much larger rise that accompanied the end of the Ice Age (about 125 meters!),

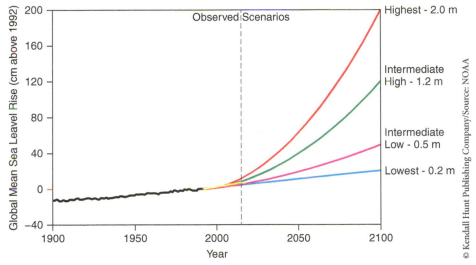

FIGURE 12.35. Four projections of sea level based on numerous computer models. The most likely scenarios point to a rise of 0.5-1.2 by the year 2100.

but stands to affect many more people who live near the coast in modern cities.

Can anything be done? That is a much more difficult question to answer. Many scientists believe that sea level rise can only be slowed or stopped by significantly limiting the amount of carbon dioxide released into the atmosphere. This would require a dramatically different makeup of energy sources to provide the energy that powers our society, one that relies much less heavily on fossil fuels than what is currently used (see chapter 11). Even if this were possible (which is unlikely within the next few decades), sea level will continue to rise as the oceans absorb more of the atmosphere's heat and receive glacial meltwater far into the future, perhaps for centuries. It appears that people will more likely have to adjust to the new risks of slightly higher ocean elevations. This will be somewhat easier for wealthier, developed countries, though there may be significant costs for coastal communities with their expensive real estate, emphasis on tourism, etc. For example, the majority of Miami, Florida, is less than five meters (16 feet) above sea level (figure 12.36). More difficult situations exist for poorer developing nations, especially those with very low-lying coastal regions and high populations, who may have to transplant thousands of people to lower-risk regions.

(a)

(b)

FIGURE 12.36. Low-lying coastal regions face a variety of difficulties with rising sea levels. Wealthy nations and cities (a) frequently have high populations and very expensive real estate, while poorer countries (b) have few resources. Risks from storms and hurricanes become more severe as sea level rises.

12.1 Circulation in the World Ocean

- Currents are masses of water that flow from one region of the ocean.
- Surface currents are formed by friction with wind. Some surface currents are linked together in circular systems called gyres.
- Circulation in the deeper parts of the ocean is controlled by water density, which is affected by temperature and salinity.
- The tidal influence of the moon can also produce currents as the tides rise and fall towards or away from shorelines.

12.2 Tides

- The gravitational pull of the moon and sun on the earth produce tides. The moon's tidal forces are stronger than the sun's. The orientation of the Earth, moon, and sun produce the Earth's tidal cycle, including spring (all aligned) and neap (perpendicular positions) tided.
- Tidal patters are a reflection of how tides are expressed at particular locations, and include diurnal, semidiurnal, and mixed tidal patterns.

12.3 Waves: Formation and Travel

- Waves are produced by friction between the wind and ocean, and are composed of oscillating (circular-moving) water particles. Waves have a crest, trough, height, and length.
- Wind speed, duration, and fetch are the factors that affect the formation of waves.
- Waves reaching the shoreline are refracted so that the wave becomes more perpendicular to the shoreline, regardless of its shape.
- Longshore transport involves both beach drift and the longshore current, which move sand along the beach and within the surf zone.

12.4 Coastal Areas and Landforms

- Coastal areas represent an important and dynamic interface between the hydrosphere, geosphere, and atmosphere.
- The shore extends from the low tide shoreline to the coast, and is shaped by tidal and storm-related processes. The shoreline is the point of contact between the ocean and the land, and migrates between the high tide and low tide shorelines.
- Coastal areas dominated by erosional processes frequently have small beaches in areas protected from direct waves. Deposition-dominated areas may display a number of sand-formed structures.
- Coastlines can be classified as emergent (rising up or extending out toward the water) or submergent (sinking downward or being flooded by the water). Both local and eustatic (global) sea level changes may be involved. Fjords and estuaries are classic features of submergent coastlines.

12.5 People, the Coast, and Rising Waters

- The beach can be viewed as a "river of sand," since longshore transport processes are constantly moving the sand along the beach. Our interactions with beaches must take this fact into consideration.

- Both hard stabilization structures (such as groins, jetties, and others) and non-structural approaches (such as beach nourishment) may be employed to manage the river of sand.

- Local and regional subsidence (sinking of loose soil and sediment), such as that seen in Louisiana and Bengladesh can result in significant losses of land to the sea.

- On a global scale, ocean elevations have been rising over the past century. Many scientists believe that the extensive use of fossil fuels, resulting in global warming, is the main reason for this sea level rise. Computer modeling forecasts continued sea level rise into the foreseeable future.

- Poorer countries with extensive low-lying coasts are at the greatest risk of sea level rise, though developed nations may also face enormous costs in protecting cities, towns, tourism, and infrastructure.

KEY TERMS

Backshore	Gyre
Barrier island	High tide
Baymouth bar	Jetty
Beach nourishment	Longshore transport
Berm	Low tide
Breakwater	Marine terrace
Coast	Mixed tidal pattern
Coastal wetland	Neap tide
Coriolis Effect	Nearshore
Current	Offshore
Diurnal tidal pattern	Rip current
Doldrums	Sea arch
Downwelling	Sea stack
Dune	Sea wall
Emergent coastline	Semi-diurnal tidal pattern
Estuary	Shoreline
Fetch	Spit
Fjord	Spring tide
Foreshore	Submergent coastline
Groin	Subsidence

Surf zone	Upwelling
Surface current	Wave
Thermohaline circulation	Wave base
Tidal bores	Wave crest
Tidal bulge	Wave-cut cliff
Tidal current	Wave-cut platform
Tidal cycle	Wave height
Tidal range	Wave refraction
Tides	Wave trough
Tied island	Wavelength
Tombolo	

REVIEW QUESTIONS

1. Why do the maps of surface currents match closely to the maps of wind direction?
2. How does the Coriolis Effect alter our perception of motion on Earth?
3. What are the key factors that drive thermohaline circulation? And how does the velocity of deep-ocean currents (part of thermohaline circulation) compare with surface currents?
4. Describe in your own words why Earth has two tidal bulges.
5. How does the orbit of the moon affect monthly tidal cycles, including the timing of spring and neap tides?
6. What are the three types of tidal patterns?
7. What factors affect the formation of waves in the open ocean?
8. Identify the labeled components of an open-ocean wave in the figure provided below.
9. Be able to sketch (without looking) the profile of the shore region in figure 12.21.
10. Why does wave refraction concentrate wave energy on headland regions?
11. Why are all the depositional landforms described in section 12.4.2 made of sand?
12. How does subsidence differ from eustatic sea level rise?

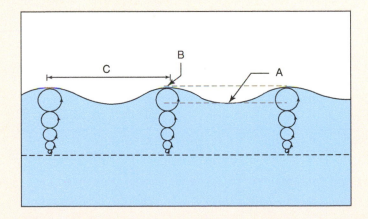

FURTHER READING

Centers for Ocean Sciences Education Excellence: http://www.cosee.net/

Climate.gov: www.climate.gov

Moran, Joseph M. *Ocean Studies: Introduction to Oceanography*, 3rd edition. American Meteorological Society.

National Oceanic and Atmospheric Administration: www.noaa.gov

Woods Hole Oceanographic Institute: http://www.whoi.edu/

REFLECT ON SCRIPTURE

After a period of teaching, Jesus left by boat with his disciples. While Jesus slept, a furious storm battered the boat, and the disciples thought they were going to die. Jesus rose, and commanded the storm to end, and the winds and waves ceased! The disciples couldn't believe it; never had they heard of anyone commanding the winds and waves. This account is found in three of the four gospels (Matthew 8, Mark 4, and Luke 8). Knowing now that winds produce waves, but that waves can continue to travel long distances after they've been generated by the wind, what does this tell you about Jesus' miracle? Does it seem more or less likely that a natural event caused/coincided with the end of the storm?

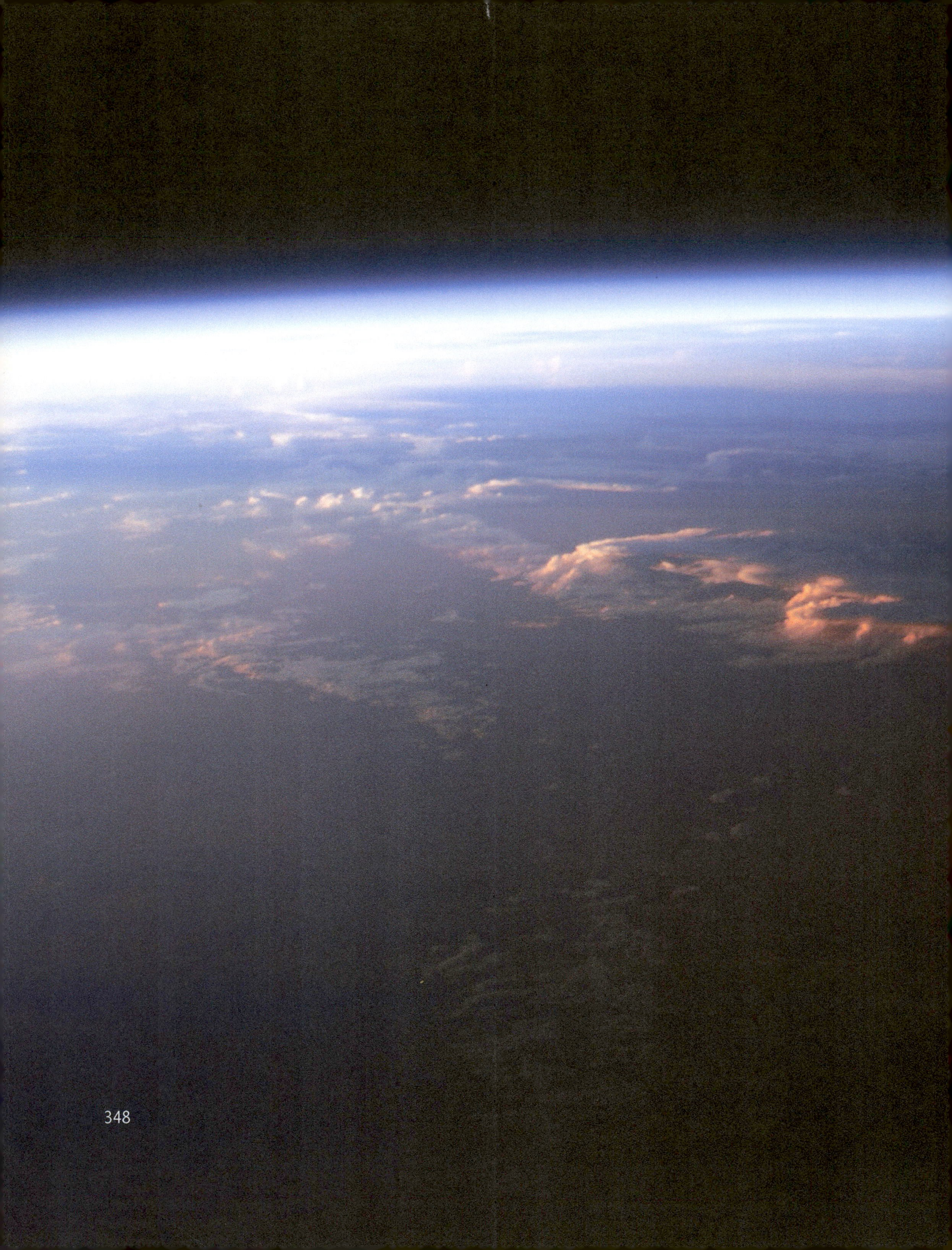

CHAPTER 13

EARTH'S ATMOSPHERE

OUTLINE:

Courtesy NASA

The thin envelop of Earth's atmosphere is 1% the radius of Earth, but is essential to sustaining life. The color of the sky is due to small particles preferentially scattering blue light. When sunlight travels large distances through the atmosphere, most of the blue is removed leaving the characteristic reds and oranges of a sunset.

Have you considered how the weather you experience compares to that in Florida, Minnesota, or Arizona? Given the broad reach of the Internet this question can be expanded to such locations as Peru, South Africa, or the Philippines. Are there similarities, and what are the differences? To answer these questions a basic understanding of Earth's atmosphere is needed. This begins with a definition of weather and climate.

13.1 WEATHER AND CLIMATE

Weather is a description of the atmosphere at a particular place and time. This description goes beyond an observation of particular events such as a blizzard, thunderstorm, or heat wave to include information useful for tracking and predicting the current and future states of the atmosphere. Weather stations satisfying criteria established by the National Oceanic and Atmospheric Administration (NOAA) regularly measure temperature, humidity, pressure, precipitation, wind speed, and direction. Although automated weather stations report this information on a continual basis, airports release hourly reports of these measurements along with cloud cover and visibility. Analysis of this data is summarized using weather maps and forecasts (figure 13.1). Weather information providers, such as Yahoo! Weather, The Weather Channel, WeatherBug, and AccuWeather, use various methods to display this information in a visually appealing manner.

Climate is the average state of the atmosphere over a region or interval of time. As a result, climate data shows less variability than weather data. By international agreement climate norms are averaged over a 30 year interval. Climate norms such as daily high and low temperatures for a particular city are used to illustrate how unusual or ordinary a day's weather is. The climate for a region is likewise averaged over time and is linked to geographic

Courtesy of NOAA

FIGURE 13.1. This daily forecast chart summarizes data gathered across the country. Pressure systems and fronts are used to highlight patterns that lead to precipitation and other significant weather events.

features such as latitude, proximity to large bodies of water, and prevalent winds. Droughts and heat waves are departures from climate norms and have a significant impact on farmers. Therefore, climate forecasts, just like weather forecasts, are valuable when determining the appropriate crop and seed hybrid to plant for the growing season. Climate statistics are not restricted to the weather parameters routinely gathered across the country, but also include frequency of hurricane landfalls, tornado occurrence by state, and number of consecutive days above 100°F. Although your favorite weather provider makes selected climate data available to you, the official repository for climate data in the United States is the National Climatic Data Center (NCDC).

13.2 THE ATMOSPHERE

Our understanding of the atmosphere has advanced significantly over the past century resulting in increasingly accurate weather forecasts. However, these advancements would not be possible without the gradual accumulation of knowledge over millennia. In 340 BC Aristotle wrote a treatise called *Meteorology* which described the phenomena of the heavens. Derived from the Greek word '*ta meteora*' (celestial phenomena or things in heaven above) the modern definition of this word is the science or study of the atmosphere. Beginning with Galileo in the 1600's meteorology has become more quantitative and has grown into the mature science that benefits us today.

13.2.1 Atmosphere's Composition

The atmosphere is composed primarily of two gases: 78% nitrogen and 21% oxygen (figure 13.2). The remaining 1% consists of argon, carbon dioxide, and a variety of trace gasses. These percentages assume that the air is dry and free from pollutants. Nitrogen gas, N_2, is relatively unreactive and provides a safe medium in which other chemical reactions can take place. Reactions involving nitrogen occur either at high temperatures or within living organisms. Lightning and combustion in automobile engines are hot enough to generate nitrogen oxides. Ammonia is also generated by nitrogen-fixing bacteria living in the root nodules of some plants. Nitrogen is converted back into its gaseous form through the decomposition of organic material.

Oxygen, O_2, on the other hand is essential for many reactions. Rusting of iron removes oxygen from the atmosphere forming iron oxide. The combustion of wood, oil and gas is a chemical reaction between the carbon compounds of the fuel and oxygen in the atmosphere. The primary products of these reactions are water vapor, carbon dioxide, and large amounts of heat. Oxygen is also removed from the atmosphere through respiration by animals and humans. Respiration is really just combustion occurring within living cells, which use enzymes to release energy in a controlled fashion. As a result, animals breathe in oxygen and exhale water vapor, carbon dioxide, and lower concentrations of oxygen. Oxygen gas is reintroduced to the atmosphere by photosynthesis. Plants use chlorophyll to convert carbon dioxide, water, and the energy of the sun to form sugars and oxygen gas.

The most significant gas in the atmosphere with respect to meteorology is water vapor. It can vary from nearly 0% over the cold surface of Antarctica to 4% in the humid jungles of the Amazon. Its importance lies in the ability to change states between a gas, liquid, and solid over a temperature range that is compatible with life. Changing states requires a significant exchange of energy and, therefore, the presence of water has a stabilizing effect on the temperature of Earth's surface.

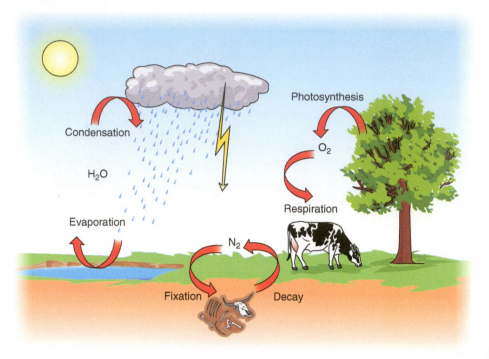

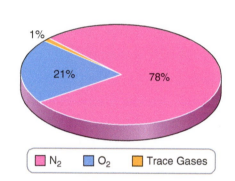

© Kendall Hunt Publishing Company

FIGURE 13.2. The predominant components of the atmosphere are nitrogen and oxygen. The remaining gases comprise less than 1% of the atmosphere. Water vapor is not included with these percentages because it varies between 0 and 4%. Several processes responsible for removing and liberating these gases in the atmosphere are also illustrated.

13.2.2 Trace Gases and Pollutants

Other gases have a much lower concentration in the atmosphere and yet have a significant impact. **Ozone**, O_3, consists of three oxygen atoms rather than two for oxygen gas. This molecule is more reactive and can damage plants and cause respiratory distress in people suffering from asthma. In larger cities *Ozone Action Days* are declared when conditions are right for ozone formation. Although ozone has negative effects near Earth's surface, higher in the atmosphere, more specifically the stratosphere,

it serves to remove harmful ultra-violet (UV) radiation. This effect plays a crucial role in protecting life on Earth's surface.

Nitrogen and sulfur oxides generated by automobile engines and coal-fired plants combine with water in the atmosphere to form **acid rain**. Acid rain leaches nutrients from soil, stresses plants, and changes the pH of streams and lakes. The environmental damage from acid rain can contribute to forest dieback as seen in figure 13.3. Incomplete combustion also generates carbon monoxide and hydrocarbons. By reducing sulfur content in

FIGURE 13.3. The effect of acid rain and other pollutants stresses trees and make them more susceptible to insect attacks. This forest is located in Ploeckenstein, Germany.

FIGURE 13.4. Soot particles, ozone, nitrogen and sulfur oxides contribute to smog. In high enough concentration these pollutants can lead to respiratory distress.

gasoline and adding catalytic converters, pollution from automobiles has been reduced. Adding smoke stack scrubbers to factories and electric power plants that use coal has also reduced the level of pollutants. These pollution standards, established by the Clean Air Act, allow the United States to enjoy better air quality than in less regulated countries such as China (figure 13.4).

The atmosphere also consists of particulate matter of varying sizes. An **aerosol** is any solid or liquid that remains suspended in the air. Large particles such as pollen, dust, and ash settle out of the air over minutes and hours. However, wind currents stir up these particles and keep them in suspension for days as seen in this satellite image of a sandstorm (figure 13.5). Smaller particles remain in the air for much longer periods of time and affect the formation of clouds. Sulfur dioxide introduced into the stratosphere by volcanic eruptions forms aerosols. These aerosols can remain in the stratosphere for years and have a cooling effect on climate.

13.2.3 Origin of the Atmosphere

Earth's atmosphere is unique among planets and moons in our solar system. Only Titan, a moon of Saturn, has an atmosphere that is composed primarily of nitrogen gas. Naturalistic models for Earth's

origin attribute the atmosphere to volcanic outgassing of a cooling planet. If that were the case, Earth would be like its nearest neighbors, Venus and Mars, and have an atmosphere of over 95% carbon dioxide, CO_2. Large oceans of water and the presence of life distinguish Earth from other objects in the solar system and allow the atmosphere to maintain a low

FIGURE 13.5. Sand storm over the Mediterranean. Sand suspended by strong winds is carried off the coast of Egypt into the Mediterranean sea. The island Cyprus can be seen in the upper right corner of the sea.

concentration of CO_2. The Genesis account of creation states that Earth's atmosphere is the purposeful creative act of God.

> *And God said, "Let there be an expanse in the midst of the waters, and let it separate the waters from the waters." And God made the expanse and separated the waters that were under the expanse from the waters that were above the expanse. And it was so. And God called the expanse Heaven. And there was evening and there was morning, the second day. (Genesis 1:6-8, ESV)*

The Hebrew word for expanse, *raqiya*, can mean firmament or vault of heaven. In the past some believed the vault of heaven was a solid top of the atmosphere. In 1874, Isaac Vail published a pamphlet, *The Waters Above the Firmament*, stating that the source of water for the forty days and nights of the Genesis Flood was the collapse of this canopy. Various forms of a canopy have been proposed from ice rings similar to those of Saturn to a vapor canopy. In 1998 Larry Vardiman reported that a vapor canopy is limited to about one meter of precipitable water. Beyond this amount the surface temperatures of Earth would be unbearable due to the heat trapping effect of this canopy. In recent years creationists have pursued other sources of the flood waters, from breaking up of the fountains of the deep to Catastrophic Plate Tectonics (see Chapter 4). Regardless of one's assumptions about the origin of Earth's atmosphere, it is universally agreed upon that our atmosphere is uniquely suited for life.

13.2.4 Density and Pressure

Earth's atmosphere is held in place by gravity and its pressure and density are dependent on altitude. **Pressure** is defined as a force per area and is equal to the weight of the atmosphere above that point. At sea level that comes to 14.7 pounds per square inch. If you were to measure the total weight of air resting on your shoulders, it would be over 1000 lbs. You don't notice this because pressure acts in all directions and cancels any need for you to support that weight.

Pressure is measured with a **barometer**. Torricelli made the first barometer from a glass tube filled with mercury and sealed at one end (figure 13.6). A **vacuum**, a location of zero air pressure, is generated at the sealed end of the tube. Air pressure at the open end pushes mercury into the column and causes it to rise towards the sealed end. While climbing a mountain, the height of the mercury column decreases indicating that air pressure is also decreasing. Through experimentation it is determined that the density of air also decreases with altitude. **Density** is the amount of mass in a volume. When the air pressure goes down, the air expands and occupies more volume, thus decreasing the density. If air density were constant, the

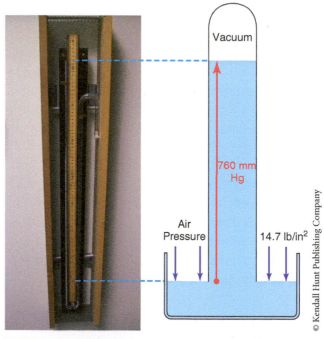

FIGURE 13.6. Air pressure is measured by using it to elevate a column of mercury in a tube. The top of the tube is sealed preventing any air from exerting a force at the top of the column. As a result, the weight of the column of suspended mercury is equal to the weight of the air pushing mercury into the column.

atmosphere would be 8 km deep and the top of Mt. Everest would be in a vacuum with an ocean of air below it.

Meteorologist measure pressure in millibars (mb) and sea level pressure is close to 1000 mb. These units are convenient because the percentage of atmosphere above you is equal to the pressure divided by 10. In Mexico City the pressure is 800 mb. That means 80% of the atmosphere is above you. This reduced pressure causes water to boil at 92 °C. As a result, food must be boiled and baked for longer periods of time. You also notice the change in air pressure with height when flying in an airplane or driving through the mountains. Your ears become uncomfortable as the pressure difference between your outer and middle ear gets larger.

13.2.5 Layers of the Atmosphere

The composition of the atmosphere is relatively uniform up to 90 km. Beyond that the lighter elements such as hydrogen and helium are present in larger percentages. This region is called the **thermosphere** and molecules interact so rarely that they do not come into thermal equilibrium. The air pressure at this level is 0.02 mb and for all practical purposes is the top of the atmosphere. The Aurora Borealis occurs in the thermosphere and is the result of charged particles from the solar wind interacting with Earth's magnetic field (figure 13.7).

Below this layer is the **mesosphere**, which extends down to an altitude of 50 km and an atmospheric pressure of 1mb. The air is dense enough in this region to cause small particles from space to heat up and glow resulting in meteors or shooting stars. Below the mesosphere is the **stratosphere**. This layer contains the ozone layer, which absorbs UV light from the sun. As a result, the temperature in the stratosphere increases with altitude, unlike the other layers of the atmosphere (figure 13.8). The lowest portion of the atmosphere is the **troposphere** and is the region where most of our significant weather occurs. The height of the troposphere at the poles is

FIGURE 13.7. Aurora Borealis. Earth's magnetic field protects it from high energy particles ejected by the sun. Near the poles where the field is strongest these light shows are common. They are seen over the continental United States when there is a high concentration of particles due to a solar flare.

©simonekesh/Shutterstock, Inc.

only 8 km because cold air is denser and, therefore, occupies less volume and height. At the equator it extends to 18 km.

Notice that these divisions of the atmosphere are based on how air temperature changes with altitude. In general rising air expands because there is less pressure. Expanding gases cool (i.e. the discharge from a fire extinguisher) and, therefore, the atmosphere is cooler at the top of a mountain than at its base. The average rate at which the troposphere cools is 6.4 °C/km. This cooling trend reverses in the stratosphere because ozone absorbs UV light and heats the air. The region where the temperature trend reverses is called the **tropopause**. This air is stable and isolates most of the weather in the troposphere from the stratosphere. Exceptions to this rule are volcanic eruptions and strong vertical motion that occurs in severe thunderstorms.

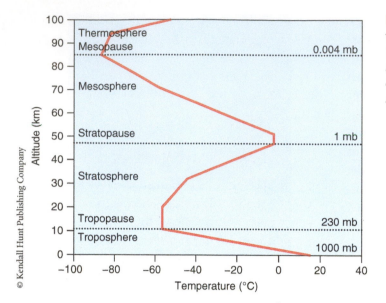

FIGURE 13.8. Layers of the atmosphere are identified by the rate at which temperature changes with altitude. Over 75% of the atmosphere is contained in the troposphere and the temperature decreases with altitude. Temperature of the stratosphere increases with height due to the absorption of UV light by ozone.

13.3 TEMPERATURE AND HEATING

Up to this point the word *temperature* has been used, but not defined. We know what hot and cold feels like, but it is relative to what we are used to. A 10 °C (50 °F) day in the summer feels cold, while that same temperature in the middle of winter feels like a heat wave. In order to avoid our subjective experience of temperature, we must use a reproducible scale based on physical principles. At its most fundamental level, **temperature** is a measure of an object's average thermal energy, both potential and kinetic energy at the microscopic scale. As the temperature of an object increases, its molecules vibrate with more speed and in general occupy more space. This is the mechanism used by thermometers that record a temperature with the height of a liquid (figure 13.9). Electrical properties also change with temperature and make digital thermometers possible.

Temperature extremes on Earth range from −89 °C (−129 °F) at Vostok Station, Antarctica to 57 °C (134 °F) in Death Valley, California. In general, warmest temperatures occur near the equator and coldest towards the poles. However, this pattern is modified by the presence of oceans and mountain ranges.

13.3.1 Types of Heating

As energy is added to an object, two things can happen. Either its temperature increases or its phase changes. In the first case, a temperature change due to added energy is called **sensible heating**. The amount of temperature change depends on the mass of the object and its composition. Water and sand

FIGURE 13.9. Bulb thermometer. As the temperature increases, the alcohol expands driving the fluid from the bulb up the column. The alcohol is dyed to make it easier to read.

provide a good example. A cup of water heats up on a stove much quicker than a gallon because it has less mass. Sand, which has a lower heat capacity, also heats up quicker than an equal amount of water. You notice this on a hot summer day at the beach when the sand is burning hot and yet the water remains cool. Both receive the same amount of energy; however, the temperature of the sand rises quicker than water. This also applies at night as the sand cools quicker than water due to its low heat capacity.

In the second case, temperature remains constant when heat is added and the material goes through a phase change. This is called **latent heating** or 'hidden' heating. In the atmosphere matter comes in three phases: solid, liquid, and gas. The best way to distinguish these phases is to consider how they behave with respect to volume and shape. Solids occur at low temperatures. Forces between the molecules of the solid keep them in a fixed position with respect to each other. Therefore, solids maintain their volume and shape. Liquids occur at higher temperatures when there is enough kinetic energy for molecules to move relative to each other. These molecules don't have enough energy to act independent of the other molecules; therefore, liquids maintain their volume, but not their shape. At even higher temperatures the molecules have enough kinetic energy to act independent of the other molecules, thus moving anywhere. As a result, gases change their volume and shape to fill the container (figure 13.10). There is much more to say about phases of matter, but we'll save that until Chapter 14 where we discuss evaporation and precipitation.

13.3.2 Heat: Transport of Energy

Heat is a peculiar word because it refers to a process rather than a substance. **Heat** is the transport of energy from a warm object to a cool object. This occurs by three different mechanisms: conduction, convection, and radiation (figure 13.11).

Conduction is the transfer of heat by the collision of molecules with its neighbors. If you place an ice cube in your hand, it feels cold. The molecules of your hand are at a higher temperature and are vibrating with more thermal energy. As the molecules of your hand bump into the less energetic molecules of the ice cube, the molecules of your hand slow down and decrease in temperature. The ice cube molecules speed up and either raise the ice cube's temperature or cause it to melt.

Convection is the vertical transfer of heat by the motion of fluids. Convection depends on the thermal property that an increase of temperature results in a decrease in density. Earth's gravity

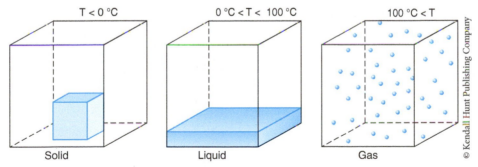

FIGURE 13.10. At temperatures below 0 °C water is a solid, which maintains both its shape and volume. Above the freezing temperature, water becomes a liquid. Its molecules are vibrating fast enough that shape can no longer be maintained and it spreads over the bottom of the container. At higher temperatures, water molecules move independent of each other and become a gas, which spreads out to fill the volume of the container.

FIGURE 13.11. Heat is transferred from hot objects to cold using three mechanisms. Conduction - Metal handles become hot through collisions of molecules with their neighbors. Convection – Warm fluids rise while the cool sinks. Radiation – Electromagnetic waves are emitted by all objects with temperature. When hot enough, the waves appear as visible light as in the fire.

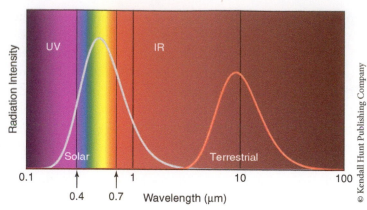

FIGURE 13.12. The sun emits light spanning the electromagnetic spectrum from the ultraviolet to the infrared. Earth's atmosphere is transparent in the visible spectrum, which is where the sun delivers most of its energy. The earth radiates only in the infrared spectrum. The atmosphere is only partially transparent at these wavelengths allowing it to trap some of this energy.

causes more dense objects to sink and less dense objects to rise. We observe this in a pan of water as it heats on the stove. If we add a drop of food color, it will circulate from the bottom to the top as warm water rises and cool water sinks. This also occurs above the ground on a sunny day leading to a rippling pattern in the air.

Radiation is the transfer of energy by electromagnetic waves. The thermal motion of molecules and their associated charged particles, generate these waves. If the molecules interact strongly with each other, they generate a continuous spectrum of wavelengths as illustrated in figure 13.12. Hot objects emit considerably more energy than cold objects and the predominant wavelength at which they emit energy is at a shorter wavelength. The sun is at a temperature of 5800 K and primarily emits its energy in the yellow to green region of the spectrum. Your body temperature is at 310 K and primarily emits in the infrared (IR) region of the spectrum. Although you can't see IR with your eye, you can feel the heat radiated from your hand by putting it close to your cheek.

13.4 THE SUN AND THE ATMOSPHERE

The sun is the source of energy for Earth and drives all the motion of the atmosphere. Apart from nuclear energy, all sources of energy used by mankind are derived from the sun, whether it is solar, wind, hydroelectric, wood, or fossil fuels. If the sun suddenly ceased to exist, Earth's surface would cool very quickly. Temperatures would fall well below freezing and the major wind patterns of the atmosphere would cease within days.

Electromagnetic radiation from the sun travels through the vacuum of space to heat Earth. This energy source delivers 1360 W/m², enough power per square meter to run an electric hair dryer. This may not seem like much, but one square kilometer could support one million hair driers operating continuously. The energy from the sun is nearly constant and yet we observe different temperatures across the world. What factors affect the temperatures we observe?

13.4.1 Daily Cycle

The most obvious change in air temperature is due to the rotation of Earth about its axis. During the day sunlight is absorbed by Earth's surface. This heating exceeds the IR cooling and the temperature increases. The sun is highest in the sky at noon and the maximum rate of heating occurs at this time. However, air temperatures continue to increase into the afternoon because solar heating continues to exceed IR cooling. Late in the afternoon, around 4:00 PM, the sun is lower in the sky and solar heating is less. At this time the temperature begins to drop. During the night only IR cooling occurs and temperatures drop even faster. Just after dawn the coldest temperature of the day is reached. If clouds are present during the night, IR radiation is trapped by the clouds and the morning temperatures are not as cold. Figure 13.13 is a graph of a weekly temperature record. Notice the pattern of highs and lows closely follow the 24 hour cycle of the day.

13.4.2 The Effect of Latitude

The next significant factor on air temperature is latitude or distance from the equator. Averaged over the course of the year, the equator receives the most solar heating and the North and South Poles the least. Near the equator at noon, the sun is directly overhead. As illustrated in figure 13.14, the sun travels through the least amount of atmosphere and the sunlight comes directly into the surface giving maximum heating. In Alaska and Canada where the latitude is 60°N, light from the noon sun comes in at a lower angle. The distance traveled through the atmosphere is greater and the intensity of sunlight is spread out over twice the area compared to the equator. As a result, Earth's surface receives less total energy from the sun over the course of the day at higher latitudes.

Incorporating the tilt of Earth, figure 13.15 plots the relative amounts of solar heating and IR cooling by latitude averaged over the course of a year. Since IR cooling depends on temperature, losses are greatest at the equator and least at the poles. Solar heating is greatest at the equator because in spite of Earth's tilt, the noon sun is always high in the sky near the equator. Notice that solar heating exceeds IR cooling from the equator up to 37° latitude both north and south. Further to the poles there is more IR cooling than solar heating. This differential heating is the driving force behind the

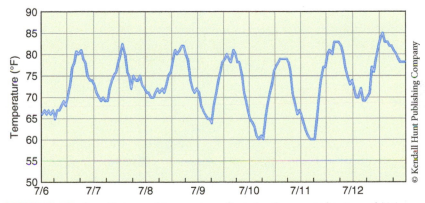

FIGURE 13.13. A weekly trace of temperatures illustrates the expected pattern of high temperatures in late afternoon and lows at sunrise. The coldest mornings occur when the skies are clear.

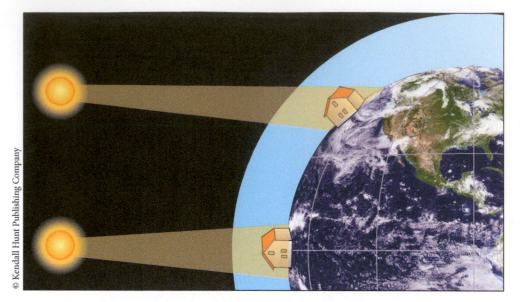

FIGURE 13.14. The most intense sunlight hits Earth's surface at noon near the equator. At higher latitudes the sun comes in at an angle causing it to travel through more atmosphere and spread over more surface area. Therefore, less energy is received and temperatures are not as high.

winds and ocean currents as energy is transported away from the tropics.

13.4.3 Seasons

Because Earth's axis is tilted by 23.5°, the latitude corresponding to maximum heating changes over the course of a year (figure 13.16). This seasonal effect of heating is due to two factors. First, the angle at which sunlight hits the surface changes thus giving less intense sunlight, as mentioned previously. Second,

the length of the day is either lengthened or shortened. The equator has 12 hours of daylight and 12 hours of night time throughout the year. However, this is not true anywhere else. March 20th marks the first day of spring, **vernal equinox**. On this day, the sun shines directly on the equator. As a result, everywhere on the planet there is 12 hours of daylight.

Three months later on June 21st the sun is directly over the **Tropic of Cancer**, located at 23.5°N. North of this location, the sun is the closest to being overhead at noon. This marks **summer solstice**, the

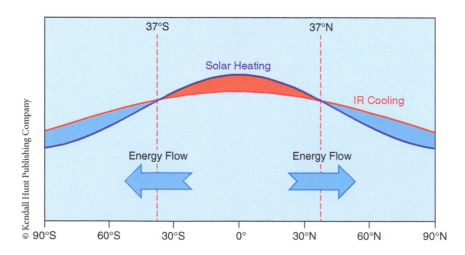

FIGURE 13.15. Near the equator heating exceeds cooling while near the poles it is the opposite. If the atmosphere were not present to transport the heat poleward, equatorial temperatures would exceed the boiling point of water during the day.

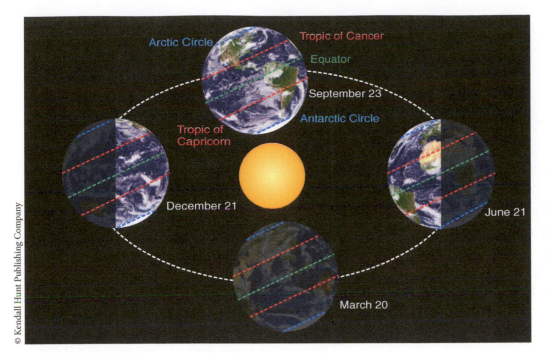

FIGURE 13.16. The Seasons. Summers are hottest because the sun is higher in the sky at noon and the days are longer. Six months later the days are shorter and the sun is lower in the sky.

first day of summer in the Northern Hemisphere. Daylight hours are greatest on this day. For example, Ohio has 15 hours of daylight and 9 hours of night time. All points north of the **Arctic Circle**, 66.5°N, are in perpetual daylight. Due to the longer days and the more direct sunlight, maximum heating from the sun is located at 30°N, corresponding to Northern Florida and Texas. In the Southern Hemisphere the daylight hours are shortened and the region is in the midst of winter. All points south of the **Antarctic Circle**, 66.5°S, are in perpetual nighttime.

On September 22nd the sun is shining directly on the equator and we have the first day of fall or **autumnal equinox**. Once again everywhere on Earth there are equal hours of daylight and night time. Winter begins on the day with the smallest number of daylight hours. **Winter solstice**, December 21st, places the North Pole in perpetual night and the sun is the lowest it will be in the noon sky. The sun shines directly on the **Tropic of Capricorn**, 23.5°S, and the Southern Hemisphere enjoys its summer.

Based on the length of days and direction of sun light, you would expect the hottest part of summer to occur on June 21st and the coldest part of winter on December 21st. However, there is a thermal lag between maximum heating and maximum temperatures. This is similar to the daily cycle, where the hottest temperatures occur at 4 PM rather than noon. Changes in vegetation, snow cover and heat retention by the ground contribute to this lag in temperatures. In general the hottest days occur in late July and August, while the coldest days occur in January and February.

Temperature lag occurs not only by season, but by geography. Since rock and land have a lower heat capacity than water, it cools faster in the fall and winter. As a result, the coldest air temperatures occur over land, away from large bodies of water. This is one reason why Bismarck, North Dakota is much colder than Seattle, Washington in January. Ocean currents, in the process of transporting heat, also modify the severity of winter weather. The Gulf Stream is a warm water current that travels north along the east coast of the United States

	Albedo
Grass	10–25%
Forest	10–20%
Bare Soil	20%
Sand	40%
Snow	80–95%
Clouds	10–90%
Water	10–60%
Asphalt	5–15%
Concrete	40–55%

FIGURE 13.17. Light colors reflect a lot of sunlight and have a high albedo. Darker colors absorb more energy from the sun and have a lower albedo. Earth's average albedo is 30%.

and heads out across the Atlantic Ocean towards Great Britain. This allows London to enjoy mild, but foggy winters, while Newfoundland and Labrador, which are at the same latitude, to suffer bitter cold winters.

13.4.4 Albedo

One final factor that impacts heating on both the large and small scale is albedo. **Albedo** is the percentage of solar radiation reflected or scattered from a surface. Objects with a high albedo reflect most of the sun's rays and absorb very little solar energy. A mirror has an albedo of 100%; however, high albedos do not require smooth surfaces. Fresh snow has an albedo of 95%, but scatters sunlight in all directions. The full moon in the night sky appears bright, but it only has an albedo of 7%, nearly the same as a chunk of charcoal. Its brightness is only apparent in contrast to the blackness of space.

The range of albedos for common surfaces is provided in figure 13.17. Snow reflects most of the incident sunlight, resulting in little heating from the sun. This reduced heating along with the energy required to melt snow, delays the onset of warm temperatures at the end of winter. Most land surfaces have albedos ranging from 10 – 30%, but even these differences can affect how warm a surface gets in the summer heat. Asphalt parking lots have an albedo less than 10%. Asphalt absorbs most of the sun's heat allowing snow to melt even when air temperatures are as low as -12 °C (10 °F). During the summer, however, asphalt becomes very hot. In contrast, the albedo of concrete is higher and results in lower summer temperatures. Similar choices are made when roofing houses. Lighter colors are used in the south while darker shingles are used in the north.

13.5 ENERGY BALANCE

The cumulative effect of all Earth's reflective surfaces is an albedo of 30%. The oceans play a major role in albedo since they cover 75% of Earth's surface. However, snow and clouds can easily modify the average albedo although they cover a smaller percentage of the globe. The total albedo, along with other heat transfer mechanisms of the atmosphere, determines Earth's average surface temperature. To maintain a constant surface temperature there must be a balance between the absorbed solar energy and the IR radiation emitted by Earth.

Figure 13.18 illustrates the transport of energy into and out of Earth's system. Treating incoming solar energy as 100%, nearly half of the sun's energy is absorbed by Earth's surface. Some is absorbed by the air and the remainder is reflected back into

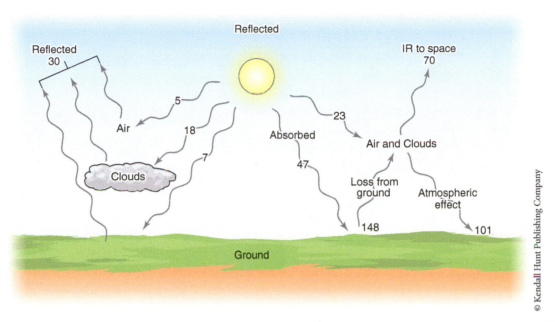

FIGURE 13.18. Energy balance of the atmosphere. Incoming radiation is either reflected or absorbed by the air, clouds or earth. Energy radiated from Earth matches the amount absorbed, otherwise the world would heat up. Some of the radiation given off by Earth is redirected back to the surface making Earth warmer than if the atmosphere were absent. The numbers are listed as percentages where incoming sunlight is 100%. Percentages greater than 100% indicate that energy given off by Earth's surface is greater than the incident solar radiation.

space. The 30% reflected is Earth's albedo, which is mostly due to the presence of clouds.

The ground loses more energy than it receives from the sun, 148%. This appears to be wrong except for the fact that Earth receives energy from a second source. The atmosphere absorbs energy from both the sun and Earth. As it loses its energy through IR radiation, some travels into outer space and the rest is redirected back to Earth. This additional heating source makes Earth warmer than if it had no atmosphere. This is called the **atmospheric effect**. Without the atmosphere Earth's surface temperature would be -18 °C (0 °F); however, with it the average temperature is 15 °C (59 °F).

13.6 CHANGING CLIMATE

Now that we know what determines Earth's temperature, the next question is how easily can this average temperature change? A century ago a Serbian scientist, Milutin Milankovic, proposed that changes in the orbit around the sun affects Earth's energy balance. If the radiant energy received at higher latitudes is reduced by 10%, it could initiate an ice age. During the Cold War concerns were raised about the possibility of a nuclear winter being triggered by the aerosols from numerous atomic bomb detonations. Current climate discussions turn the attention to the heating effect of greenhouse gases. To understand the impact of these changes in radiative forcing it is important to know how sensitive or resilient the atmosphere is to change.

13.6.1 Feedback Mechanisms

The response of the atmosphere to changes in the energy balance depends on numerous factors; however, they can be categorized as being one of two types of feedback mechanisms. A **feedback mechanism** is a series of physical processes that either enhance or inhibit a change in a physical parameter. In our case, the physical parameter is temperature. Clouds play a crucial role in moderating the temperature of Earth. If the temperature of Earth increases, we would expect evaporation from the oceans and lakes to increase. With more water vapor in the air there should be more clouds. By increasing clouds, albedo is higher and less energy gets to the surface. As a result of this chain of processes, an increase in temperature reduces heating resulting in lower temperatures. This inhibiting mechanism is called *negative feedback* and attempts to keep Earth's temperature locked into its current value (figure 13.19).

Positive feedback occurs when a change is enhanced in its effect. If Earth's temperature decreases, there is more snow cover. This increases the albedo and results in less heating. Less heating leads to lower temperatures and reinforces the initial change. If the temperature increases, the amount of snow decreases giving a lower albedo. This enhances the temperature increase and, likewise, demonstrates positive feedback.

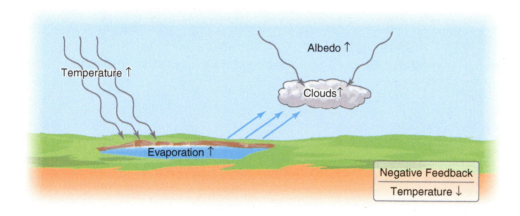

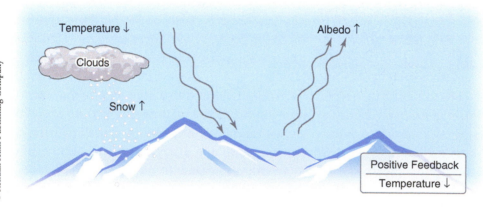

FIGURE 13.19. Feedback Mechanisms. Positive feedback reinforces an initial change while negative feedback inhibits the change.

13.6.2 The Atmospheric Effect

Although the term *greenhouse effect* is commonly used, a more accurate term is atmospheric effect. In a greenhouse heat is trapped inside a building with transparent surfaces. These surfaces prevent convection and the greenhouse remains quite warm although the temperature outside is well below freezing. However, the atmosphere does not restrict convection. What is common between the two is the radiation process. Solar radiation reaches the surface through a transparent medium, either a window or the atmosphere. The surface heats up and emits more IR radiation. The window and atmosphere are partially opaque in the IR and absorb some of the radiation as it leaves the surface. As a result, the window as well as the atmosphere reradiates some of the IR back to the surface. This wavelength dependence of transparency is the key factor.

Objects, whose absorption properties change based on wavelength, are called **selective absorbers**. As mentioned in the previous paragraph, glass is transparent in the visible spectrum and opaque in the IR. Snow demonstrates a similar effect by reflecting visible light and absorbing IR. This results in the receding snow around the base of trees as seen in figure 13.20. The trunks of trees absorb solar

FIGURE 13.20. Snow reflects sunlight, but absorbs infrared radiation. IR emitted from the trees is absorbed by snow allowing it to melt away from the trunk.

radiation coming directly from the sun as well as that which is reflected by the snow. As a result, the tree trunk heats up and emits more IR radiation. The snow closest to the tree absorbs the IR and begins to melt. After several days, the snow has noticeably receded from the trunk.

13.6.3 Greenhouse Gases

The gases of the atmosphere are also selective absorbers. The wavelength at which they absorb depends on the strength of the chemical bonds between atoms in the molecule. Figure 13.21 shows the gases' cumulative effect on the atmosphere's transparency. There are two major atmospheric windows, through which radiation is easily transmitted. The first is in the visible spectrum, which is why we can see the sun in the day and stars at night. The other window occurs in the microwave and radio wave spectrum. Although this window does not impact our weather, it allows the transmission of satellite and cell phone signals. The IR spectrum is more complicated and is dominated by absorption from oxygen, water vapor, and carbon dioxide. Other gases active in this region are methane, nitrous oxide, and ozone. Because these gases absorb IR radiation emitted from Earth's surface, they are called **greenhouse gases**. Instead of discussing each greenhouse gas separately, most scientists use a CO_2 concentration that has an equivalent effect of all the other greenhouse gases.

The scientific basis for global warming focuses on the impact of greenhouse gases on atmospheric processes. There is a well-established connection between an increase in CO_2 and absorption of IR from Earth. Figure 13.22 is a plot of CO_2 concentrations measured over past half century. These measurements were made in the middle of the Pacific Ocean and are representative of global increases of CO_2. During the summer, photosynthesis in plants removes CO_2, thus explaining the annual cycle in CO_2 concentrations.

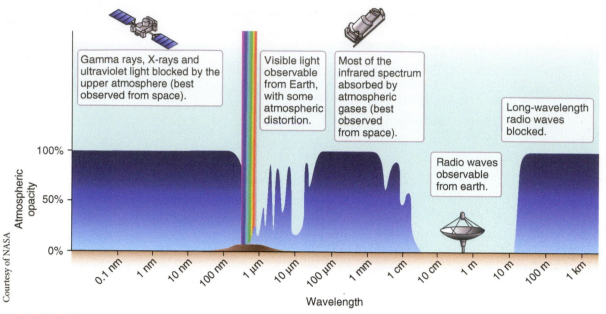

FIGURE 13.21. Gases in the atmosphere absorb radiation in the infrared range while allowing most of the visible light and microwaves to transmit. Changing the composition of the atmosphere can affect the amount of absorption and, therefore, Earth's temperature.

Water vapor also affects the amount of IR the atmosphere absorbs as illustrated by this satellite image (figure 13.23). The amount of water vapor in the atmosphere changes significantly over time and location, while the other greenhouse gases are nearly uniformly mixed and do not fluctuate much. There-fore, water vapor is treated differently than other greenhouse gases. If water vapor forms clouds low in the atmosphere, they are relatively thick. As a result, they reflect a lot of sunlight leading to net cooling. However, if these clouds are high in the atmosphere, they are relatively thin, allowing most of the sunlight through to the surface. These high clouds still trap terrestrial IR, thus leading to net warming. Currently it is determined that the net influence of all clouds leads to a net cooling of the atmosphere.

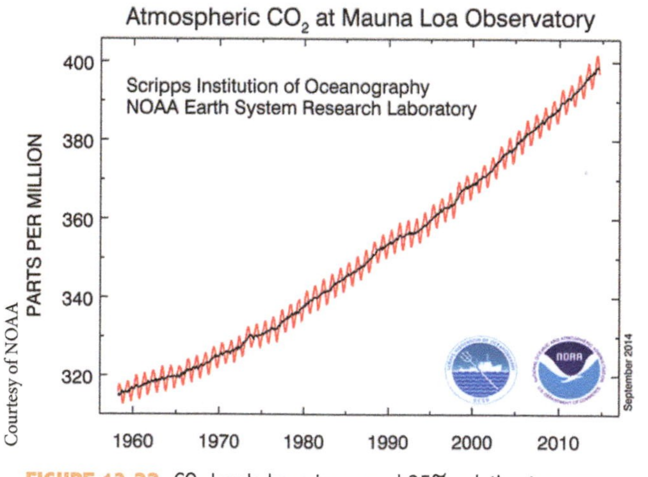

13.6.4 Global Warming

There are scientific methods of studying and model-ing Earth's systems, which can inform the debate on global warming and climate change. Modern meth-ods of collecting data and using it in computer models have developed over the last half century. Confidence in early weather forecasts was limited to days, while today it has extended to more than a week. This has come as a result of faster computers and imple-menting a better understanding of the atmosphere's physical processes. With regard to climate modeling

FIGURE 13.22. CO_2 levels have increased 25% relative to measurements in 1960. Computer models indicate that a continued increase will affect Earth's climate.

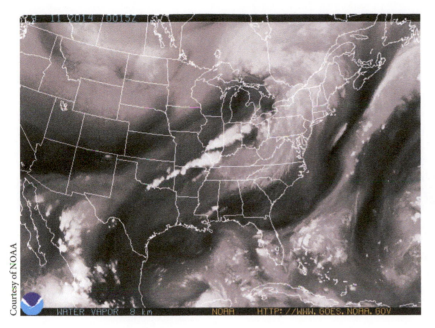

FIGURE 13.23. Since water vapor preferentially absorbs light in the 6.5 – 7.0 μm range, satellites are able to identify regions of humid air. Lighter colors indicate high levels of water vapor while black represents dry air. A cold front is easily identified by the tongue of dry air coming behind the warm humid air drawn from the Gulf of Mexico.

the challenge is to effectively implement the positive and negative feedbacks of not only the atmosphere, but also of the land, oceans, sea/land ice and biosphere. This makes the problem very complicated, but very necessary if the difficult questions of climate change are to be answered.

In spite of great progress in modeling Earth's systems, it is often hard to separate the science and the politics of global warming. When there is limited knowledge about the full complexity of a process, gaps must be filled in with assumptions. If the assumptions are wrong, there will be a bias in the results. Some people ignore climate modeling on this basis. However, even flawed models are informative if they capture the physical processes affecting the atmosphere. Many different models have been run that give a variety of results, but all indicate an increase in CO_2 which results in an increase of global temperature. The question really boils down to "which ones are the closest to reality?" If it is assumed that Earth's system is fragile to environmental change, the more extreme scenario will be promoted. Those assuming Earth is relatively robust will chose the moderate results.

Two contrasting responses to global warming are represented by the Evangelical Climate Initiative (ECI) and the Cornwall Alliance (CA). The ECI supports the Kyoto Protocol, an agreement between nations to limit CO_2 emissions to pre-1990 levels. This group of evangelicals is convinced that manmade sources of CO_2 will have a negative impact on the environment. They believe that the best way to protect the poor in climate sensitive regions both now and in the future is to restrict CO_2 emissions. In contrast, the CA opposes the Kyoto Protocol calling into question the negative impacts of CO_2 increases. In addition, the CA feels that limiting CO_2 emissions will cripple the economies of developed nations and in turn limit their ability to assist the poor impacted most by climate change. Both groups express a concern for the poor, but come to different conclusions on how to respond.

The primary question related to global warming is whether mankind is the source of and/or the solution to climate change. When the population of Earth was much smaller, it was easy to minimize the impact of humans on God's creation. However, as the population continues to rise and technology enables each person to affect his environment to a greater degree, the human factor cannot be ignored, whether it relates to global warming, pollution, deforestation, etc. Man is given dominion over the creation to exercise stewardship. With regard to global warming and climate change, stewardship should be the central issue. However, not all Christians agree on how this stewardship should be applied.

13.7 CLIMATE ZONES

In Book II of Aristotle's work *Meteorology* weather is explained by the source region of winds. The Greek work for region is *klima* from which we derive our word *climate*. Aristotle proposed three regions based on geography. Extending north from the Arctic Circle is a Frigid Zone of ice and cold. To the south of the Tropic of Cancer is the Torrid Zone, which is hot and dry. Air from these two regions clash in the Temperate Zone and gives us our weather. Aristotle's description ties climate to solar heating and provides the basis for other climate schemes.

13.7.1 Köppen Climate Classification

Modern classification systems use both temperature and precipitation patterns to define climate zones. The **Köppen climate classification** is the most widely used and is based on vegetation types. Since the types of trees, bushes, grasses and flowers depend on the length of the growing season and precipitation, they provide a good indicator of predominant weather patterns, which give rise to climate. Figure 13.24 illustrates the locations of the five predominant regions developed by Köppen: A – Tropical, B – Arid, C – Temperate, D – Continental, and E – Polar. Each climate zone consists of subgroups, which are differentiated by their vegetation type. Notice that climate zones are only defined over land surfaces.

Closest to the equator is the tropical zone, group A in the Köppen system. Temperatures are warm year round and the precipitation rates are high. Indicative of this zone are tropical rain forests, most notably those in the Amazon River basin, which can receive 280 cm (110 in) of precipitation per year. However, rain forests exist around the globe from Central America, Central Africa, India, and Southeast Asia (figure 13.25). There are subgroups of the tropical zone that take into account the effect of monsoons and other seasonal effects.

Group B represents dry regions of Earth's surface, regardless of temperature. An exceptional example of this group is the Sahara desert in North Africa, which has precipitation rates of less than 2.5 cm (1 in) precipitation per year. Arid regions with high daily temperatures include the Sahara desert, countries of the Middle East, Western Australia, and the Southwest United States. However, there are also arid regions that experience cold temperatures, such as the Gobi Desert in Central Asia, which can get to -40 °C (-40 °F) in the winter. Although it is easy to equate this group with deserts, it also includes semi-arid regions that can support grasses and shrubs (figure 13.26). These are the regions most sensitive to climate change because small changes in annual precipitation affects the ability of the land to support agriculture. During the Dust Bowl of the 1930's, a drought and loss of land productivity displaced thousands of families in the central plains of the United States and Canada. Similar conditions exist in sub-Saharan Africa which contributes to the political instability in the region.

World map of Köppen-Geiger climate classification

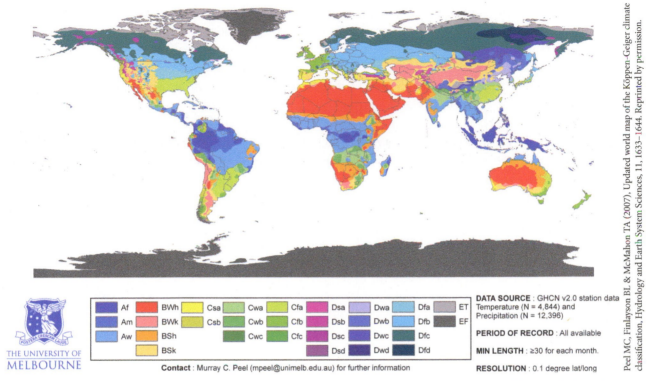

Af	BWh	Csa	Cwa	Cfa	Dsa	Dwa	Dfa	ET		
Am	BWk	Csb	Cwb	Cfb	Dsb	Dwb	Dfb	EF		
Aw	BSh		Cwc	Cfc	Dsc	Dwc	Dfc			
	BSk				Dsd	Dwd	Dfd			

DATA SOURCE : GHCN v2.0 station data Temperature (N = 4,844) and Precipitation (N = 12,396)

PERIOD OF RECORD : All available

MIN LENGTH : ≥30 for each month.

RESOLUTION : 0.1 degree lat/long

Contact : Murray C. Peel (mpeel@unimelb.edu.au) for further information

Peel MC, Finlayson BL & McMahon TA (2007), Updated world map of the Köppen-Geiger climate classification, Hydrology and Earth System Sciences, 11, 1633–1644. Reprinted by permission.

FIGURE 13.24. Köppen Climate Classification. This scheme uses vegetation to classify climate zones. As a result, it is strongly dependent on temperature and precipitation. Additional refinements of this scheme take into account the severity of winters.

Corresponding to Aristotle's temperate region is Köppen's groups C and D. Group C is a sub-tropical region located near the tropics or near large bodies of warm water. Changes between the seasons occur; however, winters are mild allowing a variety of vegetation types to thrive. As winter approaches, colder

FIGURE 13.25. Tropical rain forests are located in regions of warm year round temperatures and high precipitation. They are home to the most diverse plant and animal life on the planet.

FIGURE 13.26. Semi-arid Africa has enough precipitation to support a minimal amount of vegetation. However, shifts in climate can greatly impact the ability of these regions to support agriculture.

temperatures initiate a dormant cycle in some plants and result in a beautiful show of colors from deciduous trees (figure 13.27). Group D is further poleward and away from the moderating effect of oceans. As a result, winters are harsher thus favoring coniferous trees that maintain their foliage year round.

The last climate zone is polar, group E, and has an average temperature less than 10 °C (50 °F) during the warmest month. Corresponding to Aristotle's Frigid zone, ground is in a state of **permafrost**. Summer heating is insufficient to melt the soil to any great depth. This reduces the permeability of the soil and limits the flow of ground water. As a result, the melting of winter snow produces marshes and pools of standing water (figure 13.28). During the summer the soil at the surface is above freezing and supports small shrubs, grasses, moss and lichens. This climate zone is called **tundra**. If snow cover persists year round no vegetation can grow and this polar climate is called **ice cap**.

Sometime a sixth climate zone with a designation of H is used. This is called **highlands** and represents land that is at high elevations. Since the temperature of the atmosphere decreases with altitude, high elevations have climates similar to tundra and polar ice cap. If you have ever visited tall mountain ranges like the Rocky Mountains, Alps, or the Andes, there is an altitude at which trees no longer grow. This tree line is the starting point of the polar climate. As you progress to lower altitudes from the tree line, the temperatures get warmer and the climate zones change. Visitors to Peru can experience vegetation representative of nearly every climate group by traveling from the arid coast on the west to the headwaters of the Amazon River via the highlands of the Andes.

13.7.2 Climate Sensitivity

With an understanding of climate zones it is now possible to discuss the impact of climate change. Since climate zones are tied to precipitation rates and seasonal temperatures, any changes in these parameters will likewise shift the climate. Over the past century, the average global surface temperature has increased by 0.85 °C (1.5 °F). Figure 13.29 plots the annual fluctuation in global temperature with respect to the average during the twentieth century. There is a clear rising trend in the later half of the century. Some have pointed out that the trend stopped in the past decade and dismiss global warming; however, more time is needed to determine if the trend will resume or begin to decrease.

Apart from the debate whether the trend will increase and by how much, what are the observed

FIGURE 13.27. Deciduous forests are located in regions with plenty of rainfall and moderate winters. These forests are common in the temperate zone. In the process of shedding their foliage, these forests provide a spectacle of color in the fall.

FIGURE 13.28. Tundra has year round temperatures that are insufficient to thaw the soil below the surface. Soil in this climate region can only support small plants and shrubs.

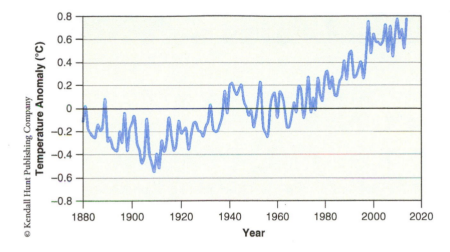

FIGURE 13.29. The average surface temperature of Earth has increased by 0.85 °C in the past century. Some claim this is part of a natural cycle while others feel this is the beginning of significant global warming

shifts in climate? While nearly nobody will complain about warmer winters in Minnesota, there may be more concerns about higher temperatures in the tropics during the summer. Shifts in temperature can increase the growing season at higher latitudes, but also make it easier for invasive animal and plant species to move into new areas. In the United States vegetative species intolerant of freezing temperatures are spreading northward, such as the mangroves of Florida. Shifts in precipitation patterns have been observed around the globe, most notably the drying of the western United States. This has resulted in less snow pack during the winter, which is a source of fresh water to major cities and for agriculture. Although this drying may be the result of long term cycles in weather patterns, such as El Niño, sustained higher global temperatures can and do have an impact on people's lives.

Increased temperature has also had an impact on polar regions. Less snow cover leads to reduced albedo, which allows for the absorption of more solar energy. This enhances the temperature increase and also results in melting more snow and ice. Although the melting of floating sea ice does not change the sea level, the run off from melting ice sheets over land does. Also, with sustained higher temperatures oceans will warm and expand in volume. Since ocean basins are not changing substantially, the height of the oceans must rise. Over the past century sea level has risen by at least 15 cm (6 inches). Melting in the

Arctic Ocean has also provided a Northwest Passage between Europe and East Asia during the summer. Countries bordering the Arctic Ocean are currently developing plans for tapping the energy resources lying under the ocean floor.

Before our discussion of climate zones, I asked where are the poor and how will they be impacted by climate change? There are still vast areas of Earth surface that are not heavily populated, such as northern latitudes, mountains, and heavily forested regions. A century ago a large percentage of the population was tied to agriculture. However, technology has allowed more land to be farmed by fewer people. As a result, over half of the world's population lives in cities and most of these cities are close to oceans and large rivers. In 2005 hurricane Katrina made landfall close to New Orleans, Louisiana, resulting in 1,833 deaths. This tragedy could have been much worse if it were not for the advanced warning and transportation infrastructure, allowing many to evacuate. Of the same intensity as Katrina, the 1970 Bhola cyclone made land fall near the Ganges River Delta in present-day Bangladesh. With high winds and a storm surge of 10 m (33 feet), 500,000 people were killed. Changes in precipitation and weather patterns will impact heavily populated areas. As Christians, we need to ask how we can best respond to the crises of today and prepare for those of the future.

CHAPTER 13 IN REVIEW: EARTH'S ATMOSPHERE

13.1 Weather and Climate

- Weather is a description of the atmosphere at a particular place and time.
- Climate is an average state of the atmosphere over a region or interval of time.

13.2 The Atmosphere

- The atmosphere consists primarily of nitrogen and oxygen gas. Other gases and pollutants have an impact on the behavior and quality of the atmosphere.
- The composition of the atmosphere is unique in the solar system and is the only location in the solar system known to be suitable for life.
- Atmospheric pressure decreases with height. It is measured with a barometer and in units of millibars.
- The atmosphere consists of the troposphere, stratosphere, mesosphere, and thermosphere. These layers are defined by how temperature changes with altitude.

13.3 Temperature and Heating

- Temperature is a measure of an object's average thermal energy.
- When energy is added to an object, it either results in a temperature increase (sensible heating) or a phase change (latent heating).
- Heat is a transfer of energy from hot to cold objects and accomplished through conduction, convection, and radiation.

13.4 The Sun and the Atmosphere

- Energy received from the sun determines the average temperature of Earth and variation of temperature is due to day/night cycles, latitude, and season.
- Albedo is the percentage of solar radiation reflected or scattered from a surface and affects how quickly an object will warm.

13.5 Energy Balance

- Earth emits just as much energy in the form of infrared radiation (IR) as it receives from solar radiation.
- The atmosphere absorbs some of this IR and reradiates it to Earth's surface resulting in higher temperatures. This is called the atmospheric effect.

13.6 Changing Climate

- Climate change may occur due to changes in Earth's orbit, composition of the atmosphere, or complex interactions between the ocean, atmosphere, land, and ice.

- Greenhouse gases increase the average surface temperature of Earth, but the amount of warming is enhanced or limited by positive and negative feedback mechanisms.
- Global warming is both a political and scientific question. Addressing the science of global warming requires a solid understanding and modeling of the entire Earth climate system.

13.7 Climate Zones

- Climate zones are identified by the types of vegetation present and are linked to temperature, precipitation, and predominant weather patterns.
- The Köppen climate classification system divides the land surface into five climate groups: tropical, arid, sub-tropical, temperate, and polar.
- Shifts in precipitation patterns, melting of polar ice, and the rise of sea level are the primary hazards of global warming.

KEY TERMS

Acid Rain	Meteorology
Aerosol	Ozone
Albedo	Permafrost
Antarctic Circle	Pressure
Arctic Circle	Radiation
Atmospheric Effect	Selective Absorber
Autumnal Equinox	Sensible Heating
Barometer	Stratosphere
Climate	Summer Solstice
Conduction	Temperature
Convection	Thermosphere
Density	Tropic of Cancer
Feedback Mechanism	Tropic of Capricorn
Greenhouse Gases	Tropopause
Heat	Troposphere
Highlands	Tundra
Ice Cap	Vacuum
Köppen Climate Classification	Vernal Equinox
Latent Heating	Weather
Mesosphere	Winter Solstice

1. What is the difference between weather and climate? Give several examples of each.
2. How are oxygen and carbon dioxide in the atmosphere connected to each other? What are the processes that generate and consume each of these gases?
3. List several pollutants of the atmosphere and how do they impact the environment?
4. Describe how the temperature, pressure, and density of the atmosphere change with altitude. How do these trends change in the different layers of the atmosphere?
5. How does the heat capacity of an object affect its rate of heating? Besides sand what other materials are you familiar with that have a low heat capacity?
6. What are the characteristic ways of distinguishing the different phases of matter?
7. Give examples of the three mechanisms of energy transport.
8. Given the relationship between the energy received from the sun and temperature, when and where are the hottest and coldest temperatures on Earth?
9. How does albedo affect the energy balance of Earth? Compare an ice free world with an ice covered Earth.
10. Is the atmospheric effect a good or a bad thing? How does it work and what role do greenhouse gases play?
11. What role does and should mankind play with regard to global warming and climate change?
12. Using a global map, identify the different climate regions. What are the challenges and benefits of the climate in that region to the people who live there?

FURTHER READING

- National Oceanic and Atmospheric Administration (NOAA) – www.noaa.gov
- Yahoo! Weather – weather.yahoo.com
- The Weather Channel – www.weather.com
- WeatherBug – weather.weatherbug.com
- AccuWeather – www.accuweather.com
- National Climatic Data Center (NCDC) – www.ncdc.noaa.gov
- The Cornwall Alliance – www.cornwallalliance.org
- Evangelical Environmental Network – www.creationcare.org
- Intergovernmental Panel on Climate Change – www.ipcc.ch

REFLECT ON SCRIPTURE

Are there any hints about the climate in the "Promised Land" during the time of Abraham or Israel's Exodus from Egypt? After reading Deuteronomy 28:12-14 & 28:23-24, consider how God used climate to work in the nation of Israel. Are there specific events from the Old Testament that are a fulfillment of this passage?

CHAPTER 14

PHENOMENA AND PROCESSES OF THE ATMOSPHERE

OUTLINE:

Courtesy of NASA

This image of a thunderstorm was taken from the International Space Station. A single strong updraft drives humid air from the surface to the tropopause. When it hits this stable layer of air it spreads out to form a large layer of cirrus clouds. The top of the updraft is seen by the rounded cumulus tops extending above the cirrus deck.

As we have learned in the Chapter 13, all motion and activity of the atmosphere is driven by energy from the sun. Since the atmosphere is relatively transparent to visible light, a large portion of that energy is deposited on Earth's surface, which results in heating. How does this heating bring about the motion of the winds? How does this result in clouds and precipitation? To begin answering these questions, we must focus on evaporation, which is second only to radiation as the means of removing energy from Earth's surface.

14.1 WATER VAPOR AND HUMIDITY

14.1.1 Evaporation

Evaporation is the vaporization of liquid water at temperatures below the boiling point. We know water boils at 100 °C and any additional heat converts liquid water into vapor. However, at lower temperatures some water molecules have enough energy to escape the surface if they are going in the right direction. As a result, a puddle of water eventually dries up without ever coming to a boil. The opposite of evaporation is **condensation**, the conversion of water vapor into liquid. At any instant of time there is competition between water molecules leaving and entering a puddle of water (figure 14.1). As more water evaporates, the amount of vapor above the puddle's surface increases. Some of these molecules in turn condense back on the surface. When a balance is achieved between the outgoing and incoming molecules, we say the water vapor has reached **saturation**. For small puddles saturation is never achieved and it completely evaporates.

The saturation value of water vapor is dependent on temperature. If the water is cold, few molecules have enough energy to leave the surface and the amount of water vapor above the surface is small. Figure 14.2 provides a table of saturation vapor pressures for different temperatures. If humid air comes in contact with a cold surface, such as a glass of ice water, the air becomes supersaturated.

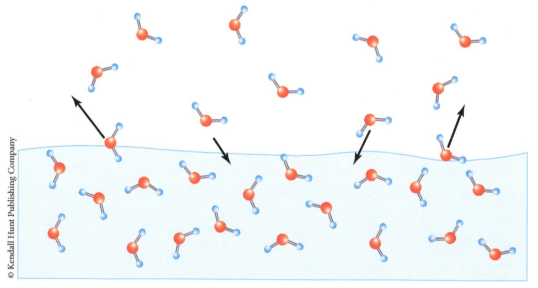

© Kendall Hunt Publishing Company

FIGURE 14.1. Evaporation and condensation are competing processes occurring over a water surface. When we use the term *evaporation*, we indicate that the net effect is a loss of liquid water and a gain of water vapor.

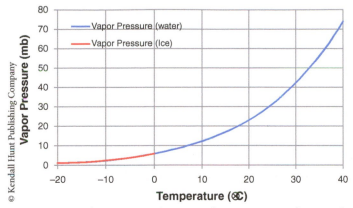

Temperature (°C)	Vapor Pressure (mb) Water (Ice)
−20	1.25 (1.03)
−15	1.91 (1.65)
−10	2.86 (2.60)
−5	4.21 (4.02)
0	6.11
5	8.72
10	12.3
15	17.0
20	23.4
25	31.7
30	42.4
35	56.2
40	73.8

FIGURE 14.2. As temperature increases, a larger percentage of water molecules evaporate to form a vapor. This effect is not linear as seen in the table and the curve.

As a result, condensation occurs and beads of sweat show up on the glass. Notice there are vapor pressures for temperatures below freezing. At these temperatures there is a competition between **sublimation**, the phase change going from solid to vapor, and **deposition**, the phase change going from vapor to solid. On a clear night, Earth's surface cools faster than the air and condensation or dew forms on the grass. If temperatures fall below freezing, deposition forms frost on the ground and windows (figure 14.3).

14.1.2 Humidity

The amount of water vapor in the air is measured in a number of ways. The easiest is to cool a glass of water until condensation forms on its surface. The temperature at which this occurs is called the **dew point temperature**. What we have done is reduce the temperature to the saturation point. Cross referencing this temperature in figure 14.2 provides the actual vapor pressure. **Vapor pressure** is a measure of the force per area exerted by the water vapor. If the dew point temperature matches the air temperature, the air is saturated and humidity is at its highest expected value. **Relative humidity** is a ratio of the actual vapor pressure compared to the saturated vapor pressure at air temperature. Therefore, relative humidity is 100% at saturation.

FIGURE 14.3. As water vapor comes in contact with surfaces below freezing, deposition occurs. It is easier for deposition to occur where frost is already present; therefore patterns extend from a central starting point.

Humidity is routinely measured using a **psychrometer**, which measures the temperature of a thermometer covered with a wet sleeve (figure 14.4). This temperature is compared to an identical thermometer without a sleeve. The difference in temperature between the two thermometers is due to evaporative cooling. Less evaporation occurs in more humid air and the little water vapor due to the cold temperatures. When this air enters the home, it warms without gaining any additional water vapor. As a result, the humidity is further from saturation and the relative humidity is low. A humidifier adds vapor to the air and makes it more comfortable in the winter by reducing the evaporation from your skin.

Relative humidity also affects our hair. At high humidity the proteins relax and the hair becomes longer. That is why people with long hair suffer the 'frizzies' on humid days. This principle is used in novelty weather forecasting houses (figure 14.5), which are really measuring relative humidity. Any device measuring relative humidity is called a **hygrometer**.

Relative humidity changes by location and time of day. After a rain shower, humidity levels can be very close to 100%. Once water puddles evaporate, relative humidity decreases as the air temperature

FIGURE 14.4. A sling psychrometer uses a dry bulb to measure air temperature and a wet bulb to measure humidity. In dry air the wet bulb experiences more evaporation resulting in a lower temperature.

Courtesy Steve Gollmer

temperature difference is small. We experience this when perspiration accumulates on our clothes on hot humid days. If we lived in Arizona where the relative humidity is 5%, evaporation occurs rapidly and our bodies stay cool and dry although the air temperature is quite warm. This cooling mechanism depletes our bodies of water and it is important to drink plenty of fluids to avoid heat cramps and sun stroke in dry climates.

14.1.3 Effects of Humidity

Relative humidity most closely matches how we respond to humidity in the atmosphere. When we enter a basement, it feels damp because the air temperature is cooler and the humidity is closer to saturation. During the winter air contains very

Courtesy Steve Gollmer

FIGURE 14.5. This novelty weather house operates on relative humidity. A hair attached to the platform lengthens or shrinks depending on humidity, moving the figurines. (This does not work inside air conditioned homes.)

increases. Since the hottest portion of the day occurs around 4 PM, this is also the time of lowest relative humidity. At night the air cools and relative humidity increases. By dawn the relative humidity reaches its highest value. Near the ground relative humidity reaches 100% and dew condenses on grass and other surfaces. Since condensation releases heat to the atmosphere, the formation of dew slows the night time cooling process. Therefore, the dew point temperature is a good predictor of morning low temperatures.

14.2 RISING AIR

Humid air is less dense than dry air and, therefore, tends to rise. Also heating at Earth's surface expands the air and contributes to convection. Once rising air moves away from the surface, the amount of heat it receives is greatly reduced. As it rises, it experiences less pressure and expands even more. The expansion process requires energy and it comes from the internal energy of the air itself. Therefore, rising air will cool as it experiences less pressure and expands to a larger volume. Conversely, the temperature of sinking air increases as the pressure rises and the volume shrinks. This relationship between pressure, volume and temperature assumes no energy enters or leaves the air and is called an **adiabatic process**. Adiabatic cooling explains why mountain tops are colder than valleys and the troposphere temperature decreases with altitude.

In addition to the adiabatic assumption we add the condition that air does not mix with its surroundings. These two conditions define a **parcel of air**. You can think of an air parcel as the air inside a balloon. The buoyancy of a hot air balloon (figure 14.6) allows it to rise above the surface of Earth. Eventually it stops rising when the buoyant force matches the force of gravity acting on the balloon and basket. If the balloon and basket are absent, the parcel of air continues to rise until its density matches the density of the air surrounding it. This is ultimately the destination of warm air rising from Earth's surface.

14.2.1 Lapse Rates

How quickly a parcel of air cools depends on whether it is dry or moist. A dry parcel cools only due to expansion of the air. This occurs at the dry **adiabatic lapse rate** of 10 °C/km. If humidity is present in the air, the air will cool until it reaches 100% relative humidity. At this point water vapor condenses into the small water droplets that form a cloud. Vertical motion results in additional condensation, which adds heat to the air parcel. Therefore, the cooling rate is only 6 °C/km and is called the moist adiabatic lapse rate. These lapse rates give you a quick method of estimating the temperature at any altitude

FIGURE 14.6. A hot air balloon rises because the air inside is less dense. Likewise, on sunny days parcels of air leave Earth's surface and rise until they reach the same temperature as the air around them.

Annette Shaff/Shutterstock.com

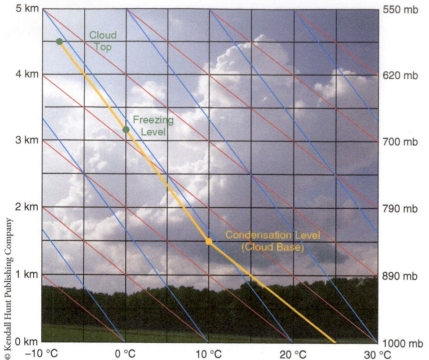

© Kendall Hunt Publishing Company

FIGURE 14.7. Dry air cools 10 °C for each kilometer it rises (red lines). Once the humidity in the air reaches the saturation point, a cloud begins to form. Beyond this point the air cools at a slower rate of 6 °C/km (blue lines). The yellow line represents the temperature of convected air starting with a surface temperature of 25 °C and a dew point of 13 °C. It saturates at 1.5 km and forms the base of the cloud.

(figure 14.7). When driving from Colorado Springs, Colorado, to the top of Pike's Peak the altitude increases by 2.5 km. Since each kilometer decreases the temperature by 10 °C, the temperature drops by 25 °C at the mountain top.

It is also known that the **dew point depression**, the difference between air temperature and the dew point, changes by 8 °C/km. This can be used to estimate the height of clouds that appear in the late morning sky. If the air temperature is 26 °C (79 °F) and the dew point is 10 °C (50 °F), the dew point depression is 16 °C. When air convected from the surface rises 2 km, the dew point depression goes to 0 °C and the relative humidity reaches 100%. The altitude where this occurs is called the **lifting condensation level** (LCL) and corresponds to the base of the cloud. In drier climates the dew point depression is greater and the cloud base is higher. Also,

cloud bases are flat because the LCL is relatively uniform.

Some information about the atmosphere can be inferred from the surface. However, to get an accurate picture of the atmosphere the National Weather Service (NWS) launches a series of balloons with weather instruments every 12 hours. This instrument, called a **radiosonde**, provides information about air pressure, air temperature, humidity, altitude, wind speed and direction. The data collected by the balloon launch is called a **sounding** and is plotted on a thermodynamic chart (figure 14.8). The solid black line is a plot of the actual air temperature at different pressure levels and is called the **environmental lapse rate**. The dashed line represents the dew point temperature. The complex structure of this plot gives information about location of clouds, stable air and the possibility of severe weather.

CHAPTER 14: Phenomena and Processes of the Atmosphere

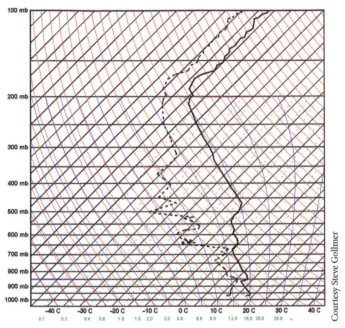

FIGURE 14.8. Data from a balloon launch is plotted on a thermodynamic chart to reveal conditions about the atmosphere's stability. Where the solid and dashed lines are close together, the humidity is near 100% indicating the presence of clouds. The temperature axis of this plot is skewed to the right. As a result, a vertical line near the surface represents a cooling rate of 6 °C/km.

14.2.2 Atmospheric Stability

The rapid development of clouds and severe weather are contingent on the stability of the atmosphere. **Stability** falls into three possible states as illustrated in figure 14.9. Just like a marble at the bottom of a bowl, a stable parcel of air returns to its starting point when disturbed. If a parcel is unstable, it will rapidly move away from the starting point like a marble balanced on an inverted bowl. Neutral stability occurs when a disturbed parcel moves away from its starting point, but not at a quickening pace.

Air that is absolutely stable returns to its starting point when pushed from its initial position. In figure 14.10a air cools according to the dry lapse rate. As it follows the red line the parcel becomes colder than the environment and is, therefore, more dense. As a result, it sinks back to its starting position.

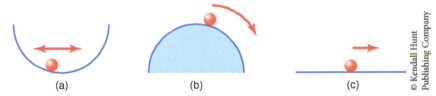

FIGURE 14.9. Three types of stability. (a) A stable object returns to its starting point when disturbed. (b) An unstable object accelerates away from it starting point. (c) A neutrally stable object doesn't return to its starting position, but it doesn't speed up either.

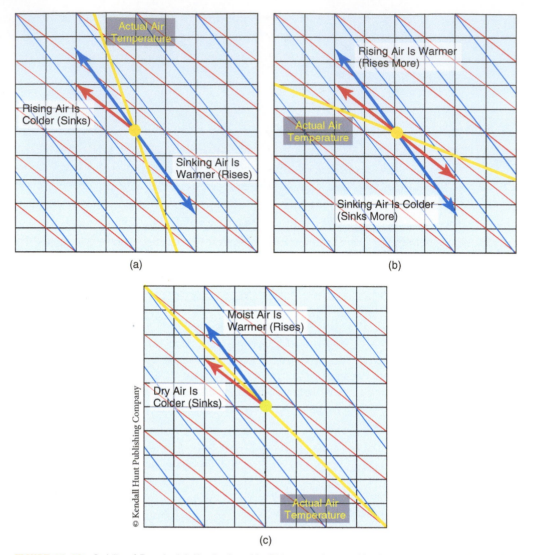

FIGURE 14.10. Stability of Parcels. (a) Absolutely stable. Rising parcels are colder than the environment and sink back to their starting position. (b) Absolutely unstable. Rising parcels are warmer than the environment and rise at a quicker rate. (c) Conditionally unstable. Rising dry parcels are stable and moist parcels are unstable. Therefore, stability depends on the humidity of the parcel.

Likewise, if it is pushed down, it becomes warmer than the environment and rises back up. If the air is moist, at 100% relative humidity, is follows the blue line and is also stable. Stability occurs after sunset when the surface begins to cool. By morning cold air lies close to the surface and temperature increases with height. This is called an **inversion**, which represents very stable air. This stable layer is disconnected from the winds higher in the atmosphere and the air is calm. This is the basis for the phrase,

"the winds go down with the sun." Cities located in valleys experience higher levels of pollution in stable air because pollutants are unable to leave the cold lower layer and mix with the rest of the atmosphere.

Figure 14.10b represents air that is absolutely unstable. Once again following either the red or blue lines for dry and moist air, you find that a rising parcel of air becomes warmer than the environment. As a result, the rising air is less dense and continues at an increasing rate of ascent. Likewise a sinking

CHAPTER 14: Phenomena and Processes of the Atmosphere

parcel becomes colder than the environment and sinks quickly. Instability leads to convection and this occurs on sunny days near the surface. If the lapse rate matches either the dry or moist lapse rate, the atmosphere is neutrally stable. But what happens between these two lapse rates?

An environmental lapse rate falling between 6 and 10 °C/km is conditionally unstable as represented by figure 14.10c. Using the diagram we see that a rising dry parcel follows the red line. The rising air is colder than the environment and is stable. However, if we have a moist parcel following the blue line, it is warmer than the environment and is unstable. As a result, the stability is conditional on whether the parcel is dry or moist.

At the beginning of the day the atmosphere is in a stable state. As the sun heats the ground, the surface air becomes unstable and sets up convection. Later in the morning the rising air reaches the lifting condensation level and a cloud begins to form. As the afternoon progresses, the cloud increases in depth. Once the cloud builds into a conditionally unstable layer, it grows explosively (figure 14.11). This growth only slows or stops when the cloud hits another stable layer of air, which is most likely the tropopause. At this point the cloud has grown into a thunderstorm. Meteorologists use atmospheric soundings to identify the conditionally unstable air and, if conditions are right, issue warnings for the development of severe weather.

14.2.3 Forced Instability

The atmosphere naturally wants to settle into a stable condition. As mentioned already, solar heating makes surface air absolutely unstable and persistent convection can push conditionally unstable air into instability. However, there are three other mechanisms which force air upward and potentially make it unstable. They are orographic forcing, fontal wedging and convergence.

Orographic forcing occurs when a predominant wind pattern pushes air over rising terrain. In the mid-latitudes the predominant wind direction is from the west. As this air encounters mountains, it is forced upward, often causing the top of the mountain to be obscured by a cloud (figure 14.12). If there is enough humidity in the air, rain falls on the windward side of the mountain. Having lost most of its water, the air passing over the mountain peak produces a drier climate on the leeward side. Also the leeward temperatures are higher because the descending air warms at the dry adiabatic lapse rate rather than the moist adiabatic rate, which occurs on the windward side.

FIGURE 14.11. Cumulus clouds build during the day due to solar heating at the surface. Once convection pushes the cloud to instability, it grows quickly in size.

FIGURE 14.12. As air is pushed over Mount Hood, a cloud forms and obscures the peak. Precipitation deposited on the windward side of the mountain leaves warm dry air to descend on the leeward side of the peak.

Frontal wedging occurs as two different types of air meet. When warm air encounters cold air, it moves upward because it is less dense. If the upward motion is strong enough, it makes the atmosphere unstable and generates severe weather. This is explained further in chapter 15. The last mechanism is **convergence**, where surface air comes in from all directions. This occurs in a low pressure system and forces surface air upward.

14.3 CLOUDS

Convection and other lifting mechanisms move energy and water vapor from the surface into the atmosphere. Stability of the atmosphere determines whether it remains close to the surface or mixes over the full depth of the troposphere. When lifted far enough, relative humidity reaches 100% and we expect water vapor to condense and form a cloud. However, for condensation to occur, enough water molecules need to interact to form a droplet. In very clean air this is unlikely to happen and relative humidities exceed 100%, a condition called **super-saturation**. However, if small particles, or aerosols, are present in the air, the water vapor easily condenses on the surface, keeping relative humidity at 100% or below. These particles are called **cloud condensation nuclei** (CCNs) and number in the hundreds of millions for every cubic meter of air. These are the dust motes you observe drifting slowly in a sun beam and comprise the dust that accumulates in your home.

If the number of CNNs is very high, it results in **haze**, a reduction in clarity when looking through the atmosphere. In dry conditions the small particles preferentially scatter blue light leaving the reddish brown hues associated with heavily polluted cities (figure 14.13a). When relative humidity is above 75%, a wet haze forms as water condenses on the surface of the CCNs. These droplets scatter blue and red light equally giving the haze a whiter appearance (figure 14.13b). If relative humidity is near 100%, the droplets grow even larger giving rise to fog.

14.3.1 Fog

Fog is a collection of small water droplets and ice crystals that reduce visibility below 1 km. Fog is

(a)   (b)

FIGURE 14.13. (a) A dry haze consists of small particles, which preferentially scatter blue light. As a result, looking through a dry haze there is a brownish orange hue. (b) With high humidity haze particles grow in size and scatter all wavelengths of light equally. Therefore, it is white in appearance.

essentially a cloud at ground level that forms due to local conditions. This occurs for various reasons and the fog type is linked to its cause. Radiation fog occurs on clear calm nights over moist surfaces (figure 14.14a). IR cooling reduces the surface temperature below the dew point and condensation occurs in the lower layer of air. This type of fog is common near river basins in the spring and fall seasons. After the sun rises, the surface air warms thus lowering the relative humidity. The fog begins to dissipate from the bottom upward giving the appearance that the fog is lifting from the ground.

Advection fog occurs when warm humid air moves over a cold surface. During the night land breezes move warm moist air over the cold California Current giving rise to fog banks (figure 14.14b). This also occurs with the passage of a warm front over snow pack during the winter. Upslope fog is the

(a)

(b)

(c)

FIGURE 14.14. (a) Radiation Fog. Clear nights with no wind allow the ground and surface air to cool quickly through the loss of IR radiation. If the ground is moist, a fog will form near the surface. (b) Advection Fog. Warm moist air moving over a cool surface results in a thick fog common in San Francisco. (c) Steam Fog. Moist warm air mixing with cool dry air can result in relative humidities greater than 100%. This type of fog is common over the hot springs in Yellowstone National Park.

result of condensation on the windward size of a mountain due to orographic forcing.

Steam fog or mixing fog is the result of a warm, moist parcel of air mixing with a dry, cool parcel. Each parcel by itself is clear because their relative humidities are below 100%. However, when they mix, the air becomes super-saturated and forms a fog. This occurs over boiling pans of water and hot springs (figure 14.14c). This effect also results in the formation of jet contrails.

14.3.2 Types of Clouds

Any reduction in visibility above Earth's surface due to water droplets or ice crystals is called a cloud. Cloud names are derived from Latin root words (figure 14.15) and allude to their appearance and altitude. Since water for clouds originates at the surface, the type of cloud is linked to weather conditions and can be used for simple forecasting.

High altitude clouds occur at temperatures colder than −40°C (−40°F) and are composed entirely of ice. They are thin in appearance and are often semi-transparent. Water vapor at this altitude originates from weather fronts and storms. High level winds move these clouds in advance of the front. Far ahead of the weather front are **cirrus** (Ci) clouds, or mare's tails which are sparse and thin in appearance (figure 14.16a). As the front approaches,

Latin	Meaning
Alto	High
Cirrus	Wisp of hair
Cumulus	Heap
Nimbo	Rain
Stratus	Layer

FIGURE 14.15. Scientific terminology often uses Latin because the meaning remain fixed unlike with modern languages. The following terms relate to cloud names.

cirrus clouds increase to form a thin layer called **cirrostratus** (Cs). If the cirrus lies next to a stable layer of air, a ripple pattern forms similar to waves on a lake (figure 14.16b).

Middle level clouds are also below freezing, but have a mix of super-cooled water droplets and ice. Named after a word meaning high, these clouds fall in a mid-range, similar to the use of the word *alto* in music. Layers of clouds at this altitude are thicker than cirrostratus and are called **altostratus** (As) These clouds give the sun a watery appearance (figure 14.16c) and prevent shadows from forming. **Altocumulus** (Ac) clouds consist of small clumps about the size of your thumbnail and have a gray appearance at their bases (figure 14.16d).

Low level clouds are above freezing and consist of water droplets. These clouds are thick enough to block the sun giving the sky a dark appearance. If the layer has no distinct features, it is called **stratus** (St). If precipitation is present, the name is changed to **nimbostratus** (Ns) (figure 14.16e). These clouds do not extend to a great vertical depth and often occur near warm and stationary fronts. Small tufts of ragged clouds below nimbostratus are called **scud**.

Clouds formed by convection have flat bases and rounded tops. Most common are fair weather **cumulus** (Cu) (figure 14.16f) that appear in the late morning and thicken as the day progresses. They have a white puffy appearance and dissipate when convection is cut off. If the atmosphere is conditionally unstable, cumulus clouds can quickly increase in height resulting in precipitation. If the cloud extends to the tropopause, the top will flatten out into a cirrus layer giving it an anvil appearance. These deeply convective clouds are called **cumulonimbus** (Cb) and generate lightning and heavy precipitation (See the figure at the beginning of the chapter).

(a) Cirrus (Ci)

(b) Cirrostratus (Cs)

(c) Altostratus (As)

(d) Altocumulus (Ac)

(e) Nimbostratus (Ns)

(f) Cumulus (Cu)

FIGURE 14.16. Cloud Types. Several of the most common cloud types are represented in these photos along with their designated abbreviation.

14.4 COLORS OF THE ATMOSPHERE

Before proceeding to precipitation it would be good to mention different optical effects occurring in the atmosphere. CCNs and aerosols scatter sunlight as it travels through the atmosphere. Given their small sizes they scatter blue light more effectively than red. As a result, the sky is seen by the blue scattered light (See the figure at the beginning of Chapter 13). During sunrises and sunsets light travels through the atmosphere at an angle allowing for more scattering. Therefore, the direct light of the sun is deficient of blue light resulting in red and orange hues (figure 14.17). Cloud droplets are much larger than CCNs and scatter all colors of light equally. As a result, the sides and tops of clouds appear white. The dark underside does not mean that the cloud's properties are different, but is due to less light traveling through the cloud.

When light scatters from small ice crystals and water droplets, an optical effect called **iridescence** occurs providing colorful pastels near the edges of thin clouds (figure 14.18). If large ice crystals are present, sunlight **refracts**, or changes its direction, as it travels through the ice crystal. Since the amount of refraction depends on the wavelength of light,

white light from the sun is separated into colors. Ice crystals have hexagonal shapes and refract light at an angle of 22°. This gives rise to a **halo**, which is red on the inside and blue on the outside. A full halo occurs when the sun is high in the sky and a thin layer of cirrostratus is present. As the sun approaches the horizon, only a portion of the halo phenomena is observed and is often called a **sundog**. Figure 14.19 illustrates a partial halo with pronounced sundogs on each side of the sun. If you are not sure you are seeing a sun dog, extend your arm and spread your hand. The angle subtended from your thumb to your little finger is about 20°, close to the angle of the halo.

A **rainbow** occurs when light travels through large drops of water. Through refraction and reflection, light changes it direction by 138° (figure 14.20). As a result, you observe a rainbow when the sun is behind you and the rain is in front of you. The angle between the anti-solar point, the location of your head's shadow, and the rainbow is 42°. Because reflection occurs within the raindrop, red appears on the outside and blue on the inside of the rainbow. For a double rainbow there is a second reflection and the order of colors is reversed (figure 14.21).

Courtesy Steve Gollmer

FIGURE 14.17. In the morning and evening sunlight travels a larger distance through the atmosphere. Small particles scatter most of the blue light leaving the reds and oranges.

Tonhom1009/Shutterstock.com

FIGURE 14.18. Thin clouds form pastel patterns as ice crystals diffract sunlight. This effect is called iridescence.

CHAPTER 14: Phenomena and Processes of the Atmosphere

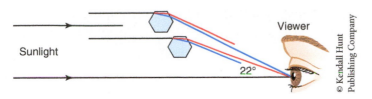

Sunlight

Viewer

22°

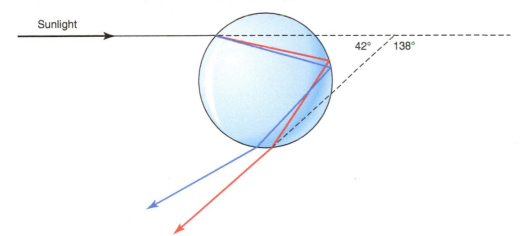

FIGURE 14.19. Light refracting through ice crystals forms a color band at 22°. Sun dogs are seen as bright color patches to either side of the sun. Halo phenomena are associated with the presence of cirrostratus clouds.

Sunlight

42°   138°

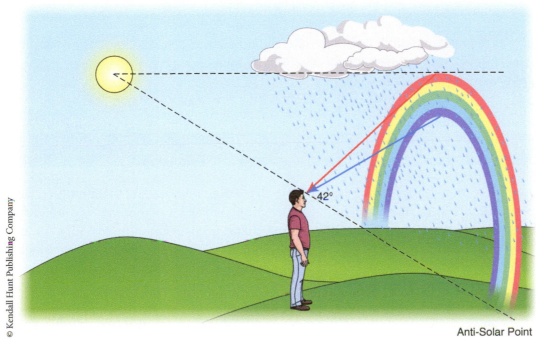

FIGURE 14.20.
Light traveling through a water droplet experiences both reflection and refraction resulting in color separation. Due to reflection within the droplet rainbows appear opposite of the sun.

42°

Anti-Solar Point

CHAPTER 14: Phenomena and Processes of the Atmosphere

FIGURE 14.21. In addition to the primary rainbow a secondary rainbow occurs at 53° due to a second reflection inside the raindrop. Since less light reaches the second reflection, this bow is always dimmer.

BOX 14.1

The rainbow is a topic of interest to creationists since it was given as a sign that God would never again destroy Earth with a global flood. What does this tell us about atmospheric conditions before the Flood? It is hard to imagine a world without rainfall and erosion, especially if sedimentary rocks are laid down between the creation and the Flood. If this is the case, then rainbows existed before the Flood, but gained an additional significance as God attached a promise to this unique phenomenon.

Genesis 2:5,6 states that before man was created "the Lord God had not caused it to rain on the land, and there was no man to work the ground, and a mist was going up from the land and was watering the whole face of the ground." This situation may have been restricted to Creation Week, since God's creation of man and the charge to cultivate the ground seems to resolve why there was no rain. It is clear that events of Creation Week and the pre-Flood world was different than what we experience today, but the question pondered by creation scientists is "how different?"

14.5 PRECIPITATION

Precipitation occurs when water droplets and ice crystals grow in size and fall to Earth. A typical cloud droplet is 0.02 mm in diameter and falls at a rate of 1 cm/s. Gentle updrafts from convection are sufficient to keep these droplets suspended in the air. As additional humidity rises from the surface, the cloud and the droplets within it grow in size. Droplets also compete with each other for the available water vapor. To sustain their size, smaller droplets need a higher vapor pressure to prevent them from evaporating (figure 14.22). Larger droplets grow at the expense of the smaller ones. Impurities in a water droplet reduce evaporation rates making it easier for the droplet to grow. Ice has a lower vapor pressure than water, so at temperatures below freezing ice crystals grow

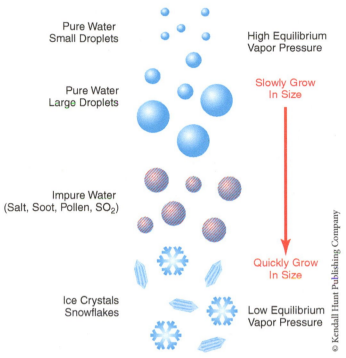

Pure Water
Small Droplets

High Equilibrium
Vapor Pressure

Pure Water
Large Droplets

Slowly Grow
In Size

Impure Water
(Salt, Soot, Pollen, SO$_2$)

Quickly Grow
In Size

Ice Crystals
Snowflakes

Low Equilibrium
Vapor Pressure

© Kendall Hunt Publishing Company

FIGURE 14.22. A higher humidity level is needed to maintain small droplets; therefore, large drops grow at the expense of the small ones. Also droplets with dissolved salts and ice crystals have the greatest advantage when competing over water vapor.

rapidly by drawing water vapor away from the liquid droplets.

14.5.1 Formation of Precipitation

As droplets grow in size, the rate at which they fall increases. Large droplets overtake the smaller ones and grow in size by merging with them. This process is called **coalescence**. By the time a droplet falls through 500 m of cloud it is 0.2 mm in size and falling at 0.7 m/s. If these droplets reach the surface, the precipitation is called **drizzle**. This type of precipitation occurs in thin clouds and is associated with warm fronts, which will be discussed in the next chapter. In thicker clouds droplets grow to larger sizes and are called **rain** when they reach the surface.

Middle level clouds are below freezing and contain ice crystals. Since ice crystals grow at the expense of their neighboring water droplets, ice crystals grow

to large sizes rapidly. Hexagonal patterns are generated as vapor deposits on the crystal and a snowflake forms. **Aggregation** is when multiple flakes stick together resulting in large clusters. If they collide with super-cooled water droplets, the water instantly freezes adding to the growing structure. If the air temperature remains below freezing, the flakes reach the ground and we call it **snow**. If temperatures near the surface are above freezing, the flakes begin to melt. They cling to form large aggregates and give us a damp heavy snow.

In tall clouds most precipitation begins as ice crystals. As snowflakes fall into warm air, they melt and become water droplets. These droplets are large and reach sizes of up to 5 mm. Figure 14.23 illustrates the relationship between air temperature and the types of precipitation. Rain and snow have already been discussed. **Freezing rain** occurs when a layer of sub-zero air lies at the surface. Rain hitting the surface instantly freezes forming a coating

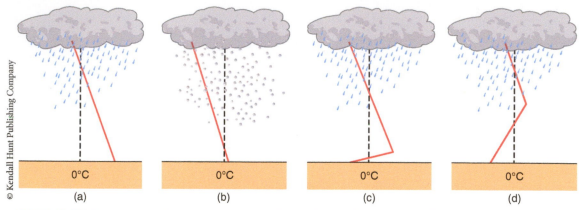

FIGURE 14.23. Temperature profiles and precipitation. (a) Rain (b) Snow (c) Freezing Rain (d) Sleet. Most precipitation starts as snow, but on its way to the surface it can melt and even refreeze. The red line represents how the air temperature changes with height. If it is to the right of the dashed line, it is above freezing.

of ice (figure 14.24). If the subzero surface layer is thick enough, rain refreezes before hitting the surface. These frozen raindrops are called **sleet** and they bounce when hitting the ground.

In a cumulonimbus cloud strongly rising air keeps precipitation suspended for long periods of time allowing it to grow to large sizes. As raindrops come in contact with large snowflakes, they freeze forming a pellet of graupel. The graupel continues to grow as it freezes additional water on its surface, a process called **riming**. If the pellet grows to the size of a pea, it is called **hail**. A strong updraft can

lift hail to the top of the cloud. Successive passes through the cloud allows hail to build up a series of growth rings (figure 14.25) and reach sizes larger than a softball.

14.5.2 Measuring Precipitation

Precipitation is measured as a depth of standing water resulting from a weather event. This amount

FIGURE 14.24. With the passage of warm fronts during the winter, warm upper air produces rain that freezes on contact with cold surfaces.

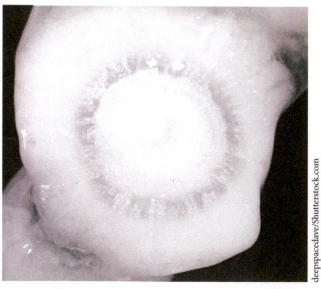

FIGURE 14.25. Growth rings indicate the number of times a hailstone cycles through the strong updraft of a thunderstorm.

CHAPTER 14: Phenomena and Processes of the Atmosphere

FIGURE 14.26. Rainfall can be quite variable; therefore, rain gauges with larger openings provide a more accurate measurement of precipitation.

Darryl Brooks/Shutterstock.com

is hard to determine because some of it soaks into the ground and the rest runs off into streams. A **rain gauge** is used to capture falling rain and provide a means of measuring the amount of standing water (figure 14.26). Rain gauges are accurate when measuring precipitation from stratus clouds since it is steady and uniform. However, cumulus clouds are smaller in size and rain falling from them is localized and sporadic. This type of rainfall is called **showers**. Precipitation from these clouds is a frustration to weather forecasters. Since there is no guarantee that everyone will get rain from a shower, the forecaster uses a probability for a chance of rain, such as 25%. However, you experience either 0% or 100% change of rain with nothing in between. This is one reason forecasters are given a bad rap.

It would appear that snowfall would be easier to measure; however, it presents its own challenges. If the ground is below freezing, snow accumulation is measured with a ruler and 10 cm of snow is equivalent to 1 cm of rain. However, 8 cm of a wet snow is closer to 1 cm of rain. Snowfall from cumulus clouds is sporadic and is called **flurries**. With the addition of wind, snow drifts form making it hard to find a representative depth of snow. A **blizzard** occurs when air temperatures are low and high winds suspend large amounts of snow particles. As a result of these challenges, the easiest way to measure snow is to capture it in a heated rain gauge and measure the depth of water.

Another method of measuring rainfall is through the use of **radar** (RAdio Detection And Ranging). The National Weather Service and a number of television broadcasters have radar systems that transmit electromagnetic waves into the atmosphere. These waves have a wavelength similar to your microwave oven and are scattered by water droplets and snow. The time it takes the wave to leave the antenna, reflect from the rain and return determines the distance to the precipitation (figure 14.27). The antenna is constantly turning so all directions are scanned. Where the rain is more intense, the signal scattered back to the antenna is stronger. Having been calibrated with surface rain gauges, radar can make estimates of rainfall rates and total precipitation. Snow doesn't reflect radar as strongly and a different calibration is used.

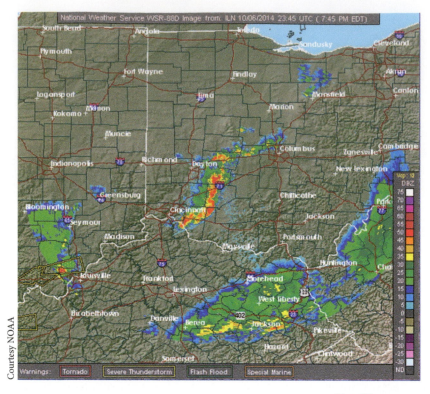

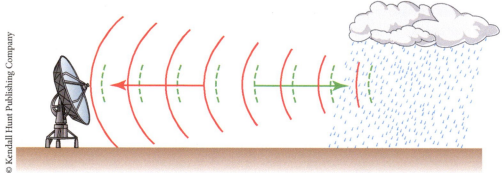

FIGURE 14.27. The strength of the reflected radar signal is related to the amount of precipitation falling through the atmosphere. Timing between the signal's transmission and reception allows a distance to be calculated.

14.6 WINDS

Up to this point we have made a connection between surface heating by the sun and convection. Humidity and stability of the air determines how clouds form and what kind of precipitation to expect. However, we have not explained how convection drives the horizontal motion of the atmosphere, which is called **advection**.

14.6.1 Thermal Circulation

To understand the connection between convection and advection let us use a simplified model illustrated in figure 14.28a. Assume the density of the atmosphere depends only on temperature and not pressure. If that

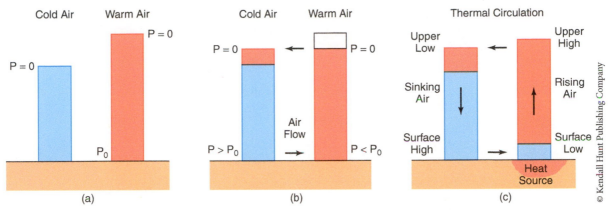

FIGURE 14.28. (a) Warm columns of air extend to greater heights than cold ones. (b) Far from the surface, warm air travels to the cold column thus increasing the surface pressure. As a result, cold surface air moves into the warm column causing it to rise. (c) As the cold air expands from surface heating the process continues.

were the case, the top of the atmosphere would be about 8 km high. Near the equator where the ground and air are warmer, the atmosphere is less dense. As a result, the top of the atmosphere is at a higher altitude. Conversely, at the poles which are much colder, the density is greater and the atmosphere is shallower. If these columns of air are placed side by side, we easily see the height difference. Since we assume that each column has the same amount of air, the surface pressure is the same. However, as you move away from the surface, the pressure difference between the columns increases.

High in the atmosphere the pressure in the cold column is lower than that in the warm. As a result, warm air flows to the cold column until 0 mb pressure occurs at the same altitude (figure 14.28b). However, by correcting the pressure difference at the top of the atmosphere, there is now more air in the cold column and the surface pressure is higher. Since we know that air flows from high pressure to low, we get advection moving surface air into the warm column. If no heating occurs at the surface, the two columns mix, thus removing the temperature difference and the circulation ceases. However, if heat is added to the base of the warm column, the newly advected cold air expands and the warm column increases in height (figure 14.28c) thus perpetuating the cycle. This motion driven by surface heating is called a **thermal circulation**.

At this point it is best to revisit the term *convergence*. As mentioned before, convergence is where air is approaching a location from all directions resulting in vertical motion. Although there are several causes for it, convergence naturally occurs at the surface of a warm column. Convection draws warm air upward leaving a surface low pressure behind. Surrounding cold air converges on this location, thus replacing the air that moved upward. Convergence also occurs high in the cold column. Sinking air leaves behind a low pressure which is filled by converging air from the warm column. The opposite of convergence is **divergence**. This occurs at the surface of the cold column and high in the atmosphere for the warm column (figure 14.29).

It is important to connect vertical motion of the thermal circulation to the stability of the atmosphere. Rising warm air moves humidity away from the surface, which eventually condenses to from clouds. As the air continues to rise, it becomes less stable and potentially generates precipitation and thunderstorms. This is the weather we associate with surface low pressure systems. Diverging air high in the warm column converges on the cold column and begins to sink. As it does, it warms and becomes more stable. This is called **subsidence**. Clouds dissipate and the possibility of precipitation is very low. Therefore, clear skies and fair weather is associated with the high surface pressure of the cold column of air.

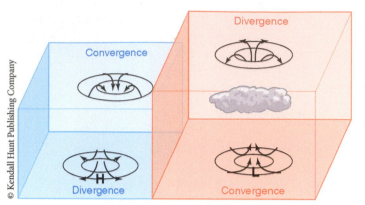

FIGURE 14.29. Convergence occurs when a surface low draws in the surrounding air, which then rises to generate clouds and precipitation. High in the atmosphere the rising air diverges or spreads out. The opposite occurs in a cold column of air where the air converges high in the atmosphere and diverges near the surface.

The effects of a thermal circulation are regularly observed near the ocean. During the day sunlight warms both the ocean and the land (figure 14.30a). Since water has a large heat capacity, its temperature increases at a slower rate than the land. As a result, a warm column of air forms over land and a cold column over the ocean. High pressure develops over the ocean and air flows to the lower pressure located inland. This air flow is called a sea **breeze**. Clouds build over land due to convection and can lead to afternoon thunderstorms. The sinking air over the ocean dissipates clouds giving clear sunny skies at the beach. During the night this thermal circulation reverses. The land cools faster than the ocean and the warm column now exists offshore. A land breeze blows out to the ocean and convection builds

clouds offshore (figure 14.30b). Just a note: winds are named by the point of origin, not their destination. Therefore, a north winds blows from north to south.

Another common thermal circulation is aided by geography. During the day solar heating on a mountain side causes air to rise vertically along the slope (figure 14.31a). This upslope wind is called a valley breeze. If the rising air is humid enough, a cloud forms on the side of the mountain and is often called upslope fog. To complete the circulation pattern, high above the valley the air sinks leading to clear skies. At night the circulation pattern reverses as IR loss causes air next to the mountain to cool and become denser (figure 14.31b). This sinking air results in a mountain breeze. As the valley fills with cool air, the warm air is pushed upward increasing the chances of cloud cover.

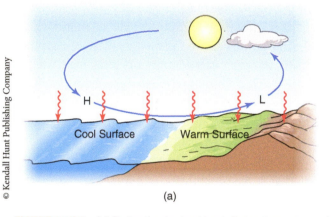

(a)

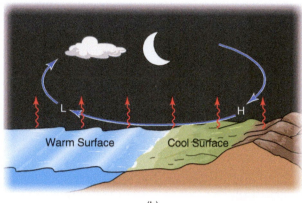

(b)

FIGURE 14.30. (a) During the day land heats faster than water and generates the rising portion of the thermal circulation. As a result, air blows in from the sea. (b) During the night the thermal circulation reverses as the land cools quicker. At night the land breeze blows out to sea.

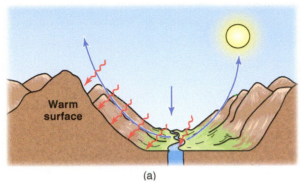
(a)

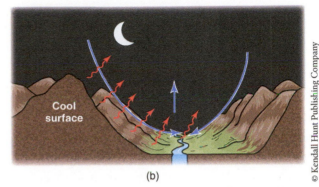
(b)

FIGURE 14.31. (a) Heating during the day allows warm air to rise along the slope of the mountain forming valley breezes. (b) During the night the circulation reverses forming a mountain breeze.

14.6.2 Forces Affecting Winds

Isaac Newton stated that an object at rest remains at rest and an object in motion remains in constant straight line motion unless acted on by unbalanced forces. The previous section made the connection between air temperature and differences in air pressure. It is these differences that lead to unbalanced forces and, therefore, generate wind. In addition to pressure difference, the Coriolis Effect and friction also introduce forces that affect the flow of wind.

Pressure has a greater impact on the wind when the difference occurs over a short distance. The ratio of pressure difference to horizontal distance is called the **pressure gradient force**. Figure 14.32

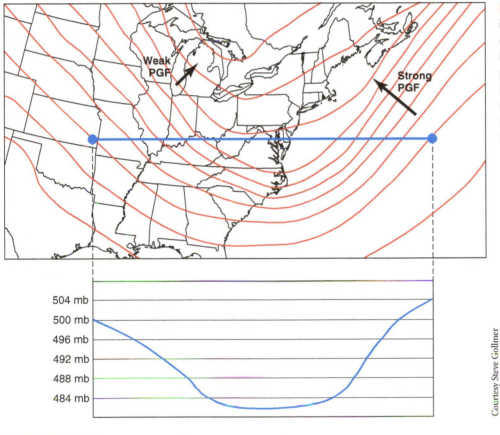

Pressure 5700 m Above Sea Level

FIGURE 14.32. Difference in pressure at any altitude exerts a force on air causing it to move from high to low pressure. Other forces may influence the actual motion of the air; therefore it may not flow perpendicular to the isobars.

Courtesy Steve Gollmer

is a contour plot of pressure over the eastern United States. The solid lines running nearly parallel to each other are **isobars** or lines of constant pressure. The pressure gradient force goes from high to low pressure and, therefore, is perpendicular to the isobars. Where the isobars are closest together, the pressure gradient force is greatest. If you think of the pressure map like an elevation map, high pressure is at the top of the hill and low pressure is in the valley. If the slope of the hill is steep, the elevation changes over a short distance and the force propelling you downhill is large. If this were the only force acting on the air, wind would always travel from high pressure to low pressure.

A second force acting on the air is the **Coriolis Effect**. In reality it is not a force, but an effect due to the rotation of Earth. To illustrate this we position ourselves at the North Pole. When a cannon ball is fired to the south (figure 14.33a), it travels in straight line motion as it flies through the air. This is the motion we would observe if we were looking down from outer space (figure 14.33b). However, we also notice that Earth is rotating on its axis in a counter-clockwise direction. If we place ourselves on Earth, we rotate with Earth. As a result, it appears that the cannon ball is veering to the right of its original path (figure 14.33c). Therefore, we conclude there is a force pushing the cannon ball to the right. This veering effect occurs no matter the location or the direction of the cannon ball's flight. The

only place the Coriolis Effect is not present is the equator. If we travel into the Southern Hemisphere, the effect is reversed and we see the ball veer to the left. Normally the Coriolis Effect is not noticeable because it shifts the landing point of a homerun baseball by only a few centimeters.

The last force is **friction**. Friction opposes motion and attempts to bring everything to a stop. Friction is proportional to the speed of the object so it is greatest for fast moving objects. When an object falls, it speeds up until the frictional force is equal and opposite to the force of gravity. This is called **terminal velocity** and affects the fall rate of raindrops and snowflakes. Air moving near Earth's surface experiences friction due to the roughness of the terrain. Water surfaces are relatively smooth allowing winds to increase to greater speeds. The presence of bushes and trees disrupts the air flow and slows the wind down. Away from the surface, the terrain has less impact and the wind speed increases. As mentioned in chapter 13, the gustiness of wind during the summer is the result of convection moving slow surface air upward and drawing fast winds towards the surface. Since convection moves slow air upward, it transmits the frictional effect of the surface higher into the atmosphere. However, once an altitude of several kilometers is reached, the frictional effect of the surface is negligible.

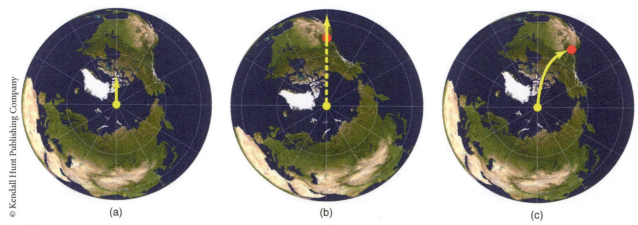

(a) (b) (c)

FIGURE 14.33. Rotation of Earth makes it appear that an object in straight-line motion is actually veering to the right in the Northern Hemisphere. This is called the Coriolis Effect.

14.6.3 Large Scale Winds

It is now possible to describe the behavior of the wind at the surface and high in the atmosphere. Of the three forces introduced above, only the pressure gradient force increases the speed of the wind. The Coriolis Effect changes its direction and friction slows it down. At the surface friction plays a significant role and wind speeds are relatively small. Sea and land breezes from thermal circulations reach speeds of 32 kph (20 mph) and can kick up dust and move small branches. At these speeds the Coriolis Effect is negligible. Higher in the atmosphere the wind speed increases as well as the impact of the Coriolis Effect. As illustrated in figure 14.34, the wind no longer runs perpendicular to the isobars,

but is turned slightly to the right. All three forces act in different directions, but they add up to zero. At even greater altitudes friction becomes less important and the pressure gradient force and the Coriolis Effect are nearly equal and opposite in direction. This occurs above the 700 mb level. If there is no friction and the isobars are straight, the winds blow parallel to the isobars and are called **geostrophic winds**.

Air flow parallel to curved isobars is called **gradient winds** and indicates an imbalance of forces. By Newton's first law air wants to move in a straight line. However, if it is rotating in a circle, there is a net force in the direction of the circle's center. Figure 14.35a illustrates the air flow around a large

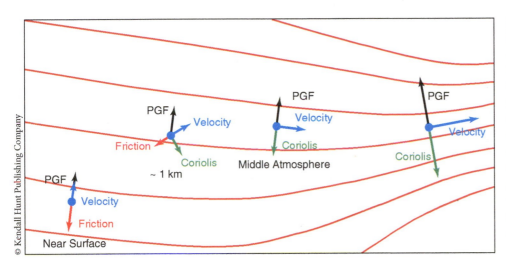

FIGURE 14.34. Geostrophic Winds. Wind speeds close to the ground are perpendicular to the isobars and are slowed by friction. Moving away from the surface, friction becomes less important and the winds veer to the right. High in the atmosphere the pressure gradient force balances the Coriolis Effect and winds move parallel to the isobars. Tighter spaced lines result in a greater pressure gradient force and stronger winds.

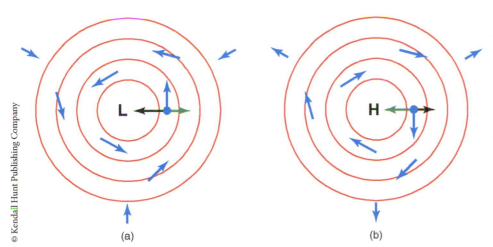

FIGURE 14.35. Gradient Winds. (a) In low pressure systems the pressure gradient force is greater than the Coriolis effect and the winds circulate counter-clockwise in the Northern Hemisphere. (b) In high pressure systems the Coriolis Effect is greater and winds circulate clockwise.

scale **low pressure system**. If the air begins at rest, it initially moves in the direction of the pressure gradient force. As the air speeds up, the Coriolis Effect veers it to the right. As a result, the air flows counter-clockwise around low pressure systems in the Northern Hemisphere. The pressure gradient force is greater than the Coriolis Effect and, therefore, the air flows in a circle. This counter-clockwise circulation is also called a **cyclone**. With a **high pressure system**, figure 14.35b, air flows away from the center of the circle. It veers to the right due to the Coriolis Effect and rotates clockwise. In this case the Coriolis Effect is greater than the pressure gradient force. This circulation, called an **anti-cyclone**, has winds rotating clockwise around a high pressure system. Near Earth's surface friction is greater and the wind speed is reduced. As a result the Coriolis Effect is weaker and the air spirals towards the center of the low pressure system and spirals outward from a high pressure system.

One final comment: Isobars are only used on surface maps representing weather data. Any data collected higher in the atmosphere is plotted on a constant pressure map and isobars become irrelevant. Instead of plotting isobars, constant height contours are used (figure 14.36). The 500 mb level in warm air is higher than in cold air because it is less dense. Therefore, greater height contours are located to the south. Similar to isobars, the pressure gradient force is perpendicular to the height contours and strongest when the contours are closest together. Combined with the Coriolis Effect, the air flow is geostrophic where the contours are straight. The geostrophic flow can develop instabilities causing warm air to extend further north than usual. This is called a **ridge**. Similarly when a cold column of air is pulled further south, it is called a **trough**. During the winter the ridge/trough pattern can become extreme resulting in warm weather in Alaska and freeze warnings in Florida.

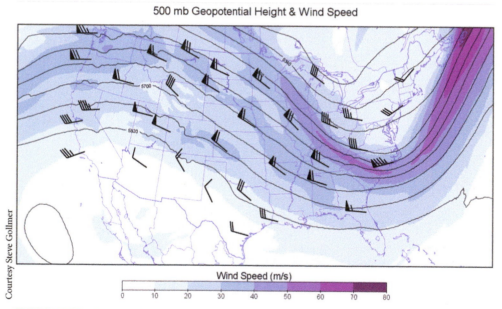

500 mb Geopotential Height & Wind Speed

Wind Speed (m/s)

0   10   20   30   40   50   60   70   80

Courtesy Steve Gollmer

FIGURE 14.36. Wind speed and direction in the middle of the atmosphere are plotted along with altitude at which the 500 mb pressure occurs. The contour lines represent constant heights that range between 5280 and 5880 m in 60 m increments. A ridge of warm air is located to the west and the east coast experiences a trough of cold air.

14.1 Water Vapor and Humidity

- Water molecules are constantly leaving and entering the surface of liquid water. If more leave than enter, then evaporation occurs. Condensation happens when more enter than leave.
- Dew point, vapor pressure, and relative humidity are different quantities used to keep track of the water vapor in the air.

14.2 Rising Air

- An adiabatic expansion of gas does not exchange energy with its surroundings and results in greater volume, lower pressure, and lower temperature.
- Rising air cools at a rate of 10 °C/km if it is dry. If it is moist (at 100% relative humidity), water condenses to form a cloud and it cools at 6 °C/km.
- If rising air becomes cooler than its surroundings, it will sink back down and remain stable. However, if it becomes warmer than its surroundings it is unstable and will rise at a more rapid rate.
- Warm air naturally rises through convection, but any air can be forced upward through orographic forcing, frontal wedging, and convergence.

14.3 Clouds

- Fog is a cloud at ground level and is named by how it forms: radiation, advection, upslope, and steam fog.
- All types of cirrus are high level clouds that consist of ice crystals.
- Low level clouds consist of water droplets and are more effective at blocking sunlight.
- Clouds that grow vertically in size are due to convection and have rounded tops giving them the name cumulus.

14.4 Colors of the Atmosphere

- Refraction of sunlight through ice crystals forms color bands call halos and sun dogs.
- Through refraction and reflection, water droplets form rainbows that are opposite of the sun relative to the observer's position.

14.5 Precipitation

- As water droplets and ice crystals fall through a cloud, they grow in size until they are large enough to reach Earth's surface. When this occurs, we call it precipitation.
- Most precipitation begins as snow high in the atmosphere. If the snow falls through a warm layer of air, it melts to form rain.

- Precipitation from cumulus clouds is sporadic resulting to rain showers. Location and intensity of showers can be determined using radar.

14.6 Winds

- Thermal circulation consists of rising air over a warm surface and sinking air over a cold surface. If the rising air is humid, it will generate clouds and precipitation.
- Sea/land breezes and valley/mountain breezes are thermal circulations that occur over 10's of kilometers.
- The pressure gradient force generates wind, but the direction of the wind is affected by friction and the Coriolis Effect.
- Geostrophic winds occur where there is little friction and a balance between the pressure gradient force and the Coriolis Effect. These winds blow parallel to isobars.
- Low pressure systems, also called cyclones, have converging air that rotates counter-clockwise around the low. Anticyclones are high pressure systems that have diverging air that rotates clockwise.

KEY TERMS

Adiabatic lapse rate	Environmental lapse rate
Adiabatic process	Evaporation
Advection	Flurries
Aggregation	Fog
Altocumulus	Freezing rain
Altostratus	Friction
Blizzard	Frontal Wedging
Breezes	Geostrophic Winds
Cirrostratus	Gradient Winds
Cirrus	Hail
Cloud condensation nuclei	Halo
Coalescence	Haze
Condensation	High pressure system, anti-cyclone
Convergence	Hygrometer
Coriolis Effect	Inversion
Cumulonimbus	Iridescence
Cumulus	Isobar
Deposition	Lifting condensation level
Dew point depression	Low pressure system, cyclone
Dew point temperature	Nimbostratus
Divergence	Orographic forcing
Drizzle	Parcel of air

Precipitation	Showers
Pressure gradient force	Sleet
Psychrometer	Snow
Radar	Sounding
Radiosonde	Stratus
Rain	Sublimation
Rainbow	Subsidence
Rain gauge	Sundog
Refraction	Super-saturation
Relative humidity	Stability
Ridge	Terminal velocity
Riming	Thermal circulation
Saturation	Trough
Scud	Vapor pressure

REVIEW QUESTIONS

1. As the temperature of water increases, how does the evaporation from a water surface change? How does this affect the amount of water vapor above the surface?
2. What happens to the relative humidity as the dew point temperature approaches the air temperature? Give examples of where and when you would expect low and high relative humidities.
3. How do the pressure, volume, and temperature of air change as it rises from the surface? How quickly does the air temperature change as it rises?
4. How is the actual temperature of air above the surface measured? What criterion is used to determine if a layer of air is stable or unstable?
5. What is the LCL and how is its location determined?
6. Describe how the temperature and relative humidity of air behaves as it is pushed up the side of a mountain and then descends on the other side.
7. What is needed to form water droplets in the sky and what factors determine how quickly they grow in size?
8. What is the difference between haze, fog, and a cloud?
9. What are the features that make it possible to distinguish between the different cloud types?
10. How is a halo different than a rainbow?
11. What are the basic types of precipitation and how does the temperature of the atmosphere affect which type of precipitation is experienced?
12. Draw a simple diagram of thermal circulation. Use it to explain how sea/land breezes and valley/mountain breezes form.

13. What is the pressure gradient force and how is it related to isobars on a weather map?

14. Pressure gradient force, Coriolis Effect and friction are all vectors: they have magnitude and direction. How does the magnitude and direction of these three relate to the velocity of the wind?

15. When looking at a weather map, how do the winds behave around a low pressure system? Around a high pressure system? Are there other features that would make it easy to identify these pressure systems?

FURTHER READING

Weather Charts at Unisys Weather – weather.unisys.com
Sounding Data at the University of Wyoming – weather.uwyo.edu/upperair/sounding.html
Cloud Appreciation Society – cloudappreciationsociety.org

We often give Gideon a hard time for his lack of faith after testing the fleece the first time. After reading Judges 6:36—40 and considering material from this chapter, how might Gideon have interpreted the fleece as a natural phenomenon rather than a miraculous sign from God? (By the way, Gideon is listed as an example of faith in Hebrews 11:32.)

CHAPTER 15

WEATHER PATTERNS AND FORECASTING

OUTLINE:

Courtesy of NASA

This image was taken from the International Space Station while Isabel was still a category 3 hurricane. It eventually made landfall on the Outer Banks of North Carolina and delivered significant damage to parts of Virginia.

In the previous two chapters we have looked at the general properties of the atmosphere and how energy from the sun is reflected, absorbed, and transferred between land, sea, and air. Convection leads to cloud formation and precipitation. Thermal wind patterns are driven by differential heating and the strength and direction of winds depend on the pressure gradient force, friction, and the Coriolis Effect. Applying this knowledge from the smallest to largest scale gives insight into the complexity of weather and an understanding of climate experienced around the world. The goal of this chapter is to do just that. In addition some weather forecasting products are introduced to remove some of the mystery behind the daily weather report.

15.1 MICROMETEOROLOGY

At the smallest scale **micrometeorology** includes phenomena lasting several minutes and extending to a kilometer in size. The layer of the atmosphere closest to Earth's surface is influenced greatly by friction and included in this sub-discipline of meteorology. The friction effect is transmitted higher in the atmosphere through swirling motion called **eddies**. Eddies are generated by sources of heat (figure 15.1) and by obstructions that redirect the wind like water swirling around a rock. On a hot summer day these eddies appear as a rippling effect over hot surfaces. At the edge of clouds eddies mix small cloud droplets with surrounding dry air in a process called **entrainment**. The droplets evaporate resulting in cool air that can form a downdraft. Entrainment is seen in figure 15.2 as

FIGURE 15.1. Smoke rising from wicks of incense illustrates the turbulent motion of air called eddies.

FIGURE 15.2. As humid air from a smoke stack mixes with cold dry air, water droplets evaporate dissipating the cloud. Pollutants coming from the stack are diluted through mixing with the surrounding air.

exhaust from a smokestack mixes with surrounding air.

As winds interact with the surface, they suspend small particles. With increased wind speed, the amount of particulate matter carried by the wind increases leading to the blinding conditions of dust storms and blizzards. When encountering an obstruction, winds slow down and deposit their suspended material. This leads to dunes and drifts as described in chapter 10. In order to control where drifts form, snow fences (figure 15.3) are erected near roads to keep them passible in winter months.

FIGURE 15.3. As winds slow down, suspended snow particles are deposited leading to drifts. By strategically positioning fences, drifts form away from the road.

15.2 GLOBAL CIRCULATION

At the largest scale, friction plays a minimal role, and global circulation patterns are mainly influenced by solar heating and the distribution of land masses. In chapter 13 it was pointed out that there is an imbalance of solar heating and IR cooling at the equator and the poles. Between 37 °N and 37 °S latitude there is net heating, while at higher latitudes IR cooling dominates. The excess heat at the equator is transported poleward by circulation of the atmosphere and the ocean. Ocean currents, as covered in chapter 12, transport warm water northward along the east coast of continents and cold water southward along the west coast. Since the atmosphere is not constrained to an ocean basin, it sets up a different pattern.

15.2.1 Circulation Cells

An ideal global thermal circulation pattern consists of a warm column of rising air over the equator and a cold column of sinking air over the poles. This one cell model (figure 15.4a) has cold air from the poles traveling towards the equator at Earth's surface and a return circulation of warm air high in the atmosphere. However, Earth's rotation generates a

Coriolis Effect that veers the air traveling south to the west and the equatorial air high in the atmosphere to the east. Since the distance between the poles is large compared to the thickness of the atmosphere, this circulation pattern is unstable. As a result, this thermal circulation breaks up into three cells (figure 15.4b). The two ends of the original circulation are the same and form what are called the **Hadley cell** and the **polar cell**. However, air flow between the two is reversed like a gear embedded between two others. The middle circulation pattern, called the **Ferrel cell**, drives surface air northward and air aloft southward. If Earth's rotation were greater, its atmosphere would break into a larger number of cells giving the banded structure observed on Jupiter.

The equator is located at the rising portion of the thermal circulation. Therefore, it has a low surface pressure causing air to converge from both the north and south. This region is called the Inter-Tropical Convergence Zone (ITCZ) and is easily identified from satellite images by a band of clouds near the equator (figure 15.5). The monotonous weather and weak surface winds at the ITCZ is why it is also called the **doldrums**. North of the equator surface air veers to the right resulting in easterly winds.

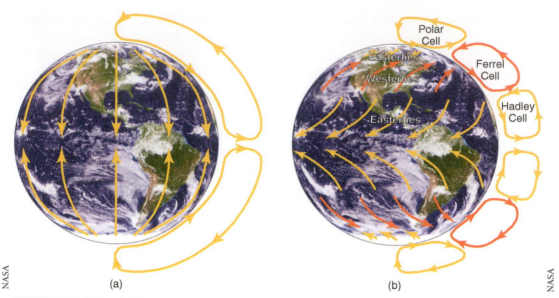

NASA

(a)

NASA

(b)

FIGURE 15.4. (a) One Cell Model — If Earth did not rotate, a large thermal circulation would be established between the equator and the pole. All surface winds would blow towards the equator. (b) Three Cell Model — The Coriolis Effect and friction cause Earth to set up a three cell circulation. This results in mid-latitude westerlies and easterlies near the equator and poles.

Courtesy of NOAA

FIGURE 15.5. In this GOES satellite IR image a band of clouds encompasses Earth just north of the equator. This is the rising portion of the Hadley cell.

These easterlies are called **trade winds** and were used by sailing vessels to travel from Europe to the New World.

At 30° latitude the Hadley and Ferrel cells meet. Air diverges from this location resulting in weak winds. Sub-tropical highs are located at the surface and sinking air leads to clear skies and little precipitation. Many arid regions correspond to this latitude including the Sahara, the Middle East, and the Australian desert. For mariners this area was called the horse latitudes and is near the Tropics of Cancer and Capricorn. Poleward of these latitudes the air veers to the east giving rise to the **westerlies**. The United States resides mostly in the Ferrel cell resulting in prevailing winds from the southwest. Sailing vessels would take advantage of these winds when returning to Europe from the Americas.

While the doldrums and horse latitudes have predictable weather, the interface between the Ferrel and Polar cells is more interesting. Convergence near 60 °N latitude forces air to rise resulting in a surface low pressure. The rising moist air becomes unstable and can generate large amounts of precipitation. There is a temperature contrast between the warm air to the south and the cold air to the north of this interface. The clash between these different air masses results in the **polar front**. Near the tropopause wind along the polar front reaches high speeds and is called the **jet stream**.

There is also a clash of westerly and easterly winds at the polar front. This sudden shift in wind direction leads to instabilities that generate patterns of ridges and troughs. The connection between instabilities in the polar front and the formation of low pressure systems was made by Norwegian meteorologists after World War I. This breakthrough greatly improved weather forecasts and is central to our understanding of **synoptic meteorology**, which spans scales of days and hundreds of miles.

15.2.2 Global Pressure and Wind Patterns

Circulation from the three-cell model is modified by the thermal circulation between oceans and continents. In the tropics continents are warmer than oceans. This weakens the sub-tropical high over land and strengthens it over the ocean. This leads to two persistent sub-tropical high pressure regions called the Pacific and Bermuda Highs (figure 15.6a). In Polar Regions the continents are colder than the ocean. Combined with the low along the polar front, the Aleutian Low and Icelandic Low become very pronounced during the winter. Given the size of the Asian continent a large high pressure system, called the **Siberian High**, establishes itself during the winter. At times this high pushes air over the North Pole thereby influencing weather in Canada and the United States. During the summer the continents warm, the Siberian High disappears and the Pacific and Bermuda Highs expand.

Figure 15.6a illustrates the circulation of the atmosphere using arrows to represent wind direction. Notice the circulation pattern is a superposition of the three-cell model and the thermal circulation between the land and ocean. The equatorial and polar easterlies are present along with the mid-latitude westerlies. However, the patterns are no longer symmetric, but include counter-clockwise rotation around the lows and clockwise rotation around the highs. Global

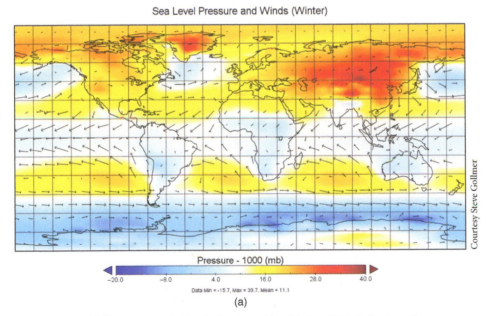

Sea Level Pressure and Winds (Winter)

Courtesy Steve Gollmer

Pressure - 1000 (mb)

−20.0   −8.0   4.0   16.0   28.0   40.0

Data Min = -15.7, Max = 39.7, Mean = 11.1

(a)

FIGURE 15.6. (a) Winter pattern showing the locations of low (blue) and high (yellow to red) pressure. The easterlies and westerlies are clearly seen as well as the rotational motion around highs and lows (compare to Figure 15.4b). (b) The summer pattern shows a disappearance of the Siberian high and an expansion of the highs over the ocean.

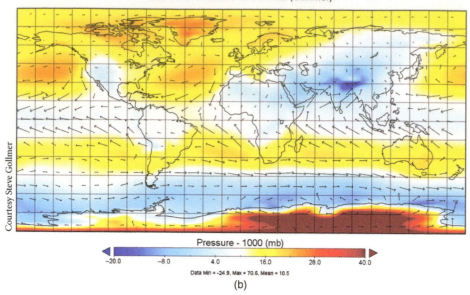

Pressure - 1000 (mb)

-20.0   -8.0   4.0   16.0   28.0   40.0

Data Min = -24.9, Max = 70.6, Mean = 10.5

(b)

Courtesy Steve Gollmer

FIGURE 15.6. *(Continued)*

circulation in the Southern Hemisphere is less complex since there is less land mass present.

The circulation pattern also shifts between seasons. Figure 15.6b shows the typical pattern during the summer. The ITCZ tracks the location of strongest solar heating and moves north. During the winter the Siberian high pushes air over the Himalayan Mountains resulting in India's dry season. However, during the summer a continental low forms and reverses the air flow. Air is now drawn from the Indian Ocean resulting in a rainy season over India as well as the rest of Southeast Asia. This seasonal shift in wind and rain patterns is called the **monsoons**.

15.2.3 El Niño

Global circulation patterns are also affected by changes in the temperature of the ocean. **El Niño** is a quasi-periodic shift of warm ocean water in the equatorial Pacific. The term quasi-periodic is used because the appearance of El Niño is not yearly like the monsoons, but occurs at intervals from two to seven years in length (figure 15.7). The effects of El Niño are more pronounced during the Peruvian winter leading to its name, which is a Spanish reference to the birth of the Christ child.

Normally the tropical Pacific Ocean has a pool of warm water near Australia and a cooler pool off the coast of Peru (figure 15.8a). This leads to a thermal circulation where warm rising air provides precipitation to Northern Australia and Southeast Asia and cool sinking air keeps Peru dry. The equatorial easterlies are strong because the surface pressure near Australia is lower than that near Peru. This pressure difference is one means of identifying El Niño and is called the **Southern Oscillation Index** (SOI). Combining these two terms leads to the acronym ENSO, which is used commonly in the meteorology literature.

During an El Niño event, the pool of warm water spreads east across the Pacific (figure 15.8b). Low pressure shifts toward the central Pacific, thus weakening the easterlies. With weaker easterlies it is easier for warm water to spread even further east and along with it precipitation. Peru experiences flooding and Peru's fishing industry is impacted as

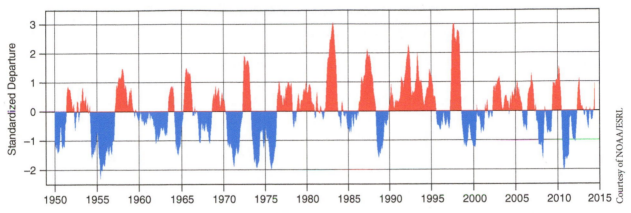

FIGURE 15.7. El Niño occurs when the index is positive and La Niña when it is negative. Notice the quasi-periodic nature of the pattern, which is due to the feedback mechanisms between the atmosphere and the ocean.

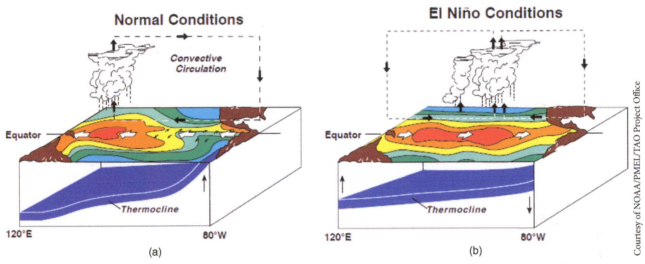

FIGURE 15.8. (a) Normally a pool of warm water occupies the western Pacific. This sets up a thermal circulation that keeps the skies over Peru clear and dry. (b) During El Niño the pool of warm water spreads across the Pacific. This change in thermal circulation shifts ocean currents and increases precipitation in Peru.

nutrient-rich cool water is cut off. The opposite of El Niño is called **La Niña** and is associated with wetter conditions in the western Pacific and cool and drier conditions in Peru.

The effects of El Niño are not limited to the tropical Pacific, but extend across the globe. Changing the position and strength of the low pressure system near Australia affects the Pacific High and the Aleutian Low. These in turn influence continental pressure systems as well as those in the Atlantic Ocean. During winter in the United States, El Niño is associated with higher levels of precipitation in California. Alaska's winters are milder while the Midwest and Southeast experience wetter and colder weather. It is also observed that hurricane activity is reduced. During La Niña Alaska experiences colder winters and the Midwest experiences warmer and drier winters.

Changes in temperature and precipitation patterns are due to the motion of large bodies of air. A large body of air that has similar properties and origination point are called **air masses**. If the air originates in Polar Regions it is cold while air from the tropics is warm. Also air residing over the ocean has higher humidity than air over the continent. The combinations of these two properties lead to four types of air masses.

Figure 15.9 illustrates the location of each air mass and the regions of the United States affected by it. Maritime tropical (mT) is warm humid air that has a profound impact on the southeastern United States. Air flow from the Gulf of Mexico keeps the southern states warm year round and provides plenty of rain. During the spring and summer, southerly winds transport maritime tropical air into the Midwest and along the Atlantic coast. Because of the high humidity, afternoons are hot and muggy and lead to the formation of thunderstorms. The severity of winter weather is moderated by this air mass. Thus this region belongs to climate group C as discussed in chapter 13.

Continental tropical (cT) air originates in the southwestern United States and Mexico and is hot and dry. Since the westerlies drive moist air from the Pacific ocean over several mountain ranges, the air becomes dry and warm due to orographic lifting. This arid region has a climate classification of group B and extends its influence into the Great Plains. When weather patterns allow the westerlies to dominate, influence of this air mass extends to the Appalachian Mountains leading to drought conditions. Continental and maritime tropical air clashes along the **dry line**, which extends from Texas to Kansas. During the spring this location provides excellent conditions for the formation of thunderstorms and tornados.

Continental polar (cP) air dominates the high latitudes. The polar climate zones of tundra and ice cap, group E, have extremely cold temperatures and dry air. This air does not remain over the poles, but extends southward over Canada and Eurasia. Since Canada and the Northern United States experience cold winters due to this air mass, these regions are given the climate designation of group D. During the summer this air mass provides relief to the Midwest

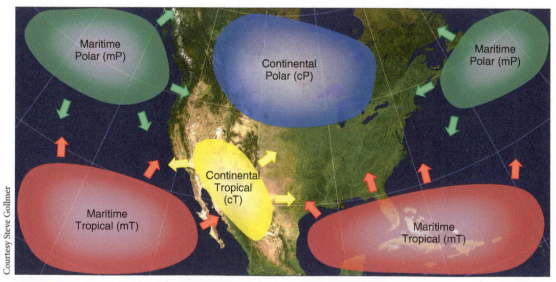

FIGURE 15.9. The clash of air masses across the continental United States gives different regions their characteristic weather and determines their climate.

by bringing cooler temperatures and less humidity. However, during the winter, outbreaks of continental polar air can push as far south as the Gulf of Mexico causing freeze damage to orange crops in Florida. This air mass clashes with maritime tropical air at the polar front and generates most of the precipitation for the Midwest.

Maritime polar (mP) air influences Alaska and the coastal regions of the Pacific Northwest. Since this air originates over the North Pacific Ocean, it is cool and humid. The ocean temperature does not change much during the year and this allows cities near the Alaska coast to enjoy 10 °C (50 °F) air during the winter. This effect keeps the temperatures in Seattle, WA moderate in even the coldest winters. Likewise during the summer, this air mass prevents Seattle from having extremely hot temperatures. The tradeoff for having moderate temperatures is cloudy days occurring over 60% of the time and 40% of the days with rain. The coastal regions of New England are also moderated by maritime polar air. However, the westerlies mix in continental polar and maritime tropical air leading to more varied weather conditions.

15.4 FRONTS

Fronts are transition zones between air masses of different temperature and humidity. On the global scale the boundary between the Ferrel and polar cells is called the polar front. It separates the cold dry air of the pole from the warmer air of the sub-tropics. The dry line is also a front, but between air masses of different humidity. Multiple fronts can appear on a weather map, but for simplicity we will focus on the polar front.

Often the polar front develops instability, which gives rise to a low pressure system, also called a synoptic scale cyclone. Since air circulates counter-clockwise, to the east of the cyclone the polar front moves northward (figure 15.10a). This results in a warm front, which is represented by a red line with regularly spaced scallops. The scallops arc on the cold air side of the front to indicate that warm air is displacing the cold air. To the west of the cyclone the polar front dips south generating a cold front. Cold air behind the front displaces warm air and is represented by a blue line with regularly spaced triangles.

If the low pressure intensifies, circulation increases. Air behind the cold front is denser and easily displaces the warm air. Therefore, the cold front advances at a faster rate than the warm front. Figure 15.10b illustrates a mature stage of the cyclone, where cold air wraps around the low pressure system and the warm sector of air resides between the two fronts. Eventually the cold front overruns the warm front generating an occluded front (figure 15.10c). This kind of front is represented by a purple line with alternating triangles and scallops. At this point the polar front is severely distorted and the cyclone is cut-off from the polar front. This process repeats itself several times a month, but can reoccur in less than a week. The cyclone moves from west to east due to the geostrophic flow high in the atmosphere. These upper level winds are sometimes called **steering winds**.

Between passages of cyclones the polar front settles down and hovers over a region. The boundary between the warm and cold air can still trigger precipitation, but not of the severity of a cold front. The polar front is now called a stationary front (figure 15.10d). It is represented by alternating symbols for a cold and warm front. The red scallops extend into the cold air and the blue triangles into the warm air.

15.4.1 Cold Front

Cold fronts exist where continental polar air displaces maritime polar or maritime tropical air. Gravity naturally forces the cold dense air under the warmer moist air causing the front to advance. In close proximity to a low pressure system, the cold front can move at speeds of 30 to 45 kph (20 to 30 mph).

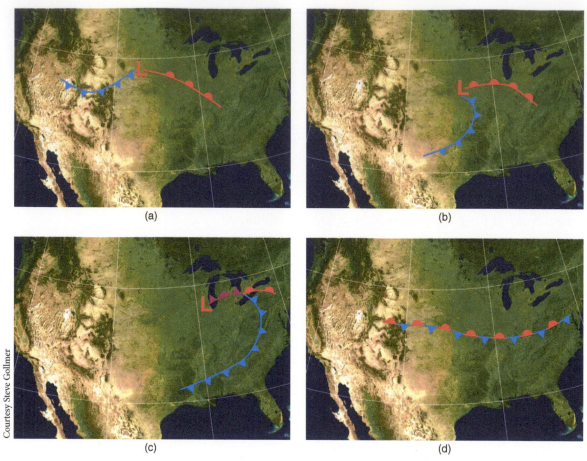

(a)   (b)

(c)   (d)

Courtesy Steve Gollmer

FIGURE 15.10. (a) Instability in the polar front causes a low pressure system to form along with its associated fronts. (b) As the low intensifies, the cold front begins to outrun the warm front. The warm humid air in the warm sector makes severe weather more likely as the cold front approaches. (c) As the cold front overruns the warm front, an occluded front forms. (d) When the polar front is stable, it does not move and is represented with a stationary front.

When viewed in profile (figure 15.11), the leading edge of the cold front is blunt causing the warm air to rise rapidly. This results in cumulus clouds near the front and generates thunderstorms in conditionally unstable air. High level winds pull clouds ahead of the front and provide a warning of its approach. Well ahead of the cold front, cirrus clouds increase in number and transition into cirrostratus. As the front gets closer, altocumulus and cumulus clouds thicken and fill the sky.

Figure 15.12 provides a summary of the effects associated with the passage of a cold front. A cold front moves from west to east and so the progression of weather corresponds to an approach of the front from the east. Initially the skies are clear, with increasing cloud cover as the front approaches. Since this air is south of the polar front, the winds are primarily westerlies. Rotation of the low pressure system modifies this direction and the winds come from the south to southwest. The air and dew point temperatures are similar indicating high humidity. Surface pressure decreases as the front approaches, reaching a minimum at the front. Most precipitation occurs at the front; however, thunderstorms can be triggered in advance of the front if the air is unstable. When the front passes, there is a sudden change in wind direction. Winds shift to the northwest as the dry polar air replaces the subtropical air. After passage of the front, precipitation drops off and the skies begin to clear.

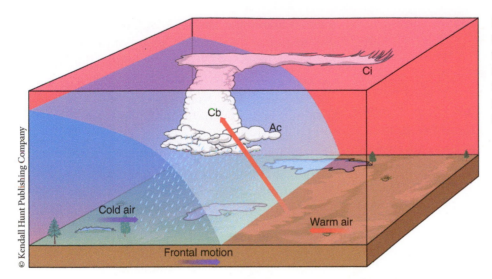

© Kendall Hunt Publishing Company

FIGURE 15.11. The rapid advance of the cold front causes air in the warm sector to rise abruptly. This makes the air unstable and can lead to severe weather. The pattern of clouds ahead of the front gives clues as to its approach.

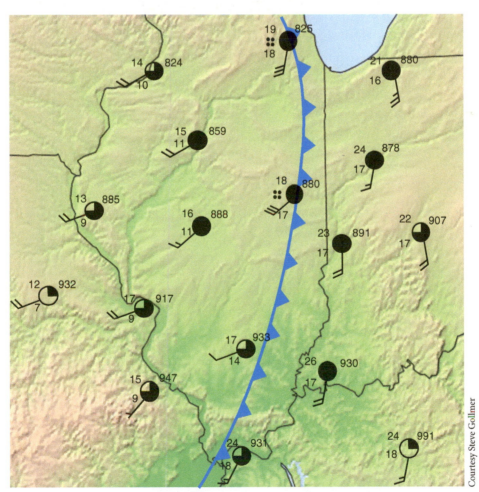

Courtesy Steve Gollmer

FIGURE 15.12. When viewed on a weather map, a cold front can be identified by several characteristics. As the front passes, temperatures drop and precipitation can occur. However, the most distinctive feature of cold front passage is the shift in wind direction.

15.4.2 Warm Front

Warm fronts travel at a slower speed than cold fronts and form a wedge shape (figure 15.13). Cold air remains at the surface, but is pushed forward by the overriding warm air. As the warm air moves over the front, there is a progression of stratiform clouds from low level nimbostratus to high level cirrus. Precipitation precedes the warm font and in the winter can begin as snow and transition into rain. Often light to moderate in intensity, precipitation can become intense if instability generates cumulonimbus clouds.

The progression of weather for a warm front is illustrated in figure 15.14. North of the front, the air is cold and dry with wind blowing from the east. As the front approaches, the dew point increases resulting in precipitation. Air temperature gradually increases and levels off as the front passes. Pressure decreases as the cyclone associated with the warm front approaches. As the front passes, pressure levels off and the wind direction shifts to the south-southwest.

15.4.3 Occluded Front

With an occlusion the cold air of a cyclone wraps around on itself. Warm air between the fronts is trapped above the surface. As a result, an occlusion looks like a warm front ahead of the transition boundary (figure 15.15). If the air behind the front is colder than the air in front of it, the surface boundary looks like a cold front. This happens when cold artic air is pulled from the north. However, if the air behind the front is maritime polar, it is warmer than the continental polar air preceding the occlusion and the surface boundary looks like a warm front (figure 15.15b). Precipitation forms in advance of the front, but can also produce thunderstorms as the cold air forces the pocket of warm air upward. As with other fronts, there is a wind shift during frontal passage.

A low pressure system and its associated fronts are easily identified in a satellite image (figure 15.16). The clouds form a comma shape as winds wrap around the cyclone in a counter-clockwise direction. The band of clouds with textured tops indicates the presence of cumulonimbus and are close to the cold front. The smoother clouds ahead of the front are cirrostratus, which are pulled forward by upper level winds. The pocket of clear sky behind the front indicates the presence of cool dry air drawn in by the low pressure system. The lowest pressures are at the center of the comma and the cold and warm fronts are either connected directly to the low or indirectly through the occluded front.

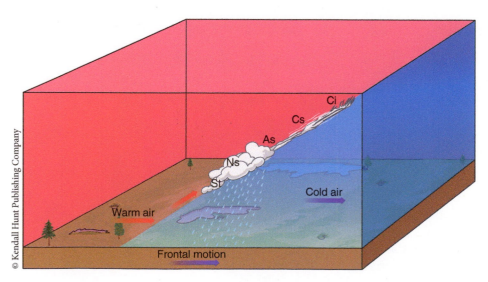

FIGURE 15.13. Warm fronts advance at a slower speed than cold fronts because the less dense air cannot displace the cold air as easily. As the warm air advances it overrides the cold air forming a wedge shape.

© Kendall Hunt Publishing Company

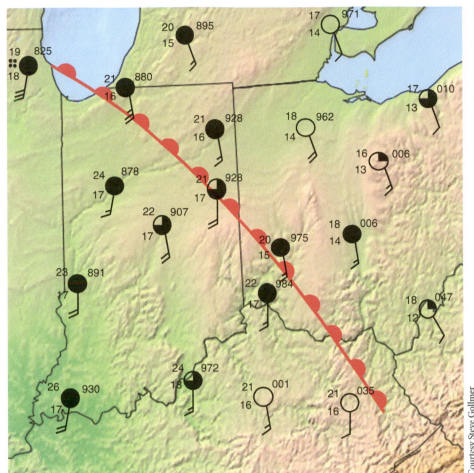

Courtesy Steve Gollmer

FIGURE 15.14. Warm fronts displace colder, dry air. As it approaches, humidity levels and temperatures increase. There is a slight shift in wind direction as frontal passage transitions polar easterlies to mid-latitude westerlies.

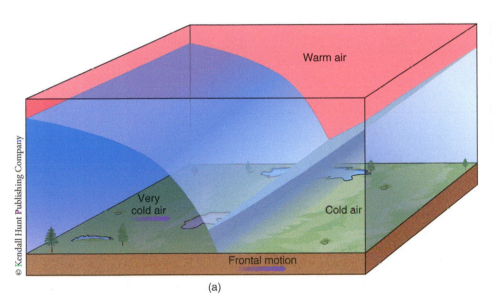

© Kendall Hunt Publishing Company

(a)

FIGURE 15.15. (a) Cold occlusion - Cold arctic air wraps around a low pressure system causing the occlusion to look like a cold front. (b) Warm occlusion - Milder Pacific air is not as cold as the continental arctic air. Therefore, the occlusion looks like a warm front.

CHAPTER 15: Weather Patterns and Forecasting

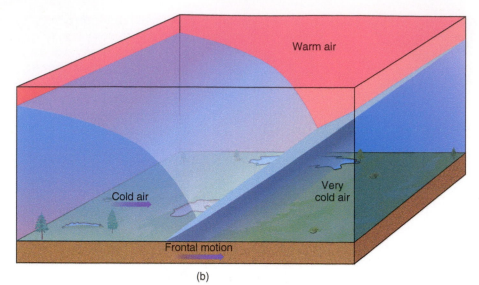

FIGURE 15.15. (Continued)

Warm air

Cold air

Very cold air

Frontal motion

(b)

FIGURE 15.16 On October 26, 2010 this system resulted in historic low pressures over the continental United States and Canada. At this point the front has already occluded. The warm sector between the warm and cold fronts is along the eastern seaboard. Although the wind speeds were not comparable in magnitude, the central pressure of this storm was equivalent to that of a category 3 hurricane.

Courtesy NOAA

15.5 THUNDERSTORMS

15.5.1 Thunderstorm Development

As mentioned in chapter 14, conditionally unstable air generates thunderstorms when it is pushed beyond its condensation point. This happens regularly during the summer in maritime tropical air masses. In the morning solar heating causes thermals to rise from the surface. By mid to late morning cumulus clouds form as the rising air reaches its condensation level. This is the first stage of a thunderstorm's life cycle and is called the **cumulus stage** (figure 15.17). The base of the cloud is flat and the top is coliform in shape as the rising thermals and turbulent mixing increase the cloud's height. The rising air is called an **updraft** and it provides a source of humidity for the growing cloud.

Once the cloud reaches a conditionally unstable layer, growth increases dramatically. As the cloud grows, its edges mix with the surrounding cool dry air. This dry air evaporates some of the cloud resulting in additional cooling. Cool air along with snowflakes and water droplets organize into a descending region called a **downdraft**. The combination of both an up and downdraft is called a **cell** and at this point the

FIGURE 15.17. In the late morning rising thermals result in the formation of a cloud. The condensation level occurs at the same altitude giving the cloud its flat base.

FIGURE 15.19. As the downdraft moves ahead of the thunderstorm, the updraft overrides the gust front forming a cloud that extends below the base of the storm. This is called a shelf cloud.

thunderstorm is in the **mature stage** (figure 15.18). During the mature stage the thunderstorm reaches the tropopause and forms the characteristic **anvil**, as high level winds pull cirrus clouds to the east of the storm. The heaviest rainfall as well as the highest instances of lightning occurs during this stage. Precipitation caught in an updraft can make multiple trips through the cloud allowing hailstones to grow to large sizes. In strong thunderstorms, updrafts can reach speeds of 96 kph (60 mph) and suspend hailstones the size of nickels or larger.

As the downdraft reaches the surface, it spreads out forming a gust front. This cold surface air forces

the updraft to the side of the cloud. As the warm updraft overrides the cold surface air, it forms a **shelf cloud** that extends below the level of the main cloud (figure 15.19). As the gust front moves away from the cloud, it causes surface turbulence, which is hazardous to landing and departing aircraft. Intense downbursts of air or **microbursts** have been responsible for a number of airplane crashes. As a result, major airports have microburst detection systems and suspend air traffic as thunderstorms approach.

The downdraft from the thunderstorm eventually cuts off the source of warm surface air and the updraft disappears. In thunderstorms not tied to a front, this condition occurs 15–30 minutes after the storm matures. This begins the **dissipation stage** and the strength and size of the storm is in decline. Outflow from the thunderstorm can spread 10's of kilometers away and facilitate the formation of other thunderstorms.

When associated with a cold front, the behavior of a thunderstorm is enhanced by frontal lifting. If the front passes in the afternoon or early evening, the intensity is greater because solar heating adds humidity and energy to the atmosphere making it less stable. The speed at which the front advances is also important. Thunderstorms die out because the inflow of warm moist air is cut off by the downdraft. However, if the speed of the front prevents this from

FIGURE 15.18. A thunderstorm cell consists of a combination of an updraft and downdraft. The top of the cloud reaches the tropopause where it flattens into an anvil of cirrus clouds.

happening, the updraft is maintained and the same cell can persist for 100's of kilometers.

15.5.2 Lightning

A common danger of thunderstorms is lightning. **Lightning** is an electrical discharge resulting from an accumulation of charge within a cloud. Freezing processes within the cloud allow preferential accumulation of negative charge at the cloud base and positive charge higher in the cloud (figure 15.20). When the accumulated charge becomes large enough, the insulating property of the atmosphere breaks down and the negative charge moves to cancel the positive charge.

A lightning discharge takes the path of least resistance. Small pockets of positive charge attract the discharging lightning bolt giving its jagged appearance (figure 15.21). Once an initial discharge occurs, it is easiest for additional charge to move

FIGURE 15.21. Taking the easiest route for eliminating charge, a lightning bolt takes a jagged path. This can lead to Earth, but also can extend between clouds and within portions of the same cloud.

through the same channel. That is why lightning appears to flicker in the same place. Although lightning can strike anywhere, it is more likely to strike tall objects. The intense heat and electric field from a lightning strike can damage electrical appliances and start fires. To protect a building **lightning rods,**

FIGURE 15.20. Precipitation processes within a thunderstorm cause negative charge to accumulate at the base of the cloud and positive towards the middle and top. When enough charge accumulates, the insulating property of the air breaks down and a lightning discharge occurs.

a series of pointed metal rods, are placed along the roof to provide a preferential location for a lighting strike. These rods are connected to a thick electric cable, which provides an alternate path for the electricity to discharge into the ground.

The high voltage and current of a lightning strike damages buildings and trees, but also kills. The annual number of deaths has decreased over time as safety procedures have been implemented. Most cloud to ground lightning occurs in close proximity of the storm. However, it has been documented that a strike can take place up to 16 km (10 mi) away. This provides the criterion for suspending outdoor sporting and water activities until 30 minutes after the last lightning strike is observed. When seeking shelter, stay away from lone trees and tall towers as they provide a preferential path for the lightning. It is best to go indoors. If driving during lightning, don't fear.

The metal frame of the vehicle provides protection to the occupants inside.

During a lightning discharge, the air is heated to high temperatures resulting in sudden expansion. This generates **thunder**, which is a shock wave that travels at the speed of sound. Although thunder causes no damage, it does generate an uneasy feeling from its suddenness and persistent rumbling. To provide some sense of calm and control, parents have children count the time between the lightning strike and the associated thunder. Since the speed of sound is 1/3 kph, dividing the time between the flash of lightning and the thunder by 3 gives the distance in kilometers. Dividing by 5 gives the distance in miles. Since the sound of thunder diminishes with distance and is reflected off surfaces, this rule doesn't work well beyond 20–25 seconds.

15.6 TORNADOES

15.6.1 Tornado Formation

A **supercell** thunderstorm is a long lasting cell that has strong rotation in its updraft. This rotating column of air, called a mesocyclone, lowers the condensation level and forms a **wall cloud** that extends below the cloud base (figure 15.22). Rotation comes from several sources, but primarily from horizontal rotation tilted upward by convection. Figure 15.23 illustrates how upper level winds generate a rolling motion of air near the surface. As convection lifts a section of rolling air, two columns of rotating air are formed; one clockwise and the other counter-clockwise. In the Northern Hemisphere counter-clockwise rotation persists because of the Coriolis Effect.

Once rotation exists in a thunderstorm, the possibility of tornado formation greatly increases. Normally warm air rising from the surface randomly collides with sinking cold air causing the flickering effect observed in the flames of a fire (figure 15.24a).

FIGURE 15.22. Extending below the base of a supercell thunderstorm, the wall cloud is a rotating structure from which tornadoes can appear.

Annette Shaff/Shutterstock.com

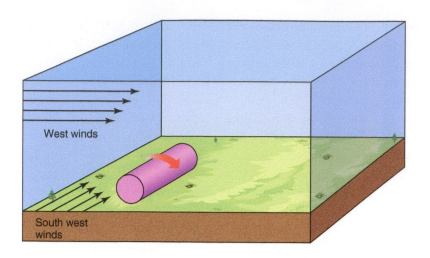

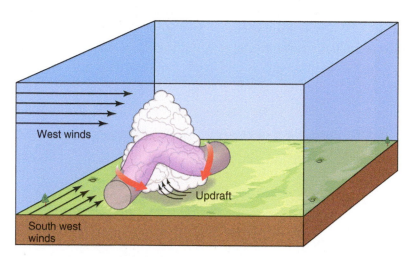

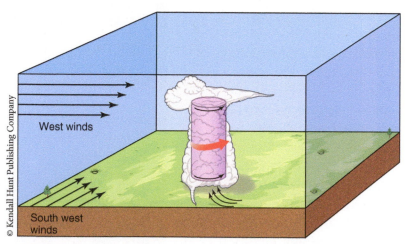

FIGURE 15.23. Strong winds above the surface can cause a tube of air to rotate along Earth's surface. Convection lifts a section of this tube converting rotation about a horizontal axis into the vertical.

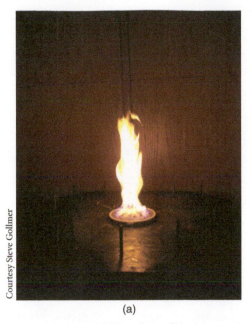

(a)

Courtesy Steve Gollmer

(b)

FIGURE 15.24. (a) Rising flames flicker because the convection is unorganized. (b) Rotating air moves less dense air to the center. This makes the convection more efficient and draws the flame into a vortex.

However, when rotation is present, the less dense air of the updraft is pushed to the center. The rate of rotation increases because the converging air conserves its angular momentum. This is the same effect that causes an ice skater to increase her rotation rate by pulling in her arms. The updraft become stronger and tightens into a vortex as illustrated in figure 15.24b. In a mesocyclone, the vortex extends below the wall cloud and when it reaches the surface it is called a **tornado**.

15.6.2 Tornado Damage

The visible portion of the tornado is composed primarily of water droplets, which is an extension of the cloud towards the surface (figure 15.25). In drier regions a visible funnel may not reach the ground; however, the flying debris at the surface indicates that a tornado is still present. Tornadoes are classified by the damage they generate at the surface. Figure 15.26 gives a summary of the enhanced **Fujita scale**, which associates wind speeds with the resultant damage. When damage is assessed, evidence of rotation is essential. If rotation is not

FIGURE 15.25. Once the rapid rotation of a funnel reaches the ground it is considered a tornado. Condensation within the funnel extends below the cloud making it visible. The debris field at the ground also gives indication as to the presence of the tornado.

present, the damage is caused by the strong outflow from a line of thunderstorms. These straight line winds, called **Derechos**, can reach speeds of more than 160 kph (100 mph) and cause damage similar to tornadoes.

Although tornadoes occur any time of day and have struck every state of the union, they are most

Scale	Wind Speed (kph)	Wind Speed (mph)	Damage
EF0	104–137	65–85	Broken branches
EF1	138–176	86–110	Peel surface off roofs
EF2	177–216	111–135	Trees uprooted, roofs removed from houses
EF3	217–264	136–165	Walls torn off, overturned cars
EF4	265–320	166–200	Houses leveled
EF5	>320	>200	Reinforced concrete heavily damaged

common in the spring near the dry line extending from Texas to Kansas. This region is often called tornado alley. High occurrence of tornados extends east of this region due to severe thunderstorms generated by the passage of cold fronts. As a result, tornado risk maps extend the high risk area to western Ohio and as far south as Alabama. As mentioned previously, when cold fronts travel fast enough, the warm moist updraft is not cut off and the thunderstorm persists for long periods of time. These conditions generate a series of supercells aligned with the front and spawn numerous tornados. One such outbreak resulted in the Tri-State tornado of 1925. This EF5 tornado lasted 3 ½ hours, traveled 352 km (219 mi) from Missouri to Indiana and killed 695 people. The Super Outbreak of 1974 occurred over 18 hours generating 148 tornados and killing between 300 and 330 people (figure 15.27). While these tornado statistics are devastating, they are dwarfed by the Bangladesh tornado of 1989, which killed more than 1,300 people.

15.6.3 Tornado Watches and Warnings

With improved forecasting and warning systems, tornado fatalities have significantly decreased in the past half century. The National Weather Service (NWS) identifies instability in the atmosphere through the use of radiosondes, launched every twelve hours. If the conditions are right for the formation of severe thunderstorms, a thunderstorm **watch** is issued. Once thunderstorms form, they are tracked through the use of radar and weather spotters. (Spotters consist of emergency workers, such as police and fire personnel, but also private citizens trained in a free storm spotter course conducted by the NWS.) At this time the watch is upgraded to a thunderstorm **warning**. Warnings are issued for locations downwind of the storm.

Tornadic activity is also identified by radar. As described in chapter 14, radar determines the intensity and distance to rainfall. Since the mesocyclone rotates counter-clockwise, it wraps precipitation and the cold downdraft around the warm updraft, thus forming a hook. This hook echo usually appears to the right rear of the storm (figure 15.28). Modern radar systems use the Doppler effect to measure wind speed and direction. Winds moving away from the radar are color coded red while approaching winds are green. Tightly spaced, strong winds going in opposite directions is indicative of mesocyclones and is adequate evidence for issuing a tornado warning, although a tornado may not be currently present.

When tornadic activity is observed, a tornado warning is issued. Tornado sirens are sounded and residents are instructed to immediately find safe shelter. The best shelter against tornadoes is in a basement under a solid counter or table. If a basement is not available, a room with no exterior

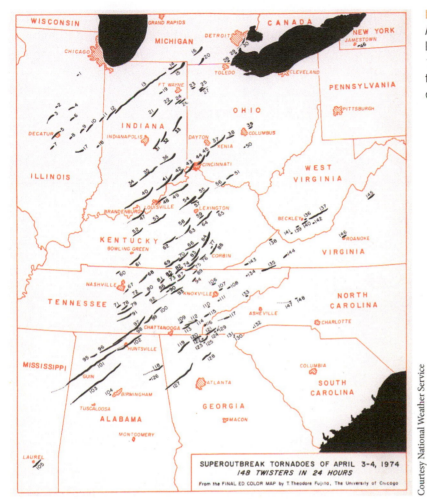

FIGURE 15.27. Super Outbreak of 1974. Associated with the passage of a cold front, this line of storms generated 148 tornadoes over 18 hours resulting in 330 deaths. Numerous tornado outbreaks have occurred over the past century.

Courtesy National Weather Service

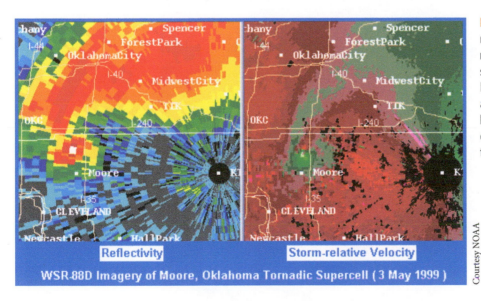

FIGURE 15.28. In a regular radar return the rotation of the mesocyclone generates a comma shaped feature called a hook echo. In modern doppler radar, wind speed and direction are used to identify locations of sudden changes in wind direction, a tell-tail signature of a tornado.

Courtesy NOAA

windows and closely spaced walls provides the best protection. This is usually in a closet under a stairwell or in a bathroom. Structures that are not securely anchored to the ground are not safe even in EF1 tornadoes. Therefore, it is best to move to an identified tornado shelter than to remain in a mobile home. If in a moving vehicle, you should stop and lie down in a ditch or culvert. It is not wise to attempt outrunning a tornado. Although the average speed of a tornado is 48 kph (30 mph), the funnel can reach speeds of 112 kph (70 mph) and travel erratically.

15.7 HURRICANES

15.7.1 Hurricane Formation

The polar front is a common source for mid-latitude cyclones; however, low pressure systems occur anywhere there is strong surface heating and conditions permitting the development of rotational motion. At higher latitudes surface temperatures are insufficient to develop strong convection. Near the equator the Coriolis Effect is negligible and unable to generate large scale rotation. However, between 5° and 20° latitude the ocean provides these conditions. Low pressure systems remaining in this region long enough will eventually intensify into hurricanes.

Hurricane season in the United States occurs between the months of June and November and impacts the east coast of the country. Many of these hurricanes originate off the coast of Africa and travel across the Atlantic Ocean (figure 15.29). During the summer ocean temperatures are above 26 °C (79 °F), which is the minimum temperature for developing and sustaining hurricanes. As the easterlies come off the coast of Africa, they form a low pressure region which enhances upward motion of the air and generates thunderstorms. Over time these storms organize due to the Coriolis Effect and the low pressure intensifies. The low frictional

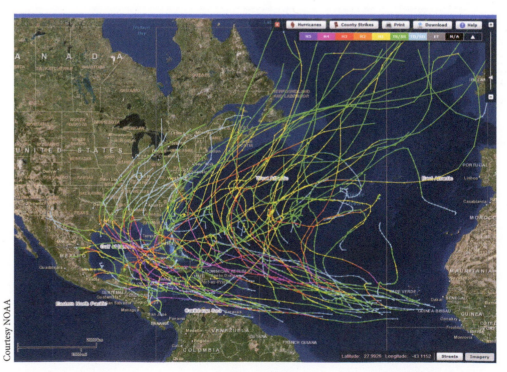

FIGURE 15.29. Warm waters off the west coast of Africa generate a large number of hurricanes that impact the eastern and gulf coasts of the United States. This image is a plot of the 77 storms that reached hurricane status between 2001 and 2010.

Courtesy NOAA

drag over the ocean allows the circulating winds to achieve high speeds.

Hurricanes begin as **tropical depressions**, which have a closed circulation around the low pressure system. As they make their way across the ocean, the winds intensify and the central pressure decreases. Once the wind speeds reach 65 kph (40 mph), the system is classified as a **tropical storm** and is given a name. Figure 15.30 is a satellite image of several storms crossing the Pacific Ocean. Each storm has a clearly defined rotation within it. If the storm drifts north into cooler water, it weakens. However, if it develops further, the wind speeds exceed 119 kph (74 mph) and the storm is called a **hurricane**. Hurricanes also develop off the coast of Mexico in the Pacific Ocean. Intense tropical storms in the western Pacific Ocean are called **typhoons** and in the Indian Ocean and South Pacific, they are called tropical cyclones.

When a storm becomes a hurricane it has a well-developed structure (figure 15.31). The lowest pressure occurs at the center of the hurricane. Not only is there convergence of air at the surface, but air is also drawn downward. This results in a cloud-free region called the **eye**. The eye is 20 – 50 km (13 – 31 mi) in diameter and has weak surface winds. However, at the edge of the eye is a region of intense thunderstorms due to the updraft of the storm. This is called the **eye wall** and contains the strongest winds and precipitation. Spiraling away from the eye wall are rain bands, which also include updrafts, but of lessening intensity. The outflow from the updraft terminates at the tropopause and forms a cirrus shield that obscures the rain bands and sometimes the eye.

15.7.2 Hurricane Damage

Hurricanes are classified by the intensity of their winds and the value of their central pressure. Developed in 1972, the **Saffir-Simpson scale** is used to predict the damage to structures due to wind speed (figure 15.32). When viewed from the coastline, winds to the left of the eye approach the coast while winds to the right recede. During landfall the circulation speed and the forward speed of the hurricane add to give higher wind speeds to the left of the storm. On the right the circulation opposes the forward speed and the overall speed is lower. In 2005 hurricane Katrina had circulation wind speeds

Courtesy NASA

FIGURE 15.30. In 2006 this unusual occurrence of three significant storms south of Japan illustrates the various levels of typhoon development. To the left is Bopha, which is the least developed. While still a tropical storm, Maria in the upper right is more developed and shows signs of a central eye. In the bottom right, Saomai has reached typhoon status.

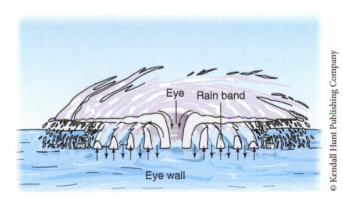

© Kendall Hunt Publishing Company

FIGURE 15.31. At the eye of the hurricane winds are mild and skies can be clear. However, the most severe portion of the storm exists at the eye wall and weakens the further you go from the eye.

Category	Pressure (mb)	Wind (kph)	Wind (mph)	Storm Surge (m)	Storm Surge (ft)
1	>980	118–153	74–95	1.2–1.6	4–5
2	965–979	154–176	96–110	1.7–2.5	6–8
3	945–964	177–209	111–130	2.6–3.8	9–12
4	920–944	210–248	131–155	3.9–5.5	13–18
5	<920	>248	>155	>5.5	>18

of 200 kph (124 mph) and an approach speed of 25 kph. To the left of the storm wind speeds were 225 kph while only 175 kph to the right. While this might not seem dramatic, the kinetic energy of the wind goes as the square of the wind speed. As a result, the energy of winds striking the Mississippi coastline was 65% greater than those in Louisiana. Once coming on shore, the storm's moisture is cut off and friction dissipates the energy of the storm.

While wind damage is significant, it is secondary to that of the storm surge (figure 15.33). **Storm surge** is the rise in water level at the coast as the hurricane makes landfall. Depending on the terrain near the coast, these waters can penetrate kilometers inland doing significant damage. In fact most of the damage and casualties from a hurricane is due to the storm surge. The low pressure near the center of the hurricane pulls upward on the ocean surface forming a dome of water. This rise in sea level is not large; however, as the hurricane makes landfall it encounters shallow waters. This amplifies the height of the storm surge in a fashion similar to a tsunami (See chapter 12). The rate at which the hurricane makes landfall and the timing with high tide impacts the size of the storm surge.

15.7.3 Hurricane Prediction

Predictions of hurricane paths and location of landfall are an ongoing area of research. In 1900 a category 4 hurricane hit Galveston, TX. Although there was warning of the approaching storm, its intensity and actual path were not well known. As a result, few evacuated before the storm surge swept over the island. Death toll from the initial storm and subsequent lack of relief came to 8,000 people or 20% of the island's population. In order to improve hurricane tracking, a squadron of planes was assigned to weather reconnaissance after World War II. This squadron was given the name Hurricane Hunters because they fly through hurricanes to accurately measure wind speed and location of the hurricane's eye. Although developing hurricanes generally travel from east to west, high pressure systems can block or modify the hurricane's trajectory. Measuring the more subtle steering winds near the storm increases the accuracy of hurricane predictions.

Hurricane prediction has also improved due to data collected from satellites and sophisticated

FIGURE 15.33. While strong winds can topple trees and damage homes, hurricane winds do not approach the strength of a tornado. However, hurricanes are more costly because they are many times larger and the storm surge can sweep homes off their foundation and drag objects miles inland.

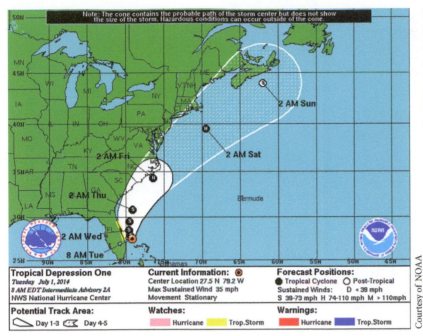

FIGURE 15.34. Since upper level winds and high pressure systems can steer hurricanes, predicting landfall has a degree of uncertainty. This uncertainty is communicated as a cone in which there is a 60-70% confidence that the hurricane will remain inside the forecast cone. Five day forecasts are given allowing communities to make the proper preparations.

computer models. The National Hurricane Center provides a number of products useful for warning the public of possible danger. Figure 15.34 is a five day forecast cone illustrating the current position and possible path of the hurricane. A cone is used because the error increases as the forecast extends to the fifth day. The predicted location of the storm is represented by a solid dot with a letter in the middle. If the letter is an 'S' it represents a tropical storm while an 'H' represents a hurricane. Colored coastlines represent the location of issued watches and warnings. **Hurricane watches** are issued several days before a storm arrives and a warning is given when the hurricane is within 24 hours of making landfall.

No two hurricanes are exactly alike because of the many factors affecting their motion and strength. Because of this it is important to take hurricane warnings seriously. Hurricanes Camille (1969) and Katrina (2005) both hit the Gulf Coast east of New Orleans, Louisiana. Camille is the second strongest hurricane to make landfall in the United States with a wind speed of 306 kph (190 mph). With a storm surge of 7.3 m (24 ft) Camille resulted in the death of 143 people and did $7.56 billion of damage (adjusted to 2005 values). By contrast Katrina made landfall as a category 3 hurricane having wind speeds of 204 kph (127 mph) (figure 15.35). The storm surge was over 4.3 m (14 ft) and resulted in the failure of the New Orleans' levee system. The outcome of this storm was the death of 1,833 people and $108 billion of damage making it the costliest natural disaster in US history. Although not as strong as Camille, Katrina's impact was greater because it had a larger physical size and made landfall closer to New Orleans. During the 36 years between hurricanes, development of coastal properties increased the cost of Katrina both financially and humanly.

FIGURE 15.35. Hurricane Katrina. Making landfall east of New Orleans, this category 3 hurricane caused levees to fail, thereby, making this storm the most costly natural disaster in the United States.

Courtesy of NASA

15.8 WEATHER FORECASTING

15.8.1 Forecasting History

The value of weather forecasts and warnings is crucial in the event of severe weather; however, accurate forecasts impact our daily lives. Early civilizations were strongly dependent on agriculture and therefore, had numerous weather and fertility gods and goddesses. Appealing to their good nature supposedly brought success in agriculture, warfare, and trade. However, they overlooked the creator God, who established the physical laws upon which all weather is dependent. When we study weather, we are not explaining away God's sovereignty, but gaining an appreciation for the complexity of His providence. Forecasters do not foresee the future, but declare what is most likely to happen barring God superseding His established laws. Therefore, weather forecasters approach their profession with confidence in the basic physical laws, but humility in their ability to predict the outcome of something as ephemeral as catching the wind.

Although it may seem an insurmountable task, weather forecasting is wildly successful. The downfall occurs when forecasters attempt to predict small details over extended periods of time. The apostle Paul forecasted the weather when he declared that "Sirs, I perceive that the voyage will be with injury and much loss, not only of the cargo and the ship, but also of our lives." (Acts 27:10) This was based, not on direct revelation from God, but on the fact that this was the season when severe storms sweep across the Mediterranean. Paul was using knowledge of climatology to comment on the safety of the voyage. It was only later in Acts 27:21-26 that Paul received direct revelation that the men on the ship would be saved.

Jesus used a common forecasting practice to point out that the Pharisees failed to see the sign of Messiah's coming. In Matthew 16:2-3 He states, "When it is evening, you say, 'It will be fair weather, for the sky is red.' And in the morning, 'It will be stormy today, for the sky is red and threatening.' You know how to interpret the appearance of the sky, but you cannot interpret the signs of the times." This statement is equivalent to our saying, "Red sky at morning, sailors take warning; Red sky at night, sailors' delight." This saying is only valid in the mid-latitudes where the prevailing winds are westerlies. After the passage of a cold front, the storm generating low pressure system is to the east. As a result, the clear skies at night give red sunsets which hearkens the approach of high pressure and fair weather (figure 15.36).

15.8.2 Forecasting Techniques

Weather forecasting uses numerous techniques which have limited success depending on location and season. The most common technique is persistence. **Persistence** assumes that the weather does not change from day to day. This technique is useful in Hawaii, where the afternoon high is in the upper 80's with a chance of afternoon showers. However, it is not as successful during the Midwest spring when the weather can change three times in the same day. **Steady-state** forecasting assumes there is a trend that continues into the future. This is useful when following a line of thunderstorm on radar (figure 15.37). Short term warnings are issued based on the progress of the storm. On the larger

FIGURE 15.36. In the middle latitudes weather tracks from west to east. Therefore, a red sunset with its associated clear skies is the expected weather the following day.

scale we expect daily high temperatures to increase as we enter summer and decrease as we approach winter. This prediction is also a forecast from **climatology**, which uses 30 year averages of weather data. **Analogue** or pattern recognition uses multiple pieces of data to recognize the recurrence of a significant weather event. The progression of cloud types in advance of a cold front is a recognizable pattern, as well as, the statement about "red skies." Finally, **numerical forecasts** use physical principles to calculate the future behavior of the atmosphere using high speed computers.

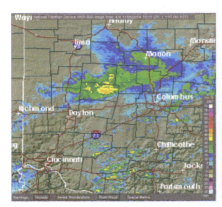

FIGURE 15.37. Storms tend to move at a constant rate. Therefore, forecasting the onset of rain is possible using radar images. The first two radar images are taken 15 minutes apart. In that time the heavier rain, indicated by yellow, advanced 15 km. Columbus, OH is 45 km further ahead and, therefore, we predict the heavier rain will arrive in 45 minutes (third image).

15.8.3 Numerical Forecasts

Numerical forecast models divide the surface of Earth into a grid and the atmosphere into layers (figure 15.38). The physical processes described in the previous two chapters are used to calculate temperature, pressure, humidity, wind speed, and direction for each location on the grid. Predictions of the future state of the atmosphere are limited only by the resolution of the model and the accuracy of the formulas representing the physical processes. Climate models use a course grid spanning the globe and run simulations centuries into the future. As computers have increased in speed, the resolution of climate models have gone from modeling the states of Illinois, Indiana, and Ohio as a single grid point to grid cells that are 111 km on a side.

Weather simulations extend only a few days into the future and, therefore, have a higher resolution. The North American Mesoscale Model (NAM) covers the continental United States and portions of Canada and Mexico with a grid 12 km on a side. This model is run every six hours and each forecast extends to 3.5 days. Data is summarized in four panel plots representing forecasts 12, 24, 36 and 48 hour into the future. Figure 15.39 gives a plot of precipitation and surface pressure. It is possible to pick out low and high pressure systems and cold fronts as they move across the country. Other useful plots are surface temperatures and winds (figure 15.40a). The 500 mb chart (figure 15.40b) shows how weather systems travel due to the mid-troposphere winds. Although forecasts beyond 5 days have limited accuracy, global patterns change more slowly and can be

Schematic for Global Atmospheric Model

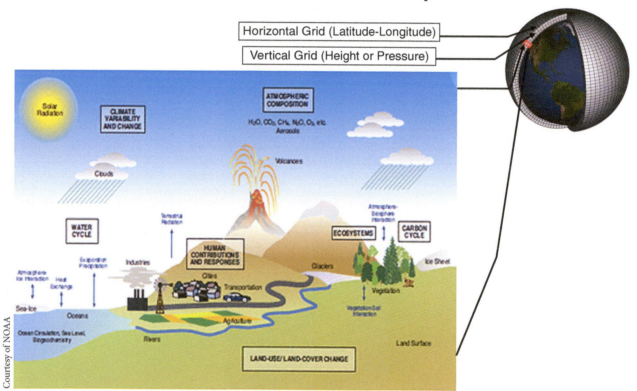

FIGURE 15.38. Breaking Earth's surface and atmosphere into a three dimensional grid, computer forecast models use the laws of physics to calculate the most likely future weather. Accuracy depends on the quality of the data provided and the amount of detail embodied in the equations used by the computer.

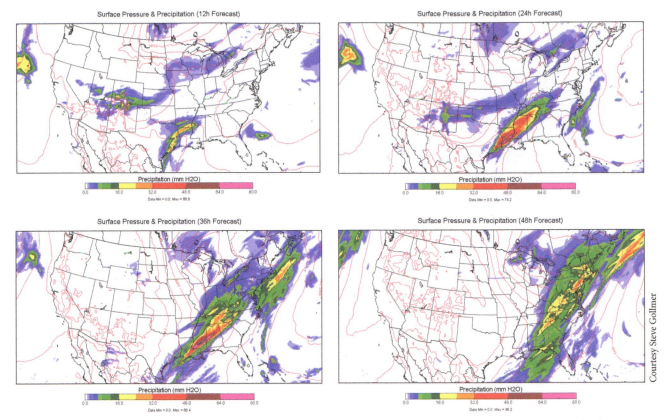

FIGURE 15.39. A common weather product provides two-day forecasts in 12 hour increments. This particular plot contains accumulated precipitation and surface air pressure predictions with a starting time of 00Z November 16, 2014.

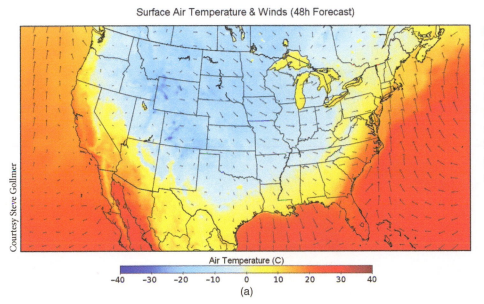

FIGURE 15.40. Besides precipitation many other forecast products are available. (A) A two-day forecast of daily high temperatures and wind speed. (B) A two-day forecast of 500 mb heights, winds, and vorticity. These winds indicate how weather systems track across the country. Vorticity is a parameter that affects convergence and divergence at the surface.

FIGURE 15.40. (*Continued*)

500 mb Height & Winds (48h Forecast)

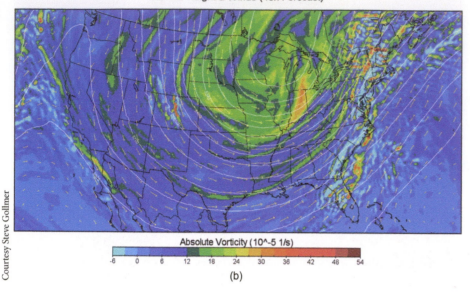

Absolute Vorticity (10^-5 1/s)

-6   0   6   12   18   24   30   36   42   48   54

Courtesy Steve Gollmer

(b)

identified with extended models. Due to non-linear interactions within the atmosphere daily forecast have a theoretical limit of two weeks. Forecasts beyond two weeks have value only in a statistical sense as they try to capture long term effects such as El Nino. 90 day climate outlooks are reported to help farmers plan their summer crops.

15.8.4 Satellite Meteorology

Alongside computer modeling, satellites have had the most profound influence on weather prediction during the 20th century. On April 1, 1960 the first successful weather satellite, TIROS-1, was launched. It demonstrated that television images of Earth provide useful information about weather patterns (figure 15.41). Since then a series of polar and geostationary satellites has been deployed for monitoring the weather. Images used regularly on TV and internet weather reports are from the **Geostationary Operational Environmental Satellite** (GOES). Geostationary satellites are placed above the equator with an orbit that takes 24 hours to complete. As a result, the satellite always looks at Earth from

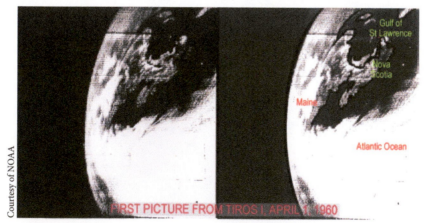

Courtesy of NOAA

FIGURE 15.41. The first successful weather satellite, TIROS-1, was launched in 1960. These first images may be of poor quality by today's standards; however, knowing the location and extent of weather systems is a valuable tool.

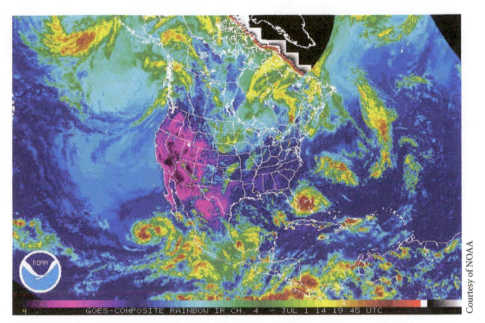

Courtesy of NOAA

FIGURE 15.42. This enhanced image combines information from both GOES east and west. The false colors highlight regions of intense convection.

the same perspective. The United States has two GOES satellites in operation at any one time. The first is positioned above 75°W longitude and looks down on the Atlantic Ocean and the Americas. The second is positioned at 135°W longitude and views the Pacific Ocean. Both provide visible and infrared images of the continental United States at ½ hour intervals. The full disk of Earth is transmitted every three hours with a resolution of 1km. Infrared composites of both satellites are generated to reveal locations of strong convective systems and cirrus cloud cover (figure 15.42).

15.1 Micrometeorology

- Phenomena lasting minutes and extending to a kilometer are part of micrometeorology. This includes smoke plumes and the lowest part of the atmosphere where friction has the largest impact.

15.2 Global Circulation

- Circulation between the equator and the pole breaks into a three-cell system: Hadley, Ferrel, and Polar Cells.
- Rising warm air at the equator results in a band of low surface pressure that corresponds to cloudiness and high precipitation.
- Sinking air at the sub-tropics results in clear skies, low precipitation, and surface high pressure systems.
- The polar front results in the most varied weather on the planet because the shear between easterly and westerly winds causes the front to generate ridges and troughs. This front generates a number of storms because polar air comes in contact with warm sub-tropical air.
- Thermal circulations between land masses and oceans complicate the circulation pattern resulting in enhanced sub-tropical oceanic highs and polar oceanic lows.
- During the winter a strong Siberian high forms that influences weather in Europe, Asia and even North America. During the summer it disappears and reverses circulation over the Himalayans resulting in monsoon rains.
- El Niño and La Niña are multi-year weather patterns that have an impact not only on the tropical Pacific Ocean, but also across the globe.

15.3 Air Masses

- Maritime tropical air masses are warm and humid. They influence the weather in the southeastern US.
- Continental tropical air masses are warm and dry. They are associated with many deserts and give the southwest US its weather.
- Continental polar air masses influence Canada and the northern US. This cold dry air is associated with severely cold weather during the winter.
- Maritime polar air masses are cool and humid. This air mass allows coastal regions in the north to maintain their moderate temperatures.

15.4 Fronts

- The surface where two different air masses meet is a front. This location can also initiate precipitation and severe weather.
- A cold front is when cold air displaces warmer air. Likewise, a warm front displaces cooler air. Warm fronts move slower than cold fronts because warm air is less dense and overrides the cold air.
- As a low pressure system moves across the country, it is preceded by a warm front followed by a cold front.

15.5 Thunderstorms

- A thunderstorm consists of a combination of an updraft and downdraft.
- A storm with a strong updraft can produce heavy rain and hail.
- The strongest thunderstorms occur in association with cold front passage in the late afternoon and evening.
- Lightning is an electrical discharge between a cloud and either the ground, another cloud or itself. The path of lightning is jagged because it takes the easiest route to its destination.

15.6 Tornadoes

- Rotation in a thunderstorm can intensify resulting in a tornado. The intensity of a tornado is rated on the Fujita scale based on the damage it generates.
- Watches indicate that conditions are favorable for the formation of serve weather, while a warning indicates that the severe weather has already developed.

15.7 Hurricanes

- Hurricanes form over warm water between 5° and 20° latitude. Low friction, strong convection and the Coriolis Effect work together to organize and intensify the storm.
- While strong winds and heavy rains do significant damage, most of a hurricane's damage is the result of the storm surge. The intensity of hurricanes is rated on the Saffir-Simpson scale.

15.8 Weather Forecasting

- Historically, weather forecasting has relied on observation of persistent trends, associated patterns and climatology. Central to modern forecasting techniques are computer models that calculate the most likely progression of weather given data collected across the globe.
- Numerous satellite systems monitor Earth's surface and atmosphere. The most familiar system is the GOES satellite, which provides half hour snap shots of data over the US.

KEY TERMS

Air mass	Downdraft
Analogue	Eddies
Anvil	El Niño
Cell	Entrainment
Climatology	Eye
Cumulus Stage	Eye Wall
Derechos	Ferrel Cell
Dissipation Stage	Fronts
Doldrums	Fujita Scale

GOES Satellite	Severe Weather Watch
Hadley Cell	Shelf cloud
Hurricane	Siberian High
Hurricane Watch	Southern Oscillation Index
Jet Stream	Steady-state
La Niña	Steering Winds
Lightning	Storm Surge
Lightning Rod	Supercell
Mature Stage	Synoptic Meteorology
Microburst	Thunder
Micrometeorology	Tornado
Monsoons	Trade Winds
Numerical Forecast	Tropical Depression
Persistence	Tropical Storm
Polar Cell	Typhoon
Polar Front	Updraft
Saffir-Simpson Scale	Wall Cloud
Severe Weather Warning	Westerlies

REVIEW QUESTIONS

1. How is friction from Earth's surface transmitted through the atmosphere? What are some features that illustrate this transmission?

2. Although the average energy received from the sun and IR energy lost by Earth is balanced, there is an imbalance on where the heating and cooling occurs. Describe this imbalance and explain how the atmosphere and ocean respond to this imbalance.

3. Using the three-cell model of the atmosphere, describe where you would expect high and low surface pressure to occur. Also describe the prevailing wind direction for each cell and where there are large and small amounts of precipitation.

4. On a world map locate regions of persistent high and low pressure systems during the Northern Hemisphere winter. Given these pressure systems where would you expect clear skies and dry weather? Cloudy skies and precipitation?

5. Describe the four types of air masses and identify regions of the US that are affected by each type.

6. What is a cold front and describe the weather you would experience as it approaches you? Do the same for a warm front.

7. What is the significance the polar front to weather in the continental US?

8. Describe the life cycle of a cumulus cloud as it develops into a thunderstorm. What features must be present in each stage?

9. What are the hazards associated with a severe thunderstorm that has an impact on life and property?

10. What is a tornado and what is necessary for its formation? What visible features are present that warn you that a tornado could form?

11. Describe the difference between a watch and a warning. Where should one take shelter during a tornado warning?

12. What conditions are necessary for the formation of hurricanes? What regions of the US are most susceptible to hurricanes and what time of year are hurricanes most likely to occur?

13. How does a hurricane inflict damage and what measures should be taken in preparation of a hurricane making landfall?

14. List different techniques of weather forecasting and give an example for each technique. How effective are these techniques?

15. What is a geostationary satellite and how is it beneficial for weather forecasting?

FURTHER READING

- NOAA's El Niño Portal – www.elnino.noaa.gov
- National Weather Service SkyWarn (Severe weather spotter training) – www.nws.noaa.gov/skywarn/
- National Hurricane Center – www.nhc.noaa.gov
- NAM Model Forecast – weather.unisys.com/nam/
- NOAA Geostationary Satellite Server – www.goes.noaa.gov

REFLECT ON SCRIPTURE

What does Ezekiel 33:1-6 have to do with weather forecasting? Ezekiel was called as a watchman to warn the people of Israel of their sin's consequences. In this passage we are reminded that a city watchman is held liable for the safety of the people in the city. In what ways does a forecaster act as a watchman? Are there times when the stakes are much higher than others? Is there a danger of too many false alarms versus failing to warn unless the outcome is obvious? This may seem like an irrelevant situation; however, forecasters have been charged and held liable for errant forecasts.

Bobboz/shutterstock.com

The planets of the solar system along with the sun. The sizes are to correct scale, but the distances are not. The planets are so far apart and from the sun that it is not possible to show them to scale in this figure.

16.1 EARTH AS A PLANET

We have already studied Earth in terms of its physical geologic composition (see Chapters 1 and 5 in particular). Now we want to consider Earth in the context of it being one of the eight planets that orbit the sun. It is helpful to begin our study of the other planets with Earth as a model, for some of the things that we learned about Earth may apply to those other planets. For instance, we know that Earth is differentiated into a core, mantle, and crust. There is evidence that other planets are at least partially differentiated as well. Many of the minerals found on Earth are also found in astronomical bodies. In another unit, we explored meteorology, the study of Earth's atmosphere. Most planets have atmospheres too, though their atmospheres are very different from Earth's atmosphere. One of the best ways to explore the planets is to employ **comparative planetology**, the study of planets by noting their similarities and differences. In this context, Earth often serves as the standard for comparison. Since we know so much about Earth's moon, it is a good basis of comparison to the satellites of the other planets.

16.2 A SURVEY OF THE SOLAR SYSTEM

What is contained in the solar system? Of course, the sun, which contains more than 99.9% of the solar system's mass, is in the solar system. But since the sun is a star, we will defer discussion of it to chapter 17. We already mentioned that there are eight planets, but what is a planet? A **planet** is a large body that orbits the sun. Millions of individual objects probably orbit the sun, but only eight are large enough to be called planets. We shall return later to the question of the minimum size for a planet. We ought to mention that the sun's gravity compels planets and other objects to orbit the sun. Orbits generally are elliptical, but planetary orbits are very close to being circular. The orbits of the planets lie in nearly the same plane, so the sun's planetary system is very flat. Earth's orbital plane, the **ecliptic**, is the plane that we use for reference.

A **satellite** is a body that orbits a planet (figure 16.1). Earth has one satellite, the moon. The moon is the name of Earth's natural satellite. Sometimes people call satellites of other planets moons, but technically, this is not proper. Here we shall endeavor to use the correct term *satellite*. Two planets, Mercury and Venus, have no satellites, but other planets have scores of satellites. Two satellites, one of Jupiter's and one of Saturn's, are larger than the planet Mercury (figure 16.2.). You may wonder why a satellite larger than Mercury is not a planet too. Size is not the issue here; it is what you orbit that matters. If these two satellites orbited the sun in their own right, they would be planets, but since they orbit planets, they are satellites.

What are the millions of smaller objects that orbit the sun? The smaller objects in the solar system generally are either **minor planets** or **comets**. For many years, the main distinction between minor planets and comets was composition. Astronomers thought that minor planets were rocky, while comets were made of various ices, such as water, dry ice, and frozen methane. However, in recent years there has been a blurring of this distinction, for many minor planets have turned out to be icy too. This difference in composition still is important, but perhaps more

FIGURE 16.1. Jupiter has four large satellites, shown here to scale with Jupiter.

FIGURE 16.2. Saturn's satellite Titan, the second largest satellite in the solar system, compared in size to Earth and moon.

important is the different sorts of orbits that comets and minor planets have. Minor planets have orbits that are similar to planetary orbits. That is, their orbits are nearly circular and lie very close to the plane of the orbits of the planets (figure 16.3). The common name for minor planets is asteroids, but astronomers prefer to call them minor planets, because it emphasizes their orbital similarity to planets. On the other hand, comets tend to have orbits very different from planets. Their orbits are very eccentric ellipses. This causes a comet to spend very little time close to the sun (a point we call **perihelion**) and much of its time far from the sun where it is very cold. Another difference is that cometary orbits generally do not lie close to the plane of the orbits of the planets, but instead are highly inclined to the solar system plane (figure 16.4). Planets and minor planets orbit the sun in the same direction, counterclockwise as viewed from above Earth's North Pole. While many comets orbit in this direction, many orbit **retrograde**, or clockwise, as viewed from above the North Pole. To many astronomers, the vast differences in orbits and composition between comets and minor planets suggest different origins and histories.

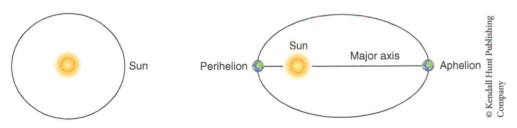

FIGURE 16.3. A comparison of a circular orbit and a very eccentric orbit.

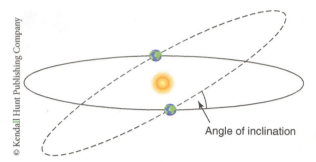

Angle of inclination

FIGURE 16.4. The orbit (dashed curve) is shown inclined to the plane of the solar system (solid curve).

16.3 THE MOON

The moon is 2,000 miles in diameter, about one quarter the size of Earth. However, the moon has only 1.3% Earth's mass. The moon orbits Earth at a distance of about 400,000 km (250,000 miles). It takes the moon a month to orbit Earth. In fact, the moon's orbital period is the basis for the month. We know from Genesis 1:16 that one purpose for the heavenly bodies is the marking of time. Earth's rotation period is the definition of the day, and the year is Earth's revolution period around the sun. The day, month, and year are the natural time units, because they have an astronomical basis. The week is not a natural unit of time, but the seven-day creation account of Genesis 1 gives us the basis for the week, and Exodus 20:11 further reinforces this. Apart from biblical creation, the origin of the week is a mystery.

Throughout the month, the moon's appearance changes. We call these changes in the moon **lunar phases** (figure 16.5). What causes lunar phases? The moon does not shine by its own light. Instead, it merely reflects the light that it receives from the sun. Some people erroneously think that Earth's shadow causes the moon's phases, but that would be a lunar eclipse, something that we will discuss in the next section. Since the moon is a sphere, the sun lights only half of the moon. The moon's phases depend upon how much of the lit half of the sun we see, and that changes as the moon orbits Earth. Let us begin by considering the moon's appearance when it passes between Earth and sun on its orbit. We call this the

new phase, because it is considered to be the start of the lunar cycle. When new, the lit half of the moon faces away from Earth, so only the dark half of the moon faces Earth. The dark half is not very bright, but, even worse, the new moon appears in the same part of the sky as the sun. This makes it impossible to see the new moon from Earth's surface. The new moon is invisible for 2–3 days.

Figure 16.6 illustrates the cause of lunar phases. A few days after new moon, the moon has moved so that a small portion of its lit half is visible in the western sky after sunset. We call the small lit sliver the *crescent phase*. More specifically, this is the waxing crescent. We call this the waxing crescent, because each successive evening the lit portion waxes, or grows, larger. About a week after new moon, we see

FIGURE 16.5. Lunar phases from waxing crescent, first quarter, waxing gibbous, full, waning gibbous, third quarter, to waning crescent.

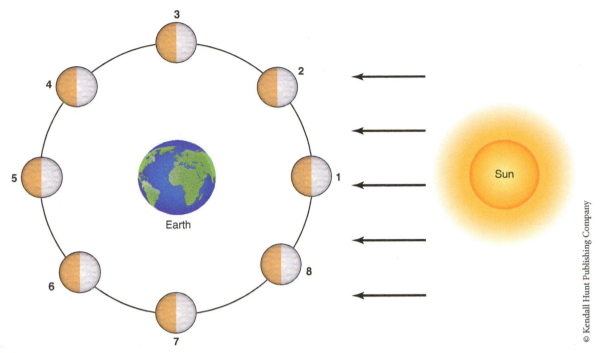

FIGURE 16.6. An Illustration of the position of the Moon at each of the lunar phases: new (1), waxing crescent (2), first quarter (3), waxing gibbous (4), full (5), waning gibbous (6), third quarter (7), and waning crescent (8). Imagine how much of the lit side of the Moon (white) that you would see from the earth at each numbered position.

half of the moon's lit side. This half-lit phase is *first quarter*. Why do we call this phase a quarter, when the moon appears half-lit? The name first quarter refers to the orbital period, not the moon's appearance. At first quarter, the moon has completed one quarter of its orbit around Earth. A few days after first quarter, the moon appears more than half-lit. We call this phase waxing gibbous. About a week after first quarter (two weeks after new), the moon is fully illuminated. This is *full moon*. The moon is now halfway through its orbit around Earth.

Through the first half of its orbit, the lit potion that we see has waxed, but now the lit portion will wane, or shrink. All of the waxing phases are visible in the early evening, but all of the waning phases are visible in the early morning. In the northern hemisphere, the waxing phases are lit on the right, while the waning phases are lit on the left. The reverse is true in the southern hemisphere. The waning phases occur in reverse order of the waxing phases. First, there is waning gibbous, then third quarter, then waning crescent, followed by new moon once again.

16.4 ECLIPSES

When the moon's phase is new, (figure 16.7) it is possible that it could pass exactly between the sun and Earth. When that happens, the moon's shadow falls on Earth. We call this a **solar eclipse**. If the alignment is just right, the sun is completely blocked out,

producing a total solar eclipse. A total solar eclipse is spectacular. The sky gets dark, and the brightest stars may become visible. Animals think that it is evening, and they proceed to go to sleep. The outer layers of the sun's atmosphere become visible. One part of the

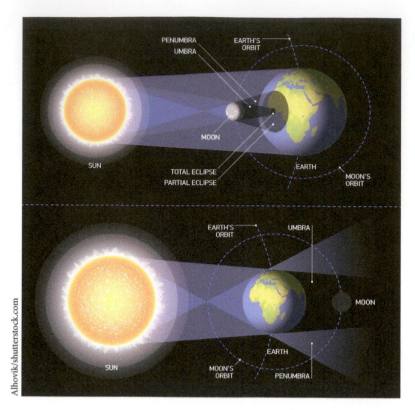

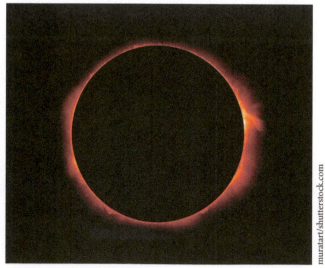

FIGURE 16.7. Illustration of how solar and lunar eclipses happen. Notice that a solar eclipse is visible over only a small part of Earth, but half of Earth (the night side) can view a lunar eclipse.

sun's atmosphere is the corona, a pearly white halo that can extend out several solar radii (figure 16.8). Near the sun's limb, or edge, blood-red loops called prominences appear (figure 16.9). During *totality*, one safely can look directly at the sun without any filters. However, totality lasts only a few minutes (figure 16.10). For more than an hour before and after totality, the moon partially blocks the sun.

FIGURE 16.8. The solar corona as seen during a total solar eclipse.

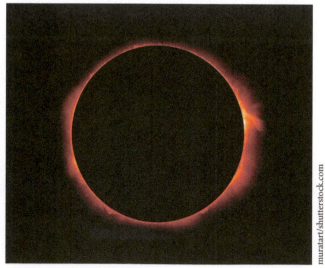

FIGURE 16.9. Solar prominences appear as loops off the limb (edge) of the sun.

FIGURE 16.10. The diamond ring effect that occurs immediately before and after a total solar eclipse.

We call this a *partial eclipse* (figure 16.11.). A partial eclipse is not safe to observe without proper filters. You ought not to try this unless you have the proper filters. The path of totality is very narrow on Earth's surface, only a couple of hundred miles wide at most. For a large area on either side of the path of totality, a partial solar eclipse is visible.

A *total solar eclipse* is very rare – one happens somewhere on Earth about every year and a half, but a total solar eclipse is visible only over a small region of Earth. On average, for a given location on Earth there is a total solar eclipse about once every four centuries. The rarity of total solar eclipses largely is because both the sun and the moon subtend an angle of ½ degree in the sky. This is because that while the sun is 400 times larger than the moon, the sun is 400 times farther away. If the moon were slightly smaller or farther away, there would be no total solar eclipses. The fact that these circumstances for total solar eclipses combining rarity and extreme beauty exist only for Earth where there are people to enjoy total solar eclipses suggest design to some people. The moon's orbit around Earth is not circular, but instead is an ellipse. The point on the orbit where the moon is closest to Earth is **perigee**. The point on the orbit farthest from Earth is **apogee**. If the moon is near apogee at the time of an eclipse, the sun' shadow fails to reach Earth, and instead there is a thin ring of the sun left visible around the moon. We call this an annular eclipse (from the word *annulus*, meaning ring) (figure 16.12). Like a partial solar eclipse, one cannot safely look directly at an annular eclipse.

At full moon, it is possible for Earth's shadow to fall on the moon. This is a **lunar eclipse** (figure 16.13). If the moon is completely immersed in Earth's shadow, there is a total lunar eclipse. However, if the moon is only partially in Earth's shadow, the eclipse

FIGURE 16.12. An annular eclipse happens when the Moon's umbra does not reach the earth. Here we include images of the partial phases before and after anularity.

FIGURE 16.11. A partial solar eclipse (here viewed through thin clouds) occurs when we are not in the Sun's umbra.

FIGURE 16.13. Montage photograph of the total lunar eclipse that occurred on April 15, 2014.

is a partial lunar eclipse. We call Earth's shadow the **umbra**. The umbra is round. Surrounding the umbra is the circular **penumbra**. The penumbra is a region where the sun's light is partially blocked. The portion of the moon that is in the penumbra is dimmed slightly, but it is difficult for the human eye to notice this. However, the umbra appears dark to the eye, and as it moves across the lunar surface, it appears as if the umbra has taken a semi-circular bite out of the moon. The circular shape of Earth's umbra was an argument that many ancients used to show that Earth was spherical, because only a spherical Earth casts a circular shadow regardless of the orientation.

During a total lunar eclipse, the moon is darkened tremendously, but the moon usually does not disappear. This is because Earth's atmosphere bends and scatters sunlight into Earth's umbra. Most of the light that is bent is red, so the moon most frequently looks reddish during a total lunar eclipse. However, there are many possible colors during a lunar eclipse. Sometimes the moon is nearly black, while other times it is golden, orange, or peach.

16.5 LUNAR FEATURES

The moon's density is 3.3 g/cm³, far less than Earth's 5.5 g/cm³. Earth's crustal and upper mantle rock have density close to the moon's 3.3 g/cm³, so the moon cannot have a large iron/nickel core as Earth has. It is possible that the moon may have a very tiny core, but that is not clear.

The most common lunar surface features are its many craters. Where did the craters come from? Since astronomers discovered craters on the moon after the invention of the telescope four centuries ago, scientists have debated that question. For much of the past four centuries, many scientists thought that volcanoes produced most lunar craters. However, others thought that craters were caused by bodies striking the lunar surfaces. In the twentieth century, most planetary scientists concluded that impacts rather than volcanoes caused the vast majority of craters on the moon. Since the 1960's, we have discovered that many other bodies in the solar system have craters too. As on the moon, planetary scientists think that most of the craters are the result of impacts. The impacting body is far smaller than the crater it produces. This is because the kinetic energy of the impacting body excavates most of the material to produce the crater. Kinetic energy is the energy of motion, and because impacting bodies are moving so fast, they possess huge amounts of kinetic energy. An impacting body stops when it slams into a surface, releasing its kinetic energy in an explosion that forms the crater. What were the impacting bodies? They probably were minor planets, with comets contributing a few craters.

The moon rotates synchronously. That is, the moon rotates and revolves at the same rate, once per month. **Synchronous rotation** causes one side of the moon to face Earth as it orbits Earth. Before 1959, when the first spacecraft passed the backside of the moon and sent back photographs, no one knew what the backside of the moon looked like. Because the lunar backside was a mystery, many people called it the dark side of the moon, meaning that it was unknown. Before Europeans had explored much of the interior of Africa in the nineteenth century, many people called Africa the Dark Continent for the same reason. Once the geography of Africa became known, the term *Dark Continent* declined. In similar fashion, the use of the phrase *dark side of the moon* declined once we mapped the lunar far side. Unfortunately, when many people now hear the term *the dark side of the moon*, they erroneously think that it refers to a side of the moon that is perpetually shrouded in darkness. However, the entire lunar surface receives light from the sun throughout the month as the moon rotates and revolves. The only exceptions might be the bottoms of a few very deep craters near the lunar poles.

Looking at the moon with the naked eye, you probably have noticed that the lunar surface has darker and lighter regions (figure 16.14). Astronomers call the lighter regions the **highlands**, because the highlands are at higher elevation than the darker regions. The darker regions are the **maria** (ma´ ree-uh), a Latin word meaning seas. This term goes back four centuries, when astronomers began studying the lunar surface with telescopes. They thought that the maria might be bodies of water, because the maria were dark and smooth. The word maria is plural; the singular is mare (ma´ ray). The maria and highlands have different color, because they are made of different kinds of rock. The highlands are made of rock similar to granite, while the maria consist of rock similar to basalt. Since basalt is denser than granite, the maria have sunk lower on the lunar mantle, while the highlands have risen higher, much as the continents float higher on Earth's mantle than the ocean basins do.

There are other differences between the lunar highlands and the maria. As previously mentioned, the maria are relatively smooth. However, the highlands are rugged, because they contain many craters. Why are there so many craters in the highlands and so few on the maria? It would seem very unlikely that

objects that struck the lunar surface to form craters somehow avoided the maria. It is more likely that the entire lunar surface was struck by impacts, so what happened to the craters on the maria? A clue comes from the observation that the lunar maria appear to be circular. The isolated maria definitely are round, but the ones that overlap look like they are the intersection of several circles. When people look at the moon through a telescope with low magnification, many of them observe that the maria look like very large craters. Astronomers think that after many of the moon's craters had already formed several large bodies struck the moon to produce extremely large craters. These craters were so large that astronomers call them impact basins. The impacts that formed the impact basins were so violent that they produced deep fractures that reached molten material in the moon. The fractures acted as conduits to bring some of the molten material to the lunar surface. The molten material filled the impact basins, and in some locations spilled onto surrounding terrain. Because the volcanic material came from deep in the moon, it has higher density than surface rocks. This caused the regions of the

© David Rives-www.davidrives.com

FIGURE 16.14. This image of the full moon shows the lunar highlands (lighter regions) and maria (darker regions).

filled impact basins to sink to lower elevation. After the lava cooled, a few impacts on the new surface produced the few craters that we see on the maria.

The front side of the moon is about equally divided between highlands and maria. However, the backside of the moon is about 95% highlands. Why are there far more maria on the front side of the moon than the backside of the moon? If the impact basins formed over many millions of years as evolutionists believe, then they ought to be more uniformly distributed on the lunar surface. Evolutionists respond that the lunar crust is thicker on the backside of the moon,

preventing volcanic overflow there. However, there is a strong correspondence between impact basins and the maria, and there are relatively few impact basins on the lunar backside. However, what if the moon is far younger than billions of years, as recent creationists believe? The impact basins would have formed in a relatively short period. Since it takes the moon a month to orbit Earth and the maria are on one side of the moon, we might conclude that the impacts that led to maria formation took place in a period of less than two weeks. Some creation scientists think that this might have happened at the time of the Flood.

16.6 TWO TYPES OF PLANETS

Table 16.1 lists a number of orbital and physical characteristics of the planets. Many of the characteristics are expressed in terms of Earth's traits. For instance, the second column lists the average distance of each planet from the sun in terms of Earths' distance from the sun. We call the average Earth-sun distance the **astronomical unit**, usually abbreviated AU. Notice that the first four planets, Mercury, Venus, Earth, and Mars, are closely spaced, but that the next four planets, Jupiter, Saturn, Uranus, and Neptune, are much more widely spaced. In fact, the first large break is between Mars and Jupiter. This

break is obvious in many other ways too, and we shall use this break to make a distinction between two types of planets. The first four planets are similar to one another, and since Earth is in this group, we call these the **terrestrial planets**, or Earth-like, planets. Earth is the largest of the terrestrial planets. Similarly, the last four planets are very similar, and Jupiter is the largest of those planets, so we call them the **Jovian planets**, or Jupiter-like, planets. (figure 16.15).

Mentioning size, the next column in the table gives the diameter of each planet in terms of Earth's diameter.

TABLE 16.1
Properties of the Planets.

Planet	Distance from sun (AU)	Diameter (# of Earths)	Mass (# of Earths)	Density (g/cm³)	Rotation period (days)	Number of satellites	Rings
Mercury	0.387	0.383	0.055	5.43	58.6	0	No
Venus	0.723	0.949	0.815	5.24	−243	0	No
Earth	1.000	1.000	1.000	5.515	0.997	1	No
Mars	1.52	0.533	0.107	3.94	1.03	2	No
Jupiter	5.20	11.2	318	1.33	0.414	16	Yes
Saturn	9.58	9.45	95.2	0.70	0.444	18	Yes
Uranus	19.3	4.01	14.5	1.30	−0.718	15	Yes
Neptune	30.2	3.88	17.2	1.76	0.671	8	Yes

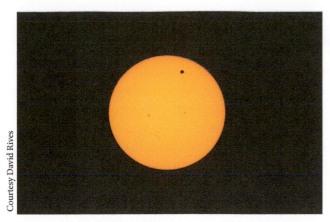

FIGURE 16.15. ·Larger spot near the top is Venus, caught here in a rare transit across the sun's face. The other spots are sunspots.

Of course, in these units Earth's size is one; notice that the other three terrestrial planets are smaller than Earth. On the other hand, you can see that the four Jovian planets are far larger than any terrestrial planet. The next column compares mass, with Earth being the standard of measure once again. You can see that, as with size, Earth is the most massive terrestrial planet, but that all the Jovian planets dwarf Earth's mass. Knowing a planet's mass and, size, we can compute its density. The next column of Table 16.1 lists the density of each planet. Density is a measure of how much matter occupies a certain volume. Density depends upon composition, so density is very important in deducing what a planet's composition likely is. As you can see, the densities of the terrestrial planets range from 3.9 to 5.5 g/cm³. This is consistent with rocky composition, so we sometimes call the terrestrial planets the rocky planets. Now look at the densities of the Jovian planets. Their densities are less than 2 g/cm³ and are far less than those of the terrestrial planets. Since rocks normally have density of at least 3 g/cm³, it is clear that the Jovian planets cannot contain much rocky material. Then what is the composition of Jovian planets? They most likely are made of hydrogen and helium. On Earth's surface, these elements are gases, so we frequently call the Jovian planets the gas giant planets. The Jovian planets have atmospheres, and the hydrogen and helium there are gaseous, but within their vast interiors, the pressure is so great that most of the hydrogen and helium there are liquid.

The next column lists the rotation periods of the planets, either in hours or in days. Venus and Uranus have negative signs; this means that these two planets rotate backwards, or retrograde. You can see that two terrestrial planets, Mercury and Venus, rotate very slowly. Mars and Earth rotate much more quickly, with either one taking about a day to rotate. However, notice that the Jovian planets all rotate very quickly, more quickly than Earth, the terrestrial planet with the shortest rotation period. This suggests another difference between the two types of planets – Jovian planets have shorter rotation periods, while terrestrial planets have longer rotation periods. The next column compares the number of satellites. There are only three satellites among the terrestrial planets, but the satellites of the Jovian planets total nearly 200, with the discovery of new ones continuing. The last column lists whether a planet has a ring system. All the Jovian planets have ring systems, but none of the terrestrial planets do, so this appears to be a distinction between the two types of planets. Table 16.2 summarizes the differences between the two types of planets.

TABLE 16.2

Comparison of the Terrestrial and Jovian Planets.

Planet type	Distance from sun	Diameter	Mass	Density	Rotation period	Number of satellites	Rings
Terrestrial	Near	Small	Small	High	Long	Few	No
Jovian	Far	Large	Large	Low	Short	Many	Yes
Pluto	39.3	0.187	0.00251	2.10	−6.39	1	No (?)
Pluto	J	T	T	J	T	J	T (?)

When Clyde Tombaugh discovered Pluto in 1930, astronomers welcomed it as the ninth planet. However, in 2006, astronomers officially decided that Pluto was not a planet. What happened, and why did astronomers make this change? The bottom row of Table 16.2 gives the relevant parameters for Pluto (figure 16.16). We can examine those traits, along with the conclusions listed in Table 16.2 to determine how Pluto stacks up as a planet. The final row of Table 16.2 lists the conclusions about Pluto. As you can see, Pluto is all over the place, with characteristics of either group. This is one reason why astronomers decided to reclassify Pluto – it did not fit in with the classification scheme of the planets. Another reason is that Pluto is just too small (figure 16.17.). Earth has 400 times more mass than Pluto, and Mercury, currently considered the smallest planet, is 22 times more massive than Pluto. For 60 years after its discovery in 1930, Pluto remained the sole known object with a planet-like orbit orbiting the sun beyond Neptune. But the early 1990's astronomers began finding many other such objects. Many of those objects have sizes comparable to Pluto, and in 2003 astronomers found one slightly larger than Pluto. If Pluto was a planet, then there is no good reason why these other large objects are not planets too. This could quickly lead to there being scores and even hundreds of planets. Astronomers thought that they ought to draw the line at some point. In 2006, the International Astronomical Union voted to define what a planet is in a manner that excluded Pluto. This was a controversial decision, and astronomers will be revisiting and updating this issue in years to come.

If Pluto is not a planet, then what is it? With the discovery of so many objects beyond the orbit of Neptune, it appears that this is another asteroid belt (the one between Mars and Jupiter has been known since the 19th century), and Pluto is just one of the larger members of this belt. There are different classifications for these bodies. Some astronomers think that they are the **Kuiper belt**, the supposed source of short period comets, so they call them KBOs (Kuiper Belt Objects). Other astronomers prefer TNO's (**Trans Neptunian Objects**),

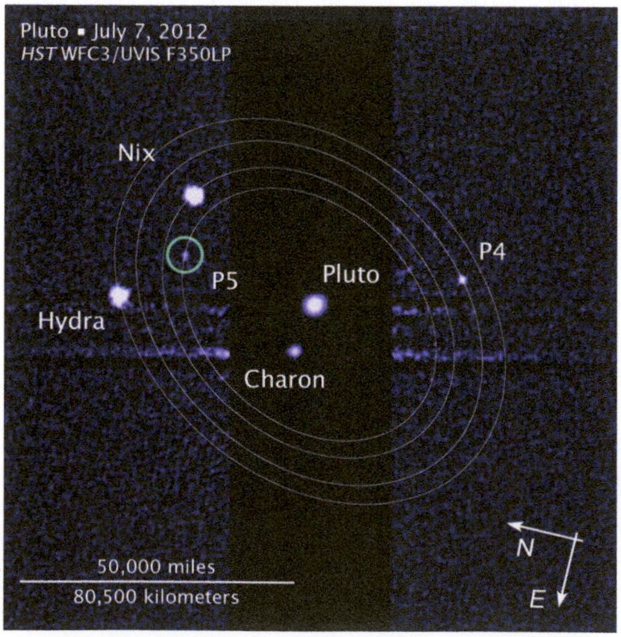

Pluto ▪ July 7, 2012
HST WFC3/UVIS F350LP

Nix

P5

Pluto

P4

Hydra

Charon

50,000 miles
80,500 kilometers

N

E

Courtesy NASA

FIGURE 16.16. Pluto has five satellites.

Courtesy NASA

FIGURE 16.17. In this size comparison you can see the difference in size between Pluto, Earth, and the Moon.

the preferred term in this text. Pluto and a few of the other larger minor planets are spherical. This is unusual, because most minor planets are odd shaped. What causes a body to be spherical? A sphere is the minimum shape that a body can have if the body has sufficient gravity. To have sufficient gravity, a body must have much mass. Most minor planets lack the mass to produce sufficient gravity to make them spherical. Astronomers recently coined a term to describe spherical minor planets such as Pluto – **dwarf planets**.

Despite common misconception, the demotion of a planet is not new. The first minor planet, Ceres, was discovered in 1801 in between Mars and Jupiter, a location where many people expected a planet to be. Astronomers called Ceres a planet for more than 40 years, even after a few more asteroid belt objects were discovered. By 1845 astronomers concluded that, much like Pluto, Ceres was simply too small to be a planet. Incidentally, since Ceres is spherical, it too is a dwarf planet. The number of dwarf planets is sure to increase in number over the years.

16.8 COMETS

Comets are small icy bodies with an admixture of very tiny solid dust particles. Their densities are very low, suggesting that they are very porous and fragile. Astronomers call this the **"dirty iceberg" theory**. Actually, what we have described is the nucleus of a comet (figure 16.18). As we previously mentioned, comet orbits are very long ellipses. The sun is near one

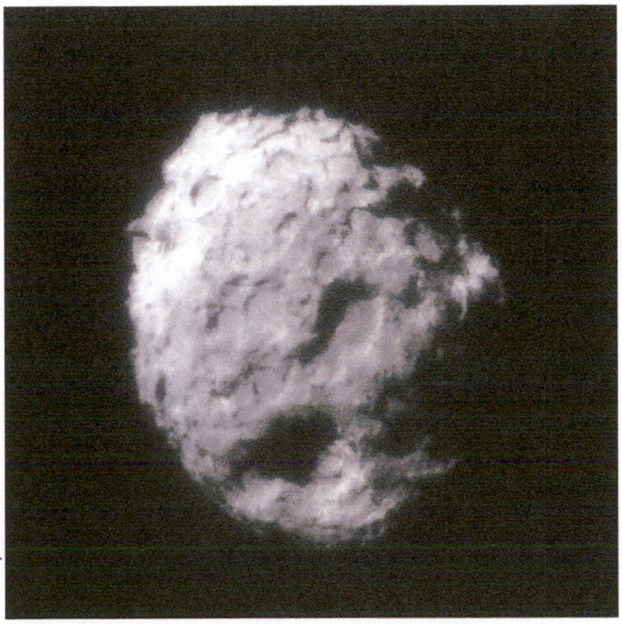

Courtesy NASA

FIGURE 16.18. The nucleus of Comet Churyumov–Gerasimenko as imaged by the Rosetta mission.

end of the ellipse. The point on the orbit nearest the sun is the perihelion, and the point farthest from the sun is **aphelion**. A comet moves very quickly when near perihelion, but it moves very slowly when near aphelion. As a result, comets spend most of the time very far from the sun. When so far from the sun, the material in a comet nucleus remains frozen. However, once each orbit when a comet nucleus approaches perihelion, the heat of the sun evaporates much of the ice directly into gas without going into a liquid phase (we call this sublimation). Much dust dislodges along with the gas. The gas expands into a large cloud around the nucleus, and excitation by solar radiation causes the cloud to glow. We call this cloud the **coma**. A comet's coma may be many tens of thousands of km across, even though the nucleus is only a few km across. Solar radiation pushes the microscopic dust particles away from the sun, producing a **dust tail**. The solar wind, an out rush of charged particles from the sun, pushes the gas away from the sun to produce a gas, or **ion, tail**. The dust tail shines by reflecting sunlight, while the ion tail glows from ionization from the sun's radiation. (figure 16.19)

The greatest brightness of a comet typically occurs shortly after it passes perihelion. However, how bright a comet appears to us on Earth depends upon its distance from Earth and how far from the

FIGURE 16.19. Comet Hale-Bopp is sometimes called the Great Comet of 1997.

comet loses a significant portion of its mass each pass near perihelion. It is obvious that a comet cannot orbit the sun many times before losing so much of its volatile material (the ices that sublime and then fluoresce to produce the bright coma and ion tail) that the comet fails to shine anymore. Indeed, some comets with periods short enough to have been observed on several returns to perihelion have grown progressively fainter over the centuries. The most famous comet, Halley's, with a period of 75–76 years, is a good example of this gradual fading. A conservative estimate of how many trips that a comet can make around the sun and still be visible is about one hundred. There is a maximum orbital size for a comet. If a comet's aphelion is too far from the sun, the gravitational perturbations of other stars can permanently remove a comet from the sun's grasp. The maximum orbital size to prevent this from happening corresponds to an orbital period of a few million years. Multiplying these two figures, we find that comets cannot be visible after they have been orbiting the sun for more than a few hundred million years. Since most scientists think (based upon various old-Earth assumptions and methods) that the solar system is 4.6 billion years old, there ought not to be any comets left.

The situation is even worse for an old solar system, for there are two other mechanisms whereby comets are lost to the solar system. The gravitational perturbations of Jupiter and the other Jovian planets frequently add energy to a comets orbit, effectively ejecting them from the solar system, never to return. Astronomers have observed ejection of comets many times, and this represents a catastrophic loss of comets. Gravitational perturbations can rob a comet of orbital energy, which results in a shorter period. This results in more frequent trips to perihelion, with even faster gradual wearing out. Another catastrophic loss of comets is collisions with planets. Astronomers have observed this once. In 1994, Comet Shoemaker-Levi IX slammed into Jupiter. Two years earlier, this comet passed close to Jupiter on its way toward perihelion. Tidal forces of Jupiter shredded the comet into more

sun it appears in our sky. On rare occasions, a comet can appear extremely bright, and its tail can extend far across the sky. However, a comet looks best in a dark sky, which can be difficult for many city dwellers to achieve. Until recent times, people viewed comets as bad omens that portended disasters. This likely resulted from what appeared to be very erratic behavior of comets. The sun, moon, and planets follow complicated motions, but they are confined to near the ecliptic and are straightforward to understand. However, comets tend to appear without warning, move across the sky in odd directions, and then disappear never to be seen again. Of course, comets follow the same sort of physical laws that the planets do. However, comets have very elliptical orbits that are highly inclined to the ecliptic, and they have longer orbital periods than the naked-eye planets. These factors result in any particular comet being visible for only a short portion of each orbit near perihelion, making its motion appear erratic.

Comet nuclei are very small (only a few km across) and hence contain very little mass. Yet, a than two dozen pieces. As those pieces moved away from perihelion, they struck Jupiter over several days, leaving behind dark marks that persisted for several days. This comet no longer exists.

Evolutionary astronomers are well aware of this problem, so they have offered two solutions to it. About 1950, the Dutch astronomer Jan Oort suggested that the sun is surrounded by a spherical cloud of comet nuclei, with the nuclei orbiting far from the sun (thousands of AU). Today we call this hypothetical distribution of comet nuclei the **Oort cloud**. The gravitational perturbations of passing stars or other objects in the galaxy were supposed to alter the orbits of these nuclei so that they fell to the inner solar system to become visible as comets. Today astronomers think that the dominant mechanism of injecting comets into the inner solar system is tides produced by the galaxy. As old comets died out, new ones would rain down to establish something close to a steady state. Oort suggested his cloud to explain the origin of long period comets, comets generally with periods greater than about 200 years. This is not an arbitrary distinction, for long period comets generally have orbits highly inclined to the ecliptic, with about half of them orbiting the sun the same direction that planets do, but half in retrograde orbits.

Short period comets, comets with periods less than about 200 years, tend to have low inclination orbits and orbit prograde, the same direction that planets orbit the sun. About the time that Oort proposed his cloud, the Dutch-born American astronomer suggested a belt that bears his name for the source of short period comets. The Kuiper belt supposedly lies just beyond the orbit of Neptune (recall that this is where Pluto is, making Pluto a member of KBO's in the estimation of many astronomers). Gravitational perturbations of the Jovian planets supposedly convert KBO's into short period comets. For about thirty years astronomers assumed that gravitational perturbations on long period comets could convert them to short period comets, so most of them assumed that the Kuiper belt was not necessary. However, by the early 1980's, astronomers had run simulations that showed that gravitational perturbations were far too inefficient to convert a significant number of long period comets into short period comets before they would be destroyed. Hence, astronomers began to accept the Kuiper belt and began to look for objects there, which resulted in the discovery of the new asteroid belt (TNO's). Today, most astronomers think that the Kuiper belt is the original source of comets, and that planetary perturbations not only produce short period comets, but that those same perturbations populate the Oort cloud from which come long period comets.

Does either the Oort cloud or the Kuiper belt exist? Even if the Oort cloud exists, it is unlikely that we will ever observe it. Are the TNO's actually KBO's? There still much room for debate on this. Many TNO's appear to be far too large to be comet nuclei. Furthermore, we know the densities of a few KBO's (Pluto is the best example). The inferred composition of TNO's do not match the composition of comets. At this time, it appears that the existence of comets still is a good argument for a recent origin of the solar system, suggesting that the solar system is far younger than billions of years.

16.9 MINOR PLANETS

The first minor planet was discovered in 1801. (figure 16.20) By the end of the 19th century, the number of minor planets had grown to about 300. By 1960, the number had grown to nearly two thousand. As of 2015, there were more than 600,000 known minor planets, and the discoveries continue unabated. Many of the ones now discovered are very small, for most of the larger ones were found long ago. The exceptions

FIGURE 16.20. The minor planet Ida has a small satellite, Dactyl.

Astronomers also classify minor planets according to composition, as inferred from color, spectroscopy, or density measurements. Minor planets in the inner solar system, those orbiting closer to the sun than the asteroid belt, tend to be made of denser material, suggesting composition similar to the terrestrial planets. Minor planets of the outer solar system, those beyond the asteroid belt, tend to be icier. This is similar to the composition of other small objects of the outer solar system, comets, and the satellites of the Jovian planets. The minor planets of the inner solar system further subdivide with regards to composition. Some primarily are made of iron and nickel. Astronomers term these M-type minor planets (the M stands for metal). Other minor planets are rockier. They are S-type minor planets, the S standing for silicate, a major component of many rocks. Some rarer minor planets are C type, with C standing for carbonaceous. Many astronomers think that C-type minor planets may be burned out comets.

to this would be the ones among the TNO's. Most minor planets are in the two belts, one between Mars and Jupiter, and the other one beyond Neptune. However, there are many found orbiting elsewhere. Astronomers often classify minor planets according to their orbits. Of particular interest are the minor planets whose orbits cross Earth's orbit, for this could result in collisions with Earth. Astronomers call these **near-Earth asteroids**, or NEA's for short.

16.10 METEORS

Occasionally, a piece of a minor planet or a comet collides with Earth. Earth's atmosphere protects us from most of these. As these fragments enter the atmosphere at an altitude of about 100 km, they interact with molecules in the atmosphere to heat the air to very high temperature. The heat is so intense that the air begins to glow around the quickly moving particle. We see the glow at night as a **meteor**, or shooting or falling star (figure 16.21). The hot air erodes the incoming particle so that smaller particles completely burn up. If an incoming object is large enough, a portion may survive to the ground. If we find a surviving piece on the ground, we call it a **meteorite**. There are three basic types of meteorites, irons, stoneys, and stoney-irons. An iron meteorite is a piece of an M-type minor planet. A stoney is a part of an S-type minor planet. The stoney-iron type are a rarer mix of the two. The ice of a comet fragment completely burns up, and the small solid pieces in a comet are too small to survive either.

There are two basic types of meteors: sporadic and shower. **Sporadic meteors** can appear anywhere in the sky at any time and travel in any direction. They generally appear alone. On a dark, clear night a typical observer can see 5–6 sporadic meteors per hour. However, throughout the year there are certain times when many more meteors than normal are visible. This is a **meteor shower**. Shower meteors do not move randomly. A shower meteor can appear anywhere in the sky, but if we trace their motions backward, they appear to diverge, or radiate, from one spot in the sky. We call the point from which they appear to diverge the radiant. We name a meteor shower by the location of the radiant. For

FIGURE 16.21. A meteor results when a fast-moving small particle of matter burns up high in Earth's atmosphere.

instance, one of the best meteor showers of the year is the Perseids, so-called because the radiant is in the constellation of Perseus. The Perseids are visible for about two weeks in August, but they peak around August 13. We can track the motions of incoming meteors to figure out what kind of orbits their parent bodies had around the sun. Sporadic meteors come from objects that have orbits similar to minor planets – nearly circular orbits that are not too inclined to Earth's orbital plane. Thus, it appears that sporadic meteors come from minor planets. On the other hand, the bodies that produce shower meteors have orbits that resemble comets. In fact, some meteor showers come from bodies that follow the orbits of known comets. What about meteor showers that are not associated with any known comet? They probably come from extinct comets. Remember that comets do not last too long.

16.11 THE ORIGIN OF THE SOLAR SYSTEM

Scientists committed to evolution think that the solar system formed from a large cloud of gas and dust that collapsed under its own gravity. Gas clouds do not spontaneously collapse like this, so the process supposedly began by some unknown mechanism. According to this theory, most of the gas fell to the center to form the sun. The small percentage of remaining material flattened into a disk, from which the planets slowly formed. The first step in this process of planetary formation was for the small dust particles in the flattened to stick together. How this happened is unknown too. Gradually, the small particles grew larger by combining. Astronomers call these hypothetical particles **planetesimals**. Eventually, some planetesimals grew large enough so that their gravity began to dominate within certain regions of the solar system. These dominant planetesimals became the seeds to form the planets.

Why are there two types of planets, the terrestrial and Jovian planets? According to the evolutionary theory, the early sun heated up and evaporated all the lighter material (mostly hydrogen) from the planetesimals in the inner solar system, close to the sun. Since those planetesimals lost the lighter material, the composition was rocky, so the planets that formed from them (the terrestrial planets) have rocky composition. Planetesimals farther from the sun kept their lighter material, so the planets that formed from them (the Jovian planets) have lighter composition. The lighter material accounted for the bulk of the mass, so when the planetesimals closer to the sun lost the light material, they lost most of their mass. This accounts for why the terrestrial planets have far less mass than the Jovian planets. While this theory can explain the differences in mass and composition between the Jovian and terrestrial planets, it cannot explain other things. For instance, it does not explain why Jovian planets rotate more rapidly than terrestrial planets.

According to the theory, not all planetesimals ended up in the planets. Some of the planetesimals orbited the newly formed planets, and many of these planetesimals coalesced to form the satellites of the planets. Other planetesimals failed to form into planets or satellites. Those planetesimals that didn't form into planets or satellites formed minor planets and comets.

Of course, we know from the Bible that the solar system did not form gradually over millions of years. Rather, according to Genesis 1:14–19 that God made the sun, moon, and stars on Day Four of the Creation Week. The word *star* has a more specific meaning today than it did in ancient times. A star was any bright object in the sky other than the sun and moon. For instance, five of the planets (Mercury, Venus, Mars, Jupiter, and Saturn) appear as bright stars in the sky. The other two planets, Uranus and Neptune, are too faint to see with the naked eye. However, through binoculars Uranus and Neptune look like faint stars. By extension, when Genesis 1:16 records that God "made the stars also," that probably refers to all of the astronomical world, including the solar system.

16.1 Earth as a Planet

- Since Earth is a planet, we can learn some things about planets in general by studying the earth.
- Comparitive planetology is a useful way to study the planets.

16.2 A Survey of the Solar System

- Besides the Sun, the solar system contains planets, satellites, and smaller solar system bodies.
- The smaller solar system bodies are minor planets and comets.
- The minor planets and comets differ in terms of their composition and their orbits around the sun.

16.3 The Moon

- The Moon is our closest neighbor in space.
- One of the God-ordained purposes of the Moon is the reckoning of time. The Moon's orbit provides the basis of our month.
- The Moon's phases are caused by the relative geometry between Earth, the Sun, and the Moon. As the Moon orbits Earth each month, the amount of the lit portion of the Moon that we see changes.

16.4 Eclipses

- There are two basic types of eclipses: lunar and solar.
- Both solar or lunar eclipses may be either partial or total.

16.5 Lunar Features

- Impacts probably formed most craters on the Moon. Craters are common features on the surfaces of many other bodies in the solar sytem.
- The maria and the lunar highlands are the basic types of surface on the Moon.

16.6 Two Types of Planets

- There are two groups of planets, the terrestrial and Jovian planets.
- The terrestrial and Jovian planets differ in at least seven ways.

16.7 What About Pluto?

- While once thought to be a planet, Pluto no longer is considered to be a planet. There are two reasons for this: Pluto is too small, and it does not classify as either a terrestrial or Jovian planet.
- Some people now consider Pluto to be a larger member of the Kuiper or the trans-Neptunian objects. Either way, Pluto now is included with a special class of dwarf planets.

16.8 Comets

- Astronomers explain comets by them being dirty icebergs.
- We classify comets as either short-period or long-period, depending upon their orbits.
- If the solar system were billions of years old, there would be no more comets left. To explain this, evolutionists invoke the Oort cloud and the Kuiper belt.

16.9 Minor Planets

- Minor planets sometimes are called asteroids.
- Minor planets have composition and orbits that are different from comets.
- Astronomers generally classify minor planets by their composition.

16.10 Meteors

- Meteors are brief streaks of light caused by bits of debris burning up high in Earth's atmosphere.
- Meteors can be either sporadic or they can occur in showers. Comet debris produces meteor showers, while sporadic meteors are debris from minor planets.

16.11 The Origin of the Solar System

- Evolutionary scientists think that the solar system gradually formed over millions of years from a cloud of gas and dust.
- The Bible teaches that God made the earth first in Day One, and that He made the planets on Day Four. Both were rapid processes, not gradual ones.

KEY TERMS

Aphelion	Highlands
Apogee	Ion tail
Astronomical unit	Jovian planets
Coma	Kuiper belt objects
Comet	Lunar eclipse
Comparative planetology	Lunar phases
Dirty iceberg theory	Maria
Dust tail	Meteor
Dwarf planets	Meteor shower
ecliptic	Meteorite

Minor planet	Retrograde
Near-Earth asteroids	Satellite
Oort cloud	Solar eclipse
Penumbra	Sporadic meteor
Perigee	Synchronous rotation
Perihelion	Terrestrial planets
Planet	Trans-Neptunian objects
Planetesimal	Umbra

REVIEW QUESTIONS

1. What are the two types of planets? Which planets are in either group? In what ways are the two groups different?
2. Why is Pluto no longer considered a planet?
3. Why do comets present a problem for a 4.6 billion year old solar system?
4. What are the two types of lunar terrain? How are they different?
5. What sort of bodies do shower meteors come from?

FURTHER READING

DeYoung, Don, and John Whitcomb, *Our Created Moon: Earth's Fascinating Neighbor*, 2nd edition. Master Books.

DeYoung, Donald B., *Astronomy and the Bible: Questions and Answers*. BMH Books.

Lisle, Jason, *Taking Back Astronomy*. Master Books.

A Pocket Guide to Astronomy. Answers in Genesis.

REFLECT ON SCRIPTURE

Read the Day Four creation account from Genesis 1, as well as Psalm 19. List purposes for heavenly bodies found in either passage. Contrast what we can learn about God from natural revelation (Psalm 19:1–6) with what we can learn about God from special revelation (Psalm 19:7–14).

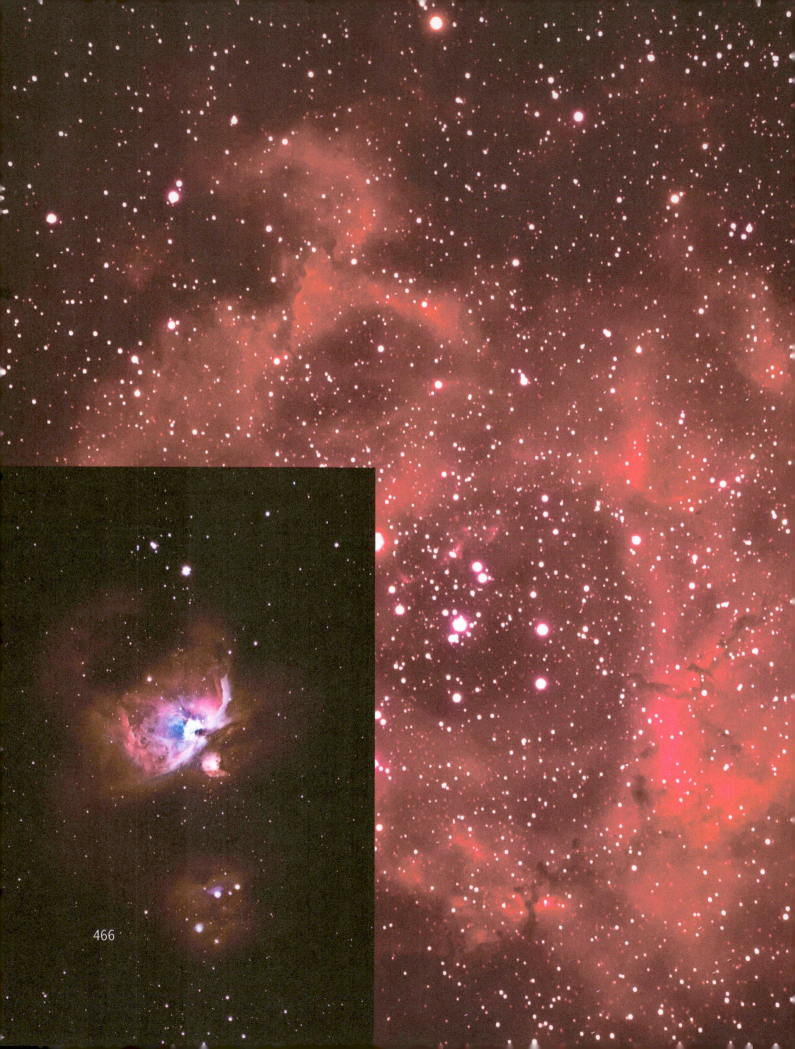

466

CHAPTER 17

BEYOND US: OUTSIDE OUR SOLAR SYSTEM

OUTLINE:

© David Rives-www.davidrives.com

The Rosette Nebula and the Nebula M42

467

In ancient times, most people thought of the sun as being very different from the stars. This probably was because the sun appears far brighter than the stars do. However, today we recognize that the sun is a star. The reason why the sun is so bright is that it is very close to us–the next nearest star is 275,000 times farther from us than the sun. Since the Bible was written in ancient times, it does not refer to the sun as a star. Instead, the Bible treats the sun, along with the moon, as objects separate from the stars. This does not mean that our modern view that the sun is a star is contrary to Scripture. Even today, everyone agrees that the sun appears very different from the stars. Besides a difference in brightness, the sun is the only star close enough that we can readily examine its surface. This means that we can study the sun in much more detail than any other star. Any conclusions that we reach about the sun could help us interpret other stars. We did a similar thing in the previous chapter when we used Earth as a template for studying other planets.

FIGURE 17.1. The sun, showing several groups of sunspots.

It is interesting to observe the sun with a telescope, but this can be very dangerous. The sun's light easily can damage human eyes, so it should only be observed with specialized equipment. There are some safe solar filters, but other solar filters are not. The safest solar filters fit over the end of a telescope, thus blocking most solar rays from entering the telescope. However, it is important to use caution when handling and storing solar filters, because damage to a filter can let harmful rays enter the telescope. The safest way to view the sun is by projecting the sun's image on a screen. The specialized equipment and know-how will provide appropriate eye protection.

The sun is a hot ball of gas (figure 17.1). Its diameter is 1.4×10^6 km, or about 109 times the diameter of Earth. The sun's mass is 2.0×10^{30} kg, about 330,000 times the mass of Earth. The **photosphere**, or visible surface of the sun, has a temperature of 5,770 K. From what we know of the physics of gases, the temperature, density, and pressure ought to increase with increasing depth inside the sun. We cannot directly measure conditions deep in the solar interior, but we can model them from what we know of physics. The best solar model tells us that the temperature in the center of the sun is more than 15 million K, and the density is about 150 g/cm³. This density is about 13 times that of lead, yet the material is a gas. This temperature is so extreme that normal atoms do not exist. Instead, all of the atoms in the sun's core are completely ionized–all of the electrons are stripped off the atoms. Therefore, the core consists of the nuclei of atoms and free electrons. This is probably true of most of the solar interior–only near the photosphere can any atoms exist.

Astronomers use spectroscopy to determine the sun's composition. The sun's spectrum has many dark absorption lines (figure 17.2). Different elements produce different lines, so astronomers can study the relative strengths of the lines to measure the composition. The sun's composition is 73%

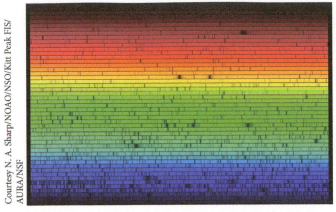

FIGURE 17.2. Spectrum of the sun showing dark absorption lines.

hydrogen, 25% helium, and 2% everything else. This composition matches that of the Jovian planets well, but not the terrestrial planets. Since we observe the spectrum of only the sun's photosphere, this composition actually is that of the photosphere. However, it seems likely that the sun's gases have mixed quite a bit, so the photospheric composition probably is close to the sun' overall composition.

The power output of the sun is immense −3.85 × 10²⁶ W. What is the source of the sun's energy? The best theory that we have is that the fusion of hydrogen into helium powers the sun. Deep in the solar core, the temperature and pressure are so great that the nuclei of hydrogen atoms (protons) slam together and become helium nuclei. It takes four protons to make one helium nucleus (During the process two protons transmute into two neutrons). The total mass of four protons is slightly more than the mass of one helium nucleus, so what happens to the remaining mass? That mass difference is converted into energy, following Albert Einstein's famous $E = mc^2$ formula, where E is the energy, m is the mass, and c is the speed of light. The speed of light is very fast, so c is a large number. Since this equation squares a large number, the amount of energy produced this way is huge. Calculations show that the sun could shine by this mechanism for about ten billion years. Of course, that does not mean that the sun is nearly that old, but only that it could last that long. For instance, most evolutionists think that the sun (as well as Earth and the rest of the solar system) is 4.6 billion years old, or about half the maximum age that the sun could have. On the other hand, biblical creationists think that the sun is only a few thousands of years old. The source of the sun's energy cannot tell us the age, but only a maximum age.

If this theory of the sun's energy source is correct, then we would expect that the sun's brightness ought gradually to increase over time. Computation shows that if the sun is 4.6 billion years old, the sun ought to have brightened by about 40%. Most evolutionists think that life arose on Earth 3.5 billion years ago. Since that time, Earth ought to have brightened by 25%. A 25% increase in solar brightness would increase the average temperature on Earth by 17 °C. The average temperature on Earth today is 15 °C, so the average temperature of Earth when life supposedly first arose would have been −2 °C. If this were true, Earth would have been almost completely frozen back then, but no one believes this. Evolutionists call this the *young faint sun paradox*. Scientists have put forth a number of explanations for this, but none is agreed upon. However, if Earth is only thousands of years old, this is not a problem.

From time to time, dark spots are visible on the photosphere. These are **sunspots** (figure 17.3). Sunspots are not actually dark, but rather they are about 1,500 K cooler than the rest of the photosphere and hence appear darker by comparison. A sunspot is very bright–it would blind you if you could

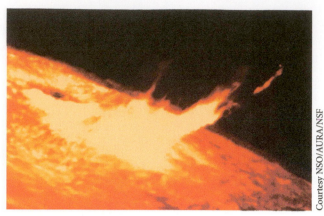

FIGURE 17.4 Solar flare.

FIGURE 17.3. Sunspots. The darker inner portion is the umbra, and the less dark outer portion is the penumbra.

look at just a sunspot with no protection for your eyes. Sunspots are regions of strong magnetic fields on the sun. Sunspots normally occur in pairs or in groups dominated by a pair of larger spots. One spot in the pair will be a north magnetic pole, while the other spot will have the opposite magnetic polarity. Since sunspots last only a few days or weeks, the appearance of sunspots can change from day to day. Additionally, sunspots will move across the face of the sun due to the sun's rotation. It takes about a month for the sun to rotate, but the equator rotates faster than the poles. The sun can do this because it is a gas throughout, so there is no reason why it should rotate at a single rate.

As sunspots fade, new ones frequently take their place. The number of sunspots visible on the sun changes over time. For several years there will be many sunspots followed by several years of few spots before returning to a time of many spots. We call this the **sunspot cycle**. The sunspot cycle averages just a little over eleven years. There was a sunspot maximum in 2013, so the next sunspot maximum will be around the year 2024. During sunspot maximum, the sun is very active. This activity takes the form of **solar flares** (figure 17.4). A solar flare is an eruption on the sun's surface, and many people mistakenly think that we can readily see solar flares (safely filtered, of course). However, solar flares produce nearly all of their energy in portions of the spectrum that are invisible to our eyes. A visible image of the sun during a solar flare would not show anything unusual.

However, a solar flare does affect us through the **solar wind**. The solar wind is a tenuous flow of charged particles outward from the sun. When a solar flare occurs, there is a gust in the solar wind. The burst of charged particles can be harmful to living things, and they can interfere with modern communications and even electrical transmission lines. However, Earth's magnetic field diverts most of the charged particles. Earth's magnetic field brings the charged particles of the solar wind close to Earth near Earth's magnetic poles. The fast-moving charged particles ionize gas in Earth's upper atmosphere. As the electrons recombine with atoms, they emit light that we call an **aurora** (figure 17.5). In the northern hemisphere people often call an aurora the northern lights, while in the southern hemisphere they go by the name southern lights. Aurorae normally are visible only at high latitudes, close to Earth's north and south magnetic poles. However, when a burst of extra charged particles in the solar wind arrives at Earth a day or two after a solar flare, aurorae may be visible at much lower latitudes.

There is a correlation between the sunspot cycle and climate. The sun will go through several decades at a time with relatively low sunspot numbers, even

FIGURE 17.5. Aurora showing green curtains. Other shapes and colors (such as red) are possible.

during maximum. At other times there may be decades of above average sunspot numbers during sunspot maximum. When sunspot numbers are high, global temperatures tend to be higher, and when sunspot numbers are low, global temperatures tend to be lower. However, correlation does not imply causation. That is, just because two events coincide is not sufficient evidence to establish a link between the two. As of yet, there is no known mechanism that can link the sunspot cycle and global temperatures on Earth.

17.2 CONSTELLATIONS

Astronomers recognize 88 **constellations**. A constellation is a group of stars that resemble the outline of some object, animal, or person. Many of the constellations are ancient, having been passed down to us, along with the lore behind them. There are many theories about who made up the constellations and why. One theory common among many Christians is that in the earliest times God revealed His complete plan of salvation through the constellations and many star names, but that the meanings became perverted over time. Proponents call this theory the "gospel in the stars." There are many problems with the gospel in the stars, so it probably is not true.

Stars move, but stars are so far away that the constellations do not appreciably change their appearance even over thousands of years. Orion (figure 17.6) is among the brightest and most recognizable of the constellations. Orion, along with the nearby Pleiades star cluster (figure 17.7), is mentioned three times in the Bible (Job 9:9, Job 38:31, and Amos 5:8). Job may be the oldest book in the Bible. Yet, when we look at Orion today, we see it much the same that Job saw it. That gives a direct connection to Job. More importantly, it ought to give us a direct connection to the God who made Orion and everything else.

FIGURE 17.6. The constellation Orion. The three bright stars in a row in the center represent Orion's belt. The two bright stars above are his shoulders, and the two bright stars below are his knees. The star near top center is Orion's head. The three stars aligned vertically below the belt outline Orion's sword.

While the constellations do not change their shapes over the years, the constellations that we see

FIGURE 17.7 The Pleiades star cluster.

in the sky continually change throughout the night. For example, the sun rises in the east, moves across the sky, and sets in the west each day, a motion caused by Earth's rotation. When the sun sets, the rotation does not stop, or else the sun would not rise the next day. As a result, most stars appear to rise in the east, move across the sky, and set in the west throughout the night. The only exceptions are those stars in the northern part of the sky. Those stars move in large circles in the north. Near the center of this motion

is Polaris, or the **North Star**. Earth's rotation points almost directly at the North Star, and the entire sky appears to slowly spin around the North Star. The North Star actually spins in a small circle, but the eye cannot notice that. This is why the North Star always appears in the north direction. In the southern hemisphere, the North Star and stars near it are not visible, but people there see a large portion of the sky in the south that we cannot see in the northern hemisphere. However, there is no bright star over the southern pole as in the northern sky, so there is no "South Star."

Throughout the year, the visible nighttime stars slowly change. This is because as Earth orbits the sun, the stars in the direction of the sun at any particular time are not visible (they are up in the day, when it is too bright for us to see stars). However, after six months, the stars that were near the sun now will be visible all night. Because of this gradual shift throughout the year, the constellations are associated with certain seasons. For instance, Orion is a winter constellation. However, in the southern hemisphere where the seasons are reversed, Orion is a summer constellation.

17.3 STARS

A single viewing of the stars at night will show that stars have different brightness. Since ancient times, astronomers have used **magnitudes** to express how bright stars are. The brightest stars are first magnitude. Stars that are a bit fainter are second magnitude, and so forth. The faintest stars that the naked eye can see on the darkest, clear night are about magnitude 6. Of course, there are stars fainter than this. With the invention of the telescope four centuries ago, astronomers realized that they had to extend the magnitude system to fainter stars. The faintest stars that the largest telescopes can now record are fainter than magnitude 30. This magnitude system is different from most measurement systems. Normally, a greater quantity goes with a larger number, but with magnitudes, the larger numbers correspond with

less light. The planets look like stars to the naked eye. The planet Venus appears far brighter than any star, so how can it be brighter than first magnitude? The answer is negative magnitudes. At brightest, Venus is −4. The full moon is about −12, and the sun is −27.

Consider a first magnitude star and a third magnitude star. The first magnitude star *appears* brighter, but is it *actually* brighter than the third magnitude star? What if the first magnitude star happened to be very close to Earth, but the third magnitude star was very far away? Then it is likely that the third magnitude star is actually brighter than the first magnitude star, but the third magnitude star merely appears fainter, because it is farther away. Astronomers call the brightness that a star appears to have in our sky the **apparent magnitude**. To express the actual brightness of a star, astronomers use **absolute magnitude**. Absolute magnitude is the apparent magnitude that a star would have at a standard distance away. The standard distance is ten parsecs (parsecs are discussed in the following section). The sun's absolute magnitude is 4.6, which is actually not that bright. The sun from the standard distance of ten parsecs would not be visible in the night sky in a city of any size, because the lights of the city would blot out the faint light of the sun.

17.4 MEASURING STARS

How do astronomers measure stellar distances? There is only one direct method, **trigonometric parallax**. Trigonometric parallax is the apparent shift in position of an object as it is observed from two different vantage points. This is a method of surveying distances on Earth. (figure 17.8). It is easy to illustrate this with one thumb held up at arm's length. When viewed with one eye, the thumb will appear in a different position than when viewed with the other eye. Now consider Earth orbiting the sun. A star's observed position on one side of the orbit will be slightly different from the star's observed position six months later on the other side of the orbit. The small shift in observed position-depends upon the baseline and the distance to the star. The baseline will be the same (the diameter of Earth's orbit) for all stars, but the distance will be different for each star. Actually, astronomers define the baseline to be one AU, the radius of Earth's orbit, so the parallax angle is half of the total measured shift.

As you might imagine, stars are so far away that the parallax of even the closest star is very small.

We divide a degree into 60 minutes of arc, and we further divide a minute of arc into 60 seconds of arc. The largest parallax angle (corresponding to the closest star) is less than one arc second (about 0.76 arc second). The math can get very messy, so astronomers define a new unit of distance, the **parsec**, abbreviated pc. The parsec is the distance that a star must be to have a parallax of one second of arc (parsec is a contraction of **par**allax of one **sec**ond of arc). A parsec is 3.26 light years. The closest star is 1.3 pc away. The equation relating parallax and distance is

$$\pi = 1/d,$$

where π is the parallax angle, and d is the distance in pc.

Until about 1990, all parallax measurements were made from the ground, and measurements made this way gave good results out to about 60 light years. At that time, the European Space Agency launched the HIPPARCHOS mission to use the prime conditions

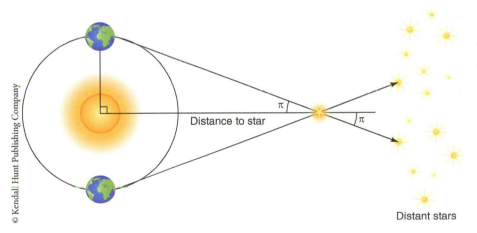

FIGURE 17.8. As Earth orbits the Sun, nearby stars appear to change position with respect to more distant stars. Apparent shift is parallax. The amount of shift is expressed by the small angle π.

Distance to star

Distant stars

above the blurring of Earth's atmosphere to measure parallax with unprecedented accuracy. That resulted in good distance measurements out to about 600 light years. New programs going into space soon will increase this to about 6,000 light years.

Trigonometric parallax is the only direct way to measure stellar distances, but there are many other indirect ways. One such method is the use of **Cepheid variables**. Many stars are variable stars, stars that vary in brightness. Cepheid variables are one type of variable stars. Cepheid variables change brightness in a very regular way over a period that can be anywhere from two days to two months. A century ago, astronomers discovered that brighter Cepheid variables had longer periods. This established a period-luminosity relation, a relationship between period and average brightness. Knowing the period-luminosity relation, an astronomer could measure the period of a Cepheid variable to determine the Cepheid variable's absolute magnitude. Knowing the average apparent magnitude, one can compute the distance. There are many other distance determination methods. Most of them rely upon finding a star's absolute magnitude, and then simple comparison to the apparent magnitude yields the distance.

Stars have different colors. However, because starlight is so dim, it is difficult for human eyes to appreciate the colors. Stellar color is a function of temperature. Red stars are the coolest. Orange stars are warmer. Yellow stars, like the sun, are warmer still. Stars much warmer than the sun are white, and the hottest stars are blue.

While astronomers can use color to gauge stellar temperature, there is a more accurate way to measure the temperatures of stars. We already mentioned that the sun's spectrum has many dark absorption lines caused by various elements. Most stars have spectra like this too. Which spectral lines are present depend not only upon which elements are present, but also upon the temperature. By studying the strengths of the various spectral lines, astronomers can deduce not only the composition of stars, but also their temperatures. About a century ago, astronomers cataloged the spectra of many stars. This classification depended upon which spectral lines were visible. This scheme used capital letters that originally were in alphabetical order. However, after several years of classification, the system was reorganized as a temperature sequence. This reorganization deleted many of the letters and shuffled the remaining letters. In order of decreasing temperature, the spectral sequence goes O, B, A, F, G, K, and M. O and B stars are blue, G stars are yellow, K stars are orange, and M stars are red. A few other specialized types reflect some odd composition differences. Each spectral type is further subdivided with Arabic numerals running from 0 to 9. For instance, the sun's spectral type is G2.

Another important stellar property is mass. The measure of any object's mass is its weight. Weight is the force of gravity, which is not the same as mass. However, weight depends upon mass, so an object's weight can be used to find its mass. How can a star be weighed? Fortunately, many stars are members of **binary star** (figure 17.9) systems. A binary star is two stars orbiting one another through their mutual gravity. By measuring the motion of the two stars, astronomers can infer what the masses of the two stars are. Binary stars are very common, so scientists can know the masses of many stars. It is good that God saw fit to make binary stars so common, because theories of how stars work depend upon stars' mass. The fact that binary stars are so common makes comparison between theory and reality much easier.

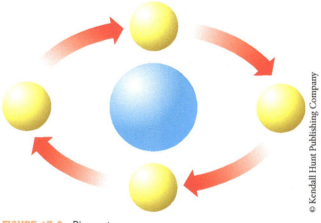

© Kendall Hunt Publishing Company

FIGURE 17.9. Binary star.

One final important stellar property is size. By size, astronomers normally refer to diameter or radius. There is only one way to directly measure stellar size–by using **eclipsing binary** stars (figure 17.10). An eclipsing binary is a binary star whose orbital plane lies close to our line of sight. As the stars revolve around one another, either star passes in front of, or eclipses, the other. The stars are so far away and so close together that individual stars are indiscernible. Rather, these binary stars appear to be a single star. However, eclipses cause the light of the star to fade and recover. By studying the way in which the light dims and brightens, astronomers can measure the sizes of the two stars. There are a few other ways that astronomers measure the sizes of stars, but they are indirect, because those methods require knowing additional information not related to size.

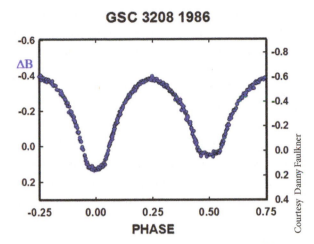

FIGURE 17.10. Light curve of the eclipsing binary star GSC 3208 1986. Blue magnitude is plotted vertically and the horizontal axis phase, a function of time. Notice that the deeper, primary eclipse is at phase equal to 0.00, while the shallower, secondary eclipse is at phase 0.50.

17.5 STAR CLUSTERS

Many stars are members of binary star systems. Other stars, like the sun, appear to be alone (except for any planets that might orbit them). However, many stars are in **star clusters** (figure 17.11). A star cluster is a collection of many stars held together by their mutual gravity. There are thousands of star clusters, but most are too faint to see with the naked eye. The brightest appearing star cluster is the Pleiades, or Seven Sisters (figure 17.12). The Pleiades are visible all night in the winter and are close to Orion. In fact, the Pleiades are mentioned along with Orion the three times that Orion is mentioned in the Bible. The Pleiades have a distinctive shape. This shape is recreated in the six stars in the Subaru symbol; Subaru is the Japanese name for the Pleiades.

FIGURE 17.11. Double star cluster in the constellation Perseus.

FIGURE 17.12. The Pleiades is a star cluster.

The Pleiades is an example of an **open star cluster**. An open star cluster has an irregular shape and contains a few dozen to a few thousand stars. On the other hand, a **globular star cluster** (figure 17.13) is spherically shaped and may contain from 50,000 to a million stars. Globular star clusters are less common than open star clusters – Earth's galaxy contains about 200 globular clusters. The nearest globular cluster is thousands of light years away, so the brightest appearing globular clusters are barely visible to the naked eye. However, through a moderate-size telescope globular star clusters can be very beautiful.

FIGURE 17.13. M13 in the constellation Hercules is an example of a globular cluster.

17.6 HERTZSPRUNG-RUSSELL DIAGRAM

A century ago, two astronomers named Hertzsprung and Russell independently discovered a relationship important in understanding stars. The **Hertzsprung-Russell diagram** (figure 17.14) is a plot of some measure of stars' intrinsic brightness versus some measure of temperature. Figure 17.14

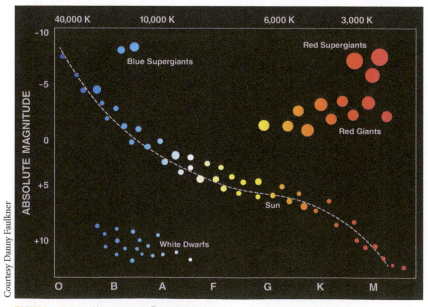

FIGURE 17.14. Hertzsprung-Russell diagram.

plots absolute magnitude versus temperature. Stellar brightness increases upward on the vertical axis, but for historical reasons, the temperature increases from right to left on the horizontal axis. Figure 17.14 shows what the Hertzsprung-Russell diagram looks like for a large group of stars. Notice that most stars fall along a roughly diagonal band from upper left to lower right. This is the main sequence. Stars to the upper left of the main sequence are hotter, brighter, and more massive than stars on the lower right of the main sequence.

A few stars are on the upper right of the diagram. Those stars are very large, because though their temperatures are low, their luminosity, or brightness, is very high. This can only be true if those stars are very large. Those stars are **giant stars**. Some of the largest stars are red giants. They are to the far right. To the lower left are very hot, faint stars. Hot stars generally are very bright, so to be so faint, these stars must be very small. These stars are **white dwarfs**, because these stars are so hot that many of them appear white. How large are giant stars? The smaller giants are only ten times or so the size of the sun. However, the largest giants can be hundreds or even thousands of times larger than the sun. The very largest stars are supergiant stars. If the sun were the size of a supergiant, the orbits of several of the planets, including Earth's, would be inside the sun. While giant stars are far larger than the sun, their masses are not that much greater than the sun's. The average density of the sun is slightly greater than the density of water. However, many giant stars have average densities thousands, or even millions of times less dense than water. How small are white dwarf stars? Typically, they are 1–2 times the size of Earth, or roughly 1–2% the size of the sun. The masses of white dwarfs are similar to that of the sun, so the densities of white dwarfs tend to be thousands of times that of water. Just one teaspoon of a white dwarf on planet Earth would weigh as much as an automobile! What causes giant stars and white dwarfs to have such strange properties? Astronomers have developed an answer to this question, a subject that is covered in the next section.

17.7 THEORIES OF STELLAR EVOLUTION

Many people mistakenly think that evolution is only about biology. However, evolutionary thinking has permeated all of human endeavors. Evolutionary thinking can be divided into three broad categories: cosmological, geological, and biological. Cosmological evolution concerns itself with the naturalistic origin of the universe (discussed in the next chapter) and the naturalistic origin of stars.

Astronomers began to develop the modern theories of stellar evolution in the 1950's. According to these theories, stars normally get their energy from the same source that the sun does–the fusion of hydrogen into helium deep inside or near their cores. Since stars have a finite amount of hydrogen, stars can shine by this mechanism only for a finite length of time. The more mass that a star has, the more hydrogen fuel that it has, but more massive stars are far brighter than less massive stars, so stars that are more massive consume energy far more quickly than lower mass stars. As stated earlier, the computed maximum lifetime of the sun is about ten billion years. Stars more massive than the sun have even shorter lifetimes, but less massive stars have longer lifetimes. For instance, the most massive stars could last only a few million years. Most astronomers think that the universe is 13.8 billion years old, so the stars that are more massive than the sun could not have been around since the beginning of the universe. The least massive stars have computed lifetimes far longer than 13.8 billion years, so some of those stars could date back to the beginning of the universe. Contrast this to the biblical creation model. According to the Bible, God made the stars on Day Four, and that was only thousands of years ago. In this view, all stars have lifetimes far longer than the age of the universe, so all stars date back to the beginning of creation.

How do evolutionary astronomers explain the existence of stars with lifetimes far less than the supposed 13.8 billion year age of the universe? They hypothesize that most stars formed long after the universe began and that star formation in an ongoing process today. They suggest that the many clouds of gas in space may hold the answer. These clouds are very thin–their density is far less than the best vacuum on Earth. However, the clouds are so large that they often contain thousands of times the mass of the sun. Gas clouds have roughly the same composition that the sun and other stars have–they are mostly hydrogen and helium, with a few percent of the remaining elements. Astronomers think that the gravity of these gas clouds help contract the clouds to form stars.

However, there is a problem with this theory. Gas clouds cannot spontaneously contract under their own gravity. While the gas clouds have feeble gravity that slightly pulls the gas inward, the clouds also have pressure that tends to push the gas apart. These forces normally balance so that gas clouds are stable against collapse. To avoid this problem, most astronomers think that an outside force jump-starts the process of collapse. Astronomers have suggested several mechanisms that could do this. One popular idea is that a supernova might explode nearby. As the shock wave of debris from the supernova passes by a cloud, the shock wave could collapse a portion of the cloud, leading to star formation. However, what is a supernova? A supernova is a large explosion that a star might undergo. Where did the star that exploded come from? If that star formed when an earlier supernova exploded, then where did the earlier star that exploded come from? This amounts to a chicken or egg sort of problem. All suggested mechanisms for star formation suffer from the same problem and require that stars first exist, so man's theories here do not really tell us where stars came from. Many astronomers now think that there was

a burst of star formation in the early universe, but there is no known mechanism to make this happen.

From time to time, news reports announce that astronomers have discovered the birth of a star. These stories are reported as if astronomers now see a star that was not there before. However, that cannot be, because the theory of star formation requires that the process takes a very long time, far longer than many human lifetimes. What really ought to be reported is that astronomers have found an object that they think is in some stage of star formation. Astronomers have developed and then modified their theories of star formation over the years as new data became available. Keep in mind that these newly discovered objects amount to snapshots that astronomers arrange within their theories.

Laying aside this objection, let us assume that stars do form somehow–what next? Astronomers think that stars end up on the main sequence, where they will remain for their computed lifetimes. For instance, the theory suggests that it took the sun about 30 million years to reach where it is on the main sequence. The sun could remain on the main sequence for ten billion years, and if the sun is 4.6 billion years, it has exhausted about half of its lifetime on the main sequence. What will happen to the sun after the ten billion years have elapsed? After ten billion years, the sun will have exhausted the store of hydrogen in its core, so it will look for some new source of energy. Theory suggests that the new source of energy will be fusion of hydrogen into helium in a thin shell around the core. This will be accompanied by a rearrangement of the structure of the sun so that the sun gradually will swell in size and cool. The sun will become a red giant star. Astronomers think that most stars eventually will do this, so this theory explains what red giant stars are–they are evolved stars that once were on the main sequence.

According to the theory, many stars, including the sun, will tap a few additional sources of energy, but those sources will be short-lived compared to the time spent on the main sequence. Red giant stars are very windy. That is, they shed matter into space.

The sun does this now, in the form of the solar wind. However, the solar wind is extremely feeble compared to the amount of gas blown off red giant stars. Some measured rates of mass loss from red giants would totally evaporate the stars in just thousands of years. Eventually, the mass loss is so great that the core of the star is exposed. The structure of the core of a red giant resembles the structure of a white dwarf, so most astronomers think that this is where white dwarfs come from. Astronomers think that the blown-off gas collects into a complex expanding shell around the newly formed white dwarf. We call these objects **planetary nebulae** (figure 17.15). A planetary nebula is a shell of glowing gas. The name is most unfortunate, because the name suggests a connection to planets. There is a superficial resemblance in that through a telescope a planetary nebula appears as a small disk, similar to the appearance of planets. White dwarfs can last for billions of years, but planetary nebulae expand into space and disappear over tens of thousands of years. Planetary nebulae have hot white dwarfs at their centers, but most white dwarfs do not have planetary nebulae around them.

Astronomers think that most stars, including the sun, eventually will end up as white dwarfs. However, according to the theory, the most massive stars have a different fate. Very massive stars are rare. Like the sun, massive stars get their energy from hydrogen fusion in their cores while on the main sequence. After leaving the main sequence to become giant stars, massive stars probably go through various other sources of energy that the sun will not. Eventually, the cores of massive stars catastrophically collapse into compact objects. See section 17.8 for more on compact objects. The collapse of the core of a massive star would release a tremendous amount of gravitational potential energy—about as much energy as the sun would produce in ten billion years. The difference is that this energy is released almost instantly as opposed to over ten billion years. This energy released so suddenly disrupts the remainder of the star to form a **supernova** (figure 17.16).

FIGURE 17.15. The planetary nebula NGC 7293

a. v. ley/Shutterstock.com

This is just one type of supernova—there are several other types. The types are described by their appearance and spectra, and astronomers explain the different types with their evolutionary theories.

Courtesy NASA

FIGURE 17.16. The bright star is the supernova 1994D, a type Ia supernova visible in the galaxy NGC 4526 in the year 1994. The galaxy is 50–60 million light years away.

A supernova rises to peak brightness very rapidly, over just a few days, and then it more gradually declines in brightness. At peak brightness, supernovae are very bright–they can be brighter than millions of bright stars. Being so bright, we can see them at great distances, even in other galaxies. One type of supernovae, the type Ia, appear to have the same brightness, so they are valuable tools in measuring distances to the galaxies in which they appear. This is an important topic of research in astronomy.

17.8 NEUTRON STARS

As discussed above, the collapse of the core of a massive star that leads to a supernova leaves behind a **compact object**. What is a compact object? A compact object is a very massive, but very small, object. Given their large mass and small size, compacts have density so high that we must use modern physics to understand them. We recognize two types of compact objects: neutron stars and black holes. A **neutron star** is a star that consists of just neutrons. The neutrons are packed to the density that exists in the nucleus of an atom. Just one teaspoon of a neutron star on Earth it would weigh more than all the automobiles ever manufactured! Neutron stars are only a few miles across, so they are extremely faint. When first predicted in the 1930's, astronomers realized that even if a neutron star were as close as the nearest stars, it would be too faint for us to see, even with the largest telescopes.

However, astronomers failed to appreciate one thing. Stars usually rotate, and so their cores must rotate too. When a spinning object contracts, its rotation speeds up. Think of a spinning ice skater as the skater pulls her arms inward. If the spinning core of a star collapses, the core must spin very rapidly. Stars have magnetic fields too. As a stellar core collapses, the strength of its magnetic field increases tremendously. Usually, the magnetic fields of astronomical bodies do not align with the rotation axes. As bodies rotate, their magnetic fields rotate too. Near a neutron star, there will be charged particles. As the neutron star spins, the magnetic field moves with it. The rapid motion of the strong magnetic field with respect to the charged particles causes the particles to emit radiation. This radiation beams along the poles of the magnetic field. Since the magnetic field spins with the star, the beam of radiation sweeps out a cone shape. If the solar system lies near the cone, it emits a flash of radiation, sort of like a search light beam (figure 17.17). The period, or length of time between flashes, is the rotation period of the neutron star. In general, the solar system will not lie along the cone swept out by a pulsar's beam, so most neutron stars cannot be found this way.

Astronomers first found neutron stars this way in 1967. In studying point radio sources, astronomers found periodic flashes, or pulses, of radio emission coming from some of the radio sources, so astronomers called them **pulsars**. A pulsar merely is a rapidly rotating neutron star that is oriented so that its periodic flashes of radiation are noticeable to observers on Earth. At first, astronomers did not know what to make of pulsars. Pulsars flashed radiation at very precise periods, about as precise as the best clocks available. For a while, astronomers even considered the possibility that they had intercepted an alien transmission. Astronomers dubbed this possibility LGM, for little green men. Eventually, astronomers figured out that a rapidly rotating neutron star made more sense. Since 1967, astronomers have discovered many more pulsars. There now are thousands of known pulsars.

Pulsars get their power from the rotational kinetic energy of the neutron star. In this way, a spinning neutron star acts as a flywheel on an engine. As a neutron star ages, its rotation ought to slow. The fastest spinning pulsars have periods of about

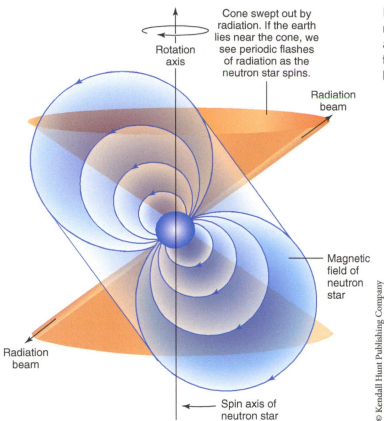

Cone swept out by radiation. If the earth lies near the cone, we see periodic flashes of radiation as the neutron star spins.

Rotation axis

Radiation beam

Magnetic field of neutron star

Radiation beam

Spin axis of neutron star

© Kendall Hunt Publishing Company

FIGURE 17.17. A neutron star (small blue circle) rapidly spins about a (vertical) axis. Beams of radiation are emitted along the axis of the neutron star's magnetic field. As the star spins, the magnetic field axis and the beams of radiation sweep out a cone.

a millisecond (a millisecond is one-thousandth of a second). The longest pulsar periods are a few seconds. Astronomers can measure the rate at which the period is changing, and from that, they can estimate the age. The pulsar in the Crab Nebula (figure 17.18) is the most famous pulsar. The Crab Nebula is a cloud of glowing gas that on photographs has a faint resemblance to a crab. The Crab Nebula's position corresponds with the position of a supernova that Chinese astronomers observed in the year 1054. Astronomers have determined the age of the Crab Nebula by measuring its expansion rate. That age is consistent with the known age of the supernova. Furthermore, astronomers can estimate the age of the pulsar by the pulsar's spin-down rate. That age is consistent with the other two ages.

Astronomers consider the Crab Nebula to be a **supernova remnant**. A supernova remnant is the expanding gas blown off the star when it

Creativemarc/Shutterstock.com

FIGURE 17.18. The Crab Nebula, a supernova remnant. One of the first pulsars discovered is inside the Crab Nebula.

exploded as a supernova. Supernova remnants get thinner as they expand, and so eventually, they disappear. However, a pulsar will last for a very long time. There are many other known supernova remnants, but, because of their relatively short lifetimes, astronomers doubt that we see most of the supernova remnants the supernovae have produced. On the other hand, astronomers are confident that they understand how supernova remnants dissipate over time. There may be a shortage of old supernova remnants that is difficult to explain in a universe that is billions of years old, but not if the universe is only thousands of years old.

17.9 BLACK HOLES

There is one other type of a compact object–a **black hole**. What is a black hole? Escape velocity is the speed that an object must have at the surface of a massive body, such as a star, to escape that body's gravity. For a given mass and size, we can use physics to calculate the escape velocity of any object. The escape velocity of a black hole exceeds the speed of light, so that not even light can escape from it. Since it gives off no light, a black hole is black. Since material objects cannot travel as fast as light, anything that falls into a black hole cannot come back out. Thus, a black hole truly is a hole in the sense that once things fall in, they cannot come out. Astronomers first suggested that black holes could exist in the eighteenth century, but no one took them seriously until the 1960's, when the term black hole was coined. There is a maximum mass that a neutron star can have, about three times the mass of the sun. Astronomers think that if a supernova leaves behind a compact object more massive than this, then the compact object must be a black hole.

A black hole is a strange object. The size of a black hole is defined by the size of its event horizon. The *event horizon* is the sphere surrounding a black hole where the escape velocity is equal to the speed of light. Anything inside the event horizon is inside the black hole, and anything outside the event horizon is outside the black hole. It is a mystery what is going on inside the black hole, though scientists can observe what is going outside of it.

How can scientists "see" something that gives off no light? Suppose that a black hole is a member of a close binary star. If the stars orbit one another closely enough, the strong tidal force of the black hole can pull matter off its companion star and on to itself. Because of angular momentum, the matter cannot fall directly onto the black hole, but rather the matter collects in a disk orbiting the black hole. Astronomers call this disk an accretion disk (figure 17.19). Astronomers have detected accretion disks around one of the stars in many close binary star systems, so this is well-understood physics. From the inner portion of the accretion disk, matter

Marc Ward/Shutterstock.com

FIGURE 17.19. An artist's conception of a black hole. The black hole is not visible, but its location is the center of the accretion disk. Notice material streaming in from the lower left, probably from the other star (out of the view here) in the X-ray binary system. Notice two jets of material perpendicular to the accretion disk coming from near the black hole.

slowly falls onto the black hole. As the matter falls toward the black hole, it heats up tremendously, and the gas gets so hot that it gives off X-rays. It is not easy to produce large amounts of X-rays, so these X-ray sources stand out. In addition, it is easy to establish whether the X-rays are coming from a close binary star. Astronomers have discovered many X-ray binaries.

When astronomers find an X-ray binary, have they necessarily found a black hole? No. Any compact object in a close binary system can produce X-rays, so the compact object could be a neutron star. How is a neutron star different from a black hole? Recall that scientists can determine the masses of the star in a binary star system, and that there is an upper limit that the mass of a neutron star can have. If the mass of the compact object exceeds the upper limit of a neutron star, astronomers assume that the compact object must be a black hole. Since the 1960's, astronomers have found many black holes this way.

The story of neutron star and black hole discovery has been filled with surprises. It tells us that the creation is far more interesting that we can imagine.

CHAPTER 17 IN REVIEW: BEYOND US: OUTSIDE OUR SOLAR SYSTEM

17.1 The Sun

- Astronomers today think of the Sun as being a star, so study of the Sun can help astronomers learn more about other stars.
- The source of the Sun's energy probably is fusion of hydrogen into helium deep in the solar core.
- Sunspots are the most obvious features on the solar surface. Sunspots vary in number over an 11 year period called the sunspot cycle.

17.2 Constellations

- Ancient astronomers divided stars into constellations, outlines of various people, animals, and objects.
- The constellations migrate across the sky during the night due to Earth's rotation.
- The constellations visible in the night sky also changes throughout the year due to Earth's revolution around the sun.

17.3 Stars

- Astronomers express the brightness of stars with magnitudes. Fainter stars have higher number magnitudes.
- Apparent magnitude is based upon how bright stars appear, but absolute magnitude measure how bright stars actually are.

17.4 Measuring Stars

- The only direct method of measuring stellar distance is trigonometric parallax. However, trigonometric parallax works for only nearby stars.
- For stars at greater distances, astronomers rely upon indirect methods, such as Cepheid variables, to find distance.
- Stars have different spectra, and their spectra usually can be classified as O, B, A, F, G, K, M.
- The only method of measuring mass is by use of binary stars.

17.5 Star Clusters

- There are two types of star clusters, open and globular clusters. There are several ways in which they differ.

17.6 Hertzsprung-Russell Diagram

- The Hertzsprung-Russell diagram is a plot of a measurement of stellar brightness verses stellar temperature.
- The main sequence is a diagonal band on the Hertzsprung-Russell diagram. Giant stars are to the upper right, while white dwarfs are to the lower left.

17.7 Theories of Stellar Evolution

- As in biology and geology, astronomers have developed evolutionary theories of how stars came about and how they change with time.
- Most astronomers think that stars supposedly form from clouds of gas and dust, though there are problems with this theory.
- Because stars expend energy and they have a finite amount of fuel, they must "die" at some point. Astronomers have developed theories of how stars die and have tied those theories to white dwarfs, black holes, and other strange astronomical objects.

17.8 Neutron Stars

- A neutron star is one type of compact object, the other being a black hole
- One way to detect neutron stars is by the effects that they have on their companions if they are in close binary stars. Such systems are X-ray binaries
- The other method of neutron star detection is if they appear as pulsars.

17.9 Black Holes

- A black hole is an object so massive and small that its gravity prevents anything, including light, from escaping.
- At this time, the only way to detect stellar black holes is when they exist in X-ray binaries.

KEY TERMS

Absolute magnitude	Open star cluster
Apparent magnitude	Parsec
Aurora	Photosphere
Binary star	Planetary nebula
Black hole	Pulsar
Cepheid variables	Solar flare
Compact object	Solar wind
Constellations	Star cluster
Eclipsing binary	Sunspot cycle
Giant stars	Sunspots
Globular star cluster	Supernova
Hertzsprung-russell diagram	Supernovae remnant
Magnitudes	Trigonometric parallax
Neutron star	White dwarfs
North star	

REVIEW QUESTIONS

1. What is the source of the sun's energy?
2. Why is the North Star special?
3. Which is brighter, a magnitude 2 star or a magnitude 4 star?
4. How big is a red giant compared to Earth? How large is a white dwarf compared to Earth?
5. What are the two types of compact objects?
6. How can we "see" a black hole?

FURTHER READING

Young faint sun paradox
Gospel in the stars
Distance determination methods

REFLECT ON SCRIPTURE

Read the passages from the Bible that mention Orion and the Pleiades, as well as Isaiah 40:26 and Psalm 147:4. Get an estimate of how many stars there are. Is God concerned with the stars? What does the number of stars and the fact that God has named them all tell you about God?

488

CHAPTER 18

MODELS OF COSMIC ORIGIN

OUTLINE:

Courtesy NASA

The Hubble Ultra Deep Field. Nearly every object in this photo is a very distant galaxy. There are approximately 3,000 galaxies in this photo, with the most distant estimated to be more than 12 billion light years away.

18.1 THE MILKY WAY

If you have looked at the sky in the summer or early autumn at night with a dark, clear sky, you have noticed a faint streak of light crossing the sky. This is the **Milky Way** (figure 18.1). The Milky Way makes a circle that goes completely around Earth, but we can see at most half of it at a time. The Milky Way is visible in the winter too, but that portion is not as bright as the summer Milky Way. Moreover, a portion of the Milky Way is not visible to most of Earth's Northern Hemisphere. When Galileo began to use a telescope four centuries ago, he scanned the Milky Way to learn its secrets. He found that the Milky Way consists of many millions of stars too faint for us to see with our eyes alone. However, the combined light of those many faint stars produces the Milky Way's glow.

Astronomers soon realized that stars are not uniformly distributed in space, but rather stars form a flat disk of stars. This is the Milky Way. The Milky Way is our **galaxy**. Like star clusters, a galaxy is a large collection of stars held together by gravity. However, galaxies contain far more stars than star clusters do. We think that the Milky Way contains about 200 billion stars. All the stars that we see are in the Milky Way. The flat disk of the galaxy is about 100,000 light years across and a few thousand light years thick. The galaxy has a bulge at its center. The nuclear bulge may be nearly 10,000 light years thick. The sun is about halfway from the center to the edge of the galaxy. The Milky Way has prominent spiral arms that connect to the nucleus. There is another major component of the galaxy—the halo. The halo is spherical and centered on the galactic center. Globular clusters are in the halo, but open star clusters are in the disk. There are some stars in the halo that are not part of globular clusters, but many of those stars are faint, so the halo does not show up readily. (figure 18.2)

FIGURE 18.1. The Milky Way.

FIGURE 18.2.

CHAPTER 18: Models of Cosmic Origin

18.2 OTHER GALAXIES

During the eighteenth century, some astronomers suggested that faint nebulae that they could see through telescopes were not clouds of gas, but rather were entire other galaxies far outside of the Milky Way. They could not see the individual stars within other galaxies, because the galaxies were much too far away. This idea became known as the **island universe theory**. Astronomers debated the island universe theory through much of the nineteenth century, but by 1880, most astronomers concluded that the Milky Way was the only galaxy. Much of this conclusion was based upon an evolutionary idea. As today, astronomers then thought that stars formed from large clouds of gas. They thought that much of the gas collapsed to the center to form the star, but leftover material flattened to a disk, from which planets might form. When astronomers saw the disks of galaxies similar to the Milky Way, they thought that they were looking at solar systems in the making. The situation changed entirely in 1924, when Edwin Hubble used the largest telescope in the world at the time to take very long-exposure photographs of the Andromeda Galaxy (sometimes called M 31) (figure 18.3). Hubble saw in his photographs very faint stars within M 31. He was able to identify them as Cepheid variables. He knew that Cepheid variables were very bright stars, so for the Cepheid variables to appear so faint, M 31 must be very far away. He even used the Cepheid variables to measure the distance to M 31. That distance placed M 31 far outside the Milky Way.

Once Hubble showed that there were other galaxies, he took the initiative to begin the study of extra-galactic astronomy. One of the first things that he did was describe and classify the different types of galaxies. Some galaxies resemble the Milky Way, with a disk and central bulge. They showed bright spiral patterns within their disks. We call these spiral arms. Hubble called galaxies with spiral arms spiral galaxies, and he divided the spiral galaxies into two types: galaxies whose spiral arms come off the bulge in the center, and galaxies whose spiral arms come off the ends of a bar passing through their centers. He called the former ones normal spirals (figure 18.4) and the latter ones barred spirals (figure 18.5). He further subdivided the two types of spirals into three subclasses, depending upon how tightly the arms were wound and how large the galactic nuclei were.

What are spiral arms? Spiral arms are the locations of very bright stars and clouds of dust and gas. The density of stars between the spiral arms is about the same as along the spiral arms, but there are

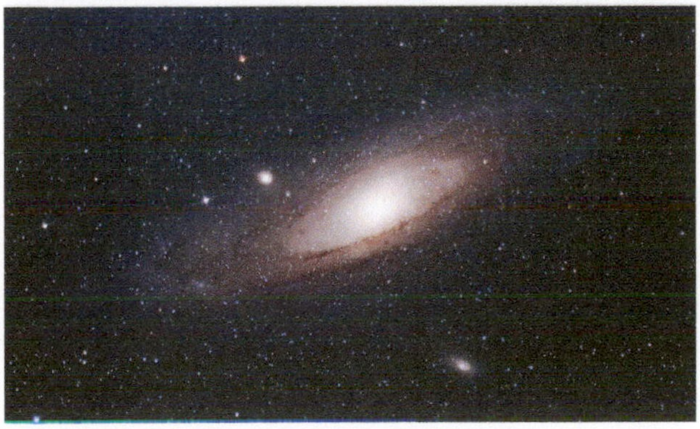

PlanilAstro/Shutterstock.com

FIGURE 18.3. M31, the Andromeda Galaxy.

Wolfgang Kloehr/Shutterstock.com

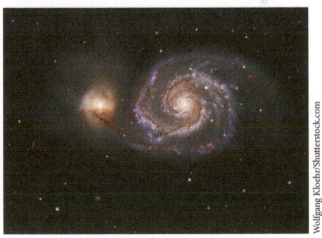

FIGURE 18.4. The Whirlpool Galaxy (M109), is a spiral galaxy with a nearby companion galaxy.

Wolfgang Kloehr/Shutterstock.com

FIGURE 18.5. M109 is a barred spiral galaxy.

few bright stars between the arms. Thus, there is less along the spiral arms than it would appear. Almost immediately, astronomers began a debate about spiral arms. The centers of galaxies orbit more quickly than the outer regions do, so some astronomers thought that spiral arms ought to eventually become so smeared as to be unrecognizable. Other astronomers suggested that spiral arms ought to start as tightly wound and un-wind with time.

Either way, if the universe is billions of years old, spiral structure of galaxies ought to have disappeared long ago. Over the past 50 years, astronomers have proposed at least three theories to explain how spiral structure could have existed for billions of years. The first theory was spiral density wave, that an acoustic wave propagates through the galaxy triggering star formation. That theory did not work very well, so eventually astronomers suggested that smaller galaxies orbiting larger galaxies stirred up spiral arms. The current thinking is that dark matter is responsible. In reality, the existence of spiral arms may be an indication that the universe is far younger than billions of years. It is interesting that very distant spiral galaxies appear nearly identical to nearby spiral galaxies. The very distant galaxies ought to be much younger, because their light presumably took billions of years to travel to Earth and hence ought to look younger than nearby galaxies of similar type. However, the similar appearance of spiral galaxies

both near and far suggests that there has been little, if any, galactic evolution.

Hubble recognized another type of galaxy: ellipticals (figure 18.6). Elliptical galaxies have an elliptical, or oval, shape without the obvious bulge that spiral galaxies have. Furthermore, elliptical galaxies lack a disk and spiral arms. Hubble further arranged elliptical galaxies into eight sub-types depending upon how circular or flattened they appeared. Many elliptical galaxies are small; we often call these common galaxies "dwarf ellipticals." However, some of the largest galaxies are giant elllipticals. Finally, Hubble found that 1–2% of galaxies defied his classification; Hubble called these irregular galaxies. Though later astronomers made refinements to Hubble's scheme of galaxy classification, his basic system remains in use today.

Hubble arranged his various types of galaxies along a tuning fork shape (figure 18.7). Hubble used his tuning fork diagram to illustrate the direction in which he thought that galaxies evolved. He thought that galaxies formed from very large clouds of gas. The forming galaxies first appeared round, but then they flattened and developed a disk with tight spiral arms that unwound to lead to irregular galaxies. However, other astronomers thought that galaxies evolved the other direction. Today, most

Courtesy NOAO/AURA/NSF

FIGURE 18.6. M87 is a giant elliptical galaxy.

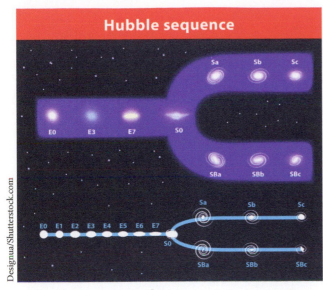

FIGURE 18.7. The Hubble tuning fork diagram shows the various types of galaxies.

astronomers do not think that galaxies evolve either way. Instead, they think that most galaxies form either as a spiral or as an elliptical and remain that way. However, there is one exception. If two or more spiral galaxies collide, astronomers think that they often transform into elliptical galaxies. Most galaxies congregate into large clusters of galaxies. The centers of many clusters of galaxies have giant elliptical galaxies at their centers. Astronomers think that these giant ellipticals formed from the mergers of many galaxies, because collisions would be so common at the centers of clusters of galaxies. This history of theories of galaxy evolution illustrates the futility of man's ideas about explaining the world apart from God.

18.3 DISTANCES TO GALAXIES

One of the most important things in studying other galaxies is finding their distances. The closest galaxies are much too far to use trigonometric parallax to find distance. Astronomers have developed many other techniques to find distances of galaxies. Many of these rely upon the use of **standard candles**. A standard candle is an object for which we know the absolute magnitude. If we measure the apparent magnitude, then, knowing the absolute magnitude, we can compute the distance. We encountered one standard candle in chapter 17–Cepheid variables. Improvements in the instruments that astronomers use have allowed astronomers to use Cepheid variables to measure the distances of galaxies to tens of millions of light years. Astronomers frequently use other standard candles. These standard candles include the brightest globular clusters within galaxies, the brightest HII regions (a type of nebula), and the brightest supergiant stars. Perhaps the most powerful standard candle that astronomers now have is the type Ia supernova. This is a particular type of supernova, and astronomers think that all type Ia supernovae have about the same

brightness when they reach their peak. If we catch a type Ia supernova in eruption in another galaxy, we can use the measured brightness to find the distance to the galaxy. Because type Ia supernovae are so bright, astronomers have used this method to measure distances of billions of light years (figure 18.8).

The most famous method of finding distances to galaxies is the **Hubble relation** (figure 18.9). In 1929,

FIGURE 18.8. A supernova can be seen as the bright white pixel (noted with the arrow) in this image of the spiral galaxy, M51 (Whirlpool Galaxy).

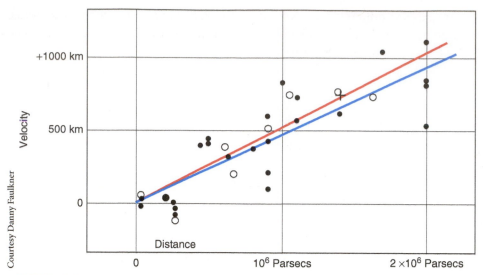

Courtesy Danny Faulkner

FIGURE 18.9. On this figure adapted from Hubble's original 1929 paper, each point represents a galaxy, with a galaxy's redshift plotted vertically and its distance plotted horizontally. Notice that the most distant galaxies have the largest velocity. This is the Hubble relation.

Hubble showed that the universe is expanding. This means that distant objects in the universe, such as galaxies, are getting farther apart. The farther away a galaxy is, the faster that it is moving away from us. Once Hubble calibrated the rate of expansion, he could use it to measure the distances to most galaxies. How does this work? We can measure the rate at which a galaxy is getting farther away from us by measuring its spectral lines. A galaxy that is moving away from us will have its spectral lines shifted toward longer wavelengths. In the visible part of the spectrum, longer wavelengths are red, so astronomers call this **redshift**. The greater the redshift, the faster that a galaxy is moving away from us. Hubble used some of the standard candles to find the distances of galaxies for which he had measured redshift. A plot of redshift verses distance produces a line. The slope of the line is the Hubble constant, and it measures the rate at which the universe is expanding. To find the distance of a galaxy for which we have measured a redshift, we divide the redshift by the Hubble constant. Over the years, Hubble altered his value for the Hubble constant, and after his death, other astronomers continued to modify the value of the Hubble constant. The Hubble constant is very important in cosmology, so we shall discuss it again in section 18.6.

18.4 ACTIVE GALAXIES

For a long time, astronomers viewed galaxies as being stable and well behaved. The reality is quite different. Astronomers call galaxies that have unusual emissions **active galaxies**. Astronomers generally group active galaxies into two broad classes: radio noisy and radio quiet. By the middle of the twentieth century, astronomers were using radio waves to study the universe. They found that some galaxies produced a lot of radio emission. A little radio emission is common for galaxies, but some galaxies produced far more power in the radio than the rest of the spectrum. Astronomers called these odd galaxies radio galaxies. There are several different types of radio galaxies. Most radio galaxies are elliptical galaxies. Other galaxies have unusual things happening in them though they are radio quiet. One of the best examples of this is the Seyfert galaxies, named for Carl Seyfert, the astronomer who described them in 1943. Seyfert galaxies are spiral galaxies with bright point-like nuclei that produce emission spectra. Most galaxy spectra have absorption lines, similar to the spectra of stars (figure 18.10).

Radio galaxies and Seyfert galaxies are two of the most common types of active galaxies. There are other types. The unusual emissions can be anywhere in their spectrum–radio, microwave, infrared, optical, ultraviolet, X-ray, or gamma-ray. The source of unusual emissions from active galaxies appears to be their nuclei, so astronomers generally focus their study on the nuclei of these galaxies. Astronomers refer to one of these objects as an **AGN**, for Active Galactic Nucleus. A common feature of AGNs is one or two jets of material shooting out from their nuclei.

Quasars are some of the more active galaxies. Quasars attracted attention in the early 1960's, because they were radio sources that appeared to be faint, blue stars. Stars do not generate appreciable radio emission, so for a while these strange objects were called "radio stars." Soon that name changed to quasi-stellar objects (QSO for short), which eventually was contracted to quasars. Quasar spectra were difficult to interpret. Finally, someone figured out that the spectra consisted of a few broad emission lines that were greatly red shifted. The first quasar discovered, 3C 273, has a redshift of nearly 17%, meaning that it is moving away from us at roughly 17% the speed of light. Assuming that this redshift is due to the expansion of the universe, 3C 273 is nearly 2.5 billion light years away. From its apparent magnitude and distance, we can compute that 3C 273 gives off one hundred times more light than all the stars in the Milky Way galaxy. (figure 18.11)

On the other hand, 3C 273 cannot be very large. We know this, because we have archival observations of 3C 273 going back more than a century, long before anyone suspected anything unusual about 3C 273. Those observations show that 3C 273 goes through variations in brightness over just a few years. To go through a cycle of brightness requires that some sort of signal propagates through an

FIGURE 18.10. M77 was one of the first galaxies classified as a Seyfert.

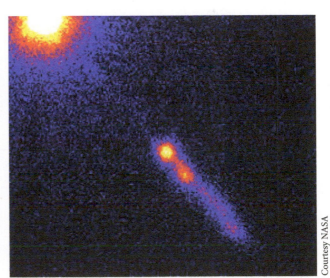

FIGURE 18.11. A jet is clearly visible in this X-ray image of the quasar 3C 273.

object to cause each part to brighten and then fade. The fastest rate that any signal can propagate is the speed of light, so the period of light variation gives us the maximum size of an object expressed in light travel time. That is, since 3C 273 varied in brightness over just a few years, 3C 273 could be no more than a few light years across. However, this produces a challenge–how can something that is just a few light years across produce such a huge amount of energy?

Before they could answer that question, astronomers needed more data. Since the discovery of the first quasar more than 50 years ago, astronomers have discovered thousands more. All of the other quasars had even larger redshifts than 3C 273, suggesting that they are even farther away. We can use the Hubble relation to find the distances of quasars, and the most distant quasars are more than 12 billion years away. In the 1960's and into the 1970's there was debate over the nature of quasars, with some astronomers questioning if quasars really are so far away. A few astronomers still doubt that quasars are very distant, but most astronomers have come to accept this. There are many reasons for believing that quasars are so distant. One reason is that quasars appear to be the cores of galaxies, which makes them the most active of AGNs. At first, the light from the galaxies that host quasars was not visible, but with larger telescopes and more sensitive cameras, astronomers were able to photograph the galaxies in which most quasars are embedded.

Assuming that quasars are the most active form of AGNs, explaining the power source of quasars likely will explain the power source of other AGNs as well. Astronomers have settled upon the theory that super massive black holes power AGNs. A super massive black hole may have millions or even billions of times the mass of the sun. Matter falling into a black hole will not go straight in. This is because nearly all matter will have some sideways motion. This sideways motion essentially is orbital motion. Orbiting bodies must conserve a quantity

called angular momentum, so as bodies fall into a black hole, their orbital motion tends to increase. The in-falling matter collides and heats because of friction. This robs orbital energy, which causes the matter to fall more deeply, which results in faster orbital motion, followed by more collisions and heating. This runaway results in very high temperatures. The in-falling matter flattens into an accretion disk, so a flattened disk of material surrounds the black hole. A complex interaction of strong magnetic fields and charged particles in the disk can produce jets of fast-moving material perpendicular to the disk. Note that the jets do not come from the black holes itself, but that the jets emanate from a region near the black hole. Matter that falls into the black hole transfers energy to power the jets. (figure 18.12)

Astronomers have used this model to explain the entire range of AGNs, including quasars. What we see in an AGN and hence how we classify it depends upon several factors. One factor is the rate at which matter falls into the black hole–a higher in-fall rate results in a brighter AGN. Also important is the thickness of the accretion disk. Some accretion disks are very thick, while others are very thin. A thick disk can better focus the jets into narrower

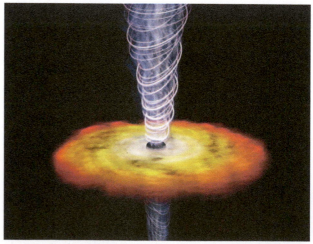

Courtesy NASA

FIGURE 18.12. In this artist's illustration of a quasar, the black hole is at the center (not visible). Notice the accretion disk and the jets perpendicular to the disk.

beams. Another factor is orientation–an AGN will look different if viewed down the axis of one of the jets than it will appear if viewed from the side.

Nor are super massive black holes confined to AGNs. It appears that virtually all large galaxies have super massive black holes lurking in their cores, including the Milky Way. The reason that we do not classify such "normal" galaxies as having AGNs is that their nuclei are reasonably well behaved. Probably very little matter is falling into the large black holes at the cores of normal galaxies. Why do we think that normal galaxies harbor super massive black holes? Near the centers of many galaxies there are objects orbiting with high speeds. Those orbits require very large masses at their center, often millions of times the mass of the sun. The orbital sizes typically are only a few light years across, yet there is no visible object there. The only things that we know of that fits this description are super massive black holes. (figure 18.13)

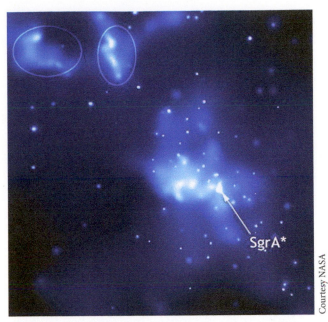

Courtesy NASA

FIGURE 18.13. Sgr A*, a radio source at the center of the Milky Way Galaxy, is visible in this X-ray image. A star called S2 orbiting very close to Sgr A* reveals that the black hole at the center of Sgr A* has a mass of about four million times that of the sun.

18.5 COSMOLOGY: THE EXPANSION OF THE UNIVERSE

Cosmology is the study of the structure of the universe. Related to cosmology is **cosmogony**, the study of the origin and history of the universe. Today many people mingle the two, but cosmology and cosmogony can and ought to be related, but separate, disciplines. A cosmology can take different forms. For instance, the geocentric and heliocentric theories are cosmologies. Our understanding of the structure of our Milky Way galaxy is a cosmology. The island universe theory was a cosmology. However, today cosmology generally refers to the structure of the universe on the largest scales.

Most stars are in galaxies. Therefore, the density of stars between galaxies is extremely low as compared to the density of stars within galaxies. Galaxies are not randomly distributed in space either. Rather, galaxies tend to be in clusters with few galaxies between clusters. The Milky Way is not part of a cluster of galaxies. Rather, the Milky Way is part of a much smaller collection of galaxies that we call the Local Group. The Local Group has three large spirals: the Milky Way, the Andromeda galaxy, and M33. In addition, there are several score small galaxies. Some of the small galaxies obviously are satellites of the three large galaxies in the Local Group. For instance, the Milky Way has two small satellites, the Large Magellanic Cloud (LMC) and the Small Magellanic Cloud (SMC). The LMC and SMC are not visible in the Northern Hemisphere, but they are fine objects in the skies of the Southern Hemisphere. They look like someone had snatched small pieces of the Milky Way and flung them into space (figure 18.14). Nearly 50 million light years away is the Virgo Cluster, the closest cluster of galaxies of any size. We call it the Virgo Cluster, because it is in the direction of the constellation Virgo. There

FIGURE 18.14. The Large Magellanic Cloud is one of two small satellite galaxies of the Milky Way.

are about a thousand galaxies in the Virgo Cluster (figure 18.15).

In the 1930's, the astronomer Fritz Zwicky studied the motions of galaxies within clusters. He found that the motions of the galaxies within cluster indicated far more mass than we could see. In many cases, clusters of galaxies contained more than ten times as much mass as contained in the galaxies that we could see in the clusters. He called this discrepancy "missing mass." By the 1970's, additional evidence for this missing mass accumulated. Eventually, astronomers changed the name of missing mass to **dark matter**

to emphasize that dark matter does not give off light. We do not know what dark matter is. Whatever it is, dark matter apparently contains far more mass than the visible matter in the universe.

Since the 1930's, astronomers have catalogued thousands of clusters of galaxies. One of the remarkable cosmological discoveries of the late twentieth century was that clusters of galaxies themselves cluster along interconnected filaments and sheets (figure 18.16). In between the filaments and sheets are large voids with very few galaxies. On a large scale, this gives the universe a lace-like appearance. Or perhaps the universe more resembles Swiss cheese with more holes than cheese. Astronomers have discussed the reason that the universe has this appearance.

One of the foundational assumptions of modern cosmology and cosmogony is that the universe is **homogeneous**, that it has the same properties throughout. One manifestations of homogeneity is that matter is uniformly distributed in space. On the

FIGURE 18.15. This is just a portion of the Virgo Cluster of galaxies.

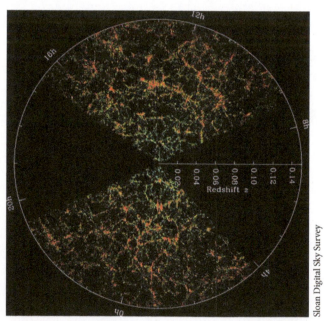

FIGURE 18.16. This three-dimensional map of the nearest 500 million light years of the universe shows several filaments and voids.

CHAPTER 18: Models of Cosmic Origin

local level, it is obvious that matter is not uniform, for we live on a planet surrounded by much nearly empty space. However, the assumption of homogeneity asserts that even on the scale of clusters of galaxies the universe can be clumpy. None of that matters if the universe is reasonably uniform on some scale much larger than clusters of galaxies or even filaments of clusters of galaxies. That is, one can assume that on a very large scale the universe may be homogeneous. However, we have mapped large portions of the observable universe, and we have yet to see homogeneity. Therefore, it is a matter of faith that at some level the universe is homogeneous. Another assumption that cosmologists often make is that the universe is **isotropic**. If the universe is isotropic, the universe has the same properties in every direction. This may appear at first to be the same as homogeneity, but homogeneity refers to being the same at every location, while isotropy refers to being the same in every direction. There have been some observations that suggest that the universe may not be isotropic either.

More than three centuries ago, Sir Isaac Newton devised the first good theory of gravity, and we still use that theory today for most situations. However, a century ago, Albert Einstein formulated the modern theory of gravity, general relativity. In Newtonian theory, massive bodies attract one another through empty space. It was a mystery how gravity accomplished this "action at a distance." General relativity addresses this mystery a little more deeply in that it treats space and time as a four dimensional entity. The presence of mass bends space-time so that objects move through this curved space-time in a manner that we perceive as the acceleration of gravity.

When Newton applied his theory of gravity to the universe as a whole, he found that gravity attracts all matter toward the center of the universe so that eventually all matter would end up in a heap at the center. Newton believed that the universe was eternal, an idea going back to the ancient Greeks. If the universe were eternal, then all matter would long ago have ended up in one place. Obviously, this has not happened, so how does one resolve this? Newton could have concluded that the universe is not eternal (as the Bible teaches), but instead he decided that the universe must be infinite in size. If the universe is infinite, then there is no center, and the universe does not collapse in on itself. This indicates that the universe is static.

When Einstein applied general relativity to the matter in the universe, he was astonished to find a large difference between his theory and Newtonian physics. A universe governed by general relativity will collapse into a heap even if it is infinite in size. He, too, could have rejected the concept of an eternal universe, but that belief was deeply ingrained. Instead, to preserve a static universe, Einstein introduced the **cosmological constant**. The cosmological constant is a repulsion that space has for itself. If the value of the cosmological constant is just right, it will balance the inward pull that the matter of the universe has.

Einstein later said that introducing the cosmological constant was his greatest blunder. Shortly after Einstein published his work on cosmology, others quickly saw another possibility. They realized that if there were no cosmological constant that the universe would either expand or contract. General relativity could not predict whether the universe is expanding or contracting, but only that these were the two possibilities. Which one is correct is a matter of observation. About this time, an astronomer named Vesto Slipher had published observations that suggested that the universe might be expanding. Edwin Hubble finally put this together in his famous paper on the expansion of the universe. In fact, Hubble did not just stumble upon his result, but instead he knew that this would be the result. However, he was the first to demonstrate a clear relationship between galaxy redshift and distance, which is what one would expect if the universe were expanding.

As the universe expands, it ought to get less dense, so the universe must have been denser in the past. We can treat the universe like a gas. As a gas expands, it cools, so the universe must cool as it expands. Therefore, the universe must have been hotter in the past. Some scientists extrapolate this trend into the ancient past and suggest that the universe began in a very dense, hot state billions of years ago. Eventually, people began to call this beginning event the **big bang**. However, this possibility flew in the face of a long-held belief–that the universe was eternal. Many people initially opposed the big bang, because it ran contrary to belief in an eternal universe, but also because it suggests that there may be a Creator.

To avoid this, some scientists held onto the idea of an eternal universe. Since an expanding universe ought to become less dense over time, how could the universe have been expanding forever? If the universe had been expanding forever, the universe ought to have zero density by now, but it clearly does not. To avoid this difficulty, some cosmologists suggested that as the universe expanded, more matter spontaneously came into existence to maintain a constant density in the universe. This idea became known as the steady state theory, or the continuous creation theory.

In the middle of the 20th century, scientists debated these two cosmological models. For a while, it appeared that the steady state theory would win out. However, in 1965, two astronomers, Arno Penzias and Robert Wilson, published their discovery of a **cosmic microwave background** (CMB) that appears to permeate the universe (figure 18.17). The universe today is transparent, but according to the big bang model, there was a time long ago when the universe was opaque. A few hundred thousand years after the big bang, the universe supposedly became transparent for the first time. The theory says that at that time, the universe had a temperature of about 3,000 K. Radiation from the hot gas that filled the universe then is just now reaching us after traveling for billions of years. However, the expansion of the universe redshifted and hence cooled that radiation so that today it has a temperature of a little less than 3 K (figure 18.18). Radiation of this type is in the microwave part of the spectrum. The steady state theory cannot explain the CMB, but the big bang model

FIGURE 18.17. Penzias and Wilson used this instrument to discover the CMB.

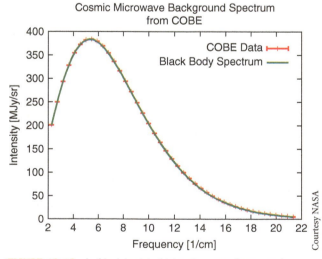

Cosmic Microwave Background Spectrum from COBE

FIGURE 18.18. In this data plot of intensity verses frequency for the CMB, the smooth curve fits the emission expected from an object with temperature a little less than 3K.

predicted it. Many people considered the CMB to be proof of the big bang, so most scientists soon abandoned the steady state theory in favor of the big bang model. For the past half century, the big bang model has been the only widely accepted model of the origin of the universe (figure 18.19).

Many Christians have embraced the big bang, seeing in it God's hand. They reason that the big bang asserts that the universe had a beginning, as does Genesis. Furthermore, many Christian apologists insist that the big bang could not have happened by itself, so the big bang gives strong evidence that there is a God. While their intention is good, there are difficulties with this approach. Does the big bang theory conform to the Bible? There are many problems. For instance, there are several textual reasons why the days of the Genesis 1 creation account were normal days. Therefore, the creation was over six normal days. Furthermore, the genealogies found in Genesis 5, 11, and elsewhere in the Old Testament show that the universe is only thousands of years old, not billions, as the big bang requires. According to the big bang model, stars began to form shortly after the big bang happened 13.8 billion years ago. Earth and the rest of the solar system did not form until about 4.6 billion years ago. Many stars supposedly formed in the 9 billion years between the big bang and the formation of Earth. However, according to Genesis, God made the stars on Day Four, three days after He created Earth. There are many other problems in the details of reconciling the big bang with Genesis creation. However, there is an overall problem. The history of science is that once widely held beliefs often are discarded in favor of new ideas. If we wed what the Bible says to the big bang, then when the big bang model is discarded, what would that do to the Bible?

The big bang has changed much over the years. For many years, astronomers thought that universe was 16–18 billion years old. This was based upon what they thought the value of the Hubble constant was. In the early 1990's, astronomers revised the Hubble constant, which immediately decreased the age of the universe, eventually to the current 13.8 billion year estimate.

Another change in the big bang model was required by the **horizon problem** and **flatness problem**. The horizon problem is the fact that the CMB has precisely the same temperature in every direction that we look. This means that portions of space in opposite directions have the same temperature. Objects normally have the same temperature after they have had an opportunity to exchange heat until they reach the same temperature. However, radiation from the CMB is just now reaching our position–how could portions of the universe in opposite directions ever have been in thermal contact to exchange heat? They could not, so why the entire universe had the same temperature was a mystery. The flatness problem involves a quantity that cosmologists indicate by the Greek letter omega, Ω. Cosmologists define Ω to be the ratio of gravitational

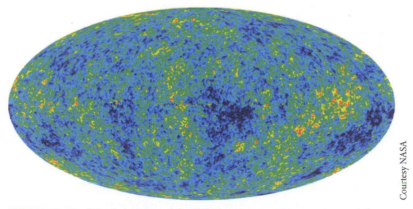

Courtesy NASA

FIGURE 18.19. This all-sky image of the CMB shows very small temperature variations.

potential energy to kinetic energy. If the value of Ω was less than one when the universe began, through then billions of years of expansion Ω ought to be zero. On the other hand, if Ω initially were greater than one, Ω now ought to be very large. However, all measurements of Ω have indicated that Ω is very close to one. How can this be, if the universe is billions of years old?

In the 1980's, cosmologists realized that **cosmic inflation** could solve the horizon and flatness problems. Inflation supposedly happened very quickly after the big bang, about 10^{-34} seconds after the big bang. Inflation refers to the universe expanding far faster than the speed of light. Inflation did not last very long–by about 10^{-32} seconds it had ended. During inflation, the universe grew far larger than it had been before. Prior to inflation, much of the universe had been in thermal contact with itself allowing for the universe to come to one temperature. Inflation pulled much of the universe out of thermal contact, but it preserved a single temperature throughout the universe. As an added benefit, whatever value Ω initially had, inflation would drive Ω almost exactly to one, from which Ω retreated from very slightly in the ensuing billions of years since inflation. Therefore, most cosmologists and astronomers assume that inflation must have happened. However, there is no evidence for inflation.

Another change in the big bang model is the adoption of dark matter. While evidence of dark matter goes back to the 1930's, it was not until the 1980's that cosmologists began to consider dark matter in their models. If most of the mass of the universe is in the form of dark matter, then it would be improper to omit dark matter from cosmological models. Another development came in the late 1990's. Astronomers found evidence that the rate of expansion of the universe is accelerating. Einstein's cosmological constant could explain this, but most cosmologists and astronomers prefer to attribute this effect to **dark energy**. The difference is that while the cosmological constant is a constant, dark energy is not. This allows the possibility that whatever is causing acceleration in cosmic expansion varies in time or in space. Another change is the consideration of the effect of string theory in the early universe. String theory is a model about how elementary particles interact. If string theory is correct, then its effects would have been very important in the early stages of the big bang when the density was so high. Therefore, big bang models today generally incorporate string theory.

Since the early 1980's, the big bang model has changed appreciably. Back then, the expansion rate and age of the universe were very different from today. No one included inflation, dark matter, dark energy, or string theory in big bang models back then, but nearly all big bang models today do. Why did the big bang model change so much? Cosmologists changed their theories to meet new problems as they developed. Many people view this as strengthening the theory. However, if a model can be changed so easily to meet any new challenge, then can the theory ever be disproved? If a model cannot be disproved, does it qualify as being science? In the early 1980's, cosmologists had complete confidence in their models, much as they have confidence in their models today, even though the models are very different. Obviously, the theories from both then and today cannot be correct. However, they both could be wrong. We cannot be sure of man's ideas, but we can be sure of God's Word.

18.7 THE LIGHT TRAVEL TIME PROBLEM

We have seen that there are galaxies that are millions and even billions of light years away. You may wonder how we can see these distant objects, if the universe is only thousands of years old. This is a profound problem, and we call it the **light travel time problem**. Recent creationists have offered several solutions to this problem. For instance, some have suggested that God created the universe with maturity. God created Adam as an adult, not a baby. In similar manner, perhaps God created the light from distant objects in transit, on its way toward Earth. However, this means that the light that we now see from most of the objects in the universe never actually left those objects. This suggests that much of the universe amounts to an illusion.

Others have suggested that the speed of light has changed. If the speed of light were infinite or nearly infinite during the Creation Week, then light from the most distant objects could have reached Earth in just thousands of years. However, there are questions about how this might have happened. The speed of light is not a free parameter, for it depends upon two fundamental constants of nature. If the speed of light were to change, the structure of matter would change.

Creation scientist Russ Humphreys has developed a creation cosmology using general relativity. In general relativity, the rate at which time passes depends upon your location. It is possible for only a few days to have passed on Earth during the Creation Week while many years elapsed elsewhere. This would permit light to have traveled to Earth from the farthest reaches of the universe while only a few days passed on Earth. Another creation scientist, John Hartnett, has used an alternate version of general relativity to create a cosmology where light similarly traveled from distant objects to Earth in a matter of days.

Jason Lisle, a creation astronomer, has argued for what he calls the anisotropic synchrony convention, or ASC for short. This proposal notes that all of our measurements of the speed of light depend upon a round trip for light. That is, we measure how long it takes light to travel from a light source to a mirror and back to the source. Dividing twice the distance between the source and mirror by the travel time yields the speed of light. However, this assumes that the speed of light is the same in either direction. If the speed of light is not the same, then our measurement is the average speed of light over the entire trip, but not the actual speed on either leg of the trip. What if the speed light is infinite one way but is half of what we think that speed of light is for the other way? We would get the same result. It is possible that the speed of light is infinite while moving toward us but half the average speed while moving away from us. In this manner, light from the most distant objects in the universe would reach us instantaneously.

Finally, another creation astronomer, Danny Faulkner, has proposed his *dasha* solution. *Dasha* is one of the Hebrew words used to describe the development of plants on Day Three of the Creation Week. God did not create plants instantly and out of nothing. Rather, He caused them to sprout out of the ground on Day Three. The Hebrew verbs used in Genesis 1:11–12 and the time constraint of this happening within one day suggest that plants rapidly grew to maturity on Day Three. That is, normal growth was accomplished very rapidly. While God made some things in the Creation Week out of nothing and perhaps instantly (Genesis 1:1), God used rapid processes too. For instance, God made man out of dust and breathed life into man. This was a process, just as the growth of plants on Day Three. There are other examples of these rapid processes during the Creation Week. Perhaps God rapidly brought the light of distant objects to Earth on Day Four.

The light travel time problem is a serious one for creationists, but creationists have responded with several different solutions to this problem. Which one is right? Perhaps none of them. However one should not be discouraged by this. There is much work yet to be done in building the creation model. The real miracle is creation itself. God brought the universe forth out of absolutely nothing. That is a difficult thing to grasp. Compared to that, getting the light here from billions of light years away is a minor problem. If God truly has the power to create the universe, then bringing the light to Earth right away is a trivial matter.

18.1 The Milky Way

- The Milky Way is our galaxy. It is round and flat, with a bulge in its center. It contains a few hundred billion stars

18.2 Other Galaxies

- In 1924, Hubble proved the island universe theory, that the Milky Way is just one of billions of galaxies, each separated by vast distances.
- Hubble later classified most galaxies as spiral or ellipticals, along with sub-types.

18.3 Distances to Galaxies

- Astronomers use a number of standard candles to estimate distances to galaxies. However, these methods are of limited use. Many of those methods do not work beyond a few tens of millions of light years
- For truly distant galaxies, astronomers use the Hubble relation to measure distances.

18.4 Active Galaxies

- AGNs are galaxies that have unusual emissions. Astronomers classify AGNs according to their properties, but astronomers think that there is one common source for most of them supermassive black holes at their cores.
- Quasars are the most energetic AGNs. Quasars are very distant, bright, and small.

18.5 Cosmology: The Expansion of the Universe

- Modern cosmology relies upon the assumption of homogeneity and isotropy, though there is evidence that neither is true.
- Hubble showed that the universe is expanding. This is another important part of modern cosmology.

18.6 Cosmogony: The Big Bang

- The big bang model has become the dominant theory of cosmogony. The evidence cited for the big bang is the presence of the CMB.
- Cosmologists have invoked inflation to explain problems with the big bang, such as the flatness and horizon problems. There is no evidence that inflation has happened.
- Cosmologists continue to invoke other factors, such as dark matter and dark energy, to explain problems with the big bang.

18.7 The Light Travel Time Problem

- The light travel time problem is a difficulty for the recent creation model. However, creation scientists have offered several solutions to this problem.

KEY TERMS

Milky Way
Active galaxies
AGN
Big bang
Cosmic inflation
Cosmic microwave background
Cosmogony
Cosmological constant
Cosmology
Dark energy
Dark matter

Flatness problem
Galaxy
Homogeneous
Horizon problem
Hubble relation
Island universe theory
Isotropic
Light travel time problem
Quasar
Redshift
Standard candle

REVIEW QUESTIONS

1. What are the basic types of galaxies?
2. How do astronomers use the Hubble relation to find distances of galaxies?
3. What is the primary evidence for the big bang?
4. What are the horizon and flatness problems? How has cosmic inflation supposedly solved these problems?
5. What is the light travel time problem? What are some of the solutions to the light travel time problem?

FURTHER READING

Faulkner, Danny. 2003. *The Universe by Design*. Master Books.

Humphreys, Russ. *Starlight and Time*. Master Books.

For more information on how astronomers measure distances, see https://answersingenesis.org/astronomy/starlight/astronomical-distance-determination-methods-and-the-light-travel-time-problem/

For more information on the *dasha* solution to the light travel time problem, see https://answersingenesis.org/astronomy/starlight/a-proposal-for-a-new-solution-to-the-light-travel-time-problem/

REFLECT ON SCRIPTURE

Look up biblical passages that speak of the God stretching the heavens. What do you think that this means? Some people think that it refers to the expansion of the universe, but others think that it may refer to something that happened on Day Two or Day Four of the Creation Week.

Compare what 2 Peter 3:10–13 and Revelation 21:1–27 tell us about the new heaven and the new Earth. Do you think that the astronomical world will be different from the one that we have now?

GLOSSARY

"A" horizon The uppermost soil horizon that is usually rich in organic matter.

aa A rough, angular, jagged type of lava.

abrasion A process of mechanical or physical weathering where a rock or particle is ground against another rock or particle.

absolute magnitude Method of expressing intrinsic brightness of stars, defined as the apparent magnitude that a star would have if it were ten parsecs away.

abyssal plain The relatively deep, broad, and flat region of the ocean underlain by oceanic crust.

Acid Rain Rain containing nitrates and sulfates. These pollutants make rain droplets acidic and can stress plants, affect the productivity of soil and change the pH of streams and lakes.

active galaxies Galaxies that have unusual emissions.

active margin A location where the edge of a continent coincides with a plate boundary.

active solar heating Structural designs that capture insolation and then move thermal energy using pumps or fans.

actualism A philosophical approach to geology which asserts that the geological processes that operated in the past were much the same as those operating today, though the scope and magnitude of events may be very different.

Adiabatic Lapse Rate The rate at which rising air cools assuming no energy transfer occurs with the surroundings. Dry air cools at a rate of 10 °C/km and moist air cools at 6 °C/km.

Adiabatic Process A process where there is no heat transfer into or out of a system.

advancing glacier When the terminus of a glacier moves forward.

advection Horizontal motion of air flow. This motion is driven by pressure differences and is in contrast to convection, which is vertical air flow due to density differences.

aerosol Small particles suspended in the air. They can consist of solid or liquid particles that have natural or human sources.

aggregation A precipitation growth process where individual ice crystals and snowflakes cling to each other.

AGN Abbreviation for "active galactic nuclei," the common term for active galaxies.

air mass A large body of air having similar properties of temperature and humidity. These properties influence the type of weather experienced by a geographic region.

albedo The percentage of radiation reflected by a surface.

alluvial fan A cone-shaped deposit at the base of a steep mountain range deposited by a river.

alluvial fans A flattened cone shaped deposit at the base of a mountain. The sediments in the fan usually accumulate be stream, mud flow and debris flow processes.

alpha decay A type of radioactive decay in which the parent isotope ejects an alpha particle (composed of two protons and two neutrons) from its nucleus.

altocumulus Middle level clouds that form globular patches. They are larger and thicker than cirrocumulus, but thinner and smaller than stratocumulus. These clouds occur at altitudes between 2400 and 6100 m.

altostratus Middle level clouds that form a uniform layer. These clouds are not thick enough to completely obscure the sun; however, transmitted sunlight is not bright enough to cast shadows. These clouds occur at altitudes between 2400 and 6100 m.

analogue A weather prediction technique that uses pattern recognition of atmospheric phenomena to make a forecast.

angle of repose The angle at which material begins to avalanche down a slope; usually about 30–33°.

angular unconformity An unconformity where the sedimentary or extrusive igneous rocks above and below the unconformity are oriented at some angle to each other.

Antarctic Circle Geographic circle located at 66.5 °S latitude. On the first day of Southern Hemisphere winter all locations south experience 24 hours of night. Likewise, on the first day of summer all locations south experience 24 hours of sunlight.

anthracite coal The highest grade of coal, sometimes called "hard coal." It has a high carbon content and burns hotter and cleaner than other types of coal.

anticline A fold in a sedimentary rock that looks like an arch.

anticyclone See high pressure system.

anvil A feature occurring at the top of a cumulonimbus cloud that consists of ice crystals. High level winds extend the cloud to one side giving it its characteristic shape.

aphanitic Refers to the fine crystal size (<1 mm) in igneous rocks.

aphelion The point on an orbit around the sun that is farthest from the sun.

apogee The point on an orbit around Earth that is farthest from the sun.

apparent magnitude A measurement of how bright a star appears.

aquiclude A rock layer that prevents the flow of groundwater or other fluid.

aquifer A rock that stores groundwater in its pore spaces.

archosaurs A group of amniotes (egg-laying vertebrates) including crocodylians, pterosaurs, dinosaurs, and birds.

Arctic Circle Geographic circle located at 66.5 °N latitude. On the first day of Northern Hemisphere winter all locations north experience 24 hours of night. Likewise on the first day of summer all locations north experience 24 hours of sunlight.

arête A sharp ridge cut into bedrock on a mountain.

artesian well A well where the water comes out of the ground under pressure. No pumping is needed.

asthenosphere A plastic-like layer in Earth about 125 km deep resting just below the lithosphere. The lithosphere slides on the asthenosphere in plate tectonics.

astronomical unit The average distance between Earth and sun, approximately 93 million miles or 150 million kilometers.

astronomy The branch of science dedicating to understanding the nature of space and the stars, planets, moons, and other objects within it.

atmosphere The mixture of gasses that surround Earth, primarily composed of nitrogen, oxygen, and argon with traces of others. It is divided into layers (e.g., the troposphere) based on temperature trends.

atmospheric effect Warming of the planet due to the selective absorption of gases in the atmosphere. Gases are transparent in the visible range and absorb in the infrared range of the spectrum. Some of the absorbed infrared is reradiated towards the surface resulting in additional heating.

atom the smallest particle of matter that cannot be split into simpler substances by chemical processes.

atomic mass The number of both neutrons and protons in the nucleus of an atom.

atomic number The number of protons in the nucleus of an atom.

aureole A rind of un-melted, but altered baked material that surrounds a pluton.

aurora Glowing light in the night sky caused by charged particles from the sun striking air particles high in Earth's atmosphere. In the northern hemisphere we call them the northern lights, while in the southern hemisphere they are the southern lights.

autumnal equinox The first day of fall when all locations on Earth experience 12 hours of sunlight and 12 hours of night time. This occurs around September 23 in the Northern Hemisphere.

"B" horizon Usually a bright colored soil horizon (yellow-brown-orange-red) found below the A horizon. Sometimes called the "subsoil." Is a place where leached materials often accumulate.

backshore The region of the shore located between the high tide shoreline and the coast.

bajada A series of overlapping alluvial fans at the base of a mountain range.

banded iron formations Layered and iron-rich sedimentary deposits in Proterozoic rocks interpreted by old-Earth geologists to indicate low-oxygen conditions during middle and late Precambrian time.

barchan dune A relatively small crescent-shaped dune, where the horns of the dune point downwind.

barchanoid ridge dune Long transverse dunes with barchans-like horns.

barrier island Narrow but long, sand-dominated islands parallel the mainland coast that are separated from it by a lagoon or sound.

base level A topographic level where water can flow no further. Sea level is the ultimate base level. Lakes can be local base levels.

base load power The minimum level of electricity that is needed day or night.

basin A regional depression that serves as an area of sediment accumulation. Also, a three-dimensional fold that is shaped like a cereal bowl, with a cross-section is a syncline.

Basin and Range Province A geographic province in the southwestern part of the United States consisting of many repeating ridges of up-faulted and down-faulted rocks.

bathymetry The science of determining the ocean's depth, including mapping of the ocean floor.

baymouth bar An elongate ridge of sand that has blocked the opening of a bay.

beach nourishment The process of transporting sand to a beach in order to replenish sediment carried away by erosion.

bed load Coarse material that is carried along the bottom of a stream bed.

bedding refers to the horizontal layers within sedimentary rocks.

bentonite A clay deposit formed from the deposition, burial, and subsequent alteration of volcanic ash.

berm A steepened ridge of sand on the shore that may have a sandy platform or terrace landward of it.

beta decay A type of radioactive decay in which the parent isotope ejects a beta particle (either negatively or positively charged) from its nucleus.

big bang The theory that the universe suddenly appeared in a very hot, dense state long ago and that the universe has expanded and cooled into the universe that we see today.

binary star Two stars orbiting one another under their mutual gravity.

Biofuels Liquid fuels derived from biomass sources.

biogeography The pattern of distribution of organisms on Earth.

biomass The use of organic material as a fuel source, including trees, waste plant material, and animal dung.

biosphere All life forms on Earth, and the interactions between them.

bituminous coal The most common type of coal, sometimes known as "soft coal."

black hole A region of space that is so dense and the gravity so strong that light cannot escape.

blizzard Severe weather characterized by high winds and reduced visibility due to falling or blowing snow.

body wave A type of earthquake wave that travels inside (and down through) Earth.

Bowen's Reaction Series A series of chemical reactions that happen within magmas leading to the crystallization of the magma. It is named in honor of N.L. Bowen.

braided stream A set of intertwining stream channels.

breakwater A hard stabilization structure built parallel to the shoreline, often to create an artificial harbor.

breccia A rock that contains clasts of angular (or sharp-edged, unrounded) material.

breeze A wind ranging between 7 and 50 kph. These winds are named for the location of their source, such as sea or land breeze.

bridal veil falls A waterfall that comes out of a hanging glacial veil that has many long and delicate streams of water, similar to a "bride's veil."

butte A spire or pinnacle-like landform.

"C" horizon Found below the B soil horizon. It is often the same color as the unweathered bedrock and will often contain pieces of the unweathered bedrock within it.

caldera A crater formed by the collapse of the summit of a volcano after an eruption.

Cambrian explosion The geologically abrupt appearance of a wide diversity of numerous animal phyla near the base of the Cambrian Period (the beginning of the Paleozoic Eon).

capillary fringe A zone just above the water table where water is drawn upwards by capillary action.

catastrophic plate tectonics (CPT) A theory of Earth surface motion in which units of lithosphere move in relation to each other rapidly during Noah's Flood.

catastrophism An approach to geology developed in the 1700s and 1800s which asserted that Earth was ancient, that Earth's geology was formed by destructive global upheavals, and that God supernaturally created organisms and ecosystems sequentially over eons of time.

cavitation A mechanical weathering process whereby rock is disintegrated by the collapse of vacuum bubbles.

cavitation A very powerful mechanical weathering process that happens when vacuum bubbles implode, causing a "sledgehammer-like" effect on a solid material.

cell A combination of an up and down draft within a convective cloud. This is easily identified by its strong reflectivity of a radar signal.

cementation a process that "glues" the individual grains of a sedimentary rock together. Cements can consist of various types like calcite or silica cement.

cepheid variables Pulsating giant or supergiant stars with periods between two days and two months. Cepheid variables follow a period-luminosity relationship that allows us to use them to measure their distances.

chain reaction A self-sustaining series of fission events.

channel deposits Sand, gravel, boulders and other rocks that occur in the bottom of a stream bed.

chemical sedimentary rock a rock that is formed when chemical reactions take place in water precipitating various minerals. Rock salt and rock gypsum would be examples.

chemical weathering A process that occurs when rocks are dissolved by various solutions or groundwater.

cinder cones A small conical-shaped volcano whose primary product is pyroclastic material.

cirque An amphitheater-like depression carved into a mountainside by a glacier.

cirrostratus High level clouds that form a thin layer and are composed of ice crystals. These clouds are semi-transparent allowing for strong sunlight transmission. Halo phenomena are associated with this type of cloud. These clouds occur at altitudes between 6000 and 12,000 m.

cirrus High level clouds that form fibrous whips sometimes called mare's tails. These clouds are composed of ice crystals and occur at altitudes between 6000 and 12,000 m.

clastic (detrital) sedimentary rock A sedimentary rock that is made from pieces or grains of other rocks. The pieces can be small (clay-sized) up to large (boulder-sized).

cleavage/cleavage plane A zone of weakness within a mineral's atomic structure that allows the mineral to be broken along a plane.

climate The accumulated statistics of weather experienced over a region for a period of time. By international agreement climatic averages are calculated over a time period of 30 years.

climatology The collection of weather observations over geographic regions and an extended period of time. Averages of these observations provide a means of forecasting expected weather for a particular region and season of the year.

cloud condensation nuclei (CCN) Aerosol particles in the atmosphere that provide a surface upon which a water droplet or ice crystal can initiate growth.

coal A brittle, carbon-rich sedimentary deposit that can be burned. It is composed of compressed and altered plant materials.

coal ash Remnant materials left over from burning coal.

coalescence A precipitation growth process where water droplets collide with each other resulting in one larger droplet.

coast A relatively flat region that begins at dunes or other permanent shore features and continues inland until there is a significant change in topography.

coastal wetland A low-lying area of land frequently flooded during high tide.

column A cylindrical cave formation that extends from the floor to the ceiling of a cave. Usually forms by a stalactite and stalagmite joining.

coma The bright cloud of gas that surrounds the nucleus of a comet when the comet is near perihelion.

comet A small solar system body with an icy composition. When near perihelion, the sun's light sublimes the ice into gas to form a coma and tail.

compact object Either a neutron star or a stellar black hole that is responsible for the X-ray emissions seen in X-ray binary stars.

compaction When sediment is turned into sedimentary rock it must be squeezed (or compacted) to drive water out and reduce the void space (pore space) within the rock.

comparative planetology The study of the planets by noting differences and similarities between planets.

competence The competence of a stream refers to the largest particle that a stream can transport under certain flow and volume conditions.

composite cones A large conical-shaped volcano which has a history of both pyroclastic and lava eruptions. They are sometimes called stratovolcanoes.

compound Substances formed by atoms that have bonded.

compressional force When two forces move toward each other.

concentrated solar power The use of mirrors to concentrate sunlight onto an object to produce thermal energy, which is often then used to generate electricity.

condensation A phase change by which vapor becomes a liquid. During this process the vapor loses energy to the surroundings. This process is the opposite of evaporation. In the atmosphere this process results in the formation of dew.

conduction The transfer of heat through the interaction of vibrating molecules. Heat flows from warmer to colder objects through physical contact.

cone of ascension A local rise in the water table formed at the bottom of a well. Usually salt water is rising upward toward the well hole.

cone of depression A local depression in the water table formed from pumping of a well. It forms because water can't flow into the well as fast as it is removed from the well.

conglomerate A type of sedimentary rock that contains rounded clasts (as opposed to the angular clasts of a breccia)

constellations A grouping of stars that suggest the outlines of various people, animals, or objects.

contact (ort thermal) metamorphic rocks When liquid rock comes into contact with pre-existing rock it often bakes or "cooks" the pre-existing rock causing some metamorphic changes to occur because of the heat.

continental arc volcanoes A linear chain of typically andesitic volcanoes formed within a continent by the subduction of a lithospheric plate (capped by ocean crust) below the continent.

continental drift A hypothesis proposed by Alfred Wegener which asserted that the continents had slowly moved away from each other and through the ocean basins.

continental glacier The largest type of glacier that covers an area > 50,000 km^2, sometimes called an ice sheet.

continental margin The edge of a continent. Typically this area is a steep slope between the shallow continental shelf and the deep abyssal plain.

continental rift A set of physical features (such as linear valleys, faults, and volcanics) formed by extensional forces acting to separate a continent into two pieces.

continental shelf The portion of a continent that is covered by ocean water.

control rods Bars placed alongside fuel rods in a nuclear reactor to slow absorb neutrons, preventing a runaway chain reaction.

convection The transfer of heat through the motion of a fluid. This is driven by gravity and differences in density. Warm, less dense fluid rises and is replace by sinking cold more dense fluid.

convergence A measure of how much substance is approaching a location. In the atmosphere this corresponds to more air entering a region than leaving it horizontally. This is the opposite of divergence.

convergent boundary A location where two plates are moving towards each other. Lithosphere is condensed or destroyed at convergent boundaries.

core The innermost part of Earth, divided into an outer layer and an inner layer and composed primarily of iron and nickel.

Coriolis Effect An apparent force due to the rotation of Earth. In the Northern Hemisphere any moving object veers to the right relative to the surface.

correlation The process of linking or matching rock units over distances in which they are not seen. Includes physical and temporal correlation methods.

cosmic inflation The hypothetical faster-than-light expansion that the universe underwent very shortly after the big bang invoked to explain the flatness and horizon problems.

cosmic microwave background (CMB) A radiation field in the microwave part of the spectrum that appears to fill the universe. The CMB is taken as the best evidence for the big bang model.

cosmogony The study of the origin and history of the universe.

cosmological constant A hypothetical repulsion that space has for itself introduced by Albert Einstein to preserve the static universe.

cosmological principle The assumption that the universe is both homogeneous and isotropic.

cosmology The study of the structure of the universe.

covalent bond A chemical bond in which two or more atoms share each other's electrons in order to fill their valence shells.

Creation Week rocks Rocks that were either called into existence early in the Creation Week (Days 1-3) or formed by processes (crystallization, erosion, metamorphism, etc.) that happened early in the Creation Week.

Creation Week The six-day period of creation and one day of rest in which God created the heavens, Earth, the seas, and all that is in them.

crest A sharp peak that separates the windward side from the leeward side of a dune.

crevasse A deep narrow crack in glacial ice.

cross-bed An angled layer in a sedimentary rock caused by avalanching of sedimentary material.

cross-bedding Angled layers horizontal sedimentary rocks formed by the migration of dunes (either above or under water).

crude oil Unrefined petroleum.

crust The uppermost layer of Earth consisting of granitic continental material and basaltic ocean floor material.

crystal form The shape a mineral takes during unrestricted growth.

cumulonimbus A thick cloud with extensive vertical development. This cloud is associated with heavy rain or snow and lightning.

cumulus A low level globular cloud with a flat base and rounded top.

cumulus stage The early stage of cumulus cloud development where only an updraft exists in the cloud. Growth of the cloud comes from warm humid air rising from Earth's surface.

current A mass of water flowing from one are to another in the ocean.

cut bank A steep river bank or small cliff formed on the outside bend of a river meander.

cyclone see *low pressure system.*

dark energy The possibility that space may repel itself in a time or spatially varying manner.

dark matter The possibility that much matter in the universe gives off no radiation, but instead reveals its presence by its gravity.

daughter isotope An element that has been produced by radioactive decay.

Day-Age view An interpretation of the Genesis 1 creation passage which asserted that each day of creation was representative of a long period of creative activity by God or equivalent to various geological ages/periods.

debris avalanche Loose material that can consist of rock, mud, ice and other debris that tumbles down a slope rapidly.

debris flow A loose collection of debris that flows down a stream bed or valley. Similar to a mudflow, but it doesn't contain as much fluid.

delta A deposit formed at the mouth of a river where it enters into another body of water.

density current A subaqueous flow of concentrated mud or sand that flows quickly along the bottom of a body of water.

density The ratio of an object's mass divided by its volume. Common units are either g/cm3 or kg/m3.

deposition A phase change by which vapor becomes a solid. During this process the vapor loses energy to the surroundings. In the atmosphere this process results in the formation of frost.

derecho Strong straight-line winds proceeding from a collection of thunderstorms usually associated with a front. These winds can produce damage comparable to small tornadoes, but do not show signs of rotational winds.

desert pavement Rocks and gravel that cover the floor of a desert.

dew point depression The difference between air temperature and the dew point temperature.

dew point temperature The temperature at which air becomes saturated with water vapor. Any reduction in temperature below this value results in the condensation of water vapor into water droplets or dew.

dewatering the removal of water from sedimentary rock as the result of compaction (squeezing).

differential weathering When one type of rock or substance weathers faster than another.

differentiation A hypothesized time early in old-Earth history in which heavy elements such as iron migrated under gravitational force to form the core, while lighter elements rose upward to form the mantle and crust.

dip slip fault A type of fault characterized by vertical movement of blocks with respect to each other.

dipole A magnetic material with a single north and south pole, such as a bar magnet or Earth's core.

dirty iceberg theory The standard model of comets, that comet nuclei contain various ices and much microscopic dust.

disappearing stream A stream which either loses surface flow over time and eventually "disappears" or one which drains into a sinkhole and begins to flow underground.

disconformity An unconformity located between two roughly parallel sedimentary or extrusive igneous units.

dissipation Stage The later stage of cumulus cloud development where only a downdraft exists in the cumulonimbus cloud. During this stage precipitation from the cloud diminishes and eventually stops.

dissolved load The chemical load of a stream. It can include dissolved rock components, pollutants, nutrients and various ions.

distributaries On a river delta the major stream splits into smaller streams just before it enters into a large body of water. The small streams are called "distributaries."

diurnal tidal pattern A shoreline tidal expression featuring a single high tide and single low tide each day.

divergence A measure of how much substance is leaving a location. In the atmosphere this corresponds to more air leaving a region than entering it horizontally. This is the opposite of convergence.

divergent boundary A location where two plates are moving away from each other. New lithosphere is produced at divergent boundaries.

doldrums A surface low pressure region that resides near the equator. This region corresponds to monotonous weather of light winds and high precipitation.

dome A three-dimensional fold that is shaped like an upside-down cereal bowl. Its cross-section is an anticline.

Dominion Mandate The role given by God to people to rule over and care for the creation.

downdraft Downward motion of air within a thunderstorm. This consists of cold air and precipitation generated within the cumulonimbus cloud.

downwelling An oceanic phenomenon in which onshore winds force water toward the coast, forcing the excess water downward and bringing warm waters to the deep ocean.

drizzle A form of precipitation that consists of water droplets smaller than 0.50 mm in diameter. These droplets fall slowly and are uniform in size.

drumlin A cigar-shaped hill that forms under moving glacial ice.

dune A structure composed of wind-blown sand and/or silt.

dust tail The portion of a comet's tail that consists of dust particles pushed away from the sun by solar radiation.

dwarf planets A relatively new designation for minor planets (asteroids) that have enough mass so that their gravity has pulled them into spherical shape.

dynamic metamorphic rock The primary ingredient of change in this type of metamorphic rock is pressure.

"E" horizon This horizon is not always present, but when it is it lies below the A/O horizons and above the B horizon. It will be extensively leached and contain minerals.

earth flow Occurs when a steep slope, typically consisting of soil, becomes saturated with water and then begins to flow relatively slowly downslope. Is significantly slower than a mudflow.

Earth science A collection of related disciplines that together give a unified understanding of the planet and its context in space. Disciplines within Earth science include astronomy, biology, geology, hydrology, meteorology, and oceanography.

Earth systems science An approach to Earth science that focuses on the interactions and feedbacks both within and among the Earth science disciplines.

eclipsing binary A binary star that we view nearly in its orbital plane so that the stars involved eclipse one another each orbit.

ecliptic The plane of Earth's orbit around the sun, or, alternately, the apparent path that the sun follows through the stars as we orbit the sun each year.

eddy A circular or swirling motion in a fluid caused by a sudden change in fluid speed.

Ediacaran fauna A fossil assemblage consisting of numerous unusual, somewhat two-dimensional organisms of unknown affinity, along with the first appearance of several animal groups, such as worms, sponges, cnidarians, and perhaps others.

El Niño An extensive warming of the central and eastern tropical Pacific ocean that results in shifts of weather patterns and precipitation. This quasi-periodic event occurs every two to seven years.

elastic rebound theory When forces build up in Earth, rocks will not bend very much. Instead they bend a little (elastic) and then they snap and break (rebound).

electron A small, negatively charged particle that orbits an atom's nucleus.

electron capture A type of radioactive decay in which the parent isotope ejects is struck by an electron. The electron fuses with a proton in its nucleus to form a neutron.

element A material of characteristic physical and chemical properties that cannot be broken down into simpler substance by chemical processes.

emergent coastline An area where land is rising up out of the ocean, or where deposition of new materials builds more land area outward into the ocean.

end moraine A deposit that forms at the terminus of a glacier.

enrichment The process of increasing the amount of U-235 above naturally occurring levels to those needed for nuclear power or weapons.

entrainment Mixing of an air current with surrounding environmental air. In a cloud this process mixes in dry air and causes the cloud to dissipate and the air to cool.

entrenched meander A bend or series of bends in a river that has formed a deep bedrock canyon.

Environmental Lapse Rate The measured rate of temperature change in the atmosphere relative to altitude.

epicenter A spot on Earth's surface just above the focus of an earthquake.

epicontinental sea An extensive covering of continental margins and/or interiors by ocean water.

equilibrium (of a glacier) When glacial ice melts just as fast as it moves forward, causing the terminus of the ice to remain in one place.

erosion Material that is picked up and transported as a result of the weathering process.

erosion The process where rock and/or soil material is picked up and transported to another location.

esker A sinuous ridge of sand and gravel that represents stream gravels that formed either under the glacial ice or within a cave in the glacial ice.

estuary A former river valley now filled with ocean water.

evaporation A phase change by which liquid becomes a vapor. During this process the liquid surface loses energy and may result in cooling.

evaporites A chemical sedimentary rock that forms as the result evaporation of water. Rock salt is one type of evaporite that can form in places like the Great Salt Lake.

experimental science Approaches to investigation that include controlled conditions in which a hypothesis is tested in a repeatable fashion.

extrusive igneous rock Rock that forms as the result of volcanic activity. The rock is *extruded* from volcanoes.

eye A circular region of light winds and possibly clear skies located at the center of a hurricane. It can range in size from 10 to 100 km.

eye wall A ring of strong updrafts and thunderstorms surrounding the eye of a hurricane. This area corresponds to the most intense winds and precipitation of the storm.

eyewitness testimony Records of event made by those who observed them.

fault A break between two masses of rocks where movement has taken place.

feedback mechanism A sequence of interactions that results in either a reinforcement (positive feedback) or inhibition (negative feedback) of the original action.

Ferrel cell A global circulation pattern set up between the rising interface of the polar cell and the sinking interface of the Hadley cell. This cell operates between 30 and 60° latitude and has westerly surface winds.

fetch The distance across a body of water over which the wind blows.

firn Granular snow that is close to turning into glacial ice.

firn limit Also referred to as the snow line; the altitude on a glacier above which snow does not melt.

fission The splitting of a large atomic nucleus into two smaller nuclei.

fjord A long and deep valley connected to the ocean that has been cut by a glacier and possesses near-vertical walls.

flatness problem The fact that the ratio of the gravitational potential energy and the kinetic energy of the universe is nearly one, though one would not expect this in a big bang universe that is billions of years old.

floodplain A relatively flat area next to a stream onto which a flooding river will overflow.

flowstone A cave formation that coats walls, ceilings and other flat areas of caves. Usually made of calcite (sometimes referred to as travertine limestone).

flurry A light snow shower lasting a short period of time.

focus The place in Earth where energy is released due to an earthquake.

fog Small water droplets suspended near Earth's surface that reduce visibility below 1 km.

fold A bend in a rock.

foliated When a rock is put under extreme pressure and new minerals begin to grow as a result, the minerals will grow perpendicular to the pressure, in the direction of least resistance. The new minerals will all line up in a particular direction, giving the rock a "layered" appearance. The layering is referred to as "foliation."

footwall (footblock) The part of a fault block that sits below the hanging wall.

foreshore The region of the shore located between the low tide and high tide shorelines.

fossil fuel Carbon-rich resources produced by the alteration of plant, animal, or other organismal remains by geological processes (especially coal, natural gas, and petroleum).

fossil The remains of a formerly living organism that is preserved in the geological record.

fracture Irregular breakage of a mineral.

fracture zone A transform boundary found within an ocean ridge between (and connecting) two spreading centers.

freezing rain A form of precipitation that consists of water droplets striking sub-zero surfaces. The precipitation instantly freezes resulting in a thickening coat of ice.

friction A force that resists motion and tends to bring objects to rest.

frontal wedging A process that lifts air vertically due to the passage of either a warm or cold front.

fronts An interface between air masses of different properties. Cold fronts occur when cold air displaces warm air. Likewise, warm fronts consist of warm air displacing cold air. If air masses have similar temperatures, but different humidity, the interface is called a dry line.

frost heaving When water freezes it expands. If underground, the expanding ice can lift objects like roads, sidewalks and buildings.

frost wedging When water freezes it expands. If the frozen water is in the crack of a rock, it can sometimes apply enough force to further break the rock apart. It is a mechanical or physical weathering process.

fuel rod Uranium-rich bars that provide fissile material for the production of heat in a nuclear reactor.

Fujita Scale A scale for classifying tornadoes according to the damage they cause and their rotational wind speed.

Fumaroles Cracks, vents, or fissures where volcanic gasses escape into the atmosphere.

galaxy A collection of millions or billions of stars held together by gravity.

Gap Theory An interpretation of the Genesis 1 creation passage which asserted that a large gap in time existed between Gen. 1:1 and 1:2, in which vast periods of ancient geological history occurred. Following a destructive event, God creates a new world in six actual days. Also called ruin-reconstruction theory.

geological column An idealized image representing the vertical (stratigraphic) relationships of rock units around the world.

geosphere The solid components of Earth.

geostrophic wind A horizontal flow of air that is parallel to straight isobars. This type of wind occurs away from Earth's surface where friction has little effect. Resulting from a balance between the pressure gradient force and Coriolis Effect, this wind influences how weather patterns move across the country.

geothermal energy The extraction of heat from below Earth's surface, used either for space heating or electricity generation.

geyser A fountain of pressurized hot water.

giant stars Stars that are much larger than normal stars.

glacial erratic A boulder that has been carried and transported by a glacier, far from its source.

glacial moraine A general term for any unsorted glacial debris; sometimes called till.

glacial striations Scratches in rock or bedrock caused by the scraping of a glacier.

glacial till A general term for any unsorted glacial debris; sometimes called moraine.

glacier A large body of moving ice.

globular star cluster A tightly bound star cluster consisting of hundreds of thousands of stars in a spherical shape.

GOES Satellites A series of geostationary satellites that transmit weather related data and images covering the United States on half hour intervals.

Two satellites are active at any one time: one covering the east coast and Atlantic Ocean and the other the west coast and Pacific Ocean.

Gondwana The southern half of Pangaea consisting of Africa, Antarctica, Australia, India, and South America.

gradient wind A horizontal flow of air that is parallel to curved isobars. This type of wind occurs in the proximity of low and high pressure systems.

granular ice Pellets of ice that if compressed further will turn into glacial ice.

Great Unconformity An extensive erosional unconformity found throughout much of the world at or near the base of the Phanerozoic rocks. It likely marks the first advance of waters over the continents during Noah's Flood.

Great Upheaval The geological events associated with Day Three of Creation Week, in which God caused the continents ("dry land") to rise out of the oceans.

greenhouse gases Gasses that primarily absorb radiation in the infrared range of the spectrum. Increasing the amount of these gases in the atmosphere results in an increased average temperature of Earth.

groin A hard stabilization structure built perpendicular to the shoreline and intended to retain sand that is eroding or entrained in longshore transport.

ground moraine Unsorted sand and gravel deposited in a sheet-like fashion by a glacier.

ground-source heat pump An energy-exchange device for space heating/cooling that uses the ground to exchange heat with the building interior.

groundwater sapping A physical or mechanical weathering process via which groundwater flowing out of a cliff face can apply pressure from the inside of the cliff face outward causing pieces of rock to splay off.

groundwater water that is found below the surface of Earth.

gyre A large, circular system of surface currents.

habit See *crystal form*.

Hadley Cell A global circulation pattern set up between the rising warm air of the equator and the sinking interface of the Ferrel cell. This cell operates between the equator and 30° latitude and has easterly surface winds called trade winds.

hail A form of precipitation that consist of balls or lumps of ice that are 5 mm in diameter or larger.

half-life The time it takes for one half of an amount of radioactive material to decay into its final, stable daughter product.

halo An optical phenomenon due to refraction through ice crystals. This results in a ring of color surrounding the sun at an angle of 22°. In dimmer light the color separation is not evident giving only a bright band of light around the moon.

hanging valley A valley formed by a tributary glacier that enters into a larger glacial trough and which sits high above the main valley floor.

hanging wall (hanging block) The part of a fault block that sits above the hanging wall.

hardness A mineral's resistance to being scratched.

haze Small particles suspended in the air that reduce clarity due to scattered light. If the particles are small enough they appear yellow against a light background. A haze does not reduce visibility as much as fog does.

heat A form of energy transfer that naturally flows from warm to cooler objects.

helictite A type of cave formation that often has bizarre shapes.

Hertzsprung-Russell diagram A plot of intrinsic brightness of stars versus temperature.

high pressure system (anticyclone) A synoptic scale pressure system that consists of closed isobars surrounding a region of high pressure. In the Northern Hemisphere air circulates in a clockwise direction around the central high.

high tide The furthest daily onshore advance of the shoreline.

highlands The heavily cratered, lighter colored regions on the moon.

historical science Investigations into events that occurred just one time in the past.

homogeneous Having the same properties throughout.

horizon problem The fact that disparate parts of the cosmic microwave background have the same temperature even though they could not have been in thermal contact.

horn A sharp mountain peak carved by a glacier.

hot spot A plume of magma originating in the lower mantle and rising upward into or through the crust.

hot springs Hot water that comes out of Earth.

Hubble relation The linear relationship between galaxy redshift and distance.

humus The part of the soil consisting of organic matter.

hurricane An intense tropical storm with wind speeds exceeding 119 kph.

hurricane watch A statement issued by the National Hurricane Center indicating that a hurricane poses a threat to an area. The statement is issued several days before landfall so residents can make preparations.

hydraulic action A physical/mechanical weathering process caused by the force of rapidly flowing water.

hydroelectric power The use of flowing/falling water to generate electricity.

hydrogen bond A chemical bond in which loose weakly charged regions of overall neutral molecules cause attraction to ions or charged regions of other molecules.

hydrologic cycle The cyclical pattern where water moves from one reservoir (e.g., the oceans, atmosphere, surface waters, glaciers, and groundwater) to another.

hydrology A branch of science that deals with the properties and flow of water.

hydrosphere All the water of Earth's surface.

hydrothermal Processes referring or related to hot water.

hygrometer A device used to measure water vapor content in the atmosphere.

hypothesis A possible explanation for some event of observation.

Ice Age A period of extensive continental glaciation during the Pleistocene Epoch.

ice cap A sheet-like glacier that has a surface area that is permanently covered in ice.

ice sheet The largest type of all glaciers >50,000 km^2 in size; sometimes called a continental glacier.

iceberg A large chunk of ice that floats in a body of water that was derived from a glacier.

ichnofossil A class of fossils that indicate the presence of an organism by the traces left behind (including footprints, trackways, burrows, resting impressions, etc.).

igneous rock Any rock that has crystallized from liquid or molten rock.

inselberg An isolated mountain that has been buried in its own eroded debris.

interface A point of contact between two different systems.

intrusive igneous rock A rock that has formed from liquid rock "intruding" into pre-existing country rock. Sometimes intrusive igneous rocks are referred to as "plutons."

inversion A location in the atmosphere where air temperature increases with altitude. This represents a region of very stable air.

ion A positively or negatively charged atom.

ion tail The tail of a comet produced by glowing ionized gas blown outward from the sun by the solar wind.

ionic bond A chemical bond in which one atom transfers one or more electrons to another atom, creating two charged particles (ions) which then are attracted by their opposite charges.

iridescence An optical phenomenon due to diffraction by water drops or ice crystals in the atmosphere. This occurs near the edges of thin clouds and results in pastel colors.

island universe theory The theory, ultimately proved to be correct, that the Milky Way is just one of very many galaxies in the universe.

isobar A line representing locations that have the same pressure value.

isostatic rebound Expansion that results after removing a very heavy object which caused compression. For example, a thick glacier might compress a landscape, when it melts expansion of the underlying rock will occur.

isostatic rebound Gradual rising or swelling of Earth's crust due to the removal of thick glacial ice or rock; a response to the removal of weight.

isotopes Atoms of an element with variable numbers of neutrons, giving them differing atomic masses.

isotropic Having the same properties in every direction.

jet stream A narrow band of strong winds in the upper atmosphere. The main jet stream is located on the polar front just below the tropopause.

jetty A pair of shore-perpendicular hard stabilization structures intended to protect a bay from sediment infill and/or wave action.

Jovian planets The four gas giant (Jupiter-like) planets.

kame A pile of sands and gravels that results from a river filling up a hole with sand and gravel in the glacial ice.

karst topography A description of the surface of Earth when it has been severely modified by solution of the underlying carbonate rock. The surface will typically have disappearing streams, sink holes, round lakes and cave openings.

kerogen Waxy organic chemical compounds found in sedimentary rocks that often have an oily smell to them. A precursor of petroleum.

kettle lake A lake that forms as the result of a block of stranded glacial ice getting left in ground moraine.

kimberlite pipe A narrow pipe-like pluton that originates from hundreds of kilometers within Earth. They are one of the major sources of diamonds.

Köppen climate classification A classification system that divides the land surface of Earth into climate zones based on the vegetation that grows there. There are five basic divisions with multiple sub-divisions based on summer and winter patterns of temperature and precipitation.

Kuiper belt Hypothetical source of short-period comets orbiting beyond the orbit of Neptune.

La Niña An extensive cooling of the central and eastern tropical Pacific ocean that results in shifts of weather patterns and precipitation. This event is considered the opposite of El Niño.

lahar Another name for a volcanic mudflow.

landslide A generic term used to describe many different types of mass wasting processes whereby large amounts of material move down a slope quickly.

late heavy bombardment A hypothesized time early in old-Earth history in which intense meteor showers destroyed most of the early surface rocks on Earth.

latent heating A transfer of heat either into or out of an object that does not result in a temperature change. Instead of changing the temperature, the heating results in a phase change.

lateral moraine A ridge of glacial till that forms along the valley edges of an alpine glacier.

Laurentia The northern half of Pangaea consisting of Asia, Europe, Greenland, and North America.

lava Liquid rock that is found above the surface of Earth.

law of lateral continuity A rule of relative dating which asserts that when sediments or extrusive igneous rocks are deposited, they remain the same composition in all directions until they either (a) contact an edge or wall to the depositional environment, or (b) thin out as energy levels decrease far from the source region.

law of original horizontality A rule of relative dating which asserts that when sediments or extrusive igneous rocks are deposited, they are spread out in broad, flat sheets.

law of superposition A rule of relative dating which asserts that in an undisturbed sequence of sedimentary rocks and extrusive igneous rocks, the oldest units are found below the younger units.

left-lateral fault In a strike slip fault when the other side has moved to the left.

lifting condensation level The level in the atmosphere where lifted air reaches a relative humidity of 100%. Water vapor begins to condense at this level forming the flat base of a cumulus cloud.

light travel time problem The difficulty in explaining light from very distant astronomical objects if the universe is only thousands of years old.

lightning An electrical discharge of high current that extends kilometers in length. Most common discharges are between a cumulonimbus cloud and the ground, but can also occur between clouds and within the same cloud.

lightning rod A grounded metal object placed at the top of a structure to provide a preferable path for a lightning discharge.

lignite coal A low-grade type of coal, sometimes called brown coal.

linear dune A large sand dune whose crests are generally parallel to wind direction. Sometimes called a longitudinal dune.

lithification The hardening process that turns sediment into sedimentary rock.

lithosphere The upper rigid portion of Earth consisting of the crust and part of the mantle. It rests on the asthenosphere and makes up Earth's plates (that move in plate tectonics).

load casts Sedimentary features that form as the result of an upper sedimentary layer pushing downward and deforming a lower sedimentary layer. Typically the features are softball to basketball in size.

loam Soil composed of sand (40%), silt (40%) and clay (20%). These are often ideal soil types.

loess Windblown silt, or silt dunes.

longshore transport The overall direction for the erosion, transport, and deposition of sediments on and near the shore.

Love wave A type of earthquake surface wave that causes the surface of Earth to move back and forth with a lateral or shearing motion.

low pressure system A synoptic scale pressure system that consists of closed isobars surrounding a region of low pressure. In the Northern Hemisphere air circulates in a counterclockwise direction around the central low.

low tide The furthest daily offshore retreat of the shoreline.

lunar eclipse When the shadow of Earth falls upon the moon.

lunar phases The change in the moon's appearance as the lit portion that we see changes when the moon orbits Earth each month.

luster The manner in which light is reflected off of a mineral surface. Categories include metallic and non-metallic (which includes subcategories such as dull, earthy, glassy, pearly, resinous, or satiny).

magma Liquid or molten rock below the surface of Earth.

magnetic reversal A 180° switch in the orientation of Earth's dipolar magnetic field.

magnitude The system of measuring star brightness that astronomers use.

mantle The part of Earth that rests between the crust and the core.

maria The darker portions of the lunar surface that do not have many craters.

marine terrace A former wave-cut platform that has been raised above the shoreline, forming a bench-like land surface.

mass wasting A generic term used to describe any large amount of material that moves down a slope, either quickly or slowly, under the force of gravity.

mature stage The stage of convective development where both an updraft and downdraft exists in a cumulonimbus cloud. During this stage precipitation rates are highest.

maturity A landscape that has maximum relief, good drainage, and a major stream that is beginning to erode horizontally (compared downward erosion) and meander on a floodplain.

meander scar The remnants of a former bend in a river often evident from above due to color changes in the soil or vegetation. It can result from an oxbow lake being filled in with debris.

meandering stream A river that winds or bends back and forth across a floodplain in a series of large loop-like patterns.

mechanical weathering Processes that disintegrate rock via physical means rather than by chemical means.

medial moraine A type of till formed in alpine glaciation where two lateral moraines join. Often is identified by dark bands in the center of an alpine glacier that parallel flow.

meltdown A catastrophic failure in a nuclear reactor in which temperatures rise too high and damage the building and/or cause other materials to explode. A meltdown may result in the dispersal of radioactive material into the surrounding area.

Mercalli intensity scale A scale (I–XII) that measures the damage done to human structures during an earthquake.

mesa A flat-topped hill.

mesosphere The layer of the atmosphere that lies between the stratosphere and the thermosphere. It extends between 50 and 90 km.

metallic bond A chemical bond in which the valence electrons are shared among neighboring atoms, but remain free to move among the three-dimensional structure of the solid.

metamorphic rock A rock that was changed from a pre-existing rock by heat and/or pressure.

metamorphism The process of change that occurs in a rock due to extreme amounts of heat and/or pressure. Rocks do not melt during metamorphism.

meteor A streak of light visible in the night sky as a small fast-moving particle enters Earth's atmosphere and burns ups.

meteor shower A large number of meteors visible when debris from a comet intersects Earth's orbit.

meteoric water Water derived from precipitation in the atmosphere. Much of this water can infiltrate into the ground and become groundwater.

meteorite Any portion of a meteor that survives to the ground.

meteorology The scientific study of the atmosphere and its processes. This includes both weather and climate.

microburst A strong downdraft from a cumulonimbus cloud that reaches the ground and extends in front of the storm. The strong wind shear from this event has resulted in aircraft accidents.

micrometeorology A branch of meteorology that studies atmospheric features and phenomena lasting minutes and extending up to a kilometer in size.

Milky Way Our galaxy, which can be seen as a broad streak in a clear, dark sky.

mineral A naturally occurring object that is a crystalline solid, generally inorganic, and has a definite chemical formula.

mineral attributes The characteristics that define a mineral.

mineral properties The unique physical characteristics that a mineral possesses, including categories for optical, shape, mass-related, mechanical, and other properties

Mining A group of activities that involve the recovery of solid metallic or nonmetallic resources from rock or loose sediment.

minor planet A small rocky body that orbits the sun. A common name for minor planets is *asteroids*.

misfit stream A stream that is quite small compared to the large valley that it flows in.

mixed tidal pattern A shoreline tidal expression featuring two high and low tides each day, but the two highs are not of the same magnitude, nor are the two lows.

Moho (Mohorovičić Discontinuity) The boundary in Earth that separates the crust from the mantle.

monsoon A wind pattern that changes direction over the year leading to a dry and rainy season.

mountaintop removal A method of coal mining in which large portions of a mountaintop are removed to allow recovery of the coal below.

mud cracks Desiccation features that typically form from water evaporating from mud.

mud pots Similar to a hot spring, but instead of water, there is mud. Often the mud will bubble as gases rise and escape from it.

mudflow A rapid flow process whereby mud flows down a river valley. Typically the mud has a consistency close to wet cement, often with less than 50% water.

mudstone A rock made mostly of clay-sized particles (<1/256 mm), but with some silt and sand components as well.

natural gas (methane) A gaseous hydrocarbon composed of one carbon atom covalently bonded to four hydrogen atoms.

natural levees A low ridge found along meandering streams. The ridge is formed during flooding as water spills out of a stream bank and deposits some of its load right next to the stream channel. Levees can also be man-made.

naturalistic evolution A theory of Earth history in which the universe and life arose by unguided processes.

neap tide The smallest tidal range of the month with the least pronounced high and low tides.

near-Earth asteroid A minor planet with an orbit very close to Earth's orbit around the sun.

nearshore The area of the ocean nearest the shore that is permanently submerged and whose bottom sediments are affected by storm wave activity.

neutron An uncharged particle in the nucleus of an atom.

neutron star A very small, dense star consisting of neutrons.

nimbostratus A low level cloud from which precipitation is falling. These clouds form a layer and give the sky a dark appearance.

nonconformity An unconformity located at the contact of a sedimentary or extrusive igneous rock and an intrusive igneous or metamorphic rock.

nonsilicate A class of minerals that do not possess a silicon-oxygen tetrahedron as their anion. Includes carbonates, halides, oxides, native elements, phosphates, sulfides, and sulfates.

normal fault A type of dip slip fault where the hanging wall moves down in relation to the footwall.

North Star The bright star on the end of the handle of the Little Dipper located near the north celestial pole so that it does not appear to move as Earth rotates.

numerical forecast A weather forecast product generated by computers using equations that describe the physical behavior of the atmosphere.

"O" horizon A soil horizon with a considerable amount of organic material, typically found at or near the surface.

oblique slip fault A type of fault that exhibits both lateral and vertical movement.

ocean ridge An elongate submarine mountain chain with a volcanic region in its center; a zone of seafloor spreading.

ocean trench Extremely deep, linear submarine regions formed by the subduction of a plate capped by ocean crust.

oceanography A branch of science that studies the world's oceans. Includes both physical and biological areas of study.

octet rule A rule of chemistry which recognizes that atoms are most stable when their valence shell is filled with eight electrons.

offshore The area of the ocean on the continental shelf that is deep enough that storm waves do not affect is sediments.

old-Earth creation A theory of Earth history in which God played an active role in creating the universe, Earth, and life (as direct creations, not through evolution) over a span of billions of years.

Oort cloud The hypothetical source of long-period comets orbiting the sun at great distances.

open star cluster A loosely bound cluster of stars that has hundreds or thousands of stars in an irregular shape.

ore A metal-rich rock or mineral resource.

organic sedimentary rock A sedimentary rock in which the particles are from organic sources, like broken pieces of shells (to make a limestone) or bits of bark and other plant material (to make coal).

orographic forcing A process that lifts air vertically due to the presence of large geographic features.

outwash deposits Sand and gravel deposits formed by rivers that carry meltwater away from a glacier.

outwash streams Rivers of meltwater that flow from a glacier.

oxbow lake A crescent-shaped lake which is the remnant of a meandering stream bend. Sometimes occurs in the shape of a horse shoe.

ozone A molecule that consists of three oxygen atoms. Its presence in the stratosphere reduces the amount of ultraviolet radiation reaching the surface.

pahoehoe A type of lava that is very smooth when it cools and looks like taffy.

paleomagnetism The record of Earth's past magnetic field information preserved in rocks.

paleontologist A scientist who studies fossils, including their biological and geological contexts.

Pangaea A hypothesized supercontinent that existed from the late Paleozoic to early Mesozoic in old-Earth timescales, or formed and rifted apart during Noah's Flood according to Catastrophic Plate Tectonics.

parabolic dune A crescent-shaped dune which usually occurs in coastal areas. Its horns often point into the wind and it is usually has been partially stabilized by vegetation.

parcel of air An imaginary mass of air that has uniform properties and does not exchange energy or mix with its surroundings.

parent isotope A radioactive atom that has not yet undergone decay.

parsec The distance that star must be to have a parallax of one second of arc (3.26 light years).

passive margin A region where the edge of a continent does not coincide with a plate boundary.

passive solar heating Structural designs that purposefully employ insolation to heat a building.

peaker plant Electrical power plants that can be quickly adjusted to serve shifting power needs during the day.

pediment An erosional ramp-like surface found at the base of a mountain range.

pegmatite A crystalline igneous rock where many of the individual mineral grains are larger than 30 mm in size.

penumbra Either a partial shadow or the less dark outer region of a sunspot.

perched water table A water table that is above (or "perched") the main water table.

perigee The point on the orbit around the sun that is closest to Earth.

perihelion The point on the orbit around the sun that is closest to the sun.

Periodic Table (of the elements) A system of categorizing elements on the basis of atomic number, electron configurations, and chemical properties.

permafrost Soil in polar regions that remains frozen year round. During the summer the surface layer can melt, but deeper layers remain frozen.

permeability The ability of water (or other fluids) to flow through a rock.

persistence A weather prediction technique that assumes that future weather will be similar to the current weather.

petroleum A liquid organic compound composed of chains of carbon bonded to hydrogen atoms, which is refined to produce various fuels, oils, and plastics.

phaneritic A crystalline igneous rock texture where the mineral grains are between 1–30 mm in diameter.

photosphere The visible surface of the sun.

planar bedding Horizontal sedimentary beds that are roughly parallel to each other.

planet A large body that orbits the sun.

planetary nebula A shell of gas around a star that the star appears to have blown off in the past.

planetesimal The hypothetical small particles that supposedly formed in the early solar system out of which planets, satellites, minor planets, and comets formed.

plate tectonics A theory of Earth surface motion in which units of lithosphere move in relation to each other over long period of geologic time. Plate tectonics combined and modified continental drift with seafloor spreading.

plate(s) A rigid, mobile unit of the geosphere composed of oceanic and/or continental crust and a portion of the uppermost mantle; also called a *lithospheric plate*.

plateau A large flat surface that is elevated above other surrounding topography.

playa A shallow desert lake which is often dry.

playa crack Similar to a mud crack, but many times larger (10's to 100's of meters in length polygons).

point bar A deposit of sand or gravel that forms in the shallow, slowly moving water on the inside of a meander bend in a stream.

Polar Cell A global circulation pattern set up between the rising interface of the Ferrel cell and the sinking air over the pole. This cell operates between 60 and 90° latitude and has easterly surface winds.

polar front The interface between the polar and Ferrel cells where cold polar easterlies meet warm westerly air brought poleward from the subtropics. This interface provides a favorable location for cyclone development.

polar wander The movement of Earth's magnetic poles with respect to the geographic poles.

polymorph A mineral that displays two or more crystal forms or mineral types while possessing the same atomic formula (e.g., diamond and graphite, both made of carbon).

porosity The amount of empty space in a rock, usually expressed as a percentage.

porphyry (porphyritic texture) An igneous rock that has two distinct grain sizes (large and small).

Precambrian An informal geochronological name given to the three eons prior to the Phanerozoic eon: the Hadean, Archean, and Proterozoic.

precipitates Sedimentary rocks that form as the result of precipitation in a body of water. Rock salt or gypsum can precipitate out of solution in special hydrothermal situations.

precipitation All liquid and solid forms of water that originate in the atmosphere and fall to the surface.

pressure Force per area. More specifically, air pressure is the weight of air above a certain point per unit area. Common units are pounds per square inch (lb/in2), inches of mercury (in Hg) and millibars (mb).

pressure gradient force A force resulting from a difference in pressure between two locations. If the pressure changes rapidly over distance, the pressure gradient force is large.

primary (P) wave A type of compressional earthquake wave that travels down into and through Earth (a body wave). It is called a "primary" wave because it travels faster than the other wave types.

principle of cross-cutting relations A rule of relative dating which asserts that geological structures or features (such as dikes, faults, or erosion surfaces) must be younger than the geological units that they affect.

principle of faunal succession A relative dating rule which asserts that fossil organisms follow one another in a definite and recognizable order within sedimentary rocks.

proton A positively charged particle in the nucleus of an atom.

proved reserve A quantity of resources that is known to exist and can be recovered economically.

psychrometer A device used to measure humidity. It uses both a dry and wet bulb thermometer to determine how close the air is to saturation.

pulsar A rapidly rotating neutron so oriented that we see periodic flashes of radiation.

pumped storage A hydroelectric system using two dams and acting like a battery. Water flows from the upper to the lower dam during the day to generate electricity. At night, water is pumped from the lower to the upper dam, replenishing it.

pyroclastic "Broken bits of fiery rock" sourced from volcanoes.

pyroclastic material Hot pieces of rock that are ejected during a volcanic eruption. It can be small (like ash) or big (like blocks and bombs)

quasar A contraction of "quasi-stellar object," a high redshift object that probably is at great distance and hence is very bright, yet is very small.

radar An instrument that transmits and measures radio waves to determine the range and reflective properties of distant objects.

radiation Electromagnetic waves emitted by any object that has a non-zero temperature on the Kelvin scale. Hotter objects emit more radiation; therefore, heat transfer occurs from warmer to cooler objects.

radioactive decay A nuclear process in which a parent atom eject material from its nucleus, resulting in the production of a different element, called the daughter element.

radioisotopic dating A variety of methods that use the measured amounts of radioactive parent and daughter products to produce a numerical age for a rock.

radiosonde An instrument package used to measure properties of the atmosphere away from the surface. Most often it is attached to a free-flight balloon and data is transmitted to a ground station using radio waves.

rain A form of precipitation that consists of water droplets 0.50 mm in diameter or larger. These droplets can have a variety of sizes and can reach diameters of 5.0 mm.

rain drop prints Sedimentary features formed from rain leaving impressions or craters in mud or sand.

rain gauge An instrument used for measuring rain amounts.

rainbow An optical phenomenon resulting in an arc of colored light due to interaction with a large number of water droplets.

RATE team A group of young-Earth creation scientists who conducted a number of experiments that challenged certain assumptions and applications of radioisotopic dating methods.

Rayleigh wave A type of earthquake wave that travels along the surface of Earth and has a rolling motion. It will move objects up and down.

recrystallization The process mineral change that takes place when a rock is heated or placed under pressure. Pre-existing minerals are usually not stable under the new pressure and temperature conditions, so the minerals change (recrystallize) to accommodate the new pressure and temperature conditions.

red beds Iron-rich sedimentary rocks (especially sandstones and shales) often inferred to be the product of arid conditions.

redshift The shift toward longer wavelengths that the spectra of very distant objects in the universe have, probably due to the expansion of the universe.

refine/refining The process of converting raw materials into finished products (e.g., crude oil into gasoline).

refraction A change in direction of a propagated wave as it moves from one medium to another.

regional metamorphic rock A type of metamorphic rock where the change has mostly occurred due to both increased temperature and pressure. These types of rocks will exhibit foliation.

rejuvenation When a meandering river cuts meandering canyons in bedrock. Initially it was thought that these developed from "old age" rivers that meandered on a flat floodplain.

relative dating A variety of methods for determining the order and sequence of geological features and events.

relative humidity The ratio of water vapor present in a volume of air compared to the saturation value at the same temperature.

relict landscape A landscape that formed by past processes and rates that are no longer operating today.

relict landscape A landscape that has undergone very little change for very long periods of time.

relief The vertical distance from the highest point in a landscape to the lowest point.

reserve a resource that has a good likelihood of existing in a resource area, and might be recovered.

residual soil An soil that forms in place from locally weathered bedrock.

resource Any material that can be used by people. Resources may be renewable or non-renewable.

retreating glacier A glacier that is melting faster than it is moving forward.

retrograde The apparent backward, east to west, motion that other planets appear to have with respect to the stars as we pass them in our orbit around the sun.

reverse fault A type of dip slip fault when the hanging wall moves up with respect to the footwall.

Richter scale A scale used to measure the magnitude of an earthquake. It is based on measurements taken from a seismogram.

ridge An elongated area of high atmospheric pressure. This is present in upper-air charts when warm air extends further north than usual in the Northern Hemisphere

rift A region of the crust (either continental or oceanic) where extensional forces allow magma to rise and new crust to form.

riming The process by which super-cooled water freezes on contact with an object. This contributes to the growth of hail stones, as well as, the buildup of ice on aircraft.

rip current A narrow offshore current along the shoreline.

ripples Small dunes or "ripples" are usually in the range of 1–4 cm in height. They form as the result of a moving air or water current and express themselves as a series of parallel humps or ridges.

river terrace A bench or series of benches in a river valley that represent former floodplains that the river has abandoned.

rock a substance made from one or more minerals.

rock avalanche Rock that tumbles quickly down a hillside.

rock creep Rock that moves slowly down a hillslope under the influence of gravity.

rock cycle A hypothetical process that shows how igneous, metamorphic and sedimentary rocks are related to each other and with the weathering process.

rockfall Rock that freefalls from a cliff face. The broken material usually accumulates at the bottom of the cliff as talus.

Rodinia A supercontinent that existed during the late Archean and early Proterozoic eons. In a young-Earth view, this may represent the original continental arrangement after Creation Week.

root wedging The breaking up of rocks by growing tree roots.

rule of inclusions A rule of relative dating which asserts that when a rock includes pieces of another rock inside it, the included pieces must be older than the rock containing them.

Saffir-Simpson scale A scale for classifying hurricanes according to their central pressure, wind speed and damage inflicted.

saltation A process where particles (usually sand) bounce along a stream bed or a desert sand dune surface. When one sand grain hits the bottom, it usually stimulates other sand grains to bounce up into a current.

sand injectite A crack that has been forcefully injected with liquefied sand; similar in shape to igneous dikes and sills.

sand waves Large sand dunes that occur on the floor of the ocean. Similar in size and shape to large sand dunes in deserts.

sandstone A rock composed primarily of sand-sized particles, 1/16-2mm in size.

satellite A body that orbits a planet.

saturation The condition where water vapor is in equilibrium with a flat surface of pure liquid water.

scarp A rock or soil face left behind after a mass movement slides away.

scientific methods A group of activities used by scientists to investigate the world, build and test hypotheses, and make predictions.

scud Ragged low-level clouds occurring below the main cloud base.

sea arch A remnant of a cave composed of a headland with rock bridge connection out to a column of limestone standing out of the ocean.

sea stack An isolated rocky spire standing out above the ocean surface.

sea wall A hard stabilization structure built directly on the shore to protect an area from waves.

seafloor spreading A hypothesis of conveyor belt-like ocean crust movement in which new crust is produced at spreading centers, moves laterally to form abyssal plains, and descends into the mantle at subduction zones.

secondary (S) wave A transverse wave type that is developed during an earthquake and travels downward into Earth. It will arrive at a seismic recording station second, just behind the P wave.

sediment Loose particles that have been derived from the weathering of rock (mud, silt, sand, etc.)

sedimentary rock Rock made of pieces (or grains) of pre-existing rock

sedimentary structures Features in sedimentary rocks that geologists can use to help them determine past conditions, rates and processes.

seismic wave Wave forms produced with energy is released from an earthquake.

seismogram A paper (or digital) record of an earthquake produced by a seismograph.

seismograph An instrument used to detect and record earthquakes. It produces a seismogram.

seismometer An instrument that is used to detect earthquakes.

selective absorber Substances that absorb radiation at one wavelength and not another. Greenhouse gases absorb in the infrared range, but not the visible range of the electromagnetic spectrum.

semi-diurnal tidal pattern A shoreline tidal expression featuring two similar-magnitude high and low tides each day.

sensible heating Heat transfer that results in a temperature change of an object.

sequence A repeating pattern of sedimentary rocks in which coarse clastics grade vertically into finer clastic and ultimately to carbonates (e.g., sandstone, shale, and limestone). Six global-scale sequences are recognized.

severe weather warning A statement issued by the National Weather Service indicating that severe weather is immanent and precautions should be taken to ensure safety. More specifically, the statement can be a severe thunderstorm or tornado warning.

severe weather watch A statement issued by the NOAA Storm Prediction Center indicating that conditions are right for the development of severe weather. More specifically, the statement can be a severe thunderstorm or tornado watch.

shale A rock composed almost exclusively of clay-sized (<1/256 mm) particles and often show fine layering.

shear force Unaligned forces pushing one part of a body in one direction, and another part of the body in the opposite direction.

shelf cloud An arc shaped cloud that forms on the leading edge of a thunderstorm's gust front. Cold air from the cloud's downdraft extends in front of the storm pushing warm humid air upward, thus forming this feature.

shield volcanoes A type of large volcano whose primary product is various kinds of lava flows. The volcano has the shape of a broad disc.

shore The area of the coastal zone including both the foreshore and the backshore.

shoreline The line of intersection between the ocean and the land.

showers Precipitation coming from convective clouds that starts and stops suddenly and varies by location.

Siberian High A high pressure region that forms over central Asia during the Northern Hemisphere winter. Influence of this feature extends into the Western Hemisphere and causes the monsoon dry season in India and Southeast Asia.

silicate A class of minerals that all possess a silicon-oxygen tetrahedron as their anion.

silicon-oxygen tetrahedron A mineral anion that includes a silicon atom bonded to four oxygen atoms, each with a single covalent bond to the silicon atom.

siltstone A sedimentary rock made predominantly of silt-sized particles.

sinkhole A hole that develops in the surface of the ground due to the ceiling of an underground cavity collapsing.

sleet A form of precipitation that consists of frozen water droplets. They are translucent and bounce when they hit the ground.

slide (also see landslide) A type of mass movement that has a relatively planar base as opposed to the curved base in a slump.

slump A type of mass movement that often has a curved base (as opposed to a planar base in a slide).

snow A form of precipitation that consists of single ice crystals that are hexagonal in shape or aggregations of crystals.

snow avalanche Snow and ice that rapidly flows or tumbles down a slope.

"Snowball Earth" A hypothesized period of global to near-global ice ages during the middle and late Proterozoic Eon.

soda straw A type of cave formation that hangs from the ceiling of a cave and is hollow. Like a drinking straw, water can flow through the hollow interior of the formation.

soil A mixture of weathered rock (sand, silt, clay) and organic debris.

soil creep A slow mass movement whereby soil flows slowly down a hillside over many years.

soil horizon A stratified layer in the soil.

soilfall When soil freefalls through the air (similar to a rock fall, except with soil).

solar eclipse When the shadow of the moon falls on Earth, blocking at least some of the sun's light.

solar flare A violent eruption on the sun's surface that ejects fast-moving charged particles, which, if they reach earth, cause magnetic storms.

solar photovoltaic panel A device in which photons of light strike a semiconductor and generate an electrical current.

solar wind A continual outrush of particles from the sun.

solifluction A special type of soil creep and flow that occurs in areas with permafrost. In the summer, a thin upper layer of soil can thaw, but it continues to rest on frozen soil. If there is any slope, the thawed soil will creep.

solution Some rocks can be dissolved by various acids. The process is referred to as "solution."

sounding A vertical measurement of the atmosphere's properties. This is done with a radiosonde and contains measurement of temperature, humidity, pressure, wind direction and speed.

Southern Oscillation Index A measure of pressure difference between the eastern and western tropical Pacific Ocean. Significantly lower pressures in the east correspond to an El Niño event.

specific gravity A unitless value that is the density of a substance divided by the density of another substance. Often used in place of density among minerals.

speleothem Any type of cave formation, usually made of travertine, or dense calcium carbonate.

spent nuclear fuel The residual radioactive material that remains after a fuel rod has been exhausted.

spheroidal weathering A description of a weathering process whereby rocks erode into round or spherical shapes.

spit A linear deposit of sand extending from the beach out into a bay or open water, and formed parallel to the prevailing longshore transport direction.

sporadic meteor A meteor that appear apart from any other meteors (not part of a meteor shower).

spreading center The area of an ocean ridge where new ocean crust is produced from rising magma.

spring deposit When water flows out of the ground and leaves some type of chemical residue behind (usually travertine limestone or tufa).

spring tide The largest tidal range of the month with the most pronounced high and low tides.

spring When water flows out of the ground.

stability A characteristic describing an object's response to small disturbances. If an object is stable, it returns to its original position while an unstable object continues to speed up in the direction of the disturbance.

stalactite A type of cave formation that hangs from the ceiling of a cave.

standard candle A bright object that we think we know the absolute magnitude so that it is useful for measuring distances in astronomy.

star cluster A collection of stars, smaller than a galaxy, bound by gravity

star dune A complex sand dune type that resembles a pointed star in appearance. It is caused by prevailing winds that change direction (often seasonally).

steady-state A forecast technique that assumes a weather system will travel in the same direction and at the same speed as it is currently moving.

steering winds Upper level winds in the atmosphere that determine how weather patterns will move across the country.

stock A discordant igneous pluton that has a surface area of <100 km² in area.

storm surge A rise in water levels along a shore due to strong winds and low surface pressure of a large storm. This incursion of water inland results in the largest amount of damage from a hurricane.

stoss slope The windward side of a sand dune.

strata Layered sedimentary rocks.

stratosphere The layer of the atmosphere immediately above the troposphere. It extends from the troposphere to about 50 km, which is the beginning of the mesosphere. Ozone in this layer absorbs ultraviolet light and results in the air temperature increasing with height.

stratus A low level cloud that forms a uniform gray layer.

streak The color of a mineral when powdered.

strike slip fault A type of fault characterized by horizontal displacement. Caused by horizontal shear forces.

strip mining A method of mining in which large amounts of overburden are removed to expose and remove mineral or fuel resources.

stromatolites Mounded structures composed of alternating layers of sediment and microbial life.

subduction/subduction zone A place in Earth's crust (usually in an ocean trench) where old oceanic crust is pulled down into Earth.

sublimation A phase change by which solid becomes a vapor. During this process the solid surface loses energy and may result in cooling. This is the opposite of deposition.

submergent coastline An area where the sea level is rising and/or erosion removes material from the land, causing the shoreline to advance inland.

subsidence A sinking of the land surface.

subsidence Descending motion of air over a large area. This occurs in high pressure systems and results in warming air and clear skies.

summer solstice – The first day of summer when the sun is highest in the sky at noon and daylight hours are the longest of the year. This occurs around June 21 in the Northern Hemisphere.

sun dog An optical phenomenon producing color spots to the left and right of the sun. This occurs near sunrise and sunset and is a truncated portion of a halo.

sunspot A dimmer, cooler region on the sun where magnetic fields are strong.

sunspot cycle An approximate eleven-year cycle over which the number of sunspots vary.

supercell A mature convective storm that contains rotation within its updraft. The structure of this storm allows it to persist for several hours and generate large hail and tornadoes.

supernova A large, brief eruption in a star that temporarily makes the star very bright.

supernovae remnant Gaseous material ejected by a supernova visible long after the supernova fades.

super-saturation A condition where water vapor content is greater than the saturation value. This occurs when there are few cloud condensation nuclei present and water droplets are difficult to form.

surf zone The area near the shore where ocean crests break. Also called "breakers".

surface current Ocean currents that flow in the upper parts of the ocean water column and are caused by friction between wind and the ocean surface.

surface wave Types of seismic waves that travel along Earth's surface. There are two kinds: Love and Rayleigh.

suspended load Material (usually clay or silt) that is carried along in the moving current of a stream and does not settle out until the water stops moving.

synchronous rotation A characteristic of most satellites, including the moon, where the satellites rotate and revolve around their planets at the same rate.

syncline A type of fold that is "U" shaped.

synoptic meteorology A branch of meteorology that studies atmospheric features and phenomena lasting days and extending to hundreds of kilometers in size. This area of study includes the motion and intensification of low pressure systems and fronts, which we most often associate with weather reports.

system A group of interacting parts that together form up a complex and independent unit.

talus creep A slow mass movement process whereby talus moves slowly towards the bottom of a slope.

talus loose material that accumulates at the base of a cliff, primarily due to rockfall.

tarn A glacial lake that forms in the bottom of a cirque.

temperature The measure of the thermal energy of molecules in an object. Hot objects have a large amount of energy while cold ones have a smaller amount of energy. Temperature is measured with a thermometer and common units are Fahrenheit (°F), Celcius (°C) and Kelvin (K).

terminal moraine A glacial end moraine that marks the farthest advance of a particular glacier.

terminal velocity The speed of a falling object when the forces of gravity and air drag balance. Density and size affect the fall rate of objects.

terminus The front edge of a glacier.

terrestrial planets The four innermost planets, including Earth, that are similar to Earth.

theistic evolution A theory of Earth history in which God created the universe, Earth, and life over a span of billions of years using natural laws and processes (e.g., biological evolution).

thermal (or contact) metamorphic rock A metamorphic rock whose primary ingredient of change is heat. This type of metamorphism usually occurs as a pluton comes into contact with pre-existing country rock.

thermal circulation Closed motion of a fluid caused by a temperature difference. Warm, less dense fluid rises and is replaced by cool, denser fluid. This motion occurs in the atmosphere from the smallest to the global scale.

thermal features Any type of hot groundwater phenomenon (springs, geysers, fumaroles, mud pots).

thermohaline circulation The large-scale flow of surface and deep-ocean waters driven by differences in temperature and salinity.

Thermosphere The layer of the atmosphere located above the mesosphere. Starting at 90 km above the surface, this layer interacts with the sun's solar wind and experiences an increase of temperature with height.

thrust fault A low angle reverse fault (about 10° or less).

thunder Sound emitted by the sudden expansion of air along the super-heated channel of a lightning discharge.

tidal bores An extremely strong onshore tidal current, in which the flood current creates a visible ridge of water.

tidal bulge the portion of the oceans that is attracted by gravity to either the sun (solar tidal bulge) or moon (lunar tidal bulge).

tidal current A rectilinear flow of water towards or away from the shore, driven by the relative strength of ebb and flood currents and following daily tidal cycles.

tidal cycle The pattern of high and low tides expressed in a location during one lunar orbit around Earth (29.5 days)

tidal range The vertical difference between high tide and low tide.

tides The daily rhythmic rise and fall of the ocean's surface elevation.

tied island A large rock or small island that is connected to the shore by a tombolo.

tombolo A ridge of sand that develops perpendicular to the beach because a large object blocks incoming waves, or because converging longshore currents combine their sediment loads.

tornado A rotating column of air extending from the base of a thunderstorm to the surface. It is often visible due to vapor condensation within the low pressure of the funnel and the debris drawn up from the ground.

trade winds Easterly winds occurring in the Hadley cell between subtropical highs and the equator.

transform boundary A location where two plates are moving past one another. Lithosphere is neither created nor destroyed at transform boundaries.

trans-Neptunian objects Any small solar system object orbiting beyond the orbit of Neptune; some people refer to these as Kuiper belt objects.

transverse dune A sand dune which has a long crest that is perpendicular to wind direction.

travertine A dense type of limestone (calcium carbonate) that usually forms in association with springs or cave deposits.

tributary A smaller river that joins into a larger river.

trigonometric parallax The only direct method of finding stellar distance from the apparent shift in position that a star undergoes as Earth orbits the sun.

Tropic of Cancer Geographic circle located at 23.5 °N latitude. On the first day of Northern Hemisphere summer the sun is directly overhead at noon.

Tropic of Capricorn Geographic circle located at 23.5 °S latitude. On the first day of Southern Hemisphere summer (Northern Hemisphere winter) the sun is directly overhead at noon.

tropical depression A mass of thunderstorms that organize into circular rotation about a central low pressure system. This occurs over the tropical ocean and has wind speeds ranging between 38 and 65 kph.

tropical storm An intensification of a tropical depression where the central pressure decreases and the wind speeds range between 65 and 119 kph. Storms of this intensity are given names.

tropopause The boundary separating the troposphere and the stratosphere. The air temperature of this layer does not decrease with height and is very stable. Thus it prevents weather in the troposphere from significantly impacting the stratosphere.

troposphere The layer of the atmosphere extending from the ground to the tropopause. In general temperature decreases with height and all weather significant to the surface is contained in this layer. Near the poles this layer is about 8 km thick while at the equator it is 18 km thick.

trough An elongated area of low atmospheric pressure. This is present in upper-air charts when cold air extends further south than usual in the Northern Hemisphere.

tsunami A large wave or sudden increase in sea level caused by a large earthquake on the ocean floor or a large landslide into the ocean.

tufa A type of limestone usually associated with spring deposits that has significant cavities and small holes within it. It will often contain fossilized remains of organisms associated with the spring water (reeds, moss).

tundra A land surface that has permafrost. During the summer the surface soil melts allowing lichens and small plants to grow.

turbidity current An rapid underwater flow of sand, silt and clay that sweeps along the bottom of a water body. It does not carry as much material as a hyperconcentrated flow.

typhoon A severe tropical storm occurring in the western North Pacific. It is comparable to a hurricane in intensity. In the South Pacific and Indian Ocean it is called a tropical cyclone.

umbra Either the total portion of a shadow where light is completely blocked, or the darker inner portion of a sunspot.

unconformity A surface within rocks that represent a break in time.

underground mining A group of mineral and fuel extraction techniques that take place under Earth's surface.

uniformitarianism A philosophical approach to geology which argues that only observed and measured processes and rates should be used to infer past geological event. Uniformitarianism is often summed up in the phrase *"the present is the key to the past."*

unloading When large amounts of rock or ice is removed from a surface.

unsaturated zone A zone above the water table that is not completely full of water. Sometimes called the vadose zone.

updraft Upward motion of air within a thunderstorm. This consists of warm humid air drawn in from Earth's surface.

upwelling An oceanic phenomenon in which offshore winds force large water away from the coast, forcing the deeper, colder, and nutrient-rich waters to rise towards the surface.

valley glacier A glacier that occupies a mountain valley. Sometimes called alpine or mountain glacier.

vapor pressure A measure of humidity based on the pressure exerted by water vapor in the air.

ventifact A rock in a desert that has been polished and shaped by windblown sand.

Vernal Equinox – The first day of spring when all locations on Earth experience 12 hours of sunlight and 12 hours of night time. This occurs around March 20 in the Northern Hemisphere.

vesicular A texture of igneous rocks describing the large numbers of holes in the rock. The holes develop as the result of volcanic gases making frothy bubbles in lava as it was cooling.

vog Similar to smog, but produced by volcanoes.

volcanic igneous rock Sometimes referred to as extrusive igneous rock because the rock was "extruded" out of a volcano.

volcanic island arc A crescent-shaped chain of volcanoes produced by the subduction of one slab of lithosphere (capped by ocean crust) below another slab of lithosphere (also capped by ocean crust).

Wadati-Benioff zone A linear trend of earthquakes near trenches that indicates the subduction of a lithospheric plate capped by ocean crust.

wadi A desert streambed which is most often dry, but on occasion will host flashfloods.

wall cloud An abrupt lowering of a cloud below the base of a cumulonimbus cloud. This occurs when a very strong updraft entering the cloud causes the condensation level to lower. This is often associated with a supercell storm and rotation is seen in the cloud.

Walther's law A rule of relative dating which asserts that in a vertical sequence of sedimentary rocks, the types of rocks found above and below each other are the same as those found adjacent sedimentary environments.

water table The place underground where the unsaturated and saturated zones meet. The water table is typically not flat, it usually contours the surface of the ground above it.

wave (ocean) An expression of energy passing through the water, produced by oscillating (circular-moving) water particles.

wave base The depth at which ocean surface waters are no longer turbulent due to the action of waves.

wave crest The highest point of an ocean wave.

wave height The difference in elevation between a wave's crest and trough.

wave refraction The tendency of ocean waves to bend towards and become parallel to the shoreline.

wave trough The lowest point of an ocean wave.

wave-cut cliff A coastal erosional feature consisting of a steep wall of rock exposing the headland.

wave-cut platform A relatively flat area in front of a wave-cut cliff.

wavelength The distance between two wave crests.

weather The state of the atmosphere at a specific location and time. This can include such observations as temperature, pressure, wind, cloud cover and precipitation.

weathering A mechanical or chemical process that breaks a larger rock down into smaller pieces or its chemical components.

westerlies The predominant west to east motion of the atmosphere in the mid latitudes.

white dwarf An Earth-sized, dense star supported by electron degeneracy pressure.

wind turbine An electricity-generating device in which wind pushes large blades, which then spin a magnet to produce an electrical current.

winter solstice The first day of winter when the sun is lowest in the sky at noon and daylight hours are the shortest of the year. This occurs around December 21 in the Northern Hemisphere.

xenolith A piece of country rock that gets incorporated into magma or lava. When the liquid rock cools, the country rock is preserved.

yardang A large rock in a desert that has been polished and shaped by windblown sand. It will often parallel the wind and resemble an overturned ship hull.

young-Earth creation A theory of Earth history in which God played an active role in creating the universe, Earth, and life in a six-day period only a few thousands of years ago.

zone of accumulation The place on a glacier where snow accumulates to form glacial ice.

zone of aeration Sometimes also called the vadose zone and is located above the water table.

zone of wastage The place on a glacier where glacial ice actively disappears either due to melting or sublimation.

INDEX

Anticline trap, 300
Anti-cyclone, 402
Anvil, 423
Aphanitic textures, 61, 62
Aphelion, 457
Aphotic zone, 15
Apparent magnitude, 472
Aquamarine, 29
Aquiclude, 246
Aquifer, 245
Archean Eon, 182
Archosaurs, 187
Arctic circle, 361
Arête, 274
Artesian spring, 246
Artesian well, 246
Asthenosphere, 14, 95, 132
Astronomical unit, 454
Astronomy, 5
Atmosphere, 10–11
Atmosphere's composition, 351, 352
Atmospheric effect, 363, 365
Atmospheric pressure, 11, 12
Atmospheric science, 5
Atmospheric stability, 383–385
Atom, 31
Atomic mass, 32
Atomic number, 31
Aureole, 77
Aurora, 470, 471
Aurora Borealis, 355
Autumnal equinox, 361
Azurite, 29

B

Backshore, 333
Bajada, 281
Banded iron formations, 182, 183
Barchan dunes, 284
Barchanoid ridge dunes, 285
Barometer, 354
Barometric pressure. *see* Atmospheric pressure
Barrier islands, 335–336

Base load power, 299
Basin, 120, 121
    and range province, 278
Bathymetry, 91
Baymouth bar, 335
Beach drift, 330
Beach nourishment, 338
Berm, 333
Beta decay, 166
Big bang model, 500–502
Binary star, 474
Bingham Canyon copper mine, 56
Biofuels, 310
Biomass, 310
Biosphere, 15–16
Bituminous coal, 72
Black holes, 482–483
Blizzard, 395
Block slides. *see* Rock slides
Body waves, 125, 132
Bonds
    chemical, 33
    covalent, 34
    hydrogen, 35
    ionic, 33–34
    metallic, 34, 35
Bowen's Reaction Series, 63, 65
Braided streams, 242
Breakwater, 338
Breccia, 67, 68
Breeze, 398
Bridal veil falls, 274
Brown coal. *see* Lignite

C

Caldera formation, 137–138
Cambrian explosion, 184, 185
Capillary fringe, 244
Carbon isotopes, 32, 33, 165
Carbon-14 decays, 167, 172
Carbonates, 41, 71
Carbonic acid, 210, 211
Caretakers of creation, 294

Fronts, 417
  cold fronts, 417
  occluded front, 420
  steering winds, 417
  warm fronts, 420
Frost heaving, 212
Frost wedging, 212
Fuel rods, 305
Fujita scale, 427
Fumaroles, 134, 253, 254

G

Gabbro, 31
Galaxy, 490
Gap Theory, 152
Garnet, 29
Gasses, 10
Gems
  precious, 47, 49–50
  semiprecious, 47
Geological column, 164, 180
Geological forces, 122
Geology, 4
Geosphere, 13–15
Geostationary Operational Environmental
    Satellite (GOES), 438
Geostrophic winds, 401
Geothermal energy, 309
Geysers, 253, 254
Giant stars, 477
Glacial erratics, 270
Glacial striations, 270
Glacial till/glacial moraine, 270
Glaciers
  alpine glaciation, 272–274
  continental glaciation,
    270–271
  continental glaciers/ice sheets, 264
  Ice Age, 265–268
  ice caps, 265
  isostatic rebound process, 265
  movement, 268–269
  valley glaciers, 265

Global circulation patterns
  circulation cells, 411–413
  El Niño, 414–415
  global pressure and wind patterns, 413–414
Global Positioning System (GPS), 103
Global warming, 366–367
Globular star cluster, 476
Glowing avalanche, 134
Gondwana, 85, 183
Gradient winds, 401
Grand Canyon rocks, 173
Granite, 30, 76
Granodiorite, 13
Graphite, 78
Grassland soil, 209
Great Unconformity, 194, 195
Great Upheaval period, 192
Greenhouse effect, 365
Greenhouse gases, 365–366
Groin, 338
Ground moraine, 271
Ground-source heat pump, 309
Groundwater
  caves, 251–253
  fumaroles, 254, 256
  geysers, 254, 256
  home uses, 248, 249
  hot springs, 254, 255
  karst topography, 253, 254
  large-scale use, 249–251
  mud pots, 254, 255
  sapping, 214, 215
  springs, 246–248
  water table, 244–245
  wells, 245–246
Gypsum, 69, 70
Gyres, 320

H

Hadean Eon, 181
Hadley cell, 411
Hail, 394
Half-time, isotopes, 168–170